NanoBioMaterials

NanoBioMaterials

Editor-in-Chief

Bhupinder Singh

Editors

Rodney J. Y. Ho

Jagat R. Kanwar

CRC Press
Taylor & Francis Group
Boca Raton London New York

CRC Press is an imprint of the
Taylor & Francis Group, an **informa** business

CRC Press
Taylor & Francis Group
6000 Broken Sound Parkway NW, Suite 300
Boca Raton, FL 33487-2742

Printed and bound by CPI Group (UK) Ltd, Croydon, CR0 4YY on a sustainably sourced paper

International Standard Book Number-13: 978-0-8153-9978-0 (Hardback)
International Standard Book Number-13: 978-1-351-13866-6 (eBook)

Library of Congress Cataloging-in-Publication Data

Names: Singh, Bhupinder, editor.
Title: NanoBioMaterials / editor, Bhupinder Singh Bhoop.
Other titles: NanoBioMaterials (Singh)
Description: Boca Raton : Taylor & Francis, 2018. | Includes bibliographical references and index.
Identifiers: LCCN 2018000548| ISBN 9780815399780 (hardback : alk. paper) |
ISBN 9781351138666 (ebook)
Subjects: | MESH: Nanostructures--therapeutic use | Biocompatible Materials--therapeutic use
| Drug Delivery Systems
Classification: LCC R857.M3 | NLM QT 36.5 | DDC 610.28--dc23
LC record available at https://lccn.loc. gov/2018000548

Visit the Taylor & Francis Web site at
http://www.taylorandfrancis.com

and the CRC Press Web site at
http://www.crcpress.com

Contents

Foreword vii
About the Series ix
Preface xiii
Editors xv
Contributors xvii

Chapter 1 Applications of Nanostructured Biomaterials 1

Shweta Sharma, Ankush Parmar, and Surinder Kumar Mehta*

Chapter 2 Nanocoutured Metallic Biomaterials and Surface Functionalization of Titanium-Based Alloys for Medical Applications 17

Jalal Azadmanjiri, Wai Hong Wong, Jagat R. Kanwar, Christopher C. Berndt, James Wang, Vijay K. Srivastava, and Ajay Kapoor**

Chapter 3 Graphene-Polymer Nanocomposites for Biomedical Applications 51

*Duo Wu, Mohammad Ramezani, and Minna Hakkarainen**

Chapter 4 Lipid-Based Nanocarriers in Lymphatic Transport of Drugs: Retrospect and Prospects 67

Vikas Rana, Sunil Kamboj, and Sheshank Sethi*

Chapter 5 Nanocarriers in Early Diagnosis of Cancer 97

Gurpal Singh, Rajneet Kaur Khurana, Atul Jain, Taranvir Kaur, and Bhupinder Singh**

Chapter 6 Dendrimers: Emerging Anti-Infective Nanomedicines 121

*Keerti Jain**

Chapter 7 Production and Utilization of Nanofibers as Promising Biomaterials 139

Yaser Dahman, Valdir Mota Junior, Yan Xuan, Simon Nagy, Sumant Saini, Jasleen Kaur, and Bilal Khan*

Chapter 8 Fibro-Porous Composite Nanobiomaterials for Enhanced Biointegration 165

*Andrew T. Wood, Dominique T. Everett, Katelynne J. Herron, and Vinoy Thomas**

Chapter 9 Nanocarrier-Mediated Therapeutic Protein Delivery 185

Akhlesh Kumar Jain, Teenu Sharma, and Sunil Kumar Jain*

Chapter 10 Nanotechnology-Based Prodrug Conjugates for Site-Specific Antineoplastic Therapy 203

*Nishi Mody, Rajeev Sharma, Surbhi Dubey, Atul Jain, and Suresh P. Vyas**

Chapter 11 Osteomyelitis: Therapeutic Management using Nanomedicines 221

Ashish Ranjan, Teenu Sharma, Atul Jain, Sumant Saini, Shantanu Bandyopadhyay, Shreeniwas Singh, and Bhupinder Singh**

Chapter 12 Nanostructured Lipid Carrier-Mediated Methotrexate Delivery Evokes Transcription Factors to Induce Selective Apoptosis in Rheumatoid Arthritis 239

Naisargee Patel, Rajeev K. Tyagi, and Neeraj K. Garg*

Chapter 13 Superparamagnetic Iron Oxide Nanoparticles for Magnetic Hyperthermia Applications 247

Yogita Patil-Sen and Vikesh Chhabria*

Chapter 14 Development of In-House Nano-Hydroxyapatite Particles for Dental Applications 263

*Sunpreet Singh and Rupinder Singh**

Index 277

Foreword

We are amidst a nanotech era, with the word, "nano" attaining vast relevance and importance. Notwithstanding the widening network of nanotechnology in fundamental terrains like water, food, shelter, energy, environment and machination, its use in therapeutics and diagnostics has particularly been paramount. Nanomedicine, nanobiomedicine, nanopharmaceuticals, nanodiagnostics, nanobiotechnology, nanobioengineering, nanobiomaterials, nanoagroceuticals, nanonutraceuticals, etc. have today become buzzwords, accordingly.

Nanotechnology *per se* embraces the study and application of various nanoscale materials, prepared by bottom-up or top-down approaches. Nanoscale biomaterials of natural or synthetic origin hold immense potential in the medical sciences. *NanoBioMaterials* clearly demonstrate close relationships between biological and material sciences, and their use in bioimaging, drug delivery, tissue engineering and many more. The introduction of nanobiomaterials in drug formulation strategies has particularly become the source of leading innovations in the past couple of decades, thus providing a distinctive platform of research to young researchers. An acute need has increasingly been felt to spell out proper guidelines and regulations, outlining the boundaries for their clinical use.

Commencing this four-volume compilation at this threshold on the subject of *Emerging Trends in NanoBioMedicine*, as a sequel to multi-volume book series on NanoBioMedicine, the chapters contributed by various experts in the field need to be applauded and welcomed. The contents of the present book volume on *NanoBioMaterials* would provide a holistic view of the varied aspects of nanobiomaterials, including their stellar merits and some key limitations, potential applications in diverse fields and their futuristic promise in biomedical sciences. It also brings forth the results of current research to those who wish to use this knowledge and know-how in the applied settings. Regulatory issues on the usage of nanobiomaterials have also been assigned due consideration. This volume provides a coherent text with focused chapters and easily accessible information in the form of well-structured and updated reference material. The treatise is likely to serve the needs of researchers and practitioners working in various domains such as nanotechnology, biotechnology, medicine, human biology, drug delivery, biosensors, engineering and technology, food technology, agriculture, herbal science, nanotoxicology and clinical research.

I firmly believe that this book volume will help to serve the unmet needs of various researchers and practitioners in harvesting the immense potential of nanomaterials in the realm of biologicals and pharmaceuticals.

Professor C.N.R. Rao
National Research Professor
Linus Pauling Research Professor &
Honorary President, Jawaharlal Nehru Centre for Adv. Scientific Research

About the Series

Implying miniaturization, the word "nano" has become a household name today. Nanotechnology encompasses the study and application of functional nanomaterials in diverse fields even at their molecular levels. Virtually, nothing remains to be nanonized at the moment, whatever the domain may be. Traversing a voyage from the origin of its idea by Richard Feynman in 1959, nanotechnology has now entered into our day-to-day living, society and culture, shaping our future in a categorical manner. On one hand, nanotechnology is being explored to develop newer nanostructures, and on the other hand, the horizons of their promising applications are being widened manifold. Not merely as an "evolution", this has lately been turning out to as a "revolution" across different scientific disciplines and industrial sectors across the globe.

Notwithstanding the promising benefits of the cutting-edge technology, its applications in healthcare are considered as paramount. This perspective assumes immense significance, as the whole world is experiencing unprecedented biomedical challenges and changes, particularly in context with "3-***D***" succession of ***D***iseases, ***D***iagnosis and ***D***rugs. In line with futuristic predictions made by Robert Freitas in 1999 in his book on 'Nanomedicine', today it is ostensibly the fastest emerging offshoot of nanotechnology. Further integration of Nanomedicine with biological sciences, such as biotechnology, biochemistry, pharmaceutical sciences, biomedical engineering, bioimaging, nanomedical robotics and biomedical technology, has spurred the genesis of a whole contemporary field of "NanoBioMedicine". Employing various nanostructures and nanodevices, the science of nanobiomedicine primarily endeavors to improve upon the patient safety, efficacy, economy and compliance of varied therapeutic, preventive and diagnostic strategies. Not merely dealing with the nuances of interactions of nanostructures with biomolecular receptors, the modern discipline of nanobiomedicine also embraces the methods of design, engineering and technological transfer of nanoformulations of drugs and biomacromolecules, and fabrication of advanced biomedical appliances. Drawing apt inputs from almost all the frontiers of science and technology, nanobiomedicine today is heralding a new era in medicare with lots of hopes and promises for the suffering society.

Nanobiomedicine, of late, has been making rapid strides, eventually providing significant impetus towards scientific enthusiasm, research efforts, industrial anticipation and regulatory attention. Of late, several nanomedicinal products have been introduced in the commercial circulation, with a few thousands more in the pipeline, and yet many more in the early stages of developmental research. These have been documented to exhibit stellar benefits not only of enhanced surface-area-per-unit volume, but also of improved aqueous solubility, target-specificity, controlled release potential, biocompatibility, stealth characteristics, precise control of particle size and ability for drug combination therapy. Such nanoscaled products principally encompass different types of nanoparticles, nanocrystals, nanoemulsions, nano-conjugates, liposomes, dendrimers, nanogels, nanocosmeceuticals, nanocapsules, nanophytomedicines, lipid nanospheres and nanocomposites. A huge proportion of these innovative nanotech-enabled products have demonstrated enormous patient benefits and plausible solutions to the erstwhile insurmountable medical problems.

Alarmingly high incidences of inadequate efficacy, safety and patient compliance reported with the current crop of drugs, have been calling for developing such newer and therapeutically superior nanomedicinal formulations to address the present patient needs and market demands. Driven by the technological updates, increased gubernatorial support, recent notifications of newer regulatory guidance for drug products containing nanomaterials and enhanced preference for targeted therapeutics, the world has now been witnessing a rapid surge in such nanomedicines. Nanopharmaceuticals, therefore, are an outsized industry today, with its market crossing a few hundreds of billions of US dollars, over 500 companies and over 80 products worldwide. In the forthcoming decades too, these nanomedicines are anticipated to massively impact almost all the terrains of patient care, personalized medicine, clinical and veterinary practice and above all, the pharmaceutical market. The ardent need in the current hour of shifting paradigms, accordingly, is to acquaint the professionals on advances in this vibrant science of nanobiomedicine, growing at a phenomenally striking pace.

A recent literature survey of the leading journals revealed a phenomenally high rate of publications in this multi- and interdisciplinary domain of nanobiomedicine. For a realm advancing as fast as this one, it is quite challenging to obtain a complete overview of the exciting developments, as most updated information on such nano-bio interfaces lies scattered in diverse journals. Owing to nonavailability of the standard treatise(s) on the subject, therefore, we have undertaken a grand task of bringing about a book series covering a gamut of subject topics spanning nanobiomedicine.

Accordingly, the current book series, "**Emerging Trends in NanoBioMedicine**", an anthology of four volumes, aspires to dexterously present the pertinent facts and figures on an array of novel subject topics. Each individual volume of the compendium is an assortment of choicest reviews and research articles, contributed by illustrious scientists and experts selectively drawn from across the world. The primary motivation behind this Herculean effort has been to bring forth high quality insight into latest research in the field as a multi-volume digest. It comprises of four volumes, *viz.*

Volume 1: NanoBioMaterials
Volume 2: NanoBioEngineering
Volume 3: NanoNutraceuticals
Volume 4: NanoAgroceuticals & NanoPhytochemicals

NanoBioMaterials, the first volume of the series, maps out an exhaustive overview of the contemporary research accomplishments of variegated nanomaterials exhibiting profound impact in improving treatment efficiency of various diseased states. This volume tacitly introduces and demonstrates close association of biology with material science, with intent for effective and safe use of nanomaterials for different biological applications. Being a promising, though burgeoning, realm of science, the treatise accentuates diverse applications of nanomaterials in the fields of drug delivery, bioimaging, tissue engineering and many more. It primarily focuses on innovations based on various nanobiomaterials, such as polymeric, metallic, lipidic, peptidal, carbon-based, and prodrug-conjugate-based, leading finally to produce next-gen biomedicinal products.

Highlighting challenges for fundamental research and opportunities for ground-breaking advances in nano-enabled bioengineering, the second volume of this series successfully unfolds the topic of **NanoBioEngineering**. Being a rational fusion of biology, engineering and nanotechnology, it presents breakthrough techniques and tools amid the existing armamentarium to utilize novel phenomena like microfluidics and nanotooling and nanomaterials like aptamers, nanocargos, nanofibers, nanocrystals, hyper-branched dendrimers, surface-engineered and superparamagnetic nanopaticles, for meeting the desired healthcare needs. The integrated volume, herein, provides a lucid account on far-reaching topics from diverse disciplines of life sciences and technology, including physics, chemistry, biology, engineering, biomedical technology and drug development.

The enormous potential of nutraceuticals in fostering human healthcare, the burning issues confronting their poor bioavailability and usage of nanotechnology principles for resolving these issues, have been adroitly deliberated upon in the third volume of the book series, **NanoNutraceuticals**. Provision of the requisite knowledge on a gamut of modern hi-tech approaches to improve their activity, especially using nanotechnology through a single textual source enshrined, this book volume can be instrumental in ameliorating the subdued popularity of nutraceuticals and lack of serious consideration by health regulatory agencies. Thus, the contents of this book volume have the capability to metamorphose the biological performance of these functional foods through their skilful delivery to the specific target using apt nanocarriers for safer and more efficacious management of the diseased states like cancer and neurodegenerative diseases, and for benefitting the human healthcare *per se*.

Volume four, **NanoAgroceuticals & NanoPhytochemicals**, is a testament to the sophisticated and unconventional extrapolation of nanotech applications in the field of agroceuticals and phytochemicals. Albeit still in its infancy, the subject matter encompassed in the book volume holds immense potential to bring forth tangible benefits to the farmers, food industry, patients and consumers. This book presents a cogent account on various nano-enabled delivery technologies of phytochemicals in agriculture to improve the storage shelf-life of crops, reduce the amount of sprayed chemicals through smart delivery of phytochemicals, sustain delivery of fertilizers, decrease the loss of nutrient(s) in fertilizers and augment percent harvesting yield through optimized water and nutrient management. The role of nanotechnology in the development of novel and innovative products out of agriculture products and herbal supplements, and their subsequent production, processing, storage and packaging, has also been explicitly highlighted in this book volume.

All these volumes put together highlight some of the most remarkable advances that have taken place recently in the vast field of nanobiomedicine, and reflect new challenges and issues that confront their further progress. The series seeks to offer the readers an apt tête-à-tête with fresh and original perspectives, epitomizing a vast multiplicity of experience, wisdom and vision, which is quite implausible to be made available collectively anywhere else.

Verily, in this context, I feel quite satiated and fascinated that the long-felt void of a repository on the much sought-after discipline of nanobiomedicine is being filled up by spearheading this current multi-volume book series. Although the field of nanobiomedicine has become too vast today, meticulous endeavor has been made to customize the contents of the book series and steer the novitiates as well as seniors on the recent and upcoming trends. Not only does this treatise embrace holistic applications of a plethora of promising nanoconstructs, especially in the fields of medical therapeutics and diagnostics, biotechnology, pharmacogenomics, human gene biology, clinical pharmacology, and so on, but it also touches upon their regulatory, safety and industrial scale-up issues.

At this juncture, I would like to take this opportunity to express my heart-felt gratitude to several people whose pooled efforts brought forth this dream of mega-series into real being. The list includes all the authors, editors, publishing house team, my research group, and research funding agencies, besides others.

First and foremost, I put across my heart-felt gratitude to the eminent and erudite authors for dexterously working on the chapters by bringing together the most comprehensive and updated information on wide-ranging subject(s). Not merely did these contributors meet with the stringent timelines, they also abided by a battery of editorial suggestions sportingly. They garnished and flavoured these chapters with apt perspectives, emanating from their own experiential wisdom. Earnest efforts have been made by them to encompass a diversity of graphic illustrations, presented immaculately as expository graphs, photographs, methodology flow charts, hierarchical tabulations, bar charts, and so on. Each chapter has been duly referenced to current literature, citing vital texts for further reading too.

As the designated Editor-in-Chief of the book series, I fall short of apposite words to express my deep appreciation to the distinguished editors of the four volumes for pouring in their invaluable and pertinent inputs from time to time. The sagacious, prompt and critical feedback provided by them as well as by the worthy reviewers of the individual chapters while spending their priceless time helped in ameliorating the quality standards of the manuscripts a great deal.

I put into record the highly painstaking, persistent and meticulous efforts of the management and staff members of the publisher, *viz.*, M/s CRC Press (Taylor & Francis Group), USA, in bringing out this mega compilation. Without their immense support and direction, it could not have been possible to bring forth this colossal series in a timely manner.

Our well-knit team of exceptionally diligent, efficient and perseverant researchers and post-docs deserve the special mention, as they were verily the spine of this massive and momentous project. Naming a few, while leaving many others, may demean their stellar contribution, for which I would always remain beholden.

I also seize this opportunity in acknowledging the University Grants Commission (UGC), New Delhi, India, for awarding this prestigious National UGC Centre of Excellence in "Applications of Nanomaterials, Nanoparticles & Nanocomposites" with focus area as "Biomedical Sciences" to the

Panjab University, Chandigarh, India. In fact, the drive to take up this marathon task for writing the explicit treatise was initiated after my assuming the charge of this Centre of Excellence, dedicated to explore and establish various nanobiomedicinal precepts. Thanks are also due to the tangible financial support provided by several other governmental agencies and pharma industrial houses to enable us to bring out tangible outcomes on diverse innovative nanostructured systems and in characterizing them. Several of such systematic works on nanopatterned structures, primarily employing Quality by Design (QbD)-based paradigms, have fetched exalted global recognitions by American Association of Pharmaceutical Sciences (USA), Select Biosciences (UK), StateEase Inc. (USA), Evonik AG (Germany), Minitab Ltd. (UK), CPhI-Asia, EBA-Oxford (UK), APTI (India) and many more.

With this, I invite and incite you to start this amusing expedition to the salubrious and picturesque world of nanobiomedicine with a missionary message, "*Think Tiny and Act Big.*" I earnestly wish, hope and pray that this four-volume book series on "**Emerging Trends in NanoBioMedicine**" will serve as a classic repertoire of essential knowledge and know-how in the domain to address the unmet needs of the scientists and practitioners from current and future generations.

Bhupinder Singh

Preface

Nanobiomaterials are nanostructured constructs of natural and synthetic materials used for living organisms. The science of nanobiomaterials employs multidisciplinary concepts and technologies arising out of biology, medicine, pharmaceutics, tissue biomedical engineering, biomaterial sciences and biochemistry. Advances in science and engineering, along with the application of nanotechnologies, are partly driven by the inventions of new nanomaterials and the generations of available materials, assembled into small, nanosized form, based upon the physicochemical principles. Some of these new nanoscale materials have contributed to our lives by providing us outcomes such as higher resolution and more energy-efficient television and computer screens, faster computers and more brilliant paints with nanopigment pellets. Advances in nanobiomaterials in healthcare can be facilitated by understanding and improving biological developments as well as their structures and functions at the nanoscale. Nanobiomaterial technology contributes a vital role in biomaterial-based pharmaceutical and the biomedical science sector. In this way, nanobiomaterials can be building blocks for bridging the gap between biology and nanotechnology.

Nanobiomaterials, the subject of this book, require consideration of nanomaterials in the biological context. The safety of nanomaterials in consumer products is typically related to their environmental impact during their manufacture, use, disposal and effects on the human body. By convention, nanobiomaterials are intended for direct human contact, both outside and inside the body. Specifically, nanobiomaterials, mostly included as a part of pharmaceutical products, are intended to assist or improve the potency and safety of drugs in the body. Nanobiomaterials serve as drug carriers or excipients in the dosage forms to address target drug localization. Biocompatible lipidic materials, for instance, are used as lipid-based vesicles and liposomes to address the side effects of antifungal, amphotericin and antitumors, doxorubicin, daunorubicin, paclitaxel and irinotecan. Additionally, biodegradable poly-lactic-glycolic polymers are used to provide long-acting nano- and micro-drug formulations. With these proven clinical applications, the search for new biocompatible and biodegradable nanomaterials has sparked keen interest in the past a few years. These concerted efforts have enabled scientific advancements in synthesis, assembly, engineering and evaluation of nanobiomaterials, as well as in developing tools and outcome matrices, validated for evaluating the likelihood of their success in clinical applications.

This book volume is intended to present an overview of the recent nanomaterials for biological applications encompassing metallic, carbon-based, lipidic, polymeric and peptidal. Chapter 1 is dedicated to metallic nanobiomaterials, demonstrating not only their promise in clinical applications, but the advancements achieved to improve their features as well. Chapter 2 is more specific to illustrate the role of metallic nanomaterials (stainless steel, cobalt and titanium) as biological hard tissue substituents like pacemaker cases, including their advantages and disadvantages, and modifications achieved to avoid the risk(s) following implantation. Graphene-based biomedical properties are lucidly described in Chapter 3, along with their applications as hydrogels and scaffolds introduced during tissue engineering. Lipid-based nanocarrier delivery systems such as liposomes, nanoemulsion and solid lipid nanoparticles are elucidated in Chapter 4, with the postulation of models to investigate their transportation via the lymphatic system. Chapter 5 unearths the enormous potential of nanotechnology-enabled diagnostic agents, such as metallic and magnetic nanoparticles, quantum dots and CNTs which by virtue of their nanometric size can access tumor sites easily, thus acting as effective biomarkers to detect cancer in a very early stage. Chapter 6 focuses on the therapeutic role of dendrimers as nanobiopolymers, particularly dendrimer-based nanohybrids and polyanionic carbosilane dendrimers, along with their applications in anti-infective nanomedicine.

Recent developments in nanofibers are highlighted in Chapter 7, involving variegated types, synthesis protocols, as well as their limitations and their applications in biomedical science.

The nanofiber chapter is followed by fibro-porous nanobiomaterial properties (Chapter 8), taking into its ambit the recent developments made to improve their bio-integration, accentuating the role of various variables to enhance their efficacy during *in vivo* applications. The next chapter (Chapter 9) illustrates successful protein delivery, focusing on maintaining bioavailability and protein stability to maximize their therapeutic efficacy, and address the challenges faced by their delivery.

Chapter 10 unravels the rationale behind the development of prodrug approach, and delineates their major applications in oncology while taking into account their ostensible limitations as well. Chapter 11 highlights the myriad nanomedical applications for osteomyelitis therapy to ameliorate treatment efficacy, represented by immense tag ability while reducing the adverse effects of traditional therapy. Chapter 12 illustrates the stellar advantages of using nanostructured lipid carriers (NLCs) in the treatment of rheumatoid arthritis, by maintaining controlled drug release, enhanced bioavailability and above all, targeted drug delivery to tissues affected by rheumatoid arthritis.

Superparamagnetic iron oxide nanoparticles are highlighted in Chapter 13, while focusing on their synthesis and fabrication protocols to enhance their features in oncology. Chapter 14 explains the use of nanohydroxyapatite (n-HAp) in the field of dentistry, along with their applications, properties and synthesis of n-HAp.

Hopefully, this book will acquaint the readership with the updated nanotech advances in order to provide the desired fillip towards biomaterial research for improving the desired safety and efficacy of drugs. Comprehension of such safety, efficacy and biocompatibility issues, tools to access scale-up, stability, predictive time-course dose-response profile in target tissues, and off-target organs, and untoward effects, could further reinforce the field of nanobiomaterials in developing newer concepts for medicinal applications.

Bhupinder Singh
Rodney J. Y. Ho
Jagat R. Kanwar

Editors

Professor Bhupinder Singh Bhoop, MPharm, PhD, DSt, FPAS, is globally acclaimed for his extensive research work on diverse nanostructured drug delivery systems, developed primarily using Quality by Design (QbD) paradigms. His work encompasses variegated nanoengineered systems, naive as well as functionalized, like SLNs, NLCs, SNELs, MWCNTs, polymeric nanoparticles, lipid-polymer hybrid nanoparticles, mixed nanomicelles, nanocapsules, nanosponges, nanobilosomes, liposomes and other vesicular systems, for a gamut of diseased states, particularly cancers, and cardiovascular, immunodeficiency and neurodegenerative disorders.
A prolific writer, he has over 280 original journal publications, 16 books, 1 directory, 52 book chapters, 5 patents and 3 technology transfers to his credit. He is on Editorial and Reviewers Board of scores of journals. A widely traveled scientist, he has delivered more than 275 invited talks across the globe, and has duly trained tens of hundreds of scientists from pharma industry on nanomedicines and QbD. He has completed 16 research projects, and guided 31 PhD's (previously and currently), 3 post-doctorates and 55 postgraduate scholars. Currently, he is serving as founder coordinator of UGC Centre for Excellence in Nano Biomedical Applications, and Coordinator, UGC Centre of Advanced Studies (Pharmaceutical Sciences), both at Panjab University, Chandigarh. He has been Chairman, University Institute of Pharmaceutical Sciences, and Dean, Faculty of Pharm Sciences, Panjab University too. His work has fetched Professor Bhoop numerous awards and accolades at global and national levels.

Professor Rodney J. Y. Ho has over 25 years of experience in leading multidisciplinary research teams and translational research projects in HIV/AIDS, cancers, multidrug transporters and analgesics, with over 130 original publications, 45 book chapters and 16 patents to his credit. His innovations include device-enabled, nose-to-brain delivery, nano-drug combination delivery, chelate-embedded nanoparticle technologies for medical imaging, and primate models to study HIV drug and virus transmission. He was also recognized among the top entrepreneurs for founding several biotechnology and biopharmaceutical companies. He was awarded the Milo Gibaldi Endowed Professor of Pharmaceutics and was an investigator at Fred Hutchinson Cancer Research Center. Currently, Dr. Ho is working as a Professor in the Department of Pharmaceutics, Bioengineering (Adjunct) and as the Director of the Targeted Long-Acting Combination Antiretroviral Therapy (TLC-ART) Program.

Professor Jagat R. Kanwar is the group leader of the Nanomedicine Laboratory of Immunology and Molecular Biomedical Research at Deakin University, Australia. He has more than 210 publications in peer-reviewed journals, 20 patents and has guided more than 25 PhD students. His work focuses on discovering novel and safe targeted nanomedicines for cancers, autoimmune disorders and inflammatory diseases. Additionally, Dr. Kanwar has also been working to investigate the underlying mechanisms involved in regeneration, apoptosis, autophagy and inflammation by targeting the production of cytokines, chemokines, oxygen radicals and matrix metalloproteinase. He has lately ventured into microfluidics, cutting-edge technologies like lab-on-a-chip techniques for cancer cells, stem cell capture, disease-specific biomarkers and exosomes.

Contributors

Jalal Azadmanjiri
School of Engineering
Faculty of Science
Engineering and Technology
Swinburne University of Technology
Hawthorn, Victoria, Australia

Shantanu Bandyopadhyay
University Institute of Pharmaceutical Sciences
UGC Centre of Advanced Studies
Panjab University
Chandigarh, India

Christopher C. Berndt
School of Engineering
Faculty of Science
Engineering and Technology
Swinburne University of Technology
Hawthorn, Victoria, Australia

and

Department of Materials
Science and Engineering
Stony Brook University
Stony Brook, New York, USA

Vikesh Chhabria
School of Pharmacy and Biomedical Sciences
University of Central Lancashire
Preston, United Kingdom

Yaser Dahman
Department of Chemical Engineering
Ryerson University
Toronto, Canada

Surbhi Dubey
Drug Delivery Research Laboratory
Department of Pharmaceutical Sciences
Dr. Hari Singh Gour Vishwavidyalaya
Sagar, Madya Pradesh, India

Dominique T. Everett
Department of Materials Science and Engineering
University of Alabama at Birmingham
Birmingham, Alabama, USA

Neeraj K. Garg
University Institute of Pharmaceutical Sciences
UGC Centre of Advanced Studies
Panjab University
Chandigarh, India

Minna Hakkarainen
Department of Fibre and Polymer Technology
KTH Royal Institute of Technology
Stockholm, Sweden

Katelynne J. Herron
Department of Materials Science and Engineering
University of Alabama at Birmingham
Birmingham, Alabama, USA

Akhlesh Kumar Jain
School of Pharmaceutical Sciences
GGV Central University
Bilaspur, CG, India

Atul Jain
UGC Centre of Excellence in Applications of Nanomaterials, Nanoparticles and Nanocomposites (Biomedical Sciences)
Panjab University
Chandigarh, India

Keerti Jain
Center of Pharmaceutics
School of Pharmaceutical Sciences
Delhi Pharmaceutical Sciences and Research University (DPRSU)
New Delhi, India

Sunil Kumar Jain
School of Pharmaceutical Sciences
GGV Central University
Bilaspur, CG, India

Valdir Mota Junior
Department of Chemical Engineering
Ryerson University
Toronto, Ontario, Canada

Sunil Kamboj
Department of Pharmaceutical Sciences and Drug Research
Punjabi University
Patiala, India

Jagat R. Kanwar
Nanomedicine-Laboratory of Immunology and Molecular Biomedical Research (NLIMBR)
School of Medicine (SoM)
Molecular and Medical Research (MMR)
Strategic Research Centre
Faculty of Health
Deakin University
Geelong, Victoria, Australia

Ajay Kapoor
School of Engineering
Faculty of Science
Engineering and Technology
Swinburne University of Technology
Hawthorn, Victoria, Australia

Jasleen Kaur
Department of Chemistry
Panjab University
Chandigarh, India

Taranvir Kaur
University Institute of Pharmaceutical Sciences
UGC Centre of Advanced Studies
Panjab University
Chandiragh, India

Bilal Khan
Department of Chemical Engineering
Ryerson University
Toronto, Ontario, Canada

Rajneet Kaur Khurana
University Institute of Pharmaceutical Sciences
UGC Centre of Advanced Studies
Panjab University
Chandigarh, India

Surinder Kumar Mehta
Department of Chemistry and Centre of Advanced Studies in Chemistry
Panjab University
Chandigarh, India

Nishi Mody
Drug Delivery Research Laboratory
Department of Pharmaceutical Sciences
Dr. Hari Singh Gour Vishwavidyalaya
Sagar, Madya Pradesh, India

Simon Nagy
Department of Chemical Engineering
Ryerson University
Toronto, Ontario, Canada

Ankush Parmar
Institute of Forensic Science and Criminology
Panjab University
Chandigarh, India

Naisargee Patel
Institute of Science
Nirma University
Sarkhej-Gandhinagar Highway
Ahmedabad, Gujarat, India

Yogita Patil-Sen
School of Physical Sciences and Computing
University of Central Lancashire
Preston, United Kingdom

Mohammad Ramezani
Department of Chemistry and Biotechnology
Swinburne University of Technology
Hawthorn, Victoria, Australia

Vikas Rana
Department of Pharmaceutical Sciences and Drug Research
Punjabi University
Patiala, India

Ashish Ranjan
Center of Veterinary Health Sciences
Oklahoma State University
Stillwater, Oklahoma, USA

Sumant Saini
University Institute of Pharmaceutical Sciences
UGC Centre of Advanced Studies
Panjab University
Chandigarh, India

Rajeev Sharma
Drug Delivery Research Laboratory
Department of Pharmaceutical Sciences
Dr. Hari Singh Gour Vishwavidyalaya
Sagar, Madya Pradesh, India

Shweta Sharma
Institute of Forensic Science and Criminology
Panjab University
Chandigarh, India

Sheshank Sethi
Department of Pharmaceutical Sciences
and Drug Research
Punjabi University
Patiala, India

Bhupinder Singh
University Institute of Pharmaceutical Sciences
UGC Centre of Advanced Studies
Panjab University

and

UGC Centre of Excellence in Applications
of Nanomaterials, Nanoparticles and
Nanocomposites (Biomedical Sciences)
Panjab University
Chandigarh, India

Gurpal Singh
University Institute of Pharmaceutical Sciences
UGC Centre of Advanced Studies
Panjab University
Chandigarh, India

Shreeniwas Singh
Department of Animal Husbandry
Jharkhand, India

Sunpreet Singh
Manufacturing Domain
School of Mechanical Engineering
Lovely Professional University
Punjab, India

Rupinder Singh
Production Engineering Department
Guru Nanak Dev Engineering College
Ludhiana, Punjab, India

Vijay K. Srivastava
Department of Mechanical Engineering Indian
Institute of Technology
Varanasi, India

Teenu Sharma
University Institute of Pharmaceutical Sciences
UGC Centre of Advanced Studies
Panjab University
Chandigarh, India

Vinoy Thomas
Department of Materials Science and
Engineering
University of Alabama at Birmingham
Birmingham, Alabama

Rajeev K. Tyagi
Institute of Science
Nirma University
Sarkhej-Gandhinagar Highway
Ahmedabad, Gujarat, India

and

Department of Periodontics
College of Dental Medicine Georgia Regents
University
Augusta, Georgia

Suresh P. Vyas
Drug Delivery Research Laboratory
Department of Pharmaceutical Sciences
Dr. Hari Singh Gour Vishwavidyalaya
Sagar, Madya Pradesh, India

James Wang
School of Engineering
Faculty of Science
Engineering and Technology
Swinburne University of Technology
Hawthorn, Victoria, Australia

Wai Hong Wong
School of Engineering
Faculty of Science
Engineering and Technology
Swinburne University of Technology
Hawthorn, Victoria, Australia

Andrew T. Wood
Department of Materials Science and Engineering
University of Alabama at Birmingham
Birmingham, Alabama, USA

Duo Wu
Department of Fibre and Polymer Technology
KTH Royal Institute of Technology
Stockholm, Sweden

Yan Xuan
Department of Chemical Engineering
Ryerson University
Toronto, Ontario, Canada

1 Applications of Nanostructured Biomaterials

Shweta Sharma, Ankush Parmar, and Surinder Kumar Mehta*

CONTENTS

1.1 Introduction ... 1
1.2 Nanobiomaterials and Their Applications ... 2
  1.2.1 Gold Nanoparticles (AuNPs) ... 2
  1.2.2 Gadolinium Nanoparticles (GdNPs) ... 3
  1.2.3 Iron Oxide/Magnetic Nanoparticles (IONPs) ... 6
  1.2.4 Ceramic Nanoparticles (CENPs) ... 7
  1.2.5 Titanium Oxide Nanoparticles (TiONPs) ... 7
  1.2.6 Silica Nanoparticles (SiNPs) ... 9
  1.2.7 Semiconductor-Based Nanoparticles/Quantum Dots (SMNPs) ... 9
  1.2.8 Organic/Carbon-Based Nanoparticles (CNPs) ... 10
1.3 Conclusions ... 11
Acknowledgment ... 12
References ... 12

1.1 INTRODUCTION

Novel innovations in the realm of biomaterial engineering have imparted exceptional asphalt to build newer materials at their nanoscale levels for applications in biological arena. Latest advances in nanotechnology have made significant impact on the development of biomaterials in order to let these interact with a living system at a cellular and subcellular level (Ernest and Shetty, 2005). Rational integration of biomedical sciences with nanotechnology has, therefore, provided impeccable impetus for the growth of nanobiomaterials to facilitate studies beyond molecular level. A gamut of nanomaterials, including polymeric, ceramic, metallic and composite particles have been explored for their diverse applications in biomedical sciences such as drug delivery, tissue engineering scaffolds, bioimaging and biosensors. Accomplishing fruitful utilization of these nanomaterials in the medical field invariably encompasses appropriate surface alterations such as entrapment, oxidation or coating to attain better stability and solubility of these nanoparticles in various aqueous media and/or biological matrices.

Lately, extensive research has surrounded the development and synthesis of nanobiomaterials. New physical properties and new advancements, both in sample preparation and device fabrication, have been notably influenced by various developments in nanoscience. Astounding strides have been undertaken using nanotechnology in the production of new surfaces, materials and structures, leading eventually to unravel innovative applications in the biomedical field (Huang and Sayed, 2010). Conventional biomedical applications like biosensors, tissue engineering, controlled release systems, intelligent systems and nanocomposite tools as part of implant design, have been grossly benefitted from nanotechnology. Chapter 1, in this context, focuses on the diverse perspectives of various nanobiomaterials and their prospective applications.

* Corresponding author.

1.2 NANOBIOMATERIALS AND THEIR APPLICATIONS

Various types of metallic nanobiomaterials are recognized with their promising biomedical applications, some of the vital ones are described below (Figure 1.1).

1.2.1 Gold Nanoparticles (AuNPs)

In the area of biomedical sciences, gold nanoparticles (AuNPs) have been explored to a large extent. AuNPs ensure inimitable physicochemical properties, complemented by their nontoxicity. A multipurpose stage has been offered by AuNPs to nanosized biological fabrication with proteins (Yeh *et al.*, 2012) and antibodies (Mukherjee *et al.*, 2007). In order to explore biological systems, AuNP bioconjugates are documented to have a primary contribution in the biomaterial design. In the range of biomedical application, multipurpose AuNPs have offered useful resources. Certain properties of AuNPs, such large surface area, electronic properties in terms of size and shape (Jackel *et al.*, 2011), biocompatibility and low toxicity (Dykman and Khlebtsov, 2011) make these as significant tools in the area of bionanotechnology. Physicochemical properties of such nanoparticles can be reformed by analyte–AuNP interactions (Uehara, 2010).

A phenomenon of surface plasmon resonance is unveiled by these nanoparticles, which vanishes the radiating light wavelength, leading subsequently to fluorescence quenching (Yeh *et al.*, 2012, Devi et *al.*, 2017). AuNPs, due to their high stability, biocompatibility and resistivity to oxidation, find wide application in biomedical sciences, including drug delivery, bioelectronics, bioimaging, biosensing and photohyperthermia (Dykman and Khlebtsov, 2012). Diverse biological moieties *viz.*

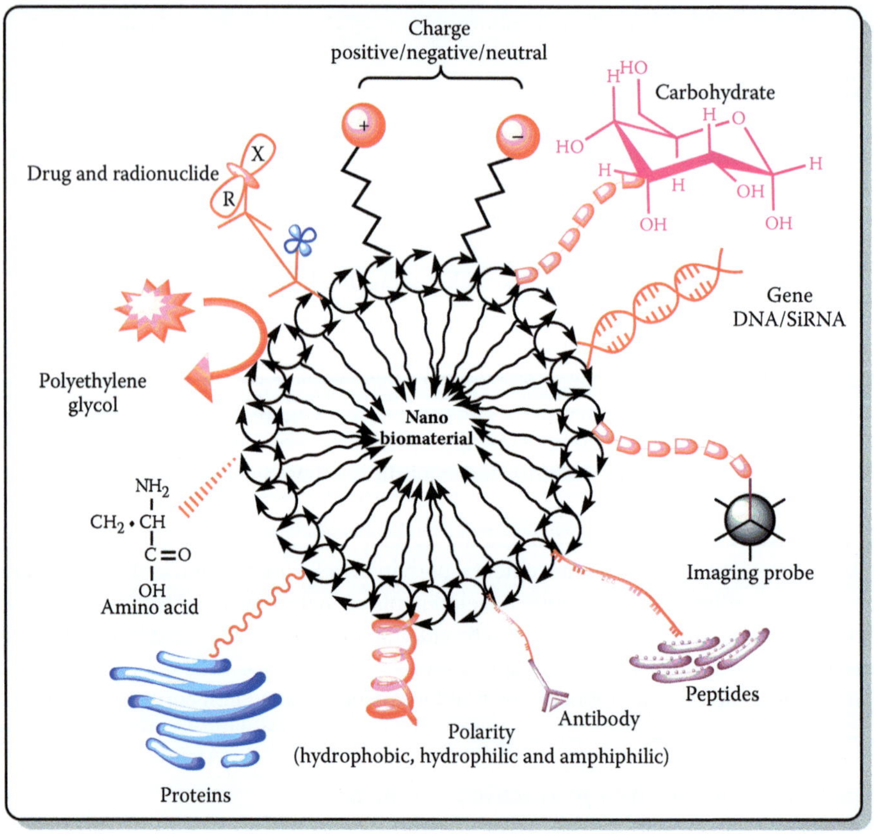

FIGURE 1.1 Biomedical applications of metallic nanobiomaterials.

antibodies, peptides, proteins and nucleic acids have been grafted on the surface of these nanoparticles for efficient and targeted delivery of active therapeutic moieties to the designated target site. This can be attributed to their unique morphology and surface properties facilitating the ease of functionalization of these nanoparticles. Bioconjugates of AuNPs have additionally turned out to be promising applicants among novel biomaterials for imaging of biological systems. AuNPs can be synthetically prepared in diverse morphologies with enhanced tunable properties (Huang and Sayed, 2010).

The process parameters involved during their fabrication tend to play vital role in governing the optical properties of these nanoparticulate systems (Jain *et al.*, 2006). Hence, it becomes mandatory to keep strong vigil on the fabrication process and optimize these in order to attain an efficient system with desired characteristics.

The diverse range of biomedical applications has been explored due to high versatility of these nanoparticles. Imperative characteristics, like enhanced surface area to volume ratio and high atomic number, have endowed these nanoparticles with contrast enhancing abilities and localized heating effect, which in turn, have felicitated the usage of these nanoparticles for theranostic application in numerous cancer therapies. Nanophotothermolysis, using lasers and gold nanoparticles, are frequently being used for damaging cancer cells along with other organisms *viz.* bacteria, viruses and fungi. Unlike other fluorescent probes, gold nanoparticles can easily be prepared and are being currently employed in the field of biomedical sciences (Amjad *et al.*, 2015). It is quite well known that cancerous cells are generally invasive in nature, and one cell leftover can lead to the regrowth of cancer. When these gold nanoparticles are exposed to extreme temperature variations, a bubble is created in the vicinity of the cells which, in turn, will burst out, sending a shockwave and leading to cell lyses (Amjad *et al.*, 2015). This facilitates the selective killing of cancerous or abnormal cells, leaving normal cells intact and unharmed. Table 1.1a highlights the biomedical applications of AuNPs.

Wang *et al.* (2012) developed a novel water-soluble fluorescent gold nanocluster (FANC) using bovine serum albumin as the template and reductant. The developed FANC possessed a negative surface charge, unique optical properties, and a higher quantum yield (~ 2%). The FANC demonstrated high colloidal stability, and can be readily conjugated with biological molecules. Table 1.1b shows the effect of FANC on expression profiles and the function of human endothelial cells and endothelial progenitor cells.

Further, the *in vitro* efficacy of these synthesized nanoparticles was envisaged and their potential cytotoxic mechanism was studied in HT-29 colon cancer cell lines. The nanoparticles portrayed an enhanced antioxidant and free radical scavenging activity *in vitro*. The outcomes of the cellular studies revealed that the nanoparticles were capable in producing potent cytotoxic effects leading to induction of apoptosis in HT-29 cancer cell lines. These findings were corroborated by Annexin V-Cy3 staining, where the morphological changes in nuclear matter was observed in accordance with the damage to the cellular membranes. This, in turn, pointed towards an enhanced targeting propensity of these developed nanoparticles towards cancerous cells (Mata *et al.*, 2016).

In another study, Zhang *et al.* (2017) carried out the facile fabrication of bovine serum albumin (BSA-PEI)–modified cationized gold nanoparticles and their intricate effect of morphology on gene delivery. It was deciphered that the nanoparticles maintained their spatial configuration leading to effective condensation of DNA. The cellular uptake experiment exhibited enhanced colocalization of these nanoparticulate systems in HepG2 cells, ascribed to the significant transmembrane endocytotic potential of the developed nanosystems. It also became apparent that the nanoparticles with a lamellar or rod-shaped geometry possessed escalated levels (~ 100-fold) of transfection *vis-à-vis* the ones with spherical morphology (Zhang *et al.*, 2017).

1.2.2 Gadolinium Nanoparticles (GdNPs)

Among the entire range of contrast agents used for magnetic resonance imaging (MRI), the gadolinium (Gd^{3+}) ion is the most frequently employed. GdNPs are anchored by gadolinium (Gd) chelates

TABLE 1.1
Biomedical Application of AuNPs and Effect of FANC

(a) Biomedical Application of AuNPs

Nanoparticles	Application	References
4-aminothiophenol and cysteamine functionalized gold nanospheres	*In vitro* cytotoxic evaluation on HeLa cells	Avilaa *et al.*, 2017
d-GNPs	Inhibitory efficacy against AChE and BChE, induction of apoptosis in HeLa cells	Dorosti and Jamshidi, 2016
Succinito Pt (IV) complex, c,c,t- [Pt $(NH_3)_2Cl_2(O_2CCH_2CH_2CO_2H)_2$] (3) PEGylated AuNPs	Cytotoxicity studies showed that the AuNPs improved the platinum uptake, increased toxicity in several cancer cell lines (HeLa, A549, MCF-7), overcome cellular resistance mechanisms associated with cisplatin	Min *et al.*, 2012
Au-PSH-Cisplatin-FA	Reduced cytotoxic effect in normal cells, enhanced cytotoxicity in folate receptor positive ovarian cancer cells	Patra *et al.*, 2010
Au@GdDTPA	These NPs attenuated X-rays exhibited an extremely high r_1 relaxivity. Au@GdDTPA exhibited a strong blood pool effect, were taken up by the lymphatic system, these NPs induced a strong magnetic resonance.	Kim *et al.*, 2010

(b) Effect of FANC on Expression Profiles and Function of Human Endothelial Cells and Endothelial Progenitor Cells

Parameters	Effect	Unchanged
mRNA level	Down regulation of VCAM and VE-Cad	Ang-1, Cx43, eNOS, PAI-1, VEGF, KDR, ZO-1, JAM-1, IL-8, MMP-9
Protein level	NA	Cx43, eNOS, PAI-1,VE-Cad, KDR, caspase-3, PCNA
Cytotoxicity	Occurs above 500 nmol/L	NA
Labeling using liposome	> 98% efficiency and 130-fold induction of fluorescence intensity at 500 nmol/L	NA
Angiogenic activity (*in vitro*)	N/A	Tube formation in matrigel
Angiogenic activity (*in vivo*)	N/A	Limb salvage.

Source: Wang HH et al., *Int J Gerontol.* 2012; 6: 1–4.
Ang-1 = angiopoietin 1; Cx43 = connexin 43; eNOS = endothelial nitric oxide synthase; FANC = fluorescent gold nanocluster; IL-8 = interleukin 8; JAM-1 = junctional adhesion molecule 1; KDR = kinase insert domain receptor; MMP-9 = matrix metallopeptidase 9; PAI-1 = plasminogen activator inhibitor 1; PCNA = proliferating cell nuclear antigen; VCAM = vascular cell adhesion molecule; VE-Cad = vascular endothelial cadherin; VEGF = vascular endothelial growth factor; ZO-1 = zona occludens 1.

as an MRI contrast agent, due to its high density (Kim *et al.*, 2013, Roux *et al.*, 2010). Alluring contrast properties are attributed to the presence of seven unpaired electrons present in the outer shell of this metal structure, which offers a longer T_2 relaxation time, thus resulting in enhanced contrast imaging property. Further, a positive contrast in MRI is offered due to relaxation of the water signal, which is catalyzed by Gd^{3+}. A number of diverse routes have been employed for the facile synthesis of gadolinium oxide nanoparticles. Among all of these synthetic reactions, the precipitation method is most commonly employed, where Gd^{3+} are allowed to react with hydroxide ions which yields Gd in precipitated manner. The precipitated hydroxides are imposed to undergo thermal dehydration resulting in the formation of Gd oxides.

GdNPs are coated with biocompatible hydrophilic ligands, acting as drug carriers in therapeutics and imaging (Kim *et al.*, 2013).

GdNPs are considered as the conventional MRI contrast agents in clinical domain, whereas some of these nanoparticles are graded as toxic (nephrogenic system fibrosis) and get accumulated in the brain. Soon after, these particles come in contact with aqueous media, flammable gases are released, which causes impulsive ignition (Wei *et al.*, 2017). The primary roles of GdNPs are; (a) the enhancement of computer memory, (b) potent X-ray material, (c) in computer tomography as scintillation materials, (d) detection of early stage cancer by targeting coated Gd nanoparticles at tumors using MRI, and (e) as a dopant in glass (Wei *et al.*, 2017).

Inferable from their ability to target specific organ or infection, lower cytotoxicity and genotoxicity, improved clearance properties, and high relaxativities (t_1 or t_2 state), these second era structures, additionally named as affinity-targeted contrast agents, have provided GdNPs with another paradigm to biomedical applications (Zhang *et al.*, 2008). Table 1.2a highlights the biomedical use of GdNPs.

An incessant attempt was made by Yousef *et al.* (2016) to thoroughly investigate the effect of radiotherapy on the survival rate of F98 glioma cells. In their study, a combinatorial approach has been employed, where X-rays are coupled with GdNPs (radiosensitizers) to enhance the cytotoxic potential of radiotherapy towards malignant cells. The line of work was also diverted towards assessment of probable biochemical changes taking place in the glioma cells during the GdNP-coupled radiotherapy via FTIR vibrational spectroscopy. Processing of the obtained data using a multivariate chemometric approach *viz.* principal component analysis (PCA) clearly revealed that presence of GdNPs leads to the generation of significant change in the biochemical parmaters (*viz.* protein and lipid profile) and genetic material (*viz.* DNA) of the glioma cells. These demarcating changes in the biochemical vitals and nuclear material might be ascribed to alteration in the cellular functions

TABLE 1.2
Biomedical Applications of GdNPs and IONPs/MNPs

Nanoparticles	Application	References
	(a) Biomedical Application of GdNPs	
Ultra-small silica-based bismuth GdNPs	Dual magnetic resonance–computed tomography image-guided radiation therapy	Detappe *et al.*, 2017
Gd-based NPs (AGuIX®)	Radiation therapy for multiple brain melanoma metastases.	Kotb *et al.*, 2016
Viral capsid–conjugated GdNPs	Higher relaxivities and can be attached with target molecules	Datta *et al.*, 2008
Carbon nanotube-encapsulated Gd^{3+}	Reduced toxicity, enhanced MRI efficacies	Bolskar, 2008
	(b) Biomedical Application of IONPs/MNPs	
Surface-modified IONPs	Assessment of safety and protein interactions in A549 cell line.	Dyawanapellya *et al.*, 2017
Epirubicin (EPI)-loaded, ultra-small, super-paramagnetic iron oxide nanoparticle (EPI-USPIONP)	Theranostic agent and chemotherapy	Yoona *et al.*, 2017
Glycyrrhizic acid-coated IONPs	Hyperthermia-based therapy	Saeedi *et al.*, 2017
Dendrimer-conjugated IONPs	Thermally activated chemotherapy	Nigama and Bahadur, 2017

as a result of GdNPs. It was also deciphered that the presence of X-rays further escalated the biochemical alterations. The results of FTIR analysis further corroborated the findings of biochemical assays. The instrumental analysis highlighted the fact that GdNPs treated F98 glioma cells demonstrated significant apoptotic changes 24 h after incubation of radiotherapy (Yousef *et al.*, 2016).

In another different approach, supramolecular self-assembly strategy was employed for the facile synthesis of a multifunctional, Gd-based nanoparticulate system encapsulating a therapeutic agent *viz.* gemcitabine-5′-monophospahte (Li *et al.*, 2016). The developed system was tested for its theranostic efficacy in MDA-MB-231 tumor bearing *in vivo* model. The present study clearly depicted that the developed systems offers a better T_1 contrast *vis-à-vis* other mainstream MRI agents. The formulated nanoparticulate system depicted enhanced retention in the designated malignant site. Further, *in vivo* studies were conducted to validate the safety, efficacy and targeting propensity of the fabricated nanosystem. It was observed on competitive analysis that the GdNPs encompassing the chemotherapeutic moiety had a more pronounced antitumor activity, leading to an effective eradication of breast cancer when compared to free drugs (Li *et al.*, 2016).

Carmona *et al.* (2017) in their studies explored the quantitative biodistribution potential of GdNPs *in vivo* via micro-PIXE imaging. It was deciphered from the study that the GdNPs were successfully diffused in a heterogeneous manner in the tumor areas. Also, the experimental results pointed that the GdNPs depicted a highly resolved and magnified imaging distribution of chemical elements (Carmona *et al.*, 2017).

1.2.3 Iron Oxide/Magnetic Nanoparticles (IONPs)

Iron oxide nanoparticles (IONPs) are quite eye-catching due to their variegated applications as catalysts and biomedicine (Fu and Ravindra, 2012). Since the last two decades, the thermophysical properties played a major role in bringing IONPs to light (Mahmoudi *et al.*, 2011, Hasany *et al.*, 2012). Varied allied domains, wherein these nanoparticles play a role in electronics and medicine, bioimaging, chemical synthesis and analysis, photonics, and cosmetics. Iron oxide nanomaterials possess a suitable surface chemistry making these convenient to be used for MRI contrast enhancement, immunoassays, tissue repair, cell separation, targeted drug delivery, hyperthermia and detoxification of biological fluids (Muller *et al.*, 2008).

Magnetic particles that are biocompatible and nontoxic are surface-coated for the accomplishment of the above-mentioned applications and permit targeted delivery in specific areas. These nanomaterials are used for the loading of drugs, enzymes, proteins and antibodies, and are directed to the specific area (*viz.* tissue, organ or tumor) with an external magnetic field (Muller *et al.*, 2008). Targeted iron oxide nanomaterials have quite favorable biomedical applications in molecular imaging, cellular labeling, apoptosis, etc. Targeted delivery of these nanomaterials *in vitro* and in animal experiments has proven to be useful in the evaluation of varied applications. Despite these, the detection of enzyme activity, proteins, oligonucleotide sequences, pathogens and enantiomeric impurities are considered among the primary applications of such nanomaterials. IONPs also play another major role in cell and *in vitro* bioseparation. A bird's eye view on the biomedical applications of IONPs have been summed up in Table 1.2b.

In the cases of drug delivery, the size of the nanoparticles plays a key role. Particles with diameter greater than 200 nm are effortlessly isolated by spleen and are ultimately discarded by the phagocytic system. These particles also exhibit significant therapeutic advantages apart in MRI. The study by Jordan *et al.* (1993) revealed that these particles absorb magnetic field energy and transform it into heat thus increasing the temperature of tumor cells in order to destroy them using hyperthermia.

Yoo *et al.* (2017) in an experiment utilized traditional coprecipitation method for the development of polyethylene glycol (PEG)-coated supramagnetic iron oxide–poly(gallol) nanoparticle complex. This activatable interpolymer complex-based sensor was specifically designed to screen varied deleterious diseases *viz.* diabetes, cancer and arthrosclerosis on the basis of quantification of oxidative stress levels. The developed sensor exhibited the phenomenon of decomplexation, when placed in an oxidative environment (ON phase) and showed enhanced T_2 relaxation. A reverse phenomenon was portrayed by the complexed nanoparticles, where diminished T_2 relaxation was observed (OFF phase). This significant declination in the T_2 level might be attributed to polymeric core, which provided foliage to the supramagnetic IONPs, ultimately resulting in the impaired interaction among the nanoparticles and the protons present in the surrounding milieu (Yoo *et al.*, 2017).

Using a different approach of surface-engineered IONPs offering triple modality *viz.* PET/NIRF/MRI were evinced by Xie *et al.* (2010). The theranostic potential of these developed multifunctional, surface-anchored nanoparticulate systems were tested *in vivo*. The outcomes of *in vivo*, *ex vivo* and histopathological studies were found to be promising interims of:

i. Effective translocation and accumulation at the designated site
ii. Enhanced retention time
iii. Reduced systemic toxicity and renal clearance
iv. Enhanced bioavailability and permeability
v. Improved half-life
vi. Enhanced permeation and retention (EPR) effect

1.2.4 Ceramic Nanoparticles (CENPs)

Currently, the nanoparticles based on ceramic materials have grown rapidly due to their noteworthy applications in biomedical sciences. Certain other applications of these materials have also been witnessed due to their remarkable properties (*i.e.* chemical inertness and heat resistance). In the field of biomedical sciences these particles act as imaging and drug delivery tools. In order to be strategies worth noticing, numerous factors (such as nature of surface, porosity, size, ratio between surface area and volume) need to be controlled. These nanoparticles are mainly composed of metal phosphates, oxides, carbides, and carbonates and metalloids of silicon, calcium, titanium, etc. Ceramic nanomaterials have proved to be of significant use in the pursuance of treating varied diseases *e.g.* cancer, glaucoma, or bacterial infection as drug delivery system, whereas some of these materials show toxicity too (Talegaonkar *et al.*, 2015).

In order to reduce the cytotoxicity of these materials in the biological systems, an attempt for the advancement of their physical-chemical properties has been made with the development of nanoscale ceramic materials like zirconia, silica, alumina, titanium oxide, and hydroxyapatite (Avti *et al.*, 2011). Certain factors of these nanoparticles, such as high stability, easy incorporation of hydrophobic and hydrophilic systems, high load capacity and a variety of administration routes, make them potent materials to control the delivery of drugs (Vega *et al.*, 2012). The new synthetic methods have abetted in improving the physiochemical properties of these ceramic nanoparticles, as a result of which, their cytotoxicity is comprehensively reduced in the biological system.

Regarding applications of CENPs in biomedicine, the controlled release of drugs has turned out to be a standout amongst the most exploited areas. In addition, these NPs can be surface-functionalized via tagging with varied organic groups for enhanced direct effects. Table 1.3a depicts various biomedical application of ceramic nanoparticles.

1.2.5 Titanium Oxide Nanoparticles (TiONPs)

The diverse crystalline structures of titanium dioxide make it a photocatalytic material, with broad dielectric and optical characteristics. Owing to the anatase stability of TiO_2, NPs at nanoscale levels like these nanoparticles have shown wide versatility in numerous fields. One of such areas is pharmacology, where these NPs have been widely used as drug eluting vehicles or excipients formulations. Also, these NPs are being utilized as a part of photodynamic treatments

TABLE 1.3
Enlisted Potentials of CENPs, TiONPs and SiNPs and SMNPs Based Bio-Imaging Studies

Nanoparticles	Application	References
	(a) Biomedical Application of Ceramic Nanoparticles	
Ceramic NPs	Near-infrared bioimaging	Soga and Kamimura 2017
Bioactive glass ceramic nanoparticle-coated poly(L-lactic acid) scaffold	Osteogenic differentiation of adipose stem cells	Mahdavi *et al.*,2017
PLLA-ceramic nanofibers	Bone regeneration	Santosa *et al.*,2017
Nano-HAP	Showed excellent results in the remineralization of teeth	Fadeel and Bennett 2010
Nanocrystalline hydroxyapatite (Nano-HAP)	Injectable paste for bone substitution, good osteoconductive properties, acts as antibody delivery agents for bone infections	Laschke *et al.*, 2007
Bioactive glass-based ceramic scaffolds	Promising for orthopedic applications in future	Chen *et al.*, 2006
	(b) Biomedical Application of Titanium Nanoparticles	
Folic acid–targeted TiO_2NPs	Cancer therapy in osteosarcoma cells	Ai *et al.*, 2017
Zinc-doped TiO_2NPs	Proangiogenesis	Nethi *et al.*, 2017
TiO_2NPs	Reduction of genotoxicity	Hackenberg *et al.*, 2017
TiO_2NPs	Prevention of skin carcinogenesis	Tyagi *et al.*, 2017
CdSe/ZnS	Maltose binding protein	Medintz *et al.*, 2005
	(c) Biomedical Application of Silica Nanoparticles	
Mesoporous silica nanoparticles	Active leukemia stem cell targeting	Beck *et al.*, 2017
Agarose and heparin encapsulated mesoporous silica nanoparticles	Biomedical silicone	Wu *et al.*, 2017
Monodispersed small silica nanoparticles	Sentinel lymph node mapping	Quan *et al.*, 2017
Ordered, large-pore mesoporous silica nanoparticles	Cellular delivery	Chiu *et al.*, 2017
Biogenic silica nanoparticles	Assessment of human lung cell viability, morphology, mitochondrial function, gene expression, ROS production	Athinarayanan *et al.*, 2017
Silica-SPIO nanoparticles	Magnetic hyperthermia agents	Saavedra *et al.*, 2010
Lanthanide-doped silica nanoparticles	Used for multiplexed immunoassays	Murray *et al.*, 2010
	(d) *In Vitro* and *In Vivo* Bioimaging Studies Using Quantum Dots	
Antimonene QDs	Near-infrared photothermal agents for effective cancer therapy	Tao *et al.*, 2017
Ti_3C_2 MXene QDs	Multicolor cellular imaging	Hue *et al.*, 2017
Ag and Mn co-doped In_2S_3/ZnS QDs	Dual modal (*in vitro* and *in vivo*) targeted imaging	Lai *et al.*, 2017
Graphene QDs	Near-infrared bioimaging and intracellular nanothermometry	Kumawat *et al.*, 2017
Fluorescent black phosphorous QDs	Evaluation of acetylcholinesterase activity	Gu *et al.*, 2017

exploiting their efficient photooxidation properties. Surgery, radiotherapy and chemotherapy are some of the therapies which are currently being used for treating cancer; however, these therapies are often associated with deleterious effects on the human health due to nonspecific cellular targeting (Vega *et al.*, 2012).

Several evidences have indicated the potential utility of NP based therapies for cancer treatment. In lieu of this, TiO_2NPs have been employed for *in vitro* studies successfully. In fact, a new chemotherapeutic agent based on TiO_2NP powder has been proposed for colon cancer treatment. The human colon carcinomas are destroyed when they are exposed to photoexcited TiO_2NPs. Furthermore, the TiO_2NPs were docked with gold and platinum which thus fundamentally lessened the survival rate of malignant or cancerous cells (Vega *et al.*, 2012).

In order to enhance the clinical applicability and significance of this deleterious disease, a few endeavors have been laid by means of utilizing the TiO_2NPs. The photocatalytic impact of TiO_2NPs is basically connected to the concentration of these NP (Vega *et al.*, 2012). The photocatalytic activation is completed by controlling light exposure at the tumor locales that have been treated with TiO_2 (Vega *et al.*, 2012). The investigations demonstrated that a critical reduction in survival rate of human colon carcinoma cells was observed when exposed with Au-or Pt-doped TiO_2 nanoparticles (Lazau *et al.*, 2007).

A novel therapeutic strategy (referred to as sonodynamic therapy) is expected to overcome the limitations of preexisting chemotherapies thereby enhancing the quality of treatment especially in case of malignant gliomas (Vega *et al.*, 2012). The TiO_2NPs acts as a photosensitizer and possess an innate ability to get activated by ultrasound. In this context, Yamaguchi *et al.* proposed TiO_2/PEG (polyethylene glycol) nanoparticles for treatment of malignant gliomas utilizing sonodynamic therapy (Yamaguchi *et al.* 2011). It was demonstrated from the experiments that TiO_2/PEG NPs did not cross the normal blood brain membrane; instead they were successfully concentrated in malignant gliomas (Yamaguchi *et al.*, 2011). Table 1.3b portrays the biomedical application of TiO_2NPs.

1.2.6 Silica Nanoparticles (SiNPs)

The diverse sorts of silica-based NPs are monoliths, rod-like particles, fibers, hollow and solid nanospheres, silica nanotubes, and mesoporous silica nanoparticles (MSNs) (Avti *et al.*, 2011). Pharmaceutically dynamic mesoporous silica atoms exhibit a few worthwhile attributes such as the automated release of potential drugs, their simplicity of disintegration and the simplicity of accessibility in the living being (Vega *et al.*, 2012). However, a major problem often accompanied with the administration of such nanoparticles is their incompatibility in the biological regimen and expanded unfavorable effects. This can be ascribed to the way that slight variations during synthesis might persist in which in turn can induce alteration in shape, size and subsequent physiochemical properties of such nanoparticles (Fadeel and Bennett, 2010). Furthermore, there are multimodal silica nanoparticles, which can be viably utilized as demonstrative markers for cancer testing and have already been approved for human testing (Vega *et al.*, 2012).

Multimodal silica nanoparticles (MSNPs) with a hydrodynamic diameter of 7 nm have been developed and are being currently used for tumor diagnosis, as conventional markers exhibit poor specificity in tumor tissues (Vega *et al.*, 2012). In addition, further surface functionalization employing polyethylene glycol (PEG), arginine-glycine-aspartic acid peptide ligands and radioactive iodine was carried out on MSNPs. It was deciphered that these nanoparticles exhibited a higher affinity and prolonged residence time in tumors and peripheral blood fluids. Further *in vivo* studies on various models of nodal metastasis in various mouse tissues publicized that these nanoparticles showed improved selectivity, increased accumulation of MSNPs in mouse melanoma xenografts. These nanoparticles were among the first to get approved for a first-in-human clinical trial (Benezra *et al.*, 2011; Vega *et al.*, 2012). The biomedical application of silica nanoparticles is listed in Table 1.3c.

1.2.7 Semiconductor-Based Nanoparticles/Quantum Dots (SMNPs/QDs)

Cadmium (Cd)-based nanoparticles are the most commonly studied nanoparticles and are represented by notation CdE, where E can be sulfide, selenide and telluride. Owing to their core shell binary structure, these nanoparticles are generally referred to as quantum dots (QDs) (Avti *et al.*, 2011). QDs (2–6 nm) represent some of the nanomaterials that significantly affect the research in many fields across the physical, chemical, and biological sciences (Algar *et al.*, 2010). Owing to their unique electro-optical properties, QDs have developed multidisciplinary interests (Murray *et al.*, 2000). In spite of the fact that QD research gained momentum in the mid-1990s, a noteworthy enthusiasm over the bioanalytical, biophysical, and biomedical research groups was ostensibly developed with the adoption of QD bioconjugates as fluorescent markers for biological imaging. When contrasted with conventional fluorescent dyes, QDs are perceived frequently for providing better illumination and photostability (Medintz *et al.*, 2005).

Based on design, QDs can be chiefly placed into two categories *viz.* "non-integrated" and "integrated". A non-integrated QD can be defined as one that is selectively introduced to a bioanalysis as a consequence of biorecognition (Algar *et al.* 2010) *e.g.* include the use of QDs as fluorescent labels in microarrays (Gerion *et al.*, 2003; Geho *et al.*, 2005), or electroactive labels in assays based on anodic stripping voltammetry (Kim *et al.*, 2006). On the other hand, an integrated QD is referred as the one, present in a system throughout a bioanalysis *e.g.* include the use of QDs for signal transduction and as a scaffold for biorecognition (Algar *et al.*, 2010).

In addition, QDs can be surface modified with biocompatible and biodegradable moieties (*viz.* polymers, chitosan, cellulose and hydroxyapatite) and can be transported into cells without affecting the core structure properties (Algar *et al.*, 2010). Other areas where these NPs have found their applications are drug delivery systems (Wang *et al.*, 2007), multiplexed gene and protein expression analysis (Zhang *et al.*, 2006) and stem cell tracking (Lei *et al.*, 2009). The biomedical application of SiNPs has been tabulated in Table 1.3d.

1.2.8 Organic/Carbon-Based Nanoparticles (CNPs)

A considerable amount of enthusiasm for material and life sciences has been developed with the production of nanosized structures. A portion of the outstanding carbon-based nanomaterials utilized as a part of biomedical applications incorporate carbon nanotubes (CNTs) or fullerene (Avti *et al.*, 2011). CNTs have attracted wide attention and have become a burning issue in bionanomaterial research, since their discovery in the 1990s (Yu and Li, 2009). These materials are cylindrical graphene structures which possess extraordinary structures and properties (large surface area, efficient thermal, electrical conductivity and mechanical strength) (Yu and Li, 2009). Numerous studies have unearthed the potential utility of CNTs in various fields; some of them are electronic devices, composite and hydrogen storage materials, catalysts, molecular adsorbents, as well as chemical and biological sensors.

Recent scenario has perceived a tremendous change in the field of biomedical application. Nowadays the application researches of CNTs in biomedicine, particularly in drug delivery, have become a new hotspot. Owing to their alluring photophysical and photochemical properties, these nanomaterials have been harnessed for various biomedical applications *viz.* MRI contrast agents, photoacoustic, cell tracking, near-Infrared and Raman imaging, pressure sensors, gene radiotracers and protein microarray and tissue engineering. (Bolskar 2008; Chen *et al.*, 2008; Sitharaman *et al.*, 2008; Jin *et al.*, 2008; Pramanik *et al.*, 2009; Mashal *et al.*, 2010)

Hartman *et al.*, (2008) exhibited the utilization of carbon nanohorns (CNHs) which are an extraordinary sort of single-walled carbon nanotubes (SWNTs). In their study, they loaded the gadolinium chloride into ultra-short SWNTs. Gadonanotubes showed relaxivity as higher as 40–90 times than commercially available contrast agents. This is attributed to the spatial confinement of paramagnetic ion clusters within the nanotubes. The relaxivity is found to be extremely sensitive to pH value in a range of 7.0–7.4. The distinction in pH values amongst tumor and normal tissues may turn out to be a great vector for early diagnosis of tumors (Hartman *et al.*, 2008).

Richard *et al.* (2008) absorbed the amphiphilic gadolinium (II) chelate on multi-walled carbon nanotubes (MWNTs). The heterogeneous suspension showed both positive and negative contrast during *in vivo* T_2 weight imaging. This extraordinary magnetic susceptibility is due to magnetic moment, caused by the motion of electrons around the circumference of nanotubes. The biomedical applications of CNTs have been tabulated in Table 1.4.

TABLE 1.4
Highlighted Biomedical Application of CNTs

Nanoparticles	Applications	References
Bacterial cellulose–functionalized MWNTs	Bone regeneration	Hernandez *et al.*, 2017
Carboxylated MWNTs	Bone substitution	Gholami *et al.*, 2017
DNA-decorated SWNTS	Surface-enhanced Raman scattering imaging of cells	Ursu *et al.*, 2017
CNTs	Gene delivery	Rius *et al.*, 2017
Graphene/SWNTs	*In vivo* bone regeneration and promotion of osteoblast function *in vitro*	Silva *et al.*, 2017
Bacterial cellulose/polyaniline/SWNTs	Biosensing	Jasim *et al.*, 2017
Platinum-folic acid-CNT complex	Selective killing of cancer cells	Dhar *et al.*, 2008

1.3 CONCLUSIONS

Continuous research in biomaterials research will essentially empower the transformation of nanoparticles, especially nanobiomaterials, from laboratory to clinical utilization. These advances will play imperative role in understanding the accompanying objectives:

(a) Targeted drug delivery of traditional and nontraditional medicines to minimize secondary systematic negative effects
(b) Right-on-time *in vivo* imaging of transforming cells
(c) Appraisal of treatment results
(d) Enhancing surgical results utilizing intraoperative imaging agents
(e) Successful programming of the immune system to target and kill pernicious cells
(f) Analytic test assays for early disease identification
(g) Tools for intrusive imaging and drug development

Chapter 1 exhibits the momentous research work and future developments on nanoparticles in the application of biomedicine through the discussion of chosen illustrations. Beyond a doubt, nanoparticles exhibit incredible potential for the leap forward of a novel era of biomedical materials. Reported studies on nanobiomaterials have prompted huge advancement with the guarantee of much more prominent advances prone to come later on. Overall, creating controlled properties, reliable and reproducible generation strategies for biocompatible nanoparticles will be essential and valuable to pave way for more noteworthy acknowledgment of commercially available nanoparticles in biomedical applications.

Chapter 1 has also demonstrated the application of this phenomena to biomedical applications, for example, bioassays and therapy. Likewise, for the successful utilization of the nanoparticles in therapeutic applications, the distribution and long-term toxic impact of nanoparticles must be thoroughly assessed, since the majority of the metallic nanoparticles like gold and silver are fairly steady and non-biodegradable. The elimination of such nanoparticles when applied for biosensing purposes must also be precisely considered. The development of nonmetal and biodegradable materials may offer an alluring alternative. Nevertheless, the capability of nanoparticles in enhancing the sensitivities of bioassays and cancer treatment efficiency recommends that the field of the nanoparticles for biomedical applications stays promising.

One truly extraordinary aspect of nanobiomaterial-based therapies is the ability to accomplish multiple objectives while simultaneously utilizing multifunctional materials. While some early types of nanoparticles are at present accessible, challenges stay for the ever progressively sophisticated materials, for example, nanobiomaterials. These hurdles are principally related with nanoparticle pharmacology, including clearance and toxicity. The interface between materials science and biomedicine is giving new and exciting avenues of research, however the reality is that issues related with nanotoxicity and *in vivo* clearance are restrictive elements. The promising elements of nanoparticles in light of biomedical performance have been featured. Given their unique properties, nanoparticles are at the cutting edge of these novel opportunities.

Nanobiomaterials have diverse applications, as we have attempted to introduce above. With the development of new production technologies, the utilization of known techniques for creating nano products and influx of new information from varied disciplines of science, new uses are being continuously devised prompting better biomaterials and implants and, therefore, higher quality of life. The following decade is relied upon to be an exceptionally energizing period for nanoscale research applications.

ACKNOWLEDGMENT

The authors are thankful to PURSE II Grant, DST, Govt. of India, for providing financial support to IFSC through PURSE grant.

REFERENCES

Ai JW, Liu B, Liu WD. Folic acid-tagged titanium dioxide nanoparticles for enhanced anticancer effect in osteosarcoma cells. *Mater Sci and Eng C*. 2017; 76: 1181–1187.

Algar WR, Tavares AJ, Krull UJ. Beyond labels: A review of the application of quantum dots as integrated components of assays, bio probes, and biosensors utilizing optical transduction. *Anal Chim Acta*. 2010; 673: 1–25.

Amjad MS, Sadiq N, Qureshi H, Fareed G, Sabir S. Nanoparticles: An emerging tool in biomedicine. *Asian Pac J Trop Dis*. 2015; 5: 767–771.

Athinarayanan J, Periasamy VS, Alhazmi M, Alshatwi AA. Synthesis and biocompatibility assessment of sugarcane bagasse-derived biogenic silica nanoparticles for biomedical applications. *J Biomed Mater Res Part B: Appl Biomater*. 2017; 105: 340–349.

Avilaa PEC, Molinaa JLP, Villavicenciob LLF, Dominguezc JPC, Chilakapatid MK, Lopezb MS. Biological response of HeLa cells to gold nanoparticles coated with organic molecules. *Toxicol In Vitro*. 2017; 42: 114–122.

Avti PK, Patel SC, Sitharaman B. *Nanobiomaterials: Current status and Future Prospects*. Nanobiomaterial Handbook: Taylor and Francis Group, LLC: (2011); pp.1–24.

Beck M, Mandal T, Buske C, Lindén M. Serum protein adsorption enhances active leukemia stem cell targeting of mesoporous silica nanoparticles. *ACS Appl Mater Interfaces*. 2017; 9: 18566–18574.

Benezra M, Medina OP, Zanzonico PB. Multimodal silica nanoparticles are effective cancer-targeted probes in a model of human melanoma. *J. Clin. Investig*. 2011; 121: 2768–2780.

Bolskar, R.D. (2008) Gadolinium endohedral metallofullerene-based MRI contrast agents. In *Medicinal Chemistry and Pharmacological Potential of Fullerenes and Carbon Nanotubes*. Springer, Dordrecht, 157–180.

Carmona A, Roudeau SE, Homel BL, Boissinot BS, Prezado Y, Ortega R. Heterogeneous intra-tumoral distribution of gadolinium nanoparticles within U87 human glioblastoma xenografts unveiled by micro-PIXE imaging. *Anal Biochem*. 2017; 523: 50–57.

Chen QZ, Thompson ID, Boccaccini AR. In vitro study of the antibacterial activity of bioactive glass-ceramic scaffolds. *Biomaterials*. 2006; 27: 2414.

Chen Z, Goodwin AP, Kattah MG, Daranciang D, Wang X, Zhang G *et al*. Protein microarrays with carbon nanotubes as multicolor Raman labels. *Nat Biotechnol*. 2008; 26: 1285–1292.

Chiu HY, Leonhardt H, Bein T. Synthesis and functionalization of ordered large-pore mesoporous silica nanoparticles for biomedical applications. *Chemie Ingenieur Technik*. 2017; 89: 876–886.

Datta A, Hooker JM, Botta M, Francis MB, Aime S, Raymond KN. Use of YbIII-centered near-infrared (NIR) luminescence to determine the hydration state of a 3,2-HOPO-based MRI contrast agent. *J Am Chem Soc*. 2008; 130: 2546.

Detappe A, Thomas E, Tibbitt MW, Kunjachan S, Zavidij O, Parnandi N *et al*. Ultrasmall silica-based bismuth gadolinium nanoparticles for dual magnetic resonance–computed tomography image guided radiation therapy. *Nano Lett*. 2017; 17: 1733–1740.

Devi RV, Sengani M, Grumezescu AM. Recent trends and methodologies in gold nanoparticle synthesis–A prospective review on drug delivery aspect. *Open Nano.* 2017; *2*: 37–46.

Dhar S, Liu Z, Thomale J, Dai HJ, Lippard SJ. Targeted single-wall carbon nanotube-mediated Pt(IV) pro drug delivery using folate as a homing device. *J Am Chem Soc.* 2008; 130: 11467–11476.

Dorosti N, Jamshidi F. Plant-mediated gold nanoparticles by Dracocephalum kotschyi as anticholinesterase agent: Synthesis, characterization, and evaluation of anticancer and antibacterial activity. *J Appl Biomed.* 2016; 14: 235–245.

Dyawanapellya S, Jagtap DD, Dandekarb P, Ghoshd G, Jain R. Assessing safety and protein interactions of surface-modified iron oxide nanoparticles for potential use in biomedical areas. *Colloids Surf B: Biointerfaces.* 2017; 154: 408–420.

Dykman L, Khlebtsov N. Gold nanoparticles in biomedical applications: Recent advances and perspectives. *Chem Soc Rev.* 2012; 41(6): 2256–2282.

Ernest H, Shetty R. Impact of nanotechnology on biomedical sciences: Review of current concepts on convergence of nanotechnology with biology. *J. Nanotechnol.* 2005; 1: 1–14.

Fadeel B, Bennett AEG. Better safe than sorry: Understanding the toxicological properties of inorganic nanoparticles manufactured for biomedical applications. *Adv Drug Deliv Rev.* 2010; 62: 362–374.

Fu C, Ravindra NM. Magnetic iron oxide nanoparticles: Synthesis and applications. *Bioinspired, Biomimetic and Nanobiomaterials.* 2012; 1(4): 229–244.

Geho D, Gurnani P, Huebschman M, Herrmann P, Espina V, Shi A *et al.* Pegylated, steptavidin-conjugated quantum dots are effective detection elements for reverse-phase protein microarrays. *Bioconjugate Chem.* 2005; 6: 559–566.

Gerion D, Kannan B, Fu AH, Parak WJ, Chen DJ, Majumdar A *et al.* Room-temperature single-nucleotide polymorphism and multiallele DNA detection using fluorescent nanocrystals and microarrays. *Anal Chem.* 2003; 75: 4766–4772.

Gholami F, Ismail S, Noor AFM. Development of carboxylated multi-walled carbon nanotubes and bovine serum albumin reinforced hydroxyapatite for bone substitute applications. *J Aust Ceram Soc.* 2017; 53 (1117).

Gu W, Yan Y, Pei X, Zhang C, Ding C, Xian Y. Fluorescent black phosphorus quantum dots as label-free sensing probes for evaluation of acetylcholinesterase activity. *Sens Actuators B Chem.* 2017; 250: 601–607.

Hackenberg S, Scherzeda A, Zappa A, Radeloffa K, Ginzkeyb C, Gehrkea T. Genotoxic effects of zinc oxide nanoparticles in nasal mucosa cells are antagonized by titanium dioxide nanoparticles. *Mutat Res Genet Toxicol Environ Mutagen.* 2017; 816–817: 32–37.

Hartman KB, Bolskar RD, Muthupillai R, Helm L, Tóth É, Merbach AE *et al.* Gd-based macromolecules and nanoparticles as magnetic resonance contrast agents for molecular imaging. *J Nano Lett.* 2008; 8: 415–419.

Hasany SF, Rehman A, Jose R, Ahmed I. Iron oxide magnetic nanoparticles: A short review. In *AIP.* 2012; 1502 (1): 298–321.

Hernandez JMG, Escalante A, Flores H, Gonzalez FJ, Gatenholm P, Toriz G *et al.* In vitro evaluation of osteoblastic cells on bacterial cellulose modified with multi-walled carbon nanotubes as scaffold for bone regeneration. *Mater Sci and Eng: C.* 2017; 75: 445–453.

Hrelescu C, Sau TK, Rogach AL, Jäckel F, Laurent G, Douillard L, Charra F. Selective excitation of individual plasmonic hotspots at the tips of single gold nanostars. *Nano Lett.* 2011; 11(2): 402–407.

Huang X, Sayed MAE. Gold nanoparticles: Optical properties and implementation in cancer diagnosis and photo thermal therapy. *J Adv Res.* 2010; 1: 13–28.

Hue Q, Zhu M, Pei Z, Li H, Wang Z, Huang Y *et al.* Photo luminescent Ti_3C_2 MXene quantum dots for multicolor cellular imaging.. *Adv Mater.* 2017; 29: 1604847.

Jain PK, Lee KS, Sayed IHE, Sayed MAE. Calculated absorption and scattering properties of gold nanoparticles of different size, shape, and composition: Applications in biological imaging and biomedicine. *J Phys Chem B.* 2006; 110: 7238–7248.

Jasim A, Ullah MW, Shi Z, Lin X, Yang G. Fabrication of bacterial cellulose/polyaniline/single-walled carbon nanotubes membrane for potential application as biosensor. *Carbohydr Polym.* 2017; 163: 62–69.

Jeong Kim T, Seok Chae K, Chang Y, Ho Lee G. Gadolinium oxide nanoparticles as potential multimodal imaging and therapeutic agents. *Curr Top Med Chem.* 2013; 13(4): 422–433.

Jin H. Where is it heading? Single-particle tracking of single-walled carbon nanotubes. *ACS Nano.* 2008; 9: 1749–52.

Jordan A, Wust P, Fählin H, John W, Hinz A, Felix R. Inductive heating of ferrimagnetic particles and magnetic fluids: Physical evaluation of their potential for hyperthermia. *Int J Hyperthermia.* 1993; 9(1): 51–68.

Khlebtsov N, Dykman L. Biodistribution and toxicity of engineered gold nanoparticles: A review of in vitro and in vivo studies. *Chem Soc Rev.* 2011; 40(3): 1647–1671.

Kim HK, Park JA, Huh MI, Jung JC, Chang Y, Kim TJ. Gd Complexes of DO3A-(Biphenyl-2,2′-bisamides) Conjugates as MRI Blood-Pool Contrast Agents. *J Mater Chem.* 2010; 20: 5411.

Kim JH, Seo KS, Wang J. Highly sensitive electrochemical determination of alfatoxin bl using quantum dots-assembled amplification labels. *IEEE Sens J.* 2006; 6: 248–253.

Kim J, Seok T, Chae K, Chang Y, Ho Lee G. (2013). Gadolinium oxide nanoparticles as potential multimodal imaging and therapeutic agents. *Curr Topics in Med Chem.* 2013; 13(4): 422–433.

Kotb S, Detappe A, Lux F, Appaix F, Barbier EL, Tran VL. Gadolinium-based nanoparticles and radiation therapy for multiple brain melanoma metastases: Proof of concept before phase I trial. *Theranostics.* 2016; 6: 418–427.

Kumawat MK, Thakur MC, Gurung RB, Srivastava R. Graphene quantum dots from mangifera indica: Application in near-infrared bio imaging and intracellular nano-thermometry. *ACS Sustain Chem Eng.* 2017; 5: 1382–1391.

Lai PY, Huang CC, Chou TH, Ou KL, Chang JY. Aqueous synthesis of Ag and Mn co-doped In2S3/ZnS quantum dots with tunable emission for dual-modal targeted imaging. *Acta Biomater.* 2017; 50: 522–533.

Laschke MW, Witt K, Pohlemann T, Menger MD. Injectable nanocrystalline hydroxyapatite paste for bone substitution: In vivo biocompatibility and vascularization. *J Biomed Mater Res Part B: Appl Biomater.* 2007; 82(B): 494.

Laurent S, Forge D, Port M, Roch A, Robic C, Vander Elst L, Muller RN. Magnetic iron oxide nanoparticles: Synthesis, stabilization, vectorization, physicochemical characterizations, and biological applications. *Chem Rev.* 2008; 108(6): 2064–2110.

Lazau C, Mocanu L, Miron I, Sfirloaga P. Consideration regarding the use of TiO_2 doped nanoparticles in medicine. *Dig J Nanomater Biostruct.* 2007; 2: 257–263.

Lei Y, H Tang, Feng M, Zou B. Applications of fluorescent quantum dots to stem cell tracing in vivo. *J Nanosci Nanotechnol.* 2009; 9: 5726.

Li L, Tong R., Li M, Kohane DS. Self-assembled gemcitabine gadolinium nanoparticles for magnetic resonance imaging and cancer therapy. *Acta Biomater.* 2016; 33: 34–39.

Mahdavi FS, Salehi A, Seyedjafari E, Sangcheshmeh AM, Ardeshirylajimi A. Bioactive glass ceramic nanoparticles-coated poly(L-lactic acid) scaffold improved osteogenic differentiation of adipose stem cells in equine. *Tissue and Cell.* 2017; 49(5): 565–572.

Mahmoudi M, Sant S, Wang B, Laurent S, Sen T. Superparamagnetic iron oxide nanoparticles (SPIONs): Development, surface modification and applications in chemotherapy. *Adv Drug Deliv Rev.* 2011; 63(1-2): 24–46.

Mashal A, Sitharaman B, Li X, Avti PK, Sahakian AV, Booske JH, Hagness SC. Toward carbon-nanotube-based theranostic agents for microwave detection and treatment of breast cancer: Enhanced dielectric and heating response of tissue-mimicking materials. *IEEE Trans Biomed Eng.* 2010; 57: 1831–1834.

Mata R, Nakkala JR, Sadras SR. Poly phenol stabilized colloidal gold nanoparticles from Abutilon indicum leaf extract induce apoptosis in HT-29 colon cancer cells. *Colloids Surf B: Biointerfaces.* 2016; 143: 499–510.

Medintz IL, Uyeda HT, Goldman ER, Mattoussi H. Quantum dot bio-conjugates for imaging, labeling and sensing. *Nat Mater.* 2005; 4: 435–446.

Min Y, Mao CQ, Chen S, Ma G, Wang J, Liu Y. Combating the drug resistance of cisplatin using a platinum pro drug based delivery system. *Angew Chem Int Ed.* 2012; 51: 6742.

Mukherjee P, Bhattacharya R, Bone N, Lee YK, Patra CR, Wang S, Kay NE. Potential therapeutic application of gold nanoparticles in B-chronic lymphocytic leukemia (BCLL): Enhancing apoptosis. *J Nanobiotechnol.* 2007; 5(1): 4.

Muller, R.N, Laurent, S, Forge, D, Port, M, Roch, A, Robic, C, Vander Elst, L. (2008). Magnetic iron oxide nanoparticles: synthesis, stabilization, vectorization, physicochemical characterizations, and biological applications. *Chem Rev.* 2008; 108(6): 2064–2110.

Murray CB, Kagan CR, Bawendi MG. Synthesis and characterization of monodisperse nanocrystals and close packed nano crystal assemblies. *Annu Rev Mater Sci.* 2000; 30: 545–610.

Murray K, Cao YC, Ali S, Hanley Q. Lanthanide doped silica nanoparticles applied to multiplexed immunoassays. *Analyst.* 2010; 135: 2132–2138.

Nethi SK, Anand NA, Oller BR, Diéguez AR, Ruiz SG, Patra CR. Design, synthesis and characterization of doped-titanium oxide nanomaterials with environmental and angiogenic applications. *Sci Total Environ.* 2017; 599–600: 1263–1274.

Nigama S, Bahadur D. Dendrimer-conjugated iron oxide nanoparticles as stimuli-responsive drug carriers for thermally-activated chemotherapy of cancer. *Colloids Surf B: Biointerfaces.* 2017; 155: 182–192.
Patra CR, Bhattacharya R, Mukherjee P. Fabrication and functional characterization of gold nano-conjugates for potential application in ovarian cancer. *J Mater Chem.* 2010; 20: 547–554.
Pramanik M, Song HK, Swierczewska M, Green D, Sitharaman B, Wang LV. *In vivo* carbon nanotube enhanced non-invasive photo acoustic mapping of the sentinel lymph node. *Phys Med Biol.* 2009; 54: 3291–3301.
Quan B, Lee C, Yoo JS, Piao Y. Facile scalable synthesis of highly monodisperse small silica nanoparticles using alkaline buffer solution and their application for efficient sentinel lymph node mapping. *J Mater Chem B.* 2017; 5: 586–594.
Richard C, Doan BT, Beloeil JC, Bessodes M, Toth E´, Scherman D. Non-covalent functionalization of carbon nanotubes with amphiphilic gd3+ chelates: Toward powerful t1 and t2 MRI contrast agents. *Nano Lett.* 2008; 8: 232–236.
Rius AC, Boase NRB, Font I, Coronas N, Perez VR, Thurecht KJ, Borrós S. In vivo fate of carbon nanotubes with different physicochemical properties for gene delivery applications. *ACS Appl Mater Interfaces.* 2017; 9: 11461–11471.
Roux S, Faure AC, Mandon C, Dufort S, Rivière C, Bridot JL, Le Pape A. Multifunctional gadolinium oxide nanoparticles: Towards image-guided therapy. *Img Med.* 2010; 2(2): 211–223.
Saavedra M, Hernández ER, Boré A, Arcos D, Regí MV, Vilaboa N *et al.* Magnetic mesoporous silica spheres for hyperthermia therapy. *Acta Biomater.* 2010; 6: 4522–31.
Saeedi M, Vahidi O, Bonakdar S. Synthesis and characterization of glycyrrhizic acid coated iron oxide nanoparticles for hyperthermia applications. *Mater Sci Eng C.* 2017; 77: 1060–1067.
Santosa D, Silva DM., Gomes PS, Fernandes MH, Santosa JD, Sencadas V. Multifunctional PLLA-ceramic fiber membranes for bone regeneration applications. *J Colloid Interface Sci.* 2017; 504: 101–110.
Sengani M, Rajeswari D. Gold nanosupplement in selective inhibition of methylglyoxal and key enzymes linked to diabetes. *IET Nanobiotechnol.* 2017; 11(7): 861–865.
Silva E, Rodrigues BVM, Vasconcellos LMR, Santos DM, Filho SPC, Marciano FR, Strano MS, Jin H. Where is it heading? Single-particle tracking of single-walled carbon nanotubes. *ACS Nano.* 2008; 9: 1749–1752.
Sitharaman B, Zakharian T. Y., Saraf A, Misra P, Ashcroft J. Water-soluble fullerene (C_{60}) derivatives as non-viral gene-delivery vectors. *Mol Pharm.* 2008; 5: 567–578.
Soga K, Kamimura M. Application of ceramic nanoparticles for near infrared bio imaging. In: Lee B, Gadow R, Mitic V. (eds.) *Proceedings of the IV Advanced Ceramics and Applications Conference.* Atlantis Press, Paris. 2017.
Tao W, Ji X, Xu X, Islam MA, Li Z, Chen S *et al.* Antimonene quantum dots: Synthesis and application as near-infrared photo thermal agents for effective cancer therapy. *Angew Chem Int Ed.* 2017; 1521–3773.
Thomas C, Kumar S, Mishra P, Talegaonkar S. Ceramic nanoparticles: Fabrication methods and applications in drug delivery. *Curr Pharm Des.* 2015; 21(42): 6165–6188.
Tyagi N, Shrivastava SK, Arora S, Omar Y, Ijaz ZM, Deshmukh SK *et al.* Comparative analysis of the relative potential of silver, zinc-oxide and titanium-dioxide nanoparticles against UVB-induced DNA damage for the prevention of skin carcinogenesis. *Cancer Lett.* 2017; 383: 53–61.
Uehara N. Polymer-functionalized gold nanoparticles as versatile sensing materials. *Anal Sci.* 2010; 26(12): 1219–1228.
Ursu EL, Doroftei F, Peptanariu D, Pinteala M, Rotaru A. DNA-assisted decoration of single-walled carbon nanotubes with gold nanoparticles for applications in surface-enhanced Raman scattering imaging of cells. *J Nanopart Res.* 2017; 19(181). DOI 10.1007/s11051-017-3876-9
Vega AIM, Quintero TG, Anita REN, Torres LSA, Castano V. Polymeric and Ceramic Nanoparticles in Biomedical Applications. *J Nanotech.* 2012; 1–10.
Wang HH, Wu YJ, Lin CAJ, Lee CH, Shen JL, Chan WH *et al.* Application of gold in biomedicine: Past, present and future. *Int J Gerontol.* 2012; 6: 1–4.
Wang L, Luhm R, Lei M. SNP and mutation analysis. *Adv Exp Med Biol.* 2007; 593: 105–116.
Wei H, Bruns OT, Kaul MG, Hansen EC, Barch M, Wiśniowska A, Cordero JM. Exceedingly small iron oxide nanoparticles as positive MRI contrast agents. *Proc Natl Acad Sci U S A.* 2017; 28: 114(9): 2325–2330.
Wu F, Xu T, Zhao G, Meng S, Wan M, Chi B *et al.* Mesoporous silica nanoparticles-encapsulated agarose and heparin as anticoagulant and resisting bacterial adhesion coating for biomedical silicone. *Langmuir.* 2017; 33: 5245–5252.
Xie J, Huang J, Lee S, Wang J, Gao J, Li X *et al.* PET/NIRF/MRI triple functional iron oxide nanoparticles. *Biomaterials.* 2010; 31: 3016.

Yamaguchi S, Kobayashi H, Narita T. Sono dynamic therapy using water-dispersed TiO_2-polyethylene glycol compound on glioma cells: Comparison of cytotoxic mechanism with photodynamic therapy. *Ultrasonics Sonochem.* 2011; 18: 1197–1204.

Yeh YC, Creran B, Rotello VM. Gold nanoparticles: Preparation, properties, and applications in bionanotechnology. *Nanoscale.* 2012; 4(6): 1871–1880.

Yoo E, Nardacci LE, Beaman DJ, Drinnan CT, Lee C, Fishbein KW *et al.* Activatable inter polymer complex-super paramagnetic iron oxide nanoparticles as magnetic resonance contrast agents sensitive to oxidative stress. *Colloids Surf B: Biointerfaces.* 2017; 158: 578–588.

Yoona J, Choa SH, Seonga H. Multifunctional ultra-small super paramagnetic iron oxide nanoparticles as a theranostic agent. *Colloids Surf A: Physicochem Eng Asp.* 2017; 520: 892–902.

Yousef I, Seksek O, Gil S, Prezado Y, Susoe JS, Rovira IM. Study of the biochemical effects induced by X-ray irradiations in combination with gadolinium nanoparticles in F98 glioma cells: First FTIR studies at the Emira laboratory of the SESAME synchrotron. *Analyst.* 2016; 141: 2238–2249.

Yu CAI Shao, Lie KJ. Advance in research on carbon nanotubes as diagnostic and therapeutic agents for tumor. *Chinese J Anal Chem.* 2009; 37: 1240–1246.

Zhang P, Li B., Du J, Wang Y. Regulation the morphology of cationized gold nanoparticles for effective gene delivery. *Colloids Surf B: Biointerfaces.* 2017; 157: 18–25.

Zhang T, Gerion D, Ding L, Elboudwarej O, Cooke PA, Gray JW *et al.* Cellular effect of high doses of silica-coated quantum dot profiled with high throughput gene expression analysis and high content cellomics measurements. *Nano Lett.* 2006; 6: 800.

Zhang Y, Venugopal JR, Turki AE, Ramakrishna S, Su B, Lim CT. Bio composites containing natural polymers and hydroxyapatite for bone tissue engineering. *Biomaterials.* 2008; 29: 4314.

2 Nanocoutured Metallic Biomaterials and Surface Functionalization of Titanium-Based Alloys for Medical Applications

Jalal Azadmanjiri, Wai Hong Wong, Jagat R. Kanwar, Christopher C. Berndt, James Wang, Vijay K. Srivastava, and Ajay Kapoor**

CONTENTS

2.1 Introduction 17
2.2 Metallic Biomaterial Implants 20
  2.2.1 Historical Progress on Metallic Biomaterials 22
  2.2.2 Crucial Necessities in Design of Metallic Biomaterials 23
  2.2.3 Current Metallic Implants and Their Behavior in a Physiological Environment 25
2.3 Surface Treatment of Metallic Biomaterials 28
  2.3.1 Physical Surface Treatments 28
    2.3.1.1 Magnetron Sputtering Deposition 28
    2.3.1.2 Pulsed Laser Deposition 29
    2.3.1.3 Lithography Process 30
    2.3.1.4 Plasma Spraying 30
  2.3.2 Mechanical Surface Treatments 31
  2.3.3 Chemical Surface Treatments 32
    2.3.3.1 Atomic Layer Deposition 33
    2.3.3.2 Silanization and Langmuir-Blodgett Deposition 33
    2.3.3.3 Layer-by-Layer Assembly Approach 35
    2.3.3.4 Electrochemical Method: Electrochemical Anodization 35
2.4 Conclusions 42
Acknowledgment 42
References 42

2.1 INTRODUCTION

Biomaterials usually are identified as synthetic products that are utilized for the replacement of hard or soft tissues in the human body. These become close to the blood when inserted into the human body through orthopaedic surgery (Bauer *et al.*, 2013). Their biocompatibility being the most important characteristic, biomaterials can either be inorganic or organic in nature. The field of implantology has acquired various strategies whether in terms of the utilized implantation materials or modifications

---

* Corresponding authors.

of these materials to attain the best case of implantation (Teo *et al.*, 2016). Several materials like metals, ceramics and polymers have been underpinned for biomedical applications. However, the applied material should not only be bioinert, meanwhile it should also be biocompatible (Williams, 1998; Ratner, 2001). An effective implant needs to match the replaced tissue in the view of shape and inner structure. Hence, the design of an implant should cover the physical, chemical, biological and morphological properties of exposed surface to living organs (Ratner, 2001). On the other hand, surface treatments of an implanted material not only ban the negative impact on the host tissue, but they also improve the connection between these two components (Chen and Thouas, 2015). The actual behavior of an implant is categorized to be "incompatible", "biotolerant", "bioinert" and "bioactive" (Geetha *et al.*, 2009). "Incompatible" means the dissolution of implant substances into the living members in toxic concentration that leads to the rejection of the implant. "Biotolerant" term means the release of substance from the surface of an implant. However, it is in nontoxic concentration and leads to isolation of the implant from the surrounding tissue. "Bioinert" means that it is applicable when there is no dissolution from the surface of the implant. "Bioactive" also means a good interaction in between the implant along with the living tissue which almost successfully leads to the proliferation and differentiation of the cells of living organs. Obviously, the living organ is naturally capable of regenerating itself, but the implant evidently does not have such capability (Geetha *et al.*, 2009). Consequently, the success of any implant relies on the synergism resulted from various biomedical materials combinations. It is stated that hierarchical structure accompanies some materials whether it was natural or synthetic which in turn provides the rise to specific properties of a bulk material (Lakes, 1993). Likewise, the assembly of bone, as illustrated at Figure 2.1a–e, has more than one length scale that starting through nanoscale to macroscale (Rho *et al.*, 1998; Weiner and Wagner, 1998; Vaughan *et al.*, 2012). Bone is a natural composite material of hierarchical architecture that is mainly created of hydroxyapatite (HA) crystals and organic collagen proteins. Despite such hierarchical building, a bone shows high degree of toughness and resistivity to fractures and it becomes light for flexible movement (Wu *et al.*, 2016).

Biomaterials might be utilized inside various areas of the body system, for instance, it is used as artificial parts of the heart, blood vessel stents and implants in shoulder, hips, knees, ears, elbows, and dental structure (Williams, 1998; Ratner, 2001; Bauer *et al.*, 2013). In essence, the actual demand for hard tissue implants like spinal, hip and knee substitution acquires a high percentage compared to other kinds of implants. Human joints are subject to arthritis like osteoporosis and osteoarthritis. This degenerative disease unfavourably causes either pain or damaging the function of the joints, hence, implantation is required. Figure 2.2 illustrates kind of implants used for the total substitution of knee and hip (Geetha *et al.*, 2009). However, in some cases, revision surgery might be needed due to failure of implantation. Figure 2.3

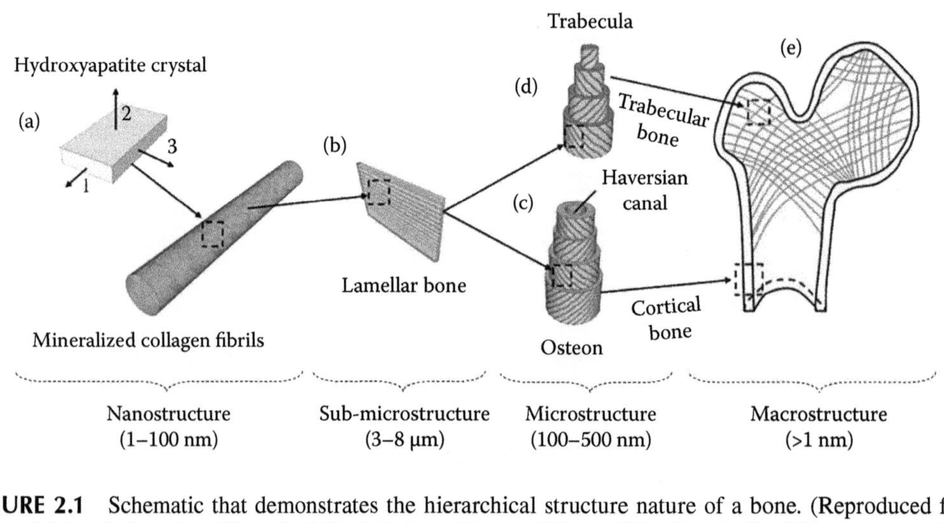

FIGURE 2.1 Schematic that demonstrates the hierarchical structure nature of a bone. (Reproduced from data published in Vaughan TJ *et al.*, *J Mech Behav of Biomed Mater.* 2012; 12: 50–62. With permission.)

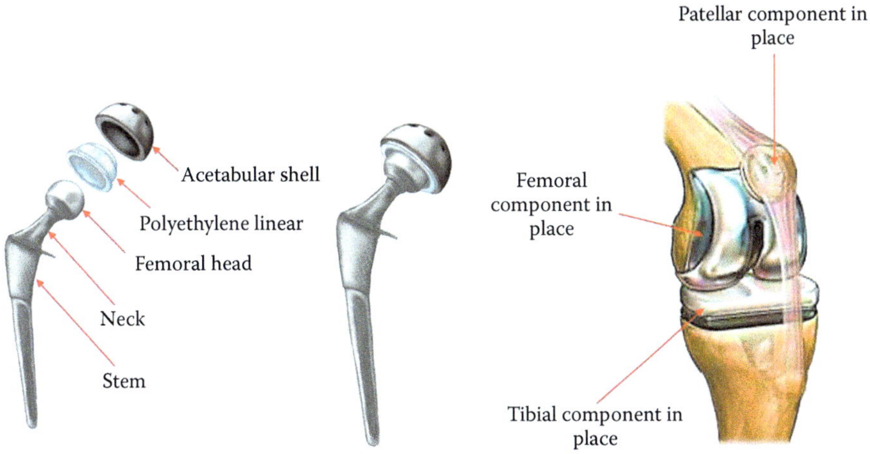

FIGURE 2.2 Implant for the complete substitution of hip and knee.

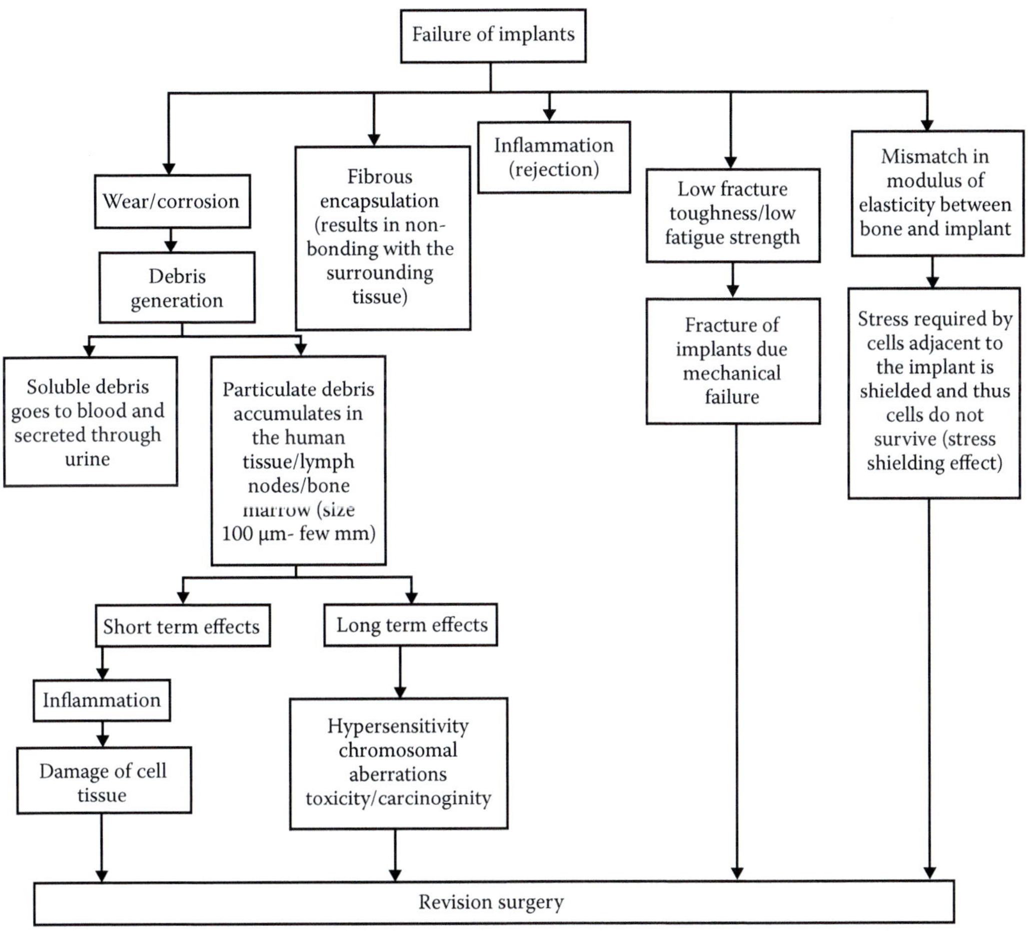

FIGURE 2.3 Different reasons for implant failure. (Reproduced from data published in Geetha M *et al.*, *Prog Mater Sci.* 2009; 54: 397–425. With permission.)

shows some various causes that collapse an implant (Geetha *et al.*, 2009). The principle of Chapter 2 is introducing main metallic implants, essential requirements in plot of those materials and major techniques that have been done with respect to the field of implantation utilizing metallic biomaterials. On top of that, some substantial physical, mechanical and chemical surface modification techniques and strategies undertaken for the optimization of the surface of implants are reviewed. Consequently, it outlines the recent works for fabricating metallic biomaterials, nanotubes and the corresponding cell responses.

2.2 METALLIC BIOMATERIAL IMPLANTS

A metallic biomaterial is broadly utilized in the manufacturing of health-related devices for hard tissue substitution such as pacemaker cases, screws, external fixators, bone plates, dummy heart valves, and dental implants (Ratner *et al.*, 2004; Balazic *et al.*, 2007; Cho *et al.*, 2017). Fabrication methods of metal-based medical devices are cutting, forming and other alternative operations (*i.e.*, electron beam melting and water jet cutting). So far, metallic biomaterials are showing excellent fracture toughness, fatigue strength and tensile strength compared to polymer and ceramic biomaterials. Hence, these biomaterials are chosen as implant materials, owing to reliable mechanical performance and high corrosion resistance, otherwise which can lead to release of number of harmful toxins, when come in contact with human body fluids (Breme and Biehl, 1998; Ratner *et al.*, 2004; Sumita *et al.*, 2004; Balazic *et al.*, 2007; Prasad *et al.*, 2017). The four main categories of metallic implant biomaterials are stainless steel (SS)-, cobalt (Co)- and titanium (Ti)-based alloys, and valve metals that are included magnesium alloy, nickel-titanium (NiTi) shape-memory alloys and, last but not least, tantalum alloy (Blackwood, 2003; Azadmanjiri *et al.*, 2016a, 2017a).

Stainless Steel: Iron-based alloys that consist of a high chromium (Cr) percentage (11–30 wt%) are known as SS (Chen and Thouas, 2015). The chemical composition of different kinds of SS determines whether it belongs to Cr or chromium-nickel (Cr-Ni) group. SS can also be categorised into martensitic, austenitic, ferritic and duplex (austenitic plus ferritic) base on their characteristic microstructure. Martensitic SS is mainly used as dental and surgical due to it hardness (up to 97 HRB) while application of ferritic SS in the medical device is limited. SS is commonly used as an implant material is austenitic SUS 316L (Chen and Thouas, 2015).

Cobalt-based alloys: Initially, Co alloys were utilized in dental implants due to their superior resistance to degradation, which slowly spread into surgical application (Sury and Semlitsch, 1978). The Co alloys that are suitable for biomedical applications are classified into cast or wrought alloys.

Two cobalt-chromium-molybdenum (CoCrMo) and cobalt-nickel-chromium-molybdenum (CoNiCrMo) have been employed for biomedical applications. Cr and Mo both are added to enhance corrosion resistance and strength of the alloy (Henriques *et al.*, 2015; Martinez-Nogues *et al.*, 2016). Despite similar abrasive characteristics of cast CoCrMo and wrought CoNiCrMo alloys, the CoNiCrMo alloy is more recommended for load-bearing applications, due to poor frictional ability of CoCrMo alloys with the other materials and even itself (Park and Kim, 2013). Since CoNiCrMo alloy has Ni in its composition, the release of Ni ions has been evaluated in Ringer's solution (it is a made solution from various salts that dissolved in water for the aim of producing an isotonic solution relative to the bodily fluids of an animal) at 37°C and then compared with 316L SS. It has been surprisingly demonstrated that although the concentration of Ni in CoNiCrMo alloy is 3 times higher than the SS, however, the rate of Ni release is almost the same for both of them (3×10^{-10} g/cm^2/day) (Enkelhardt *et al.*, 2012). Furthermore, *in vitro* investigations proved that Co particulate influences osteoblast-like cells which, in turn, harmfully prohibits the proliferation of type I collagen, osteocalcin and alkaline phosphatase in the cell culture medium. Despite of that implication, Cr and CoCr alloy particulates have no substantial toxicity on the cell lines (Granchi *et al.*, 1996). The elastic modulus of CoCr alloys, that ranges from 220 to 234 GPa, becomes disadvantageous factor, it being higher than the other materials such as SS. Consequently, the distribution of load to bone in artificial joint substitution is not properly adapted which might result in failure implant (Schmalzried *et al.*, 1996). Therefore, stainless steel can be considered for fabricating Ni-free, C-free Co alloys that have been of great interest (Chiba *et al.*, 2005).

Valve metals: Valve metals such as Tantalum (Ta), Zirconium (Zr), Magnesium (Mg) and Niobium (Nb) have effectively contributed to the biomedical field either as alloying elements with Ti or base metals for biomedical applications. Ta has shown acceptable biocompatibility since it has been bioactive (Balla *et al.*, 2010). However, its poor characteristics from the viewpoint of mechanical requirements, restricts its utility in biomedical applications. For instance, it has high density (16.6 g/cm^3) compared to other metallic biomaterials. Zr has good corrosion resistance in various media. Zr possesses biocompatibility due to a self-formed oxide layer, which can even be formed in air or oxygenated electrolyte. This layer is able to decrease corrosion and increase osseointegration (Hanawa, 2004; Zander and Köster, 2004; Hanawa, 2012). A knee implant made of Zr has been commercially approved because of the acceptable wear resistance of zirconia (zirconium dioxide, ZrO_2) fabricated on the surface of the implant (Patel and Spector, 1997; Good *et al.*, 2005; Stojilovic *et al.*, 2005). Mg is an important material for biomedical field, especially orthopaedic surgery, in terms of its biodegradable behavior (Song, 2007). Mg becomes a suitable candidate for bimetallic implantation due to its good mechanical properties. For instance, Mg has low density (1.74 g/cm^3) as well as low elastic modulus (45 GPa) which is close to natural bone (Witte *et al.*, 2005; Staiger *et al.*, 2006). However, Mg is prone to corrosion forming Mg^{2+} ions accompanied by H_2 gas evolution. Although Mg^{2+} is principal component in the human body especially bones, but the gas evolution resulting from the chemical reactivity of Mg is major concern during the surgery (Saris *et al.*, 2000; Hartwig, 2001). Therefore, Mg requires considerate surface treatment for enhancing both the corrosion resistivity and the bioactivity. Nb demonstrated good characteristics which largely enable to become interesting candidate for biomedical applications. Nb has pronounced ductility and also good corrosion resistivity in several aggressive mediums because of the creation of consistent oxide layer on its surface and low Young's modulus (Asselin *et al.*, 2007; O'Brien *et al.*, 2008). However, the low strength of Nb alloys becomes critical factor in terms of its engagement in biomedical devices (Niendorf *et al.*, 2007).

Titanium-based alloys: Ti is a low-density (4.5 g/cm^3) metal as it only approximates 60% of the density of iron (Fe) (7.9 g/cm^3) and nearly half of the density of Co (8.9 g/cm^3). This metal can be strengthened using alloying elements and deformation processing. Allotropic transformation take place when Ti is heated to approximately 885°C which change its HCP (α-phase) crystal structure into BCC (β-phase) crystal structure. Based on the final microstructure of Ti, it can be group into α alloy, α+β alloys and β alloys. ASTM and BS7252/ISO 5832 are the material specification that used to define Ti and its alloys internationally (Chen and Thouas, 2015). The considering factors while selecting Ti alloys for implementation are osseointegration, density, shear strength, biocompatibility and corrosion resistance (Rack and Qazi, 2006).

Ti has natural phenomenon analogous to the other valve metals for creating a surface layer of titanium dioxide (TiO_2) which greatly promotes it for the biomedical applications. TiAl4V alloy is the most usual Ti-based alloy for biomedical implantation. Aluminium (Al) constitutes 5.5–6.5% while vanadium (V) constitutes 3.5–4.5%. Al is being added for stabilizing α-phase whilst V is added for stabilizing β-phase. Due to the fact that Al and V have allergic influence on the cells, an interest of fabricating alternatives of Ti-based alloys has been in progress. Therefore, $Ti_6A_{17}Nb$ and $Ti_5A_{12.5}Fe$ have been developed where they do not have V (Niinomi, 2002). Further, $Ti_{15}Sn_4Nb_2Ta_{0.2}Pd$ and $Ti_{15}Zr_4Nb_2Ta_{0.2}Pd$ have been developed as Al-free and V-free Ti-based alloys (Okazaki *et al.*, 1995). From the view of the mechanical characteristics, β-Ti alloys considerably have low Young's modulus than α and α+β Ti-based alloys. Consequently, focus on fabricating low Young's modulus β-Ti alloy is the main research area. Hence, $Ti_{13}Nb_{13}Zr$, $Ti_{12}Mo_6Zr_2Fe$, $Ti_{15}Mo$, $Ti_{15}Mo_5Zr_3A_1$, $Ti_{35}Nb_7Zr_5Ta$ and $Ti_{29}Nb_{13}Ta_{4.6}Zr$ (TNTZ) have been all developed (Niinomi, 2003). TiNi alloy (Nitinol) has been one of the most notable Ti alloys due to its novel properties such as, shape memory effect (SME) and superelasticity (Prymak *et al.*, 2005). The elastic modulus of Nitinol (38–48) is much lower than standard 316L SS. Therefore, Nitinol has undertaken various medical applications such as orthodontic wires, stents and bone implantation. The biocompatibility of Nitinol implant in soft tissues has been confirmed both *in vitro* and *in vivo* investigations (Ryhanen *et al.*, 1998; Assad *et al.*, 1999). Since Nitinol contains nickel (Ni) in its composition, the expectation of Ni ions release becomes obvious. However, it has been reported that the concentration of Ni ion release from Nitinol is below the critical concentration which induces allergic effect (Kapanen *et al.*, 2001). On the other hand, the fabrications of Ni-free superelastic Ti alloys, which are β-type alloys, are acquired great interest. A few examples are TiNbSn, TiMoGa, TiMoGe, TiMoAl, TiTa, TiNb, TiNbAl, TiScMo, TiMoAgSn and TiNbTaZr (Niinomi *et al.*, 2005). Hence, it can be observed that the choice of both suitable alloying elements for Ti and appropriate surface treatment evidently influence the overall properties of the alloy which in turn determine the validity of the alloy for being implanted.

2.2.1 Historical Progress on Metallic Biomaterials

There are records of medical implants manufactured from metallic materials during the 19th century. Requirements for approaches to bone repair, such as internal bone fracture fixation of larger bones, were the driving forces for the development of metallic implants during the early 19th. The success of implanting metallic products, for instance, bone pins and spinal products, manufactured from iron (Fe), gold (Au) or silver (Ag) had been contributed by the implementation of Lister's aseptic surgical method in 1860 (Niwa, 2004; Chen and Thouas, 2015). During 1969, Clemson University, South Carolina, hosted the first meeting regarding biomaterials and these materials gained the recognition since then (Geetha *et al.*, 2009). Ever since, metallic biomaterials are playing an important role in form of the temporary devices (*i.e.*, fracture plates), permanent implants (*i.e.*, whole joint replacements) as well as dental implants (Chen and Thouas, 2015).

SS was considered as a practicable implant material in the early days due to it high availability and easy process of alloying. Cr, Mo and Ni are used as the alloying elements for the alloys like 316L (ASTM F138). These alloys are mainly used to develop temporary devices and are used in total joint replacement surgery. However, there is a limitation in total joint replacement due to the high modulus elasticity in SS which is around 200 GPa. Hence, due to this SS limitation, researchers aimed to look for new alloys that have more wear and corrosion resistivity, higher fatigue strength and more biocompatibility. This is when Co-based alloys came into action. Wrought alloys were used in building prosthetic stem- and load-bearing components met the requirements of the researchers, however, Co-Cr-Mo alloys (ASTM F75 and F799) still had a large modulus (~210 GPa) and less biocompatibility, which are undesirable characteristics for an implant material (Long and Rack, 1998).

Ti alloys acquired much attention in the industry in 1970s due to its excellent tissue compatibility, higher corrosion resistance, lower modulus of elasticity and excellent specific strength (Nag and Banerjee, 2012). TiO_2 that readily forms a layer of nascent oxide in the atmosphere, had excellent osseointegration properties but lacks material strength, which limits its usage in application like hip cap shells, pacemaker cases and heart valve cages (Lee *et al.*, 2002; Nag and Banerjee, 2012). In order to improve strength for total joint replacement, Ti alloy Ti-6Al-4V ELI (ASTM F136) was developed. This extra low interstitial alloy contained 6 wt% of Al and 4 wt% of V with elastic modulus of ~110 GPa which is just half of SS and Co-Cr-Mo alloys. The additional comparison mechanical properties between these alloys are showing in Table 2.1. There are some problems like adverse tissue reactions and cytotoxicity because of the existence of V (Li *et al.*, 2010; Chen and Thouas, 2015). Therefore, other alloying elements such as Nb and Fe were used to replace with V to establish the other type of Ti-6Al-7Nb (Semlitsch *et al.*, 1992; Chen and Thouas, 2015) and Ti-5Al-2.5Fe alloys (Chen and Thouas, 2015).

There are some studies stating that the released V and Al ions caused long term health issues, e.g., Alzheimer's disease, peripheral neuropathy and osteomalacia (Nag and Banerjee, 2012). These issues lead the researchers' focus into a condition where they crucially needed to develop a new alloy that included brilliant biocompatibility without detrimental reaction, superior corrosion in human body, perfect mechanical strength and fatigue resistance (Chen and Thouas, 2015). Besides, the new alloy must have low modulus elasticity, wear resistance and low density. Unfortunately, researchers do not have

TABLE 2.1
Data Comparing the Mechanical Properties of Various Alloys

Materials	Modulus of Elasticity (GPa)	Ultimate Tensile Strength (UTS) (MPa)	Yield Strength
316L stainless steel	200	465–950	170–750
CoCrMo alloys	200–230	600–1795	275–1585
Ti alloys	105	785	692
Ti-15Mo-5Zr-3Al	75	882–975	870–968
Cortical bone	10–40	90–140	–

Source: Long M and Rack HJ, *Biomater.* 1998; 19: 1621–1639.

many choices in choosing alloying elements because the only elements that are not harmful to humans are Ti, Mo, Nb, Ta, Zr, Fe and tin (Sn) (Nag and Banerjee, 2012). Among all the listed elements, Ta is the only alloying element that shows osseocompatibility similar to Ti. However, as it is already mentioned, the high density of Ta is preventing it to be primary alloying additive (Nag and Banerjee, 2012).

Researchers also investigated and analysed the Mo- and Nb-based alloys, to develop a novel alloy. Mo-based β-Ti alloy [Ti-11.5Mo-6Zr-4.5Sn] that is also known to βIII was developed which had the same phase transformations occurring as binary Ti-Mo. This alloy has low yield strength, high ductility and high toughness when quenched at room temperature (Feeney and Blackbur, 1970; Nag and Banerjee, 2012). The combination of these properties and the Fe effect of Ti-Mo formed Ti-12Mo-6Zr-2Fe that shows excellent yield strength and modulus values (Wang *et al.*, 1993; Nag and Banerjee, 2012). Tang *et al.* (2000) developed Nb-β-Ti alloys, based on the system like Ti-Nb-Ta, Ti-Nb-Ta-Zr, Ti-Nb-Ta-Mo, and Ti-Nb-Ta-Sn. Among the entire chosen alloy, tensile strength and elongation of Ti-29Nb-13Ta-4.6Zr alloy was better than or equivalent to conventional Ti alloys. Recently, research efforts on NiTi-shape memory alloys and growth of new Mg-based alloys for bone tissue engineering are expanding (Kuroda *et al.*, 1998; Tang *et al.*, 2000; Nag and Banerjee, 2012).

2.2.2 Crucial Necessities in Design of Metallic Biomaterials

Mechanical efficiency: A biomaterial that is using biomedical utilizations has to be powerful and tough in order to replace of human body sections. The important mechanical properties that require being considered during the selection and developing biomaterials are: Young's modulus, ultimate tensile strength (UTS) and toughness. There are many types of metallic biomaterials that are used in the industry, however, SS, Co-based alloy and Ti-based alloys are the main metallic biomaterial alloys, due to their mechanical properties that have been shown in Table 2.1 (Long and Rack, 1998; Chen and Thouas, 2015). Even these biomaterials are commonly used as implants but their Young's modulus is significantly greater compared to the natural bone that differ from 4 to 30 GPa, depend on the type of bone (Geetha *et al.*, 2009). This high Young's modulus leads to "stress shielding effect" where the constructed implant tends to sustain most of the load which preventing the needed stress reaching to the adjacent remodelling bone and cause biological responses. The stress shieling effect causes: (i) decrease of bone's thickness, (ii) mass losses of bone, (iii) atrophy, and (iv) increase of osteoporosis (Nag and Banerjee, 2012; Chen and Thouas, 2015). When the mentioned biological issues happen, the constructed implant will be loosened and then additional revision surgery will be required (Sumner and Galante, 1992; Nag and Banerjee, 2012). Therefore, it is necessary to have a biomaterial that has high strength and modulus of elasticity that closes to the normal bone. The stress distribution among the constructed implant and normal bone can definitely develop once a biomaterial with similar modulus of elasticity to bone is used to manufacture an implant. The implantation time frame also can be extended and reoperation can be avoided (Nag and Banerjee, 2012).

Fatigue limit is one of the important factors that need to be considered within developing an alloy, as well as during applying them as an implant. Fatigue limit is the range of cyclic stress that endured by materials before it fails as fatigue failure. Generally, implants are produced using microstructural inhomogeneity of metallic alloys, depending on the implant position, shape and cyclic stress that is applied to the implant (Long and Rack, 1998; Nag and Banerjee, 2012). For instance, knee joints and constructed hip joints are subjected to cyclic stress due to human beings walked a few thousand steps a day at 1 Hz rate (Chen and Thouas, 2015). Constructed hip joints experienced a larger loading stress than the patient's body weight. Because the positioned joint is out of perpendicular alignment, when patient start to walk, the patient's body weight is supported just by one leg at a time when taking a step ahead or back. The approximated loading stress that suffers by a single leg is ~50 MPa (Chen and Thouas, 2015). Dental implants also have the same experience because of chewing motion and non-osseous tissue implants. The fatigue mechanical working status of some implants is listed in Table 2.2 (Sumita and Hanawa, 2003; Chen and Thouas, 2015). Fatigue ruptures can easily happen through regular service situations without any caution or indicator which is harmful to the patient. Biomaterials that employ to create implants are estimated to endure millions of cyclic deformations through the assistance period and fracture resistance (Chen and Thouas, 2015).

Corrosion resistance: Corrosion resistivity is another major factor that influences the longevity of an implant in human body. The environment of the body is very different compared to the ambient condition, thus, a metal with excellent corrosion resistivity in the air may not carry out well in human body (Chen and Thouas, 2015). A human body fluid in normal condition has ~0.9% saline, solutions of mostly Na^+, Cl^-, amino acids, a range of soluble proteins and other traces ions. The pH value of these fluids in 1 atmosphere are 7.2–7.4 at 37°C, that is very close to neutral, however, it drops to 3–4 when inflammation is caused by surgery or injury (Sumita and Hanawa, 2003; Chen and Thouas, 2015). The surroundings become aggressive for an implant once inflammation incorporated with variation in ionic strength in relation to high blood pressure or because of ion deposits. The low corrosion resistance of an implant would not persist in the aggressive environment and metal debris ions release into the body and caused allergic and toxic reactions (Geetha *et al.*, 2009).

Wear resistance: Wear is serious issue in any joint replacements. Therefore, wear resistance is important during the service period of an implant. Implant loosening in a joint system usually is happening at a low wear resistance or high friction coefficient. Aseptic loosening is the common issue that causes by wearing and can be occurred in any material. This happen when microparticles are sorted at the bearing couples of a joint replacement (Lennox *et al.*, 1987; Lombardi *et al.*, 1989; Chen and Thouas, 2015). The actual quantity of microparticles appears to enhance because of long-term wear and a series of biological reactions. The environment becomes acidic when enzymes and metabolites are releasing during the reaction. Bone and implant ablation that take place during aseptic loosening is because of the produced enzymes, acidic chemicals, ion and debris (Papageorgiou *et al.*, 2008; Chen and Thouas, 2015).

TABLE 2.2
Fatigue Mechanical Working Conditions of Some Implants

Implants	Expected Total Number of Loading of a 65-year-old Patient for the Life Time (*i.e.*, 20 Years)	Loading Frequency (Hz)
Joint	10^7	1
Tooth filling	10^9	1
Pacemaker	10^7	1

Source: Chen Q and Thouas GA, *Mater Sci and Eng R: Rep.* 2015; 87: 1–57.

Osseointegration: The procedure of growing a new bone and bone recovery is described as osseointegration, which is actually the essential requirement in orthopaedics. An implant loosening is due to the inability of an implant surface incorporate with adjacent bone and surrounding tissues. In this case micromotions and fibrous are formed between an implant and bone (Nasab *et al.*, 2010; Chen and Thouas, 2015). Therefore, it is essential to produce an implant with suitable surface that will connect to adjacent bone. The factors that perform a significant role in improving osseointegration are surface roughness, surface chemistry and surface topography (Geetha *et al.*, 2009).

Biocompatibility: Biocompatibility is the measure achievement of a biomaterial dependent on how properly human body respond to a biomaterial. There are not any precise measurements and accurate description for biocompatibility, however, a response between a host tissue and fluid with an inserted implant determine the acceptability of the biomaterial (Ratner *et al.*, 2004). Biocompatibility of materials is influenced by the materials degradation in the body environment and host response caused by the material. The requirements of biocompatibility are complicated and rigorous. Therefore, a desirable biomaterial is predicted to be non-toxic and do not cause allergic reaction as it used in close contact with living tissue (Geetha *et al.*, 2009). The requirements of biocompatibility for different health-related applications is varied because of the various chemical compositions in human body (Chen and Thouas, 2015).

2.2.3 Current Metallic Implants and Their Behavior in a Physiological Environment

Metallic implants are famous and widely employed in the biomedical field, because they performed well in their roles to improve the life's quality of a patient. Table 2.3 lists the three main (SS, Co and Ti) metallic alloys and their application that are commonly seen in this field.

In fact, corrosion is the major chemical aspect that might accompany metallic implant when inserted into the living organ. Corrosion means the degradation and depreciation of the metal to oxides or hydroxides or other components as a result of reacting metallic implant with the tissue fluid. Tissue fluid in the human body represents an aggressive medium to the metallic implant where it has water, dissolved oxygen, proteins, chloride, hydroxide and other ions. Corrosion occurs due to the ionization of the atoms of metal to ions (*i.e.*, $Ti \rightarrow Ti^{2+} + 2e^-$) which, in turn, react with oxygen or other species of the tissue fluid. Then, it forms an unstable component which can be easily flaked off. As a result, the release of corrosion products into the living organ may cause various consequences such as inflammation and cell damage which results in failure of implantation and consequently it requires revision surgery. However, in some cases the creation of the protective oxide layer onto the surface of a metal which in turn somewhat impedes more dissolution. In principle, the stability and the adherence of a passive film formed onto the surface of particular biomedical material rely on the working conditions, such as the type of the electrolyte, the working period, pH of the electrolyte and temperature of the medium (Epelboin *et al.*, 1979). Metallic biomaterials such as SS, Co-based alloys, valve metals and Ti-based alloys are spontaneously capable of forming oxide passive film onto its surface whether in air or aqueous medium. For instance, TiO_2 is fabricated on the surface of Ti-based alloys while Cr_2O_3 is formed on the surface of SS and Co-based alloys (Milosev *et al.*, 2000; Milosev and Strehblow, 2000; Schultze and Lohrengel, 2000; Milosev and Strehblow, 2003). Localized corrosion contributes greatly to biometallic implants where the total mass loss is small. On the hand, the whole system can be totally damaged. Although the discovery of localized corrosion is far less than that of general corrosion, the consequences of localized corrosion is much more dangerous than general corrosion. Figure 2.4 simply indicates the surface of biometallic implant after being subject to localized corrosion where the pits started to grow up at certain places on the surface because of the segregation of the passive film at these sites. For more details on localized corrosion and the fundamentals of formation, see Schmuki (2002), while we below briefly discuss the behavior of the most common biometallic materials in physiological medium and the corresponding corrosion phenomena.

Stainless steels: the body fluid contains chloride (Cl^-) in its composition with a concentration of 103 $mmol^{-1}$ (Oyane *et al.*, 2003), which in turn influences the pitting potential of SS. The maximum

TABLE 2.3
Three Main Metallic Alloys and Their Applications

Material	Implants
Stainless Steel	
Type 304	Trachea tube and preformed dental crown
Type 316L	Bone screw, pins, nail bone plates, wires, Harrington spine instrumentation, hydrocephalus drainage valve and intramedullary bone nails
Type 304 and 316	Retention pins for dental amalgam and preformed endodontic post and core
Type 316 and 316 L	Mandibular wire mesh prostheses, stapedial prostheses for middle ear repair and wire mesh
Types 304, 316 and 316L	Sutures, electrodes and lead wires
Type 302, 303, 304, and 305	Fixed orthodontic appliances
17-7PH, 17-7PH (Nb), PH-15-7Mo and type 301, 304, 316, 316L, 420 and 431	Neurosurgical aneurysm and microvascular clips
Orthinox	Total joint prostheses
Stainless steel	Balloon-expandable stent and Arzbaecher pill electrode
Stainless steel wire formed in a zigzag configuration of 5–10 bends	Self-expanding stent
Co-based	
Co-28Cr-6Mo (Cast)	Permanent implant
Co-28Cr-6Mo (Forged)	Permanent implant
Co-28Cr-6Mo (Wrought)	Permanent implant
Co-Cr-Ni-Mo-Fe (Wrought)	Permanent implant
Co-35Ni-20Cr-10Mo (Wrought)	Permanent implant
Co-35Ni-20Cr-10Mo (Forged)	Permanent implant
Co-Ni-Cr-Mo-W-Fe (Wrought)	Short-term implant
Co-20Cr-15W-10Ni (Wrought)	Short-term implant
Ti-based	
α *microstructure*	
CP-Ti grade 1 CP-Ti grade 2 CP-Ti grade 3 CP-Ti grade 4	Dental implants, implantable infusion drug pump, maxillofacial and craniofacial implants, staples for spinal surgery and screws
α-β *microstructure*	
Ti-6Al-4V Ti-6Al-4V ELI	Parts for orthodontic surgery, artificial heart valves, components in high-speed blood centrifuges and bone fixation devices
Ti-6Al-7Nb	Femoral prosthesis steams, spinal components, fracture fixation plates, fasteners, nails, rods, screws and wires
Ti-3Al-2.5V	Tubing and intramedullary nails

Source: Chen Q and Thouas GA, *Mater Sci and Eng R*: *Rep*. 2015; 87: 1–57.

pitting resistivity in physiological media has been investigated in several works (Tracana *et al.*, 1995; Walczak *et al.*, 1998). The most common types of corrosion that affect SS implant are pitting, crevice, fatigue, stress corrosion cracking, fretting and galvanic corrosion. These types of corrosions all lead to defect in the implant material which, in turn, breaks the function of the implant. The corrosion of 316L SS in bioenvironment has been extensively studied with particular interest in the release of metallic ions such as Ni^{2+} in the living organism (Cater and Hicks, 1956; Heath *et al.*, 1971; Case *et al.*, 1994; Pereira *et al.*, 1995; Morais *et al.*, 1998). For instance, the effect of release of

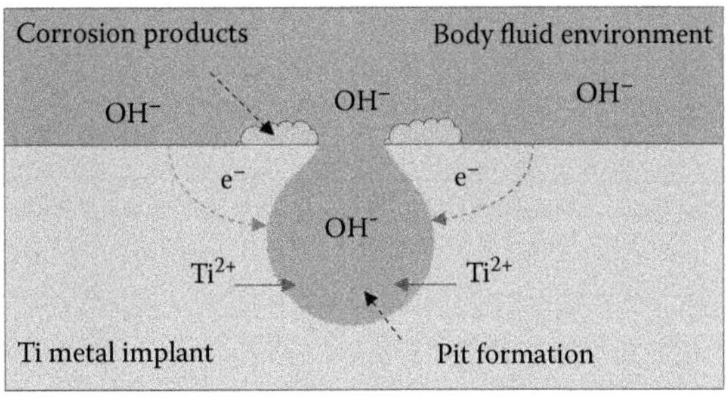

FIGURE 2.4 Schematic of localized corrosion attack for implanted biomaterial in a living organ.

the ions from SS slurry on the cell proliferation of rat bone marrow has been investigated (Morais *et al.*, 1998). It leads to a certain concentration of the ions released from the corrosion product, and it is able to negatively impact the cell culture of rat bone marrow (Morais *et al.*, 1998). The surface coating of SS by biocompatible elements such as hydroxyapatite (HA) has been attempted to increase the corrosion resistance of SS which in turn limits the release of ions (Sivakumar *et al.*, 1994; Fathi *et al.*, 2003). Comparative studies on the electrochemical behavior of both HA-coated stainless steel and bare SS in Ringer's solution showed that the open circuit potential of the HA-coated stainless steel is shifted to the positive direction which denotes enhancement of the corrosion resistance (Sivakumar *et al.*, 1994; Fathi *et al.*, 2003). Furthermore, two layers of HA/Ti-coated SS substantially improved the corrosion resistance of stainless rather than the bare SS, in 0.9 wt% NaCl at 37°C (Fathi *et al.*, 2003). However, surface coating of the SS may promote crevice corrosion which unfavourably leads to severe degradation of the implant during long-term use.

Cobalt alloys: The Co-Cr-Mo alloy is considered as an important biomedical material because of its good strength, wear, toughness and corrosion resistances. It has been used as the surface layer of the total joint and knee replacement (Black, 2005). The mixed oxides layer from both Co and Cr and enhance the corrosion resistance of Co-based alloys. The passive range of Co-based alloys is wide (-300 to 600 mV) and the breakdown passivity of Co-based alloys in Hank's solution has not been reported (Milosev and Strehblow, 2003). Co-Cr-Mo alloy demonstrated good anti-corrosion behavior in 0.15 M NaCl and exhibited thicker self-fabricated oxide layer on the top of its surface after heat and nitric acid (HNO_3) treatment which was responsible for the improvement of the surface characteristics (Strandman and Landt, 1982; Smith *et al.*, 1991; Codaro *et al.*, 2000). Spectroscopic investigations on the surface of the CoCr alloy have been performed, utilizing different surface analysis methods such as X-ray photoelectron spectroscopy (XPS), Auger electron spectroscopy (AES) and surface-enhanced Raman spectroscopy (SER) (Li *et al.*, 1999; Hanawa *et al.*, 2001; Milosev and Strehblow, 2003). It has been investigated that an oxide thin film with 3.1 nm thickness on the surface of CoCrMo alloy, consist of Cr_2O_3, $Cr(OH)_3$ and Cr_2O_3, $Cr(OH)_3$, CoO and MoO is fabricated, in Hank's solution (Milosev and Strehblow, 2003). CoCrMo alloy releases metals ions such Cr^{2+}, Ni^{2+} and Mo^{2+} which apparently act as toxic materials to living tissues (Shettlemore and Bundy, 2001). Table 2.4 shows the effects of some elements (Ni, Cr, Mo, C and W) on Co in respect of mechanical properties and corrosion resistance (Chen and Thouas, 2015).

Titanium alloys: Ti and Ti-based alloys have good mechanical characteristics. Among all of the Ti-based alloys, Ti-6Al-4V has been broadly used in biomedical applications. Ti exhibited high pitting potential due to the formation of TiO_2 layer on its surface which, in turn acts as barrier layer (Hanawa and Ota, 1991; Khan *et al.*, 1996). The corrosion resistivity of Ti and Ti-based alloys has been proved, however, the thickness of the passive oxide layer changes with time. For instance, the existence of calcium phosphates has been reported in the thin film formed on Ti implant in physiological media (Healy and Ducheyne, 1992; Esposito *et al.*, 1999). The higher affinity of Ti to oxygen greatly improves the regeneration of oxide passive thin film. However, it has been found that the regeneration of that oxide passive layer is easier in case of saline water, versus that of physiological

TABLE 2.4
Effect of Ni, Cr, Mo, C and W Elements on Co Properties

Elements	Effect on Mechanical Properties	Effect on Corrosion Resistance
Ni	Increase castability, enhance solid-solution strengthening	Increase corrosion resistance
Cr	Enhance wear resistance	Cr_2O_3 to corrosion resistance
Mo	Enhance solid-solution strengthening	Increase corrosion resistance
C	Increase castability, enhance wear resistance	–
W	Decrease corrosion fatigue strength, enhance solid-solution strengthening	Decrease corrosion resistance

Source: Chen Q and Thouas GA, *Mater Sci and Eng R: Rep.* 2015; 87: 1–57.

medium (Sukenik *et al.*, 1990). This is most likely because of the H_2O_2 formation in the physiological medium close to the surface of Ti implant (Tengvall *et al.*, 1989; Pan *et al.*, 1996). β-Ti alloys have shown better corrosion resistance as well as wear corrosion characteristics in phosphate buffer solution. For instance, the corrosion resistance of Ti-Nb-Zr is ascribed to the inert oxide film of Nb and Zr oxides formed onto surface of the alloy (Khan *et al.*, 1999; Lopez *et al.*, 2001; Yu *et al.*, 2001). The porous nature of Ti displayed lower corrosion characteristics than that of blank Ti and 316L Stainless steel, which is due to the fact that pores are susceptible to localized corrosion (Seah *et al.*, 1998). Another study on Ti-Ni alloys proved that porous Ti-Ni has a lower corrosion resistance than solid Ti-Ni (Li *et al.*, 2003).

2.3 SURFACE TREATMENT OF METALLIC BIOMATERIALS

The surface of a metallic biomaterial plays an important role for having a better interaction between the physiological medium and the surface of the implant. Hence, surface treatment of the implant can develop several properties such as wear and corrosion resistance, bioactivity, biocompatibility and osseointegration, since the surface of the manufactured implant lacks these. Surface modification has been approached by various strategies. In section 2.3, the notable surface modifications are briefly explained as three main categories of physical, mechanical and chemical methods.

2.3.1 Physical Surface Treatments

The main goal of physical surface treatments is to fabricate the coating layer onto the surface of substrate based on thermal or kinetic or electric energy. This section briefly shows main different methods for physical surface treatments.

2.3.1.1 Magnetron Sputtering Deposition

There are many sputter deposition techniques such as reactive sputtering, ion-beam sputtering, ion-assisted deposition, magnetron sputtering, high target utilization sputtering and gas flow sputtering. Magnetron sputtering is the most common sputtering technique for fabricating thin film coating (Figure 2.5) (Azadmanjiri *et al.*, 2014, 2017b). Magnetron sputtering is a physical vapour deposition process that can be hired to fabricate multilayered materials. The atoms are generated in the vapour phase using bombardment of a solid source with energetic particles. After that, a single layer of material is deposited onto the substrate. The current experimental work employs repeated magnetron sputtering to fabricate disparate layers (Srivastava *et al.*, 2016).

The microstructure and properties of the coated layer depends on its composition. The structure and composition can be manipulated by adjusting the process parameters. Sputtered thin film manufactured by magnetron sputtering has excellent chemical and microstructural uniformity. Almost any metal target material can be sputtered without decomposition as well. The water-cooled

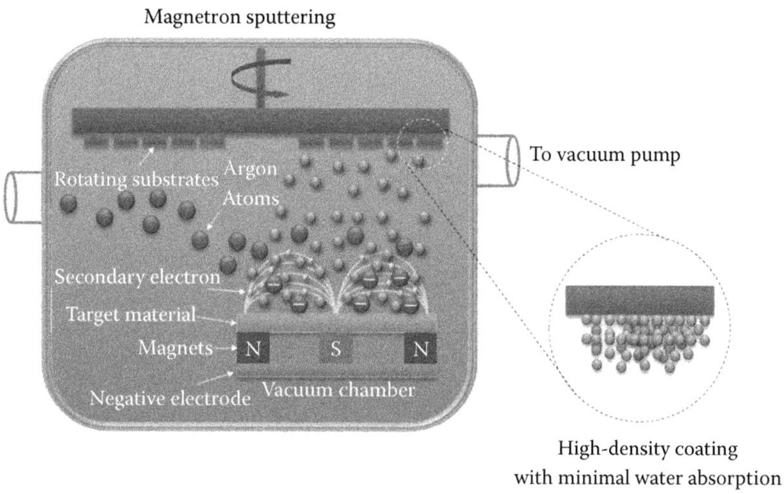

FIGURE 2.5 Schematic that depicts depositing a thin film by magnetron sputtering.

target minimizes the effects of radiated heat so that thermal microstructural effects are diminished. However, slow deposition speed, low adhesion, and low density are several drawbacks for magnetron sputtering compared to other technologies.

2.3.1.2 Pulsed Laser Deposition

Pulsed laser deposition (PLD) is a physical gas-phase technique for depositing thin film. PLD is become a popular technique to fabricate high-purity nanostructure materials of any composition. In PLD techniques, the target is vaporized from the target using a pulsed laser in an ultra-high vacuum chamber or in the existence of a reagent gas. A schematic of a PLD method for the synthesis of nanostructure materials is demonstrated in Figure 2.6. Target atoms may be either evaporated under the vacuum condition and deposited directly on the substrate; or the target atoms may interact in the presence of a reagent gas to form the favourable compound; *i.e.*, oxide in the case of oxygen, carbide for methane, nitride for nitrogen, ammonia, etc. The elemental combination and size dispensation of nanoparticles can be altered by changing experimental parameters, such as composition of the inert gas and the reagent gas in the chamber and varying the temperature gradient and laser pulse power.

PLD is a fast technique (*i.e.*, 1.3 monolayer for Fe deposition per minute, using Nd: YAG laser (Chubunova *et al.*, 1994) and the stoichiometry of the deposited film is very close to the target

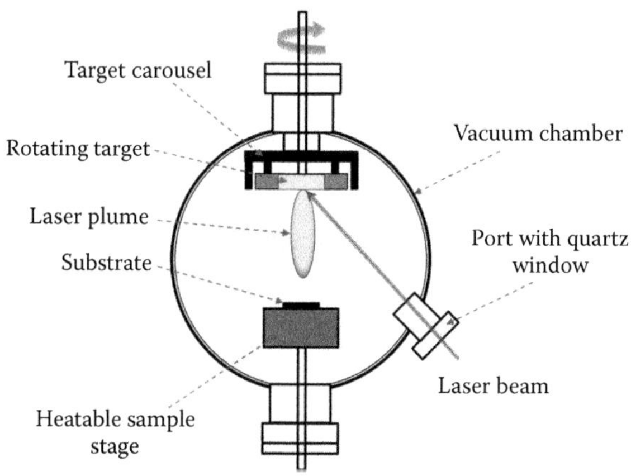

FIGURE 2.6 Schematic of the pulsed laser deposition (PLD) technique.

material, hence, it is possible to fabricate stoichiometric thin film from a single alloy bulk target (Krebs *et al.*, 2003). The properties of the fabricated thin film can also be controlled in terms of crystalline structure, thickness and composition (Krebs *et al.*, 2003). PLD does not need expensive or corrosive precursors, and large volume targets are not necessary with this technique compared to the other thin film methods, such as magnetron sputtering. However, clusters and droplets may be observed at the surface of the deposited films that lead to higher roughness and this is a main drawback for optical, electrical (Morintale *et al.*, 2010) or other applications of the deposited film.

2.3.1.3 Lithography Process

Lithography is an easy, inexpensive and fast physical method for fabricating self-assembled 2D nanostructured materials (NSMs) on different substrates. Lithography consists of many alternative types of surface preparation in which a design is transferred from a photomask or reticle onto a substrate surface that might enable multiple copies to be made from one exposure (Tiwari *et al.*, 2012). The two nanofabrication categories for lithography are (i) conventional methods such as e-beam lithography and focused ion beam (FIB) lithography (Tiwari *et al.*, 2012); and (ii) unconventional methods such as soft nanoimprint lithography (Stewart *et al.*, 2009), nanosphere lithography (Haes *et al.*, 2005), colloidal lithography (Zheng *et al.*, 2008), nanoimprint lithography (Lucas *et al.*, 2008) and solution-phase synthesis (Lee *et al.*, 2009).

Lithography has been used extensively by researchers in fields of physics, chemistry and biology (Lee *et al.*, 2009). The unconventional approaches of lithography are more preferable (Tiwari *et al.*, 2012). Zemła *et al.* (2009) demonstrated selective protein adsorption to polymer patterns formed by self-organization (blended films of polystyrene/poly methyl methacrylate and polystyrene/plyethylene glycol) and soft lithography (micromolded bilayers of partially brominated polystyrene/polyethylene glycol), as shown in Figure 2.7.

The selective protein microarrays have many applications in biology, biomedical devices, biosensor technology and tissue engineering (Zemła *et al.*, 2009). Lithography can produce high-quality prints. However, it is unsuited for economically producing small print runs which is in a few orders per day.

2.3.1.4 Plasma Spraying

The plasma spraying technique relies on melting and spraying a material on a substrate utilizing an electric field. The temperature, density and velocity of plasma beams play an essential role in the fabrication of coating. It has been illustrated that almost all materials are subject to melting

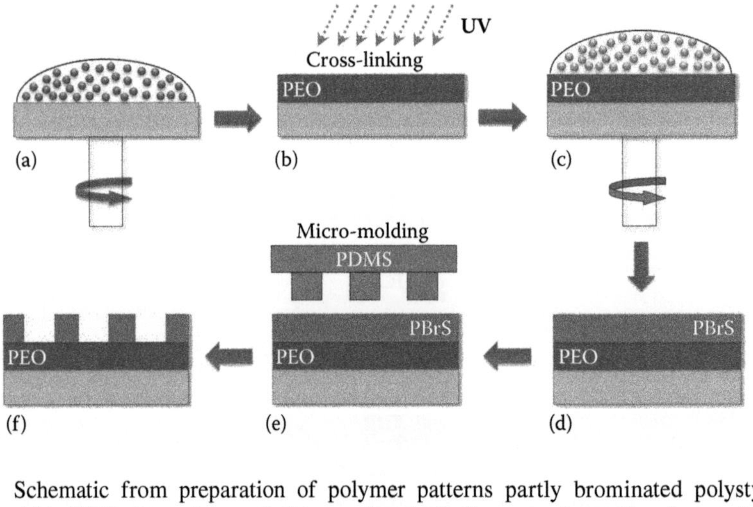

FIGURE 2.7 Schematic from preparation of polymer patterns partly brominated polystyrene (PBrS)/poly(ethylene oxide) (PEO) formed by soft lithography. (a–d) Present spin-casting from selective solvents for deposition bilayers of PEO and PBrS. (e and f) Demonstrate the polymer patterns by solvent-assisted micromolding technique, (PDMS= polydimethylsiloxane). (Reproduced from data published in Zemła J *et al.*, *Biomacromolecules*. 2009; 10: 2101–2109. With permission.)

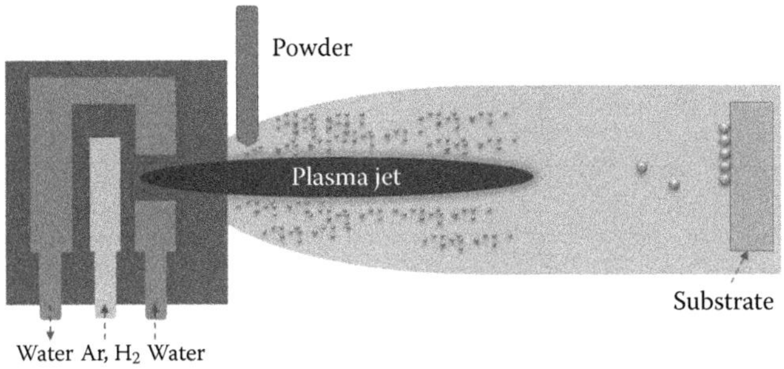

FIGURE 2.8 Schematic of the plasma spray process technique.

in a plasma jet (Figure 2.8). The utilized gases for plasma spraying are either single atom gases like argon (Ar) and helium (He) or double atom gases like N_2 and H_2. A mixture of them is sometimes used to accurately verify inert atmosphere. Prior to undertaking plasma spray, the substrate is cleaned and rinsed in organic solvents as well as roughened to enhance the interaction between both the coating and the substrate. A plasma spray technique has various advantages such as high deposition rate (80 g/min), thick deposition layer (5 mm) and low total cost. Furthermore, the deposited layer has rough surface which has potential for several fields such as the implantation of biomaterials for bone. Ceramic materials such as Al_2O_3, ZrO_2 and TiO_2 have been deposited onto Ti substrate by plasma spray technique in order to be employed for orthopaedic field. However, the bioinertness of Al_2O_3 and ZrO_2 limits its use as a consequence of their weak interaction with the bone. Therefore, bioactive ceramic materials have been attempted by plasma spray technique to be coating for Ti. HA constitutes almost 70% of bone composition. Therefore, the strategy of the plasma spray deposition of artificial HA on Ti has been of great interest. HA easily connects and bonds to bone. However, it has weak bond with Ti due to mismatch of their thermal coefficients (13.3×10^{-6} K^{-1} for HA and 8.4–8.8×10^{-6} K^{-1} for Ti). Moreover, HA is prone to degradation in simulated body fluid. As a result, the bond between HA and substrates can be easily broken which, in turn, causes loosening of the implant, as well as create risk of debris accumulation (Degroot *et al.*, 1987; Collier *et al.*, 1993). The crystallinity of deposited HA coating by the plasma spray technique is of a major concern. Therefore, further deposition treatments are suggested such as furnace heating (Lynn and DuQuesnay, 2002), laser treatment (Cheang *et al.*, 1996) and water vapour treatment (Cao *et al.*, 1996). A 50-μm thickness of plasma spray deposition layer is recommended in order of avoiding fatigue failure while keeping good biocompatibility. Calcium silicate (CaO-SiO_2) coating has been attempted by plasma spray technique onto Ti substrate and it has been reported that CaO-SiO_2 has bioactivity and biocompatibility where it promotes the proliferation of osteoblast-like cells (De Aza *et al.*, 2000; Liu *et al.*, 2002). However, the bond strength of calcium silicate coating with Ti (39 MPa for $CaSiO_3$ and 38.9 for CaO-SiO_2) is still weak despite it is higher than that of HA coating (Liu *et al.*, 2002). Efforts have been made to strengthen the bond strength of calcium silicate coating utilizing composites of that material with either ZrO_2 (Liu and Ding, 2002) or TiO_2 (Liu and Ding, 2003). Along the same line, Ti coating utilizing the plasma spray technique has been pursued on Ti substrates (Vercaigne *et al.*, 1998). It has been pointed out that alkali and heat treatment for the plasma spray Ti coating significantly boosted the osseointegration.

2.3.2 Mechanical Surface Treatments

Mechanical surface treatment is essential whether in preparing the surface for further surface modification methods like the physical or chemical processes of modification in terms of machining, grinding and polishing (Lucchini *et al.*, 1996), or to be directly involved in implantation after mechanical treatment in the view of adapting the roughness and topography of the surface

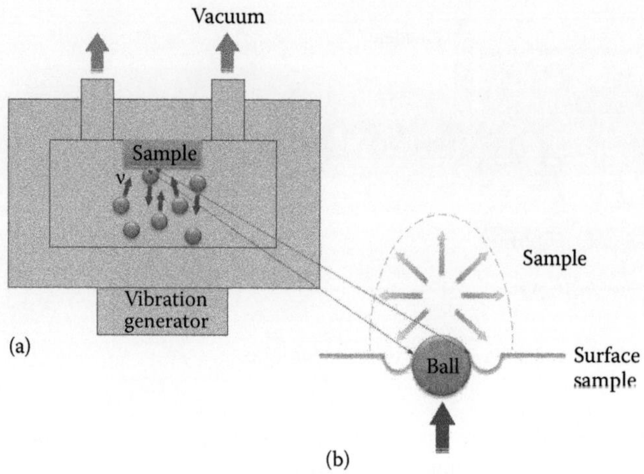

FIGURE 2.9 (a) Schematic view that illustrates the SMAT setup; (b) the plastic deformation exposed to the surface that is induced by the repeated multidirectional efficacy of the flying balls.

(Azadmanjiri *et al.*, 2015). Blasting is a method of altering the surface roughness and refining and removing the contaminations of the surface (Kern and Thompson, 1994; Wennerberg *et al.*, 1996). Abrasive particles such as alumina (Al_2O_3), Titania (TiO_2) and HA particles are employed for blasting the surface of biometallic material because of the concern of the stability in chemical solutions as well as its biocompatibility. However, it has been announced that some segments of that particles might still be attached to the surface of implant after blasting, causing cell damage when inserted into living organ, especially in case of alumina particles (Thompson and Puleo, 1996; Wennerberg *et al.*, 1996). The use of biocompatible and osteoconductive materials like HA or Tricalcium phosphate [TCP, Ca_3 $(PO_4)_2$] demonstrated significant osseointegration (Mueller *et al.*, 2003). Surface mechanical attrition treatment (SMAT) was also utilized in biomedical field in order to enhance the surface characteristics by creating a nanostructure in the macroscale (Azadmanjiri *et al.*, 2015). In a short prescription, spherical balls target the surface of the samples (metallic biomaterial) due to the vibration by a high-power ultrasound. Some parameters such as frequency of vibration, SMAT time, working distance and diameter of balls are effective and involve during SMAT process. It has clearly reported that the SMATed surface behaves better than the un-SMATed surface whether in terms of corrosion resistance or cell proliferation (Azadmanjiri *et al.*, 2015). Additionally, the hardness of SMATed surface is greater than that of un-SMATed surface. Figure 2.9 show a schematic of SMAT setup and plastic deformation exposed to the surface. For a detailed discussion on mechanical surface modifications, see Azadmanjiri *et al.* (2015).

2.3.3 Chemical Surface Treatments

Chemical approaches play a significant role in developing surface treatment with nanostructured materials, by mixing materials and solutions at a molecular level; hence providing good chemical homogeneity in comparison to physical methods (Tiwari *et al.*, 2012). Additionally, chemical processing design and synthesis of novel materials can be settled into the final end productions (Tiwari *et al.*, 2012). However, toxic reagents and solvents are frequently used for the synthesis of nanostructure materials (Dahl *et al.*, 2007), and the chemical methods are relatively slow to actualize. There are many chemical deposition methods used for fabricating nanostructure materials, such as sol-gel methods (Azadmanjiri and Ebrahimi, 2004; Azadmanjiri and Ebrahimi, 2006; Azadmanjiri, 2008), *in situ* polymerization (Azadmanjiri *et al.*, 2011, 2012a), Silanization and Langmuir-Blodgett (LB) techniques (ClementeLeon *et al.*, 1997; Azadmanjiri *et al.*, 2012b), layer-by-layer (LbL) assembly

(Lee *et al.*, 1988; Keller *et al.*, 1994), self-assembly (Sellinger *et al.*, 1998) and atomic layer deposition (ALD) (Lee *et al.*, 2013) processes. Some of important ones will be explained below.

2.3.3.1 Atomic Layer Deposition

ALD or atomic layer epitaxy (ALE) is a chemical deposition technique for creating a significant surface area, high-quality and thin film layers with atomic scale accuracy (Knez *et al.*, 2007; Azadmanjiri *et al.*, 2016c, 2018a). Figure 2.10 demonstrates an ALD technique where the thin film layers cyclically develop for a binary substance from gaseous precursors. ALD is actually a surface-controlled layer-by-layer thin film deposition approach. It has the ideal potential for generating thin films where thickness and film structure can be governed at the atomic level. ALD can generate a broad range of film materials with high density, low impurity and precise control of thickness. Additionally, a low deposition temperature can mitigate undesirable outcomes on sensitive substrates.

The ALD technique has been extensively considered by researchers in nanotechnology areas because of its control over the deposited layer thickness of materials and the potential for the adjustment of chemical and physical properties in the nanoscale array (Knez *et al.*, 2007). There is an extensive review on metal and nitride thin films created by ALD for semiconductor device applications by Kim (2003). The surface chemistry of the trimethylaluminum [Al $(CH_3)_3$] (TMA) and H_2O process by ALD is elaborated by Puurunen *et al.* (2005). Furthermore, the mechanical properties of some metal oxides such as TiO_2 (Sun *et al.*, 2012; Azadmanjiri et al., 2018b) and Al_2O_3 (Dingemans *et al.*, 2010) have been explored systematically. Different groups have made deposition bio-based multilayer barrier films by ALD (Vartiainen *et al.*, 2016). The slowness of deposition is the major restriction of the ALD technique; usually a portion of a monolayer is deposited in each cycle. However, a very thin layer is required for next-generation biomaterial devices and thus the slowness of ALD methods may not be a critical issue.

2.3.3.2 Silanization and Langmuir-Blodgett Deposition

Surface treatment, to modify the connections of fluids with them, is a crucial step in creating beneficial microfluidics devices, especially those utilized in biological applications. Silanization as a

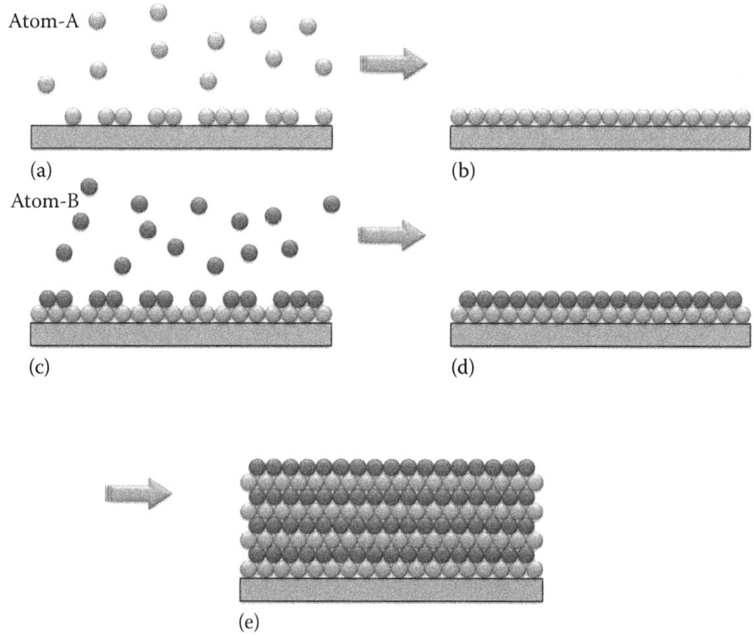

FIGURE 2.10 Schematic demonstration of an ALD technique where the thin film layers cyclically increase for a binary compound from gaseous precursors.

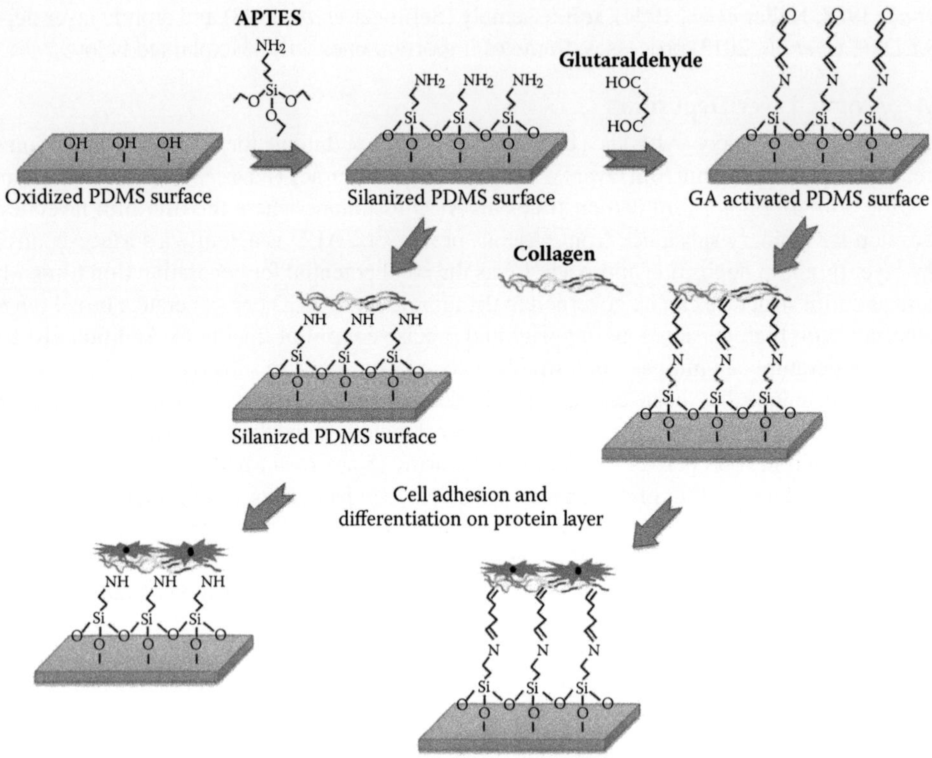

FIGURE 2.11 Schematic demonstration of PDMS surface modification by APTES with/without GA and protein crosslinking and subsequent cell culture. (Reproduced from data published in Chuah YJ *et al.*, *Biomater. Sci.* 2015; 3: 383–390. With permission.)

common term is applying to the formation of organosilane monolayers on substrates. Glass *et al.* (2011) studied various organosilane deposition protocols and some application notes have been provided as a practical resource to a broader range of techniques for all users.

Silanization is a low-coat and efficient chemical covalent coating route for addressing a material surface by means of self-assembly with organofunctional alkoxysilane molecules. HA, bioglass, TiO_2 and many other metal oxide surfaces can be silanized due to rich hydroxyl groups, which attack and displace the alkoxy groups on the silane, hence, forming a covalent –Si–O–Si– bond. There are numerous kinds of commercially existing silane combining agents which usually are simple to react with hydroxylated surface and present active groups, such as amino groups and carboxyl groups, to the surface (Qiu *et al.*, 2014). Figure 2.11 takes polydimethylsiloxane [PDMS, $CH_3(Si(CH_3)_2O)_nSi(CH_3)_3$] as an example to demonstrate chemical structure of the surface modification by 3-aminopropyltriethoxysilane (APTES, $C_9H_{23}NO_3Si$) with/without glutaraldehyde (GA), adopted by protein immobilization using collagen, and examined the stability of the confluent cell layer and the potency of adhered mesenchymal stem cells (MSCs) for up to three weeks.

Silanization is a simple and effective route; however, the reaction conditions such as concentration of the silane and reaction time must be carefully controlled to prevent from forming thick polymerized silane networks on the surface. Otherwise, in some conditions, the bond between silane and the surface can also subject to hydrolysis (Wasserman *et al.*, 1989).

The LB method is a technique for producing mono- and multi-molecular films, which is implemented by transferring Langmuir films to the surface of a solid substrate (Figure 2.12). Langmuir films consist of a monolayer of amphiphilic compounds formed on the surface of a liquid. LB method enables the fabrication of closely packed, well-ordered and organized multilayered films from oriented monolayers and allow for the immobilization of several functional molecules onto

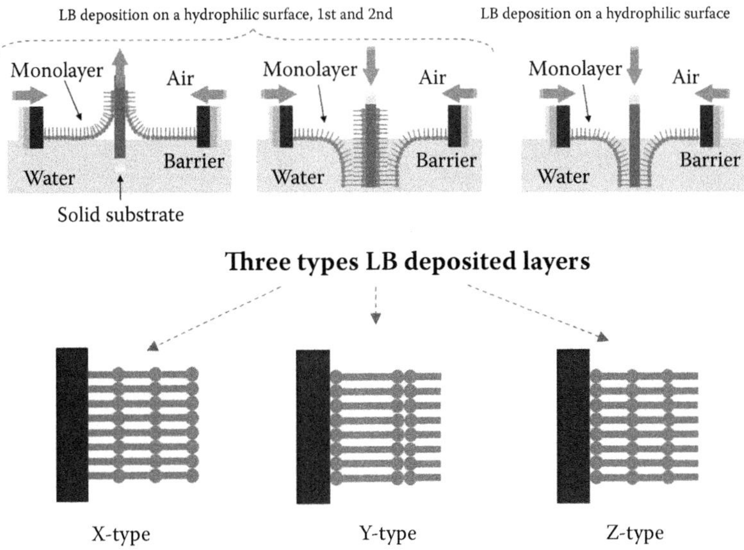

FIGURE 2.12 Schematic demonstrating the LB technique for producing mono- and multimolecular films and three types of LB deposited layers on a hydrophilic surface.

surfaces (Borges and Mano, 2014). However, LB is an expensive and time-consuming method that requires specialized instrumentation. It also requires specific and limited molecules (*i.e.*, amphiphilic molecules) to prepare the films. Additionally, a weak molecular interaction between the film and the solid support is a disadvantage for LB (Love *et al.*, 2005).

2.3.3.3 Layer-by-Layer Assembly Approach

The LbL assembly method includes the consecutive adsorption of supplementary multivalent molecules on a surface layer. Interaction between the molecules and deposited layers could be either electrostatic or non-electrostatic (Borges and Mano, 2014). LbL assembly is an effective, reproducible, robust, easy and flexible method to produce highly ordered nanoscale structures or types with expanded functionalities and activities. This method may be employed to fabricate nanostructured polymeric multilayer thin films and nanocomposites with specific thicknesses, structures, properties, compositions and functionalities on any substrate (Borges and Mano, 2014). In order to synthesize a high-quality thin film and avoid any contaminations from the previous adsorption step – as well as prevent weakly adsorbed layers – washing steps are required between the deposition of each molecule layer (Hoogeveen *et al.*, 1996). The LbL method generally occurs through a different of deposition techniques such as spraying, dip-coating, spin-coating and perfusion that have been extensively reviewed in previous works (Chiarelli *et al.*, 2001; Michel *et al.*, 2005; Li *et al.*, 2012). The architecture and properties of the deposited films by LbL technique can be accurately controlled at the nanoscale level (Borges and Mano, 2014). There are several molecular interactions between thin film layers for the LbL assembly technique that have been overviewed by J. Borges and J.F. Mano (Borges and Mano, 2014). Figure 2.13 shows a schematic for an LbL self-assembly of a multilayer coating by sequential adsorption of oppositely charged polyelectrolytes.

2.3.3.4 Electrochemical Method: Electrochemical Anodization

Since the nanoporous structure greatly improves cell adsorption, attachment and osteogenesis of the cells, nanotubes fabricated by electrochemical anodization have been of great interest. Recent review articles have thoroughly discussed the fabrication of nanotubes as well as the cell behavior (Macak *et al.*, 2007; Roy *et al.*, 2011; Brammer *et al.*, 2012; Minagar *et al.*, 2012; Tan *et al.*, 2012; Minagar *et al.*, 2013; Wang *et al.*, 2014). Therefore, nanotube fabrication and the cell activity are briefly discussed

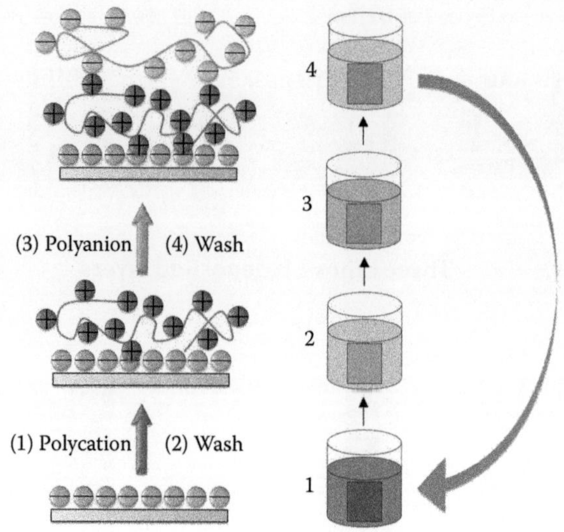

FIGURE 2.13 Schematic that demonstrates the LbL self-assembly of a multilayer coating using consecutive adsorption of oppositely charged polyelectrolytes.

here. Table 2.5 displays the recent work for fabricating metallic biomaterial nanotubes as well as the corresponding cell response. Numerous research strategies have been appointed for the fabrication of nanotubes onto the surface of metallic biomaterials. Nanotube structures verify the proper surface area for the cells where it is able to be interlocked inside the tubes and it can conveniently spread into the distances located between the tubes. Electrochemical anodization has been extensively applied and it is not only limited to Ti and Ti-based alloys. However, this technique requires an innovative strategy if it is being used for Ti and its alloys. Because Ti surface is capable to adsorb oxygen extensively. Hence, the formed oxide layer on top surface of Ti prevents a steady penetration of fluoride anions (F^-) during anodization and formation of uniform nanotubes. Figure 2.14 simply depicts the electrochemical anodization setup where Ti (or Ti-based alloy, Zr and Ta) is connected to the anodic side while platinum (Pt) is connected to the cathode side. Ti nanotubes formed in fluorine-based electrolyte through three consecutive steps since anodization is held (Figure 2.14). First, the oxidation of the Ti to TiO_2 takes place to form compact oxide thin film. After that, the adsorption of fluoride anions (F^-) on the surface of Ti and reaction with the oxide layer results in the fabrication of porous thin film. Finally, the formed soluble complex $[(TiF_6)^{-2}]$ dissolves progressively from the surface of Ti into the bulk of solution where TiO_2 nanotubes are then fabricated. In fact, some factors such as time of anodization, applied voltage, the type of electrolyte and annealing conditions affect the geometry and topography of the fabricated nanotubes. The fabrication of nanotubes from aqueous electrolytes reduces both the voltage and time compared to the organic electrolyte (*i.e.*, ethylene glycol, polyethylene glycols) when wider nanotubes are required (Macak *et al.*, 2005; Wan *et al.*, 2009). For instance, TiO_2 nanotubes with a diameter up to 100 nm and length up to 4 μm has been fabricated at 20 V from an aqueous electrolyte containing 1M $(NH_4)_2SO_4$ and NH_4F, while high voltage (60 V) and longer time (18 h) were required for fabricating TiO_2 nanotubes with diameter up to 100 nm and 45 μm in length from organic electrolyte containing NH_4F, H_2O and ethylene glycol (EG, $C_2H_6O_2$). The length of nanotubes does not play significant role with the cell activity (Bauer *et al.*, 2009). It has been reported that the diameter of the fabricated nanotubes has a critical impact on osteogentic cell proliferation (Park *et al.*, 2007, 2009). It is pointed out that 15-nm diameter TiO_2 nanotubes constitute the borderline of cell activity beyond which cell apoptosis begins. In contrast, cell osteogenesis was highly activated in case of 100-nm diameter TiO_2 nanotubes (Das *et al.*, 2009). This leads to the presumption that the material–cell interaction mechanism is controlled by some stimuli that might not be clearly grasped. However, the difference of cells in the same cell line, the nanotubes' diameter,

TABLE 2.5
Metallic Biomaterials Nanotubes, Characteristics, Synthesis Condition and Cell Activity

Material	Nanotube Characteristics	Synthesis Condition	Cell Activity	Ref.
Ti	80 nm (diameter), length (200–6000 nm), wall thickness (7–34 nm)	30 V, 60 min, 60°C	Coating of the PLGA on titanium nanotubes favors the adhesion and proliferation of osteoblasts and improves the biocompatibility of the system.	(Wang *et al.*, 2017)
Ti	70 nm (diameter), nanograined wall, 18 μm (length)	HF, 20 V, 120 min, no annealing on nanograined wall TiO_2	Enhanced cell adhesion due to the interlocking of cells into nanotubes, human mesenchemal stem cells (hMSCs) were employed for stem cell test.	(Azadmanjiri *et al.*, 2016)
Ti	Group 1 (Two-step anodization): 100 nm (diameter), 54 μm (length) Group 2 (One-step anodization): 100 nm (diameter), 4 μm (length)	0.25 wt% NH_4F + 2 vol% + EG Step 1: 30 V, 1 h Step 2: 30 V, 3 h 0.25 wt% NH_4F + 10 vol% + EG, 30 V, 3 h, annealing to be crystallized as well as hydrophilic	The fabricated TiO_2 on surface of Ti by one-step anodization has a rougher surface and sharper edges than that fabricated by two-step anodization. Moreover, the surface energy obtained by two-step anodization (γ = 58.19 ± 2.32 mJm^{-2}) is higher than that of one-step one (γ = 45.56 ± 1.13 mJm^{-2}). Therefore, fabricated TNTs on the surface displayed higher biocompatibility and cell viability than that of both one-step and CP-Ti. Consequently, the topography of the surface might possibly have impact on the behavior of the cell, SaOS2 osteoblast-like cells were used for cell culture.	(Wang *et al.*, 2014)
Ti	70 nm (diameter), 500 nm (length)	0.25 wt% NH_4F + 2 vol% H_2O+EG, 60 V, 30 min	Three surface treatment processes were consecutively performed to produce microporous/nanoporous Ti starting from micro-arc oxidation. Then deposition of Ti thin film and finally anodic oxidation. The as-fabricated hierarchical combination of microporous and nanoporous surface showed super hydrophilicity of contact angle (θ = 0). Further, the hierarchical surface promoted cell osteogenesis and protein adsorption compared to dense Ti or even microporous surfaces.	(Han *et al.*, 2014)
Ti6Al4V	25 nm and 35 nm (diameter)	0.5 wt% HF, 20 V, 60 min, annealing to be crystallized	First, laser melting creates a porous surface which after that undertook anodization. Nanotubes were created on a microsurface. Heat-treatment at 600°C destroyed the nanostructure resulting in crystal of microstructure instead.	(Yavari *et al.*, 2014)

(*Continued*)

TABLE 2.5 (CONTINUED)
Metallic Biomaterials Nanotubes, Characteristics, Synthesis Condition and Cell Activity

Material	Nanotube Characteristics	Synthesis Condition	Cell Activity	Ref.
Ta coated Ti	100 nm (diameter), 10 nm (wall thickness), 300 nm (length)	Acetic acid + 0.5 wt% HF 20 V, 30 min, Ti nanotubes were annealed to be crystallized and after that coated by Ta (20-nm layer thickness) utilizing sputter technique	Human osteoblast cells respond better in case of Ta-coated TiO_2 nanotubes than flat Ti and bare TiO_2 nanotubes which was due to the physico-chemical effect in terms of the surface energy ($\theta = 4$ for Ti nanotubes while $\theta = 0$ for Ta-coated TiO_2 nanotubes) and isoelectric field. Ta-coated TiO_2 nanotubes demonstrated better cell viability and faster rate of cellular matrix of mineralization (30%) than that of bare TiO_2 nanotubes.	(Frandsen *et al.*, 2014)
Ti coated Ti-6Al-4V	30 nm (diameter), 10 nm (wall thickness)	0.5 wt% HF, 5 V	Nano/micro hybrid structure was fabricated on the surface which seemingly signalled to better cell response, protein adsorption and gene expression in comparison to the as-sprayed Ti.	(Xie *et al.*, 2014)
Ti-6Al-4V	50 nm and 100 nm (diameter)	0.3 wt% NH_4F + 2 vol% H_2O + EG, 20 V and 60 V, 4 h annealing to be crystalized at different temperatures 300, 450 and 600°C for 1.5 and 3 h	Nanotubes created at 60 V and annealed at 600°C for 30 h interestingly demonstrated the highest degree of hydrophilicity as well as roughness. Furthermore, the 60 V nanotubes were able to keep their hydrophilicity up to 11 days without annealing while the aging time extended 60 days as a consequence of annealing at 600°C. It was obvious that voltage and annealing temperature empowered the hydrophilicity due to the formation of crystalline phase which is responsible for delaying hydroxylation/dehydroxylation equilibrium. This implication paves the way for anti-aging biocompatible Ti implants.	(Hamlekhan *et al.*, 2014)
Ti50Zr	3–120 nm (diameter), 3–32 nm (wall thickness)	0.1–0.5 wt% NH_4F + 1 M $(NH_4)_2SO_4$, 5–20 V, 2 h, annealing to be crystallized and more hydrophilic as well	The surface energy of fabricated titania-zirconia-zirconium titanate nanotubes composite (TiO_2-ZrO_2-$ZrTiO_4$) on the surface ($\gamma^S = 73.64$ mJm^{-2}) is higher than that of nanotubes fabricated on pure Ti ($\gamma^S = 68.51$ mJm^{-2}) which promotes Ti50Zr alloy to be acceptable biomedical material than pure Ti.	(Minagar *et al.*, 2014)

(*Continued*)

TABLE 2.5 (CONTINUED)
Metallic Biomaterials Nanotubes, Characteristics, Synthesis Condition and Cell Activity

Material	Nanotube Characteristics	Synthesis Condition	Cell Activity	Ref.
TiCu deposited on Ti	80 nm (diameter)	0.25 wt% NH_4F + 1 vol% H_2O + EG	Stoichiometric ratio of TiO_2/CuO nanotubes was fabricated on the surface which led to better antibacterial activity and biointegration of the implant surface with the cells. Even so, the concentration of Cu in the thin film should be less than 1 vol% with respect to cytotoxicity of Cu. *Escherichia coli* (*E. coli*) and *Staphylococcus aureus* (*S. aureus*) were used for testing the antibacterial activity of the fabricated Ti-Cu nanotubes while Murine preosteoblast cells (MC3T3-E1) were used for cell culture.	(Hang *et al.*, 2014)
Ti-coated zirconia knee implant	50 nm (diameter), 200 nm (length)	0.25 wt% NH_4F + 2 vol% + EG 20 V, 15 min, annealing to be crystallized	TiO_2 nanotubes created onto the surface of zirconia implant exhibited both higher wettability and cell osseointegration than that of the bare zirconia implant after 24 h of cell culture. MC3T3-E1 mouse osteoblast cells were used.	(Frandsen *et al.*, 2013)
Ti	100 nm (diameter), 1 μm (length)	1 M H_3PO_4 + 0.3 wt% HF, 20 V, 60 min	TiO_2 nanotubes created on the surface provided appropriate cell osteogenesis than that of bare Ti. *In vitro* investigations matched the *in vivo* atmosphere by preparing conditional medium of MC3T3-E1 cells (osteoblasts cells-CM) which contains paracrine.	(Kim *et al.*, 2013)
Ti	120 nm (diameter), 800 nm (wall thickness)	1 M Na_2SO_4 + 0.5 wt% NaF 20 V, 4.5 h, annealing to be crystallized	Nanotubular structure of Ti enhanced both corrosion properties and biocompatibility of Ti, where the corrosion rate of TiO_2 nanotubes [0.64 millimetres per year (mpy)] in simulated body fluid is much lower than that of solid Ti surface (6.42 mpy).	(Balakrishnan and Narayanan, 2013)
Ti	30 nm and 100 nm (diameter)	0.5% HF, 20 V, 30 min, annealing to be crystallized	The antibacterial activity of TiO_2 nanotubes against bacteria strains such as *E. coli* and *S. aureus* was evaluated in the presence of curcumin. It was found that curcumin enhanced the bacteria attachment to the surfaces of TiO_2 nanotubes and bacterial biofilm was created on the surface; despite that it has an antimicrobial effect on the bacteria itself.	(Oh *et al.*, 2013)

(*Continued*)

TABLE 2.5 (CONTINUED)
Metallic Biomaterials Nanotubes, Characteristics, Synthesis Condition and Cell Activity

Material	Nanotube Characteristics	Synthesis Condition	Cell Activity	Ref.
Ti-30Ta	100 nm (diameter)	HF + H_2SO_4 + 5% dimethyl sulfoxide (DMSO), 35 V, 40 min, annealing to be crystallized	Nanotubes created on the surface display the lowest contact angle compared with the untreated Ti-30Ta, alkali and heat-treated Ti-30Ta, alkali and heat-treated Ti-30Ta with SBF coating and ion beam etched Ti-30 Ta.	(Capellato *et al.*, 2013)
Zr	–	1 mol/L H_3PO_4, 3–30 V, 60 min	Nanoporous structure exhibited better corrosion resistivity and it further promoted osseointegration *in vivo*.	(Sanchez *et al.*, 2013)
Zr	20 nm and 70 nm (diameter), 7 µm (length)	0.75 M NH_4F + 1 M $(NH_4)_2$ SO_4, 10–25 V, 30 min	The wettability of the as-fabricated ZrO_2 nanotubes are accordingly controlled by the diameter of the nanotubes as long as the layer thickness of nanotubes is constant with the annealing. For instance, it is more hydrophilic than intact Zr (67.4° for intact Zr and 18.2° for ZrO_2 nanotubes) while it becomes 8.7° after annealing. However, the aging time dramatically affects the wettability of nanotubes whether they are annealed or not. It becomes 38.6° for as-fabricated nanotubes and 28.2° for the as-annealed nanotubes. Then, it is clear that wettability can be adjusted according to the requirement of the orthopaedic implant.	(Wang *et al.*, 2013)
Zr		1 mol/L H_3PO_4, 3–30 V, 60 min	Nanoporous zirconium was fabricated by anodization. Meanwhile, phosphates ions were incorporated into the thus-anodized surface which in turn promoted the formation of HA when immersed in SBF.	(Sanchez *et al.*, 2011)
Zr	40 nm (diameter) and 10 µm (length)	First step: 0.75 M NH_4F + 1 M $(NH_4)_2$ SO_4, 20 V, 30 min Second step: 0.1 M NH_4F + 1 M $(NH_4)_2$ SO_4, 20 V, 15 min, annealing at 300°C for 6 h to be crystallized	Two-step anodization resulted in fabricating nanotubular surface of ZrO_2, which demonstrated the superhydrophilic surface compared to dense Zr. ZrO_2 nanotubes showed compatible topography for cell adhesion and proliferation. Furthermore, the alkaline phosphate activity test proved the high ratio of Ca and P deposited by the cells on ZrO_2 nanotubes which almost cover the tubular structure as proved by SEM.	(Frandsen *et al.*, 2011)
Ti	>100 nm (diameter)	H_3PO_4 + HF, 20 V, 1 h, stirring while anodized	Showed good cell adhesion which remained for almost 7 days.	(Das *et al.*, 2009)

(*Continued*)

TABLE 2.5 (CONTINUED)
Metallic Biomaterials Nanotubes, Characteristics, Synthesis Condition and Cell Activity

Material	Nanotube Characteristics	Synthesis Condition	Cell Activity	Ref.
Ti	15–100 nm (diameter)	H_3PO_4 + HF, 1–20 V, 1 h	15 nm was the critical diameter for the nanotubes beyond which cell fate occurred. Osteoblast and osteoclast cells were used for cell culture.	(Park *et al.*, 2009)
Ti	30–100 nm (diameter)	Acetic acid/HF, 5–20 V, annealing to be crystallized	Bone forming ability increased because of increasing the nanotubes' diameters which led to higher integrin clusters and focal contact formation.	(Brammer *et al.*, 2009)
Ti	30–100 nm (diameter)	Acetic acid/HF, 5–20 V, 30 min, annealing to be crystallized	Cell adhesion noticeably appeared while 30 nm did not show significant cell differentiation. Human mesenchyme stem cells were used for the cell test.	(Oh *et al.*, 2009)
Zr	50 nm (diameter), 20 μm (length)	0.5 wt% M NH_4F + 1 M $(NH_4)_2SO_4$, 20 V, 3 h	Bone-like appetite was fabricated onto ZrO_2 nanotubes when immersed in SBF for 30 days.	(Guo *et al.*, 2009)
Ta	2.3–11.1 μm (length)	HF + H_2SO_4 (1:9) + 5 vol% dimethyl sulfoxide (DMSO), 10–20V, 20 min	Tantala nanotubes have superior functionality than that of dense Ta in terms of cell adhesion, proliferation and differentiation. Human foetal osteoblasts (hfoB1.19) were employed for cell culture.	(Ruckh *et al.*, 2009)
Ti	51 nm (diameter), 50 nm (wall thickness), 600 nm (length)	H_2SO_4 + NaF + Citric acid 20 V, 4 h, annealing to be crystallized	Showed significant cell division within 5 days where Osteoblastic precursor cell (OPC1) used for the cell proliferation test.	(Kim *et al.*, 2008)
Ti	15–100 nm (diameter)	H_3PO_4 + HF, 1–20 V	100 nm diameter exhibited cell fate while excellent cell osseointegration observed for 15 nm diameter. Rat mesenchyme stem cells were utilized for cell culture.	(Park *et al.*, 2007)
Ti	80 nm (diameter), 400 nm (length)	HF, 20 V, 35 min, annealing to be crystallized and then loaded with antibiotics	Continuous increase of cell adhesion up to 3 weeks' time.	(Popat *et al.*, 2007)
Ti	70 nm (diameter), 15 nm (wall thickness), 250 nm (length)	HF, 20 V, 60 min, annealing to be crystallized	Obvious cells adhesion due to interlocking of cells into nanotubes, MC3T3-E1 cells were employed for cell test.	(Oh *et al.*, 2006)
Ti	70 nm (diameter), 15 nm (wall thickness), 250 nm (length)	HF, 20 V, 60 min, alkali treatment (NaOH) and then annealing to be crystallized	Soaking in SBF and MC3T3-E1, Alkali treatment enhanced cell adhesion.	(Oh and Jin, 2006)

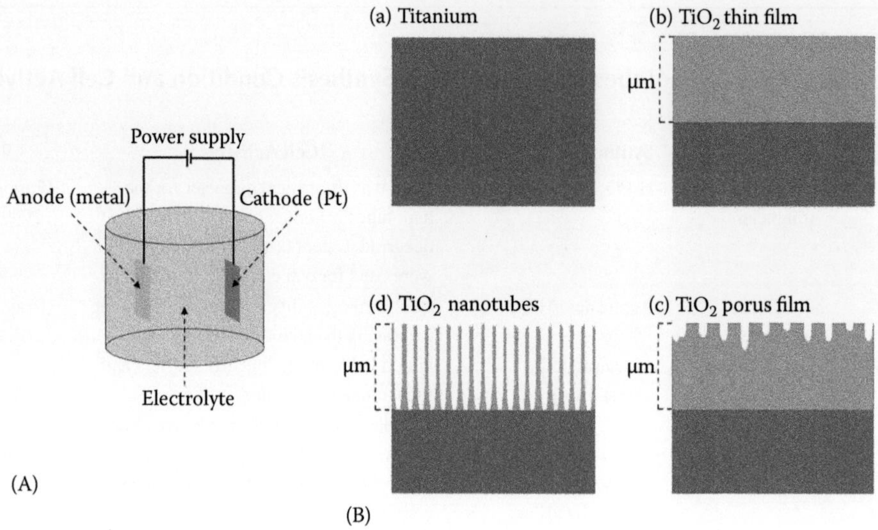

FIGURE 2.14 Schematic illustration of the electrochemical anodization: (A) experiment setup; (B) mechanism of fabrication.

the conditions of nanotubes' fabrication and the conditions of cell culture all constitute important factors which might significantly alter that mechanism of interaction and the focal contact formation.

2.4 CONCLUSIONS

Metallic nanomaterials have been used in the fabrication of medical implants for a long time. Demands for bone repair, such as the internal fracture fixation of long bone, are the factors that drive the development of metallic implants. Metallic biomaterials also play an important role in most of the temporary devices and permanent implants, as well as dental implants. The main categories of metallic implant biomaterials are SS, Co-based alloys, Ti-based alloys and miscellaneous others. Metallic materials provide better mechanical properties (*i.e.*, excellent fracture toughness, fatigue strength and tensile strength) compared to polymer and ceramic biomaterials and, in some cases, high corrosion resistance that leads to the reduced release of harmful toxins. Wear resistance, osseointegration and biocompatibility are also the essential requirements to be associated with metallic biomaterials. The main issue of Ti-based implants is chemical etching under a physiological environment and mechanical damage because of corrosion of implant during its long-term implantation in the human body. Corrosion leads to ion release and wear debris that cause inflammation of the surrounding tissues and eventually leads to implant failure. The development of new techniques for surface modification is one of the most research areas to resolve the problems and further improve the biological performance of implants.

ACKNOWLEDGMENT

The authors acknowledge the financial support through the Australia-India Strategic Research Fund (AISRF) ST060048.

REFERENCES

Assad M, Lemieux N, Rivard CH, Yahia LH. Comparative in vitro biocompatibility of nickel-titanium, pure nickel, pure titanium, and stainless steel: Genotoxicity and atomic absorption evaluation. *Bio-Med Mater and Eng.* 1999; 9: 1–12.

Asselin E, Ahmed TM, Alfantazi A. Corrosion of niobium in sulphuric and hydrochloric acid solutions at 75 and 95 degrees C. *Corr Sci.* 2007; 49: 694–710.

Azadmanjiri J. Structural and electromagnetic properties of Ni-Zn ferrites prepared by sol-gel combustion method. *Mater Chem and Phy.* 2008; 109: 109–112.

Azadmanjiri J, Berndt CC, Kapoor A, Wen C. Development of Surface Nano-Crystallization in Alloys by Surface Mechanical Attrition Treatment (SMAT). *Crit Rev Soli Sta Mater Sci.* 2015; 40: 164–181.

Azadmanjiri J, Berndt CC, Wang J, Kapoor A, Srivastava VK, Wen CE. A review on hybrid nanolaminate materials synthesized by deposition techniques for energy storage applications. *J Mater Chem A.* 2014; 2: 3695–3708.

Azadmanjiri J, Berndt CC, Wang J, Kapoor, Srivastava VK. Nanolaminated composite materials: structure, interface role and applications. *RSC Adv.* 2016b; 6: 109361–109385.

Azadmanjiri J, Ebrahimi SAS. Influence of stoichiometry and calcination condition on the microstructure and phase constitution of NiFe2O4 powders prepared by sol-gel autocombustion method. *Physica Status Solidi C: Conferences.* 2004;3414–3417.

Azadmanjiri J, Hojati-Talemi P, Simon GP, Suzuki K, Selomulya C. Synthesis and electromagnetic interference shielding properties of iron oxide/polypyrrole nanocomposites. *Polym Eng and Sc.* 2011; 51: 247–253.

Azadmanjiri J, Seyyed Ebrahimi SA. The effects of pH and citric acid concentration on the characteristics of nanocrystalline NiFe2O4 powder synthesized by a sol-gel autocombustion method. *Phy Met and Metall.* 2006; 102: S21–S23.

Azadmanjiri J, Simon GP, Suzuki K, Selomulya C, Cashion JD. Phase reduction of coated maghemite (γ-Fe2O3) nanoparticles under microwave-induced plasma heating for rapid heat treatment. *J Mater Chem.* 2012a; 22: 617–625.

Azadmanjiri J, Suzuki K, Selomulya C, Amiet A, Cashion JD, Simon GP. The use of plasma treatment for simultaneous carbonization and reduction of iron oxide/polypyrrole core/shell nanoparticles. *J Nanopart Res.* 2012b; 14.

Azadmanjiri J, Wang PY, Pingle H, Kingshott P, Wang J, Srivastava VK *et al.* Enhanced attachment of human mesenchymal stem cells on nanograined titania surfaces. *RSC Adv.* 2016a; 6: 55825–55833.

Azadmanjiri J, Srivastava VK, Kumar P, Nikzad M, Wang J, Yu A. Two- and three-dimensional graphene-based hybrid composites for advanced energy storage and conversion devices. *J Mater Chem A.* 2018a; 6: 702–734.

Azadmanjiri J, Wang J, Berndt CC, Kapoor A, Zhu DM, Ang ASM, Srivastava VK. Influence of charged defects on the interfacial bonding strength of tantalum-and silver-doped nanograined TiO_2. *Phys Chem Chem Phys.* 2017a; 19: 11881–11891.

Azadmanjiri J, Wang J, Berndt CC, Kapoor A, Zhu DM, Ang ASM, Srivastava VK. Thantalum-and silver-doped titanium dioxide nanosheets film: Influence on interfacial bonding structure and hardness of the surface. *Ind Eng Chem Res.* 2016c; 56: 434–439.

Azadmanjiri J, Wang J, Berndt CC, Wen C, Srivastava VK, Kapoor A. Structural and mechanical properties of magnetron-sputtered Al-Au thin films. *App Phys A.* 2017b; 123: 71.

Azadmanjiri J, Wang J, Berndt CC, Yu A. 2D layered organic-inorganic heterostructures for clean energy. *J Mater Chem A.* 2018b; 6: 3824-3849.

Balakrishnan M, Narayanan R. Synthesis of anodic titania nanotubes in Na2SO4/NaF electrolyte: A comparison between anodization time and specimens with biomaterial based approaches. *Thin Solid Films.* 2013; 540: 23–30.

Balazic M, Brojan M, Bombac D, Caram RJ, Kosel F, Kopac J (2007). Titanium and Titanium Alloy Applications in Medicine. *Surface Engineered Surgical Tools and Medical Devices.* Jackson MJ, Ahmed W. (eds.) Boston, MA, Springer US, 533–576.

Balla VK, Bose S, Davies NM, Bandyopadhyay A. Tantalum–A Bioactive Metal for Implants. *Jom.* 2010; 62: 61–64.

Bauer S, Park J, Faltenbacher J, Berger S, von der Mark K, Schmuki P. Size selective behavior of mesenchymal stem cells on ZrO2 and TiO2 nanotube arrays. *Integrative Bio.* 2009; 1: 525–532.

Bauer S, Schmuki P, von der Mark K, Park J. Engineering biocompatible implant surfaces Part I: Materials and surfaces. *Prog in Mater Sci.* 2013; 58: 261–326.

Black J. *Biological Performance of Materials Fundamentals of Biocompatibility, Fourth Edition.* Hoboken, Hoboken: CRC Press, 2005.

Blackwood DJ. Biomaterials: Past successes and future problems. *Corr Rev.* 2003; 21: 97–124.

Borges J, Mano JF. Molecular interactions driving the layer-by-layer assembly of multilayers. *Chem Rev.* 2014; 114: 8883–8942.

Brammer KS, Frandsen CJ, Jin S. TiO2 nanotubes for bone regeneration. *Trends in Biotech.* 2012; 30: 315–322.

Brammer KS, Oh S, Cobb CJ, Bjursten LM, van der Heyde H, Jin S. Improved bone-forming functionality on diameter-controlled TiO2 nanotube surface. *Acta Biomat.* 2009; 5: 3215–3223.

Breme J, Biehl V. (1998). *Metallic Biomaterials. Handbook of Biomaterial Properties.* Black J, Hastings G. (eds.) Boston, MA, Springer US, 135–144.

Cao Y, Weng J, Chen JY, Feng JM, Yang ZJ, Zhang XD. Water vapour-treated hydroxyapatite coatings after plasma spraying and their characteristics. *Biomater.* 1996; 17: 419–424.

Capellato P, Riedel NA, Williams JD, Machado JPB, Popat KC, Claro APRA. Surface Modification on Ti-30Ta Alloy for Biomedical Application. *Eng.* 2013; 05 (09): 7.

Case CP, Langkamer VG, James C, Palmer MR, Kemp AJ, Heap PF *et al.* Widespread dissemination of metal debris from implants. *J Bone and Joint Surg-British Vol.* 1994; 76B: 701–712.

Cater WH, Hicks JH. The recent history of corrosion in metal used for internal fixation. *Lancet.* 1956; 271: 871–873.

Cheang P, Khor KA, Teoh LL, Tam SC. Pulsed laser treatment of plasma-sprayed hydroxyapatite coatings. *Biomat.* 1996; 17: 1901–1904.

Chen Q, Thouas GA. Metallic implant biomaterials. *Mater Sci and Eng R: Rep.* 2015; 87: 1–57.

Chiarelli PA, Johal MS, Casson JL, Roberts JB, Robinson JM, Wang HL. Controlled fabrication of polyelectrolyte multilayer thin films using spin-assembly. *Adv Mater.* 2001; 13: 1167–1171.

Chiba A, Kumagai K, Takeda H, Nomura N. Mechanical properties of forged low Ni and C-containing Co-Cr-Mo biomedical implant alloy. Pricm 5: The Fifth Pacific Rim International Conference on Advanced Materials and Processing, Pts 1–5. Zhong ZY, Saka H, Kim TH *et al.* Zurich-Uetikon, Trans Tech Publications Ltd., 2005, 475–479, 2317–2321.

Cho YD, Seo YJ, Lee YM, Rhyu IC, MoRyoo H, Ku Y. An overview of biomaterials in periodontology and implant dentistry. *Adv Mater Sci Eng.* 2017, 1–7.

Chuah YJ, Kuddannaya S, Lee MHA, Zhang Y, Kang Y. The effects of poly(dimethylsiloxane) surface silanization on the mesenchymal stem cell fate. *Biomater Sci.* 2015; 3: 383–390.

Chubunova EV, Khabelashvili ID, Lebedinskii YY, Nevolin VN, Zenkevich A. Direct observation of silicide growth at Fe-Si interface during pulsed laser deposition. *Thin Solid Films.* 1994; 247: 39–43.

ClementeLeon M, Agricole B, Mingotaud C, Gomez Garcia CJ, Coronado E, Delhaes P. Toward new organic/inorganic superlattices: Keggin polyoxometalates in Langmuir and Langmuir-Blodgett films. *Langmuir.* 1997; 13: 2340–2347.

Codaro EN, Melnikov P, Ramires I, Guastaldi AC. Corrosion behavior of a cobalt-chromium-molybdenum alloy. *Russian J Electrochem.* 2000; 36: 1117–1121.

Collier JP, Surprenant VA, Mayor MB, Wrona M, Jensen RE, Surprenant HP. Loss of hydroxyapatite coating on retrieved, total hip components. *J Arthro.* 1993; 8: 389–393.

Dahl JA, Maddux BLS, Hutchison JE. Toward greener nanosynthesis. *Chem Rev.* 2007; 107: 2228–2269.

Das K, Bose S, Bandyopadhyay A. TiO2 nanotubes on Ti: Influence of nanoscale morphology on bone cell-materials interaction. *J of Biomed Mater Res Part A.* 2009; 90A: 225–237.

De Aza PN, Luklinska ZB, Anseau MR, Hector M, Guitian F, De Aza S. Reactivity of a wollastonite-tricalcium phosphate Bioeutectic (R) ceramic in human parotid saliva. *Biomater.* 2000; 21: 1735–1741.

Degroot K, Geesink R, Klein C, Serekian P. Plasma sprayed coatings of hydroxyapatite. *J Biomed Mater Res.* 1987; 21: 1375–1381.

Dingemans G, Beyer W, Van De Sanden MCM, Kessels WMM. Hydrogen induced passivation of Si interfaces by Al2 O 3 films and SiO2/Al2O3 stacks. *App Phys Lett.* 2010; 97.

Enkelhardt A, Nes CS, Faur N. Comparative study of the biomechanics of materials used in stenting. Advanced Materials and Structures Iv. Nicoara M, Raduta A, Opris C. Stafa-Zurich, Trans Tech Publications Ltd., 2012; 188: 76–81.

Epelboin I, Keddam M, Mattos OR, Takenouti H. Dissolution and passivation of Fe and Fe-Cr alloys in acidified sulfate medium – influences of Ph and Cr content. *Corr Sci.* 1979; 19: 1105–1112.

Esposito M, Lausmaa J, Hirsch JM, Thomsen P. Surface analysis of failed oval titanium implants. *J Biomed Mater Res.* 1999; 48: 559–568.

Fathi MH, Salehi M, Saatchi A, Mortazavi V, Moosavi SB. In vitro corrosion behavior of bioceramic, metallic, and bioceramic-metallic coated stainless steel dental implants. *Dental Mater.* 2003; 19: 188–198.

Feeney JA, Blackbur MJ. Effect of microstructure on strength, toughness, and stress-corrosion cracking susceptibility of a metastable beta titanium alloy (Ti-11.5Mo-6Zr-4.5Sn). *Metall Trans.* 1970; 1: 3309–3318.

Frandsen CJ, Brammer KS, Noh K, Connelly LS, Oh S, Chen LH *et al.* Zirconium oxide nanotube surface prompts increased osteoblast functionality and mineralization. *Mater Sci & Eng C-Mater for Bio App.* 2011; 31: 1716–1722.

Frandsen CJ, Brammer KS, Noh K, Johnston G, Jin S. Tantalum coating on TiO2 nanotubes induces superior rate of matrix mineralization and osteofunctionality in human osteoblasts. *Mater Sci & Eng C-Mater for Bio App*. 2014; 37: 332–341.

Frandsen CJ, Noh K, Brammer KS, Johnston G, Jin SH. Hybrid micro/nano-topography of a TiO2 nanotube-coated commercial zirconia femoral knee implant promotes bone cell adhesion in vitro. *Mater Sc & Eng C-Mater for Bio App*. 2013; 33: 2752–2756.

Geetha M, Singh AK, Asokamani R, Gogia AK. Ti based biomaterials, the ultimate choice for orthopaedic implants - A review. *Prog in Mater Sci*. 2009; 54: 397–425.

Glass NR, Tjeung R, Chan P, Yeo LY, Friend JR. Organosilane deposition for microfluidic applications. *Biomicroflu*. 2011; 5: 036501-036501-036507.

Good V, Widding K, Hunter G, Heuer D. Oxidized zirconium: A potentially longer lasting hip implant. *Mater & Des*. 2005; 26: 618–622.

Granchi D, Ciapetti G, Savarino L, Cavedagna D, Donati ME, Pizzoferrato A. Assessment of metal extract toxicity on human lymphocytes cultured in vitro. *J Biomed Mater Res*. 1996; 31: 183–191.

Guo LM, Zhao JL, Wang XX, Xu RQ, Lu ZM, Li YX. Bioactivity of zirconia nanotube arrays fabricated by electrochemical anodization. *Mater Sci & Eng C-Biom and Supr Sys*. 2009; 29: 1174–1177.

Haes AJ, Chang L, Klein WL,Van Duyne RP. Detection of a biomarker for Alzheimer's disease from synthetic and clinical samples using a nanoscale optical biosensor. *J Amer Chem Soc*. 2005; 127: 2264–2271.

Hamlekhan A, Butt A, Patel S, Royhman D, Takoudis C, Sukotjo C *et al*. Fabrication of Anti-Aging TiO2 Nanotubes on Biomedical Ti Alloys. *Plos One*. 2014; 9: 10.

Han CM, Kim HE, Koh YH. Creation of hierarchical micro/nano-porous TiO2 surface layer onto Ti implants for improved biocompatibility. *Sur & Coat Tech*. 2014; 251: 226–231.

Hanawa T. Metal ion release from metal implants. *Mater Sci & Eng C-Bio and Supr Sys*. 2004; 24: 745–752.

Hanawa T. Research and development of metals for medical devices based on clinical needs. *Sci and Tech of Adv Mater*. 2012; 13: 15.

Hanawa T, Hiromoto S, Asami K. Characterization of the surface oxide film of a Co-Cr-Mo alloy after being located in quasi-biological environments using XPS. *App Sur Sci*. 2001; 183: 68–75.

Hanawa T, Ota M. Calcium-phosphate naturally formed on titanium in electrolyte solution. *Biomater*. 1991; 12: 767–774.

Hang RQ, Gao A, Huang XB, Wang XG, Zhang XY, Qin L *et al*. Antibacterial activity and cytocompatibility of Cu-Ti-O nanotubes. *J Biomed Mater Res Part A*. 2014; 102: 1850–1858.

Hartwig A. Role of magnesium in genomic stability. *Muta Res-Fund and Molec Mech of Muta*. 2001; 475: 113–121.

Healy KE, Ducheyne P. Hydration and preferential molecular adsorption on titanium invitro. *Biomater*. 1992; 13: 553–561.

Heath JC, Freeman MAR, Swanson SAV. Carcinogenic properties of wear particles from prostheses mede in cobalt-chromium alloy. *Lancet*. 1971; 1: 564–570.

Henriques B, Bagheri A, Gasik M, Souza JCM, Carvalho O, Silva FS *et al*. Mechanical properties of hot pressed CoCrMo alloy compacts for biomedical applications. *Mater & Des*. 2015; 83: 829–834.

Hoogeveen NG, Cohen Stuart MA, Fleer GJ, Böhmer MR. Formation and stability of multilayers of polyelectrolytes. *Langmuir*. 1996; 12: 3675–3681.

Kapanen A, Ryhanen J, Danilov A, Tuukkanen J. Effect of nickel-titanium shape memory metal alloy on bone formation. *Biomater*. 2001; 22: 2475–2480.

Keller SW, Kim HN, Mallouk TE. Layer-by-layer assembly of intercalation compounds and heterostructures on surfaces: Toward molecular "beaker" epitaxy. *J Amer Chem Soci*. 1994; 116: 8817–8818.

Kern M, Thompson VP. Effect of sandblasting and silica-coating procedures on pure titanium. *J Dent* 1994; 22: 300–306.

Khan MA, Williams RL, Williams DF. In-vitro corrosion and wear of titanium alloys in the biological environment. *Biomater*. 1996; 17: 2117–2126.

Khan MA, Williams RL, Williams DF. The corrosion behaviour of Ti-6Al-4V, Ti-6Al-7Nb and Ti-13Nb-13Zr in protein solutions. *Biomater*. 1999; 20: 631–637.

Kim H. Atomic layer deposition of metal and nitride thin films: Current research efforts and applications for semiconductor device processing. *J Vac Sci & Tech B*. 2003; 21: 2231–2261.

Kim MJ, Lim HJ, Lee BG, Kim JH, Choi J, Kang HG. Establishment of Validation Methods to Test the Biocompatibility of Titanium Dioxide. *Bull Kor Chem Soc*. 2013; 34: 1857–1863.

Kim SE, Lim JH, Lee SC, Nam SC, Kang HG, Choi J. Anodically nanostructured titanium oxides for implant applications. *Electrochem Acta*. 2008; 53: 4846–4851.

Knez M, Nielsch K, Niinistö L. Synthesis and surface engineering of complex nanostructures by atomic layer deposition. *Adv Mater.* 2007; 19: 3425–3438.

Krebs H-U, Weisheit M, Faupel J, Süske E, Scharf T, Fuhse C *et al.* (2003). Pulsed Laser Deposition (PLD) — A Versatile Thin Film Technique. *Advances in Solid State Physics.* Kramer B. (ed.) Berlin, Heidelberg, Springer Berlin Heidelberg, 505–518.

Kuroda D, Niinomi M, Morinaga M, Kato Y, Yashiro T. Design and mechanical properties of new beta type titanium alloys for implant materials. *Materials Science and Engineering a-Stru Mater Proper Microstruc and Proc* 1998; 243: 244–249.

Lakes R. Materials with structural hierarchy. *Nature.* 1993; 361: 511–515.

Lee BH, Yoon B, Abdulagatov AI, Hall RA, George SM. Growth and properties of hybrid organic-inorganic metalcone films using molecular layer deposition techniques. *Adv Func Mater.* 2013; 23: 532–546.

Lee CM, Ho WF, Ju CP, Lin JHC. Structure and properties of Titanium-25 Niobium-x iron alloys. *J Mater Sci-Mater in Med.* 2002; 13: 695–700.

Lee H, Kepley LJ, Hong HG, Mallouk TE. Inorganic analogues of Langmuir-Blodgett films: Adsorption of ordered zirconium 1,10-decanebisphosphonate multilayers on silicon surfaces. *J Amer Chem Soci.* 1988; 110: 618–620.

Lee S, Mayer KM, Hafner JH. Improved localized surface plasmon resonance immunoassay with gold bipyramid substrates. *Analy Chem.* 2009; 81: 4450–4455.

Lennox DW, Schofield BH, McDonald DF, Riley LH. A histologic comparison of aseptic loosening of cemented, press-fit, and biologic ingrowth prostheses. *Clin Ortho and Rela Res.* 1987: 171–191.

Li J, Zhou L, Li ZC. Corrosion behaviors of a new titanium alloy TZNT for surgical implant application in Ringer's solution. *Rare Metals.* 2010; 29: 37–44.

Li Y, Wang X, Sun J. Layer-by-layer assembly for rapid fabrication of thick polymeric films. *Chem Soc Rev.* 2012; 41: 5998–6009.

Li YH, Rao GB, Rong LJ, Li YY, Ke W. Effect of pores on corrosion characteristics of porous NiTi alloy in simulated body fluid. *Mater Sci & Eng C-Biom and Supr Sys.* 2003; 363: 356–359.

Li YS, Wang K, He PX, Huang BX, Kovacs P. Surface-enhanced Raman spectroelectrochemical studies of corrosion films on implant Co-Cr-Mo alloy in biosimulating solutions. *J Ram Spectr.* 1999; 30: 97–103.

Liu XY, Ding CX. Plasma sprayed wollastonite/TiO2 composite coatings on titanium alloys. *Biomater.* 2002; 23: 4065–4077.

Liu XY, Ding CX. Plasma-sprayed wollastonite 2M/ZrO2 composite coating. *Sur & Coat Tech.* 2003; 172: 270–278.

Liu XY, Tao SY, Ding CX. Bioactivity of plasma sprayed dicalcium silicate coatings. *Biomater.* 2002; 23: 963–968.

Lombardi AV, Mallory TH, Vaughn BK, Drouillard P. Aspetic loosening in total hip-arthroplasty secondary to osteolysis induced by wear debris from titanium-alloy modular femoral heads. *J Bone and Join Surg-Amer Volu.* 1989; 71A: 1337–1342.

Long M, Rack HJ. Titanium alloys in total joint replacement – A materials science perspective. *Biomater.* 1998; 19: 1621–1639.

Lopez MF, Gutierrez A, Jimenez JA. Surface characterization of new non-toxic titanium alloys for use as biomaterials. *Surf Sci.* 2001; 482: 300–305.

Love JC, Estroff LA, Kriebel JK, Nuzzo RG, Whitesides GM. Self-assembled monolayers of thiolates on metals as a form of nanotechnology. *Chem Rev.* 2005; 105: 1103–1169.

Lucas BD, Kim JS, Chin C, Guo LJ. Nanoimprint lithography based approach for the fabrication of large-area, uniformly oriented plasmonic arrays. *Adv Mater.* 2008; 20: 1129–1134.

Lucchini JP, Aurelle JL, Therin M, Donath K, Becker W. A pilot study comparing screw-shaped implants – Surface analysis and histologic evaluation of bone healing. *Clin Oral Impl Rese.* 1996; 7: 397–404.

Lynn AK, DuQuesnay DL. Hydroxyapatite-coated Ti-6Al-4V Part 2: The effects of post-deposition heat treatment at low temperatures. *Biomater.* 2002; 23: 1947–1953.

Macak JM, Tsuchiya H, Ghicov A, Yasuda K, Hahn R, Bauer S *et al.* TiO2 nanotubes: Self-organized electrochemical formation, properties and applications. *Curr Opin in Soli St & Mater Sci.* 2007; 11: 3–18.

Macak JM, Tsuchiya H, Schmuki P. High-aspect-ratio TiO2 nanotubes by anodization of titanium. *Ang Chem-Inter Ed.* 2005; 44: 2100–2102.

Martinez-Nogues V, Nesbitt JM, Wood RJK, Cook RB. Nano-scale wear characterization of CoCrMo biomedical alloys. *Trib Inter.* 2016; 93: 563–572.

Michel M, Izquierdo A, Decher G, Voegel JC, Schaaf P, Ball V. Layer by layer self-assembled polyelectrolyte multilayers with embedded phospholipid vesicles obtained by spraying: Integrity of the vesicles. *Langmuir.* 2005; 21: 7854–7859.

Milosev I, Metikos-Hukovic M, Strehblow HH. Passive film on orthopaedic TiAlV alloy formed in physiological solution investigated by X-ray photoelectron spectroscopy. *Biomater.* 2000; 21: 2103–2113.

Milosev I, Strehblow HH. The composition of the surface passive film formed on CoCrMo alloy in simulated physiological solution. *Electr Acta.* 2003; 48: 2767–2774.

Minagar S, Berndt CC, Gengenbach T, Wen C. Fabrication and characterization of TiO2-ZrO2-ZrTiO4 nanotubes on TiZr alloy manufactured via anodization. *J Mater Chem B.* 2014; 2: 71–83.

Minagar S, Berndt CC, Wang J, Ivanova E, Wen C. A review of the application of anodization for the fabrication of nanotubes on metal implant surfaces. *Acta Biomater.* 2012; 8: 2875–2888.

Minagar S, Wang J, Berndt CC, Ivanova EP, Wen C. Cell response of anodized nanotubes on titanium and titanium alloys. *J Biome Mater Res Part A.* 2013; 101: 2726–2739.

Morais S, Sousa JP, Fernandes MH, Carvalho GS, de Bruijn JD, van Blitterswijk CA. Effects of AISI 316L corrosion products in in vitro bone formation. *Biomater.* 1998; 19: 999–1007.

Morintale E, Constantinescu C, Dinescu M. Thin films development by pulsed laser-assisted deposition. *Physics AUC.* 2010; 20: 43–56.

Mueller WD, Gross U, Fritz T, Voigt C, Fischer P, Berger G *et al.* Evaluation of the interface between bone and titanium surfaces being blasted by aluminium oxide or bioceramic particles. *Clin Oral Impl Res.* 2003; 14: 349–356.

Nag S, Banerjee R. Fundamentals of Medical Implant Materials. *Mater for Med Dev.* 2012; 23: 11.

Nasab MB, Hassan MR, Sahari BB. Metallic biomaterials of knee and hip – A review. *Trends in Biomater and Artif Organs.* 2010; 24: 69–82.

Niendorf T, Canadinc D, Maier HJ, Karaman I, Yapici GG. Micro structure-mechanical property relationships in ultrafine-grained NbZr. *Acta Mater.* 2007; 55: 6596–6605.

Niinomi M. Recent metallic materials for biomedical applications. *Metall and Mater Trans a-Phys Metall and Mater Sci.* 2002; 33: 477–486.

Niinomi M. Recent research and development in titanium alloys for biomedical applications and healthcare goods. *Sci and Tech of Adv Mater.* 2003; 4: 445–454.

Niinomi M, Hanawa T, Narushima T. Japanese research and development on metallic biomedical, dental, and healthcare materials. *JOM.* 2005; 57: 18–24.

Niwa S. Recent trends and development of biomaterials for artificial bone. *Mater Jpn.* 2004; 43: 186–192.

O'Brien B, Stinson J, Carroll W. Development of a new niobium-based alloy for vascular stent applications. *J Mech Behav of Biomed Mater.* 2008; 1: 303–312.

Oh S, Brammer KS, Li YSJ, Teng D, Engler AJ, Chien S *et al.* Stem cell fate dictated solely by altered nanotube dimension. *Proc of the Nat Acad of Sci of the USA.* 2009; 106: 2130–2135.

Oh S, Daraio C, Chen LH, Pisanic TR, Finones RR, Jin S. Significantly accelerated osteoblast cell growth on aligned TiO2 nanotubes. *J Biomed Mater Rese Part A.* 2006; 78A: 97–103.

Oh S, Jin S. Titanium oxide nanotubes with controlled morphology for enhanced bone growth. *Mater Sci & Eng C-Biom and Supr Sys.* 2006; 26: 1301–1306.

Oh S, Na IY, Choi KH. The Effect of Curcumin Against in vitro Adhesion of Implant Device-associated Bacteria on Nanosized Titanium Dioxide. *J of Nano Res.* 2013; 23: 83–90.

Okazaki Y, Ohota M, Ito Y, Tateishi T. Corrosion resistance of implant alloys in pseudo physiological solution and role of alloying elements in passive films. *Nip Kinz Gak/J of the Jap Inst of Met.* 1995; 59: 229–236.

Oyane A, Kim HM, Furuya T, Kokubo T, Miyazaki T, Nakamura T. Preparation and assessment of revised simulated body fluids. *J Biomed Mater Res Part A.* 2003; 65A: 188–195.

Pan J, Thierry D, Leygraf C. Electrochemical impedance spectroscopy study of the passive oxide film on titanium for implant application. *Electrochim Acta.* 1996; 41: 1143–1153.

Papageorgiou I, Shadrick V, Davis S, Hails L, Schins R, Newson R *et al.* Macrophages detoxify the genotoxic and cytotoxic effects of surgical cobalt chrome alloy particles but not quartz particles on human cells in vitro. *Muta Res – Fund and Molecu Mech of Muta.* 2008; 643: 11–19.

Park J, Bauer S, Schlegel KA, Neukam FW, von der Mark K, Schmuki P. TiO2 Nanotube Surfaces: 15 nm - An Optimal Length Scale of Surface Topography for Cell Adhesion and Differentiation. *Small.* 2009; 5: 666–671.

Park J, Bauer S, von der Mark K, Schmuki P. Nanosize and vitality: TiO2 nanotube diameter directs cell fate. *Nano Lett.* 2007; 7: 1686–1691.

Park JB, Kim YK. *Metallic Biomaterials.* Published: Boca Raton; 2013, Crc Press-Taylor & Francis Group.

Patel AM, Spector M. Tribological evaluation of oxidized zirconium using an articular cartilage counterface: A novel material for potential use in hemiarthroplasty. *Biomater.* 1997; 18: 441–447.

Pereira ML, Abreu AM, Sousa JP, Carvalho GS. Chromium accumulation and ultrastructural-changes in the mouse-liver caused by stainess-steel corrosion products. *J Mater Sci-Mater in Medic.* 1995; 6: 523–527.

Popat KC, Eltgroth M, LaTempa TJ, Grimes CA, Desai TA. Decreased Staphylococcus epidermis adhesion and increased osteoblast functionality on antibiotic-loaded titania nanotubes. *Biomater.* 2007; 28: 4880–4888.

Prasad K, Bazaka O, Chua M, Rochford M, Fedrick L, Spoor J, Symes R, Tieppo M, Collins C, Cao A, Markwell D, Ostrikov K, Bazaka K. Metallic Biomaterials: Current Challenges and Opportunities Materials. 2017; 10: 884.

Prymak O, Bogdanski D, Koller M, Esenwein SA, Muhr G, Beckmann F *et al.* Morphological characterization and in vitro biocompatibility of a porous nickel-titanium alloy. *Biomater.* 2005; 26: 5801–5807.

Puurunen RL. Surface chemistry of atomic layer deposition: A case study for the trimethylaluminum/water process. *J App Phys.* 2005; 97: 112–118.

Qiu Z-Y, Chen C, Wang X-M, Lee I-S. Advances in the surface modification techniques of bone-related implants for last 10 years. *Regene Biomater.* 2014; 1: 67–79.

Rack HJ, Qazi JI. Titanium alloys for biomedical applications. *Mater Sci and Eng C.* 2006; 26: 1269–1277.

Ratner BD (2001). *A Perspective on Titanium Biocompatibility. Titanium in Medicine: Material Science, Surface Science, Engineering, Biological Responses and Medical Applications.* Berlin, Heidelberg, Springer Berlin Heidelberg, 1–12.

Ratner BD, Hoffman AS, Schoen FJ, Lemons JE (2004). *An Introduction to Materials in Medicine. Biomaterials Science.* Burlington, Burlington: Elsevier Science, 37–40.

Rho JY, Kuhn-Spearing L, Zioupos P. Mechanical properties and the hierarchical structure of bone. *Medi Eng & Phy.* 1998; 20: 92–102.

Roy P, Berger S, Schmuki P. TiO2 Nanotubes: Synthesis and Applications. *Ang Chem-Inter Ed.* 2011; 50: 2904–2939.

Ruckh T, Porter JR, Allam NK, Feng XJ, Grimes CA, Popat KC. Nanostructured tantala as a template for enhanced osseointegration. *Nanotech.* 2009; 20: 7.

Ryhanen J, Kallioinen M, Tuukkanen J, Junila J, Niemela E, Sandvik P *et al.* In vivo biocompatibility evaluation of nickel-titanium shape memory metal alloy: Muscle and perineural tissue responses and encapsule membrane thickness. *J Biomed Mater Res.* 1998; 41: 481–488.

Sanchez AG, Ballarre J, Orellano JC, Duffo G, Cere S. Surface modification of zirconium by anodisation as material for permanent implants: *In vitro* and *in vivo* study. *J Mater Sci-Mater in Medi.* 2013; 24: 161–169.

Sanchez AG, Schreiner W, Duffo G, Cere S. Surface characterization of anodized zirconium for biomedical applications. *App Surf Sci.* 2011; 257: 6397–6405.

Saris NEL, Mervaala E, Karppanen H, Khawaja JA, Lewenstam A. Magnesium – An update on physiological, clinical and analytical aspects. *Clin Chim Acta.* 2000; 294: 1–26.

Schmalzried TP, Peters PC, Maurer BT, Bragdon CR, Harris WH. Long duration metal-on-metal total hip arthroplasties with low wear of the articulating surfaces. *J Arthro.* 1996; 11: 322–331.

Schmuki P. From Bacon to barriers: A review on the passivity of metals and alloys. *J Sol Sta Electrochem.* 2002; 6: 145–164.

Schultze JW, Lohrengel MM. Stability, reactivity and breakdown of passive films. Problems of recent and future research. *Electroch Acta.* 2000; 45: 2499–2513.

Seah KHW, Thampuran R, Teoh SH. The influence of pore morphology on corrosion. *Corr Sci.* 1998; 40: 547–556.

Sellinger A, Weiss PM, Nguyen A, Lu Y, Assink RA, Gong W *et al.* Continuous self-assembly of organic-inorganic nanocomposite coatings that mimic nacre. *Nature.* 1998; 394: 256–260.

Semlitsch MF, Weber H, Streicher RM, Schon R. Joint Replacement Components Made of Hot-Forged and Surface-Treated Ti-6Al-7Nb Alloy. *Biomater.* 1992; 13: 781–788.

Shettlemore MG, Bundy KJ. Examination of in vivo influences on bioluminescent microbial assessment of corrosion product toxicity. *Biomater.* 2001; 22: 2215–2228.

Sivakumar M, Mudali UK, Rajeswari S. Investigation of failures in stainless-steel orthopedic implant devices-fatigue failure due to improper fixation of a compression bone plate. *J Mater Sci Lett.* 1994; 13: 142–145.

Smith DC, Pilliar RM, Metson JB, McIntyre NS. Dental implant materials.2. Preparative procedures and surface spectroscopic studies. *J Biomed Mater Res.* 1991; 25: 1069–1084.

Song GL. Control of biodegradation of biocompatable magnesium alloys. *Corr Sci.* 2007; 49: 1696–1701.

Staiger MP, Pietak AM, Huadmai J, Dias G. Magnesium and its alloys as orthopedic biomaterials: A review. *Biomater.* 2006; 27: 1728–1734.

Stewart ME, Yao J, Maria J, Gray SK, Rogers JA, Nuzzo RG. Multispectral thin film biosensing and quantitative imaging using 3D plasmonic crystals. *Analy Chem.* 2009; 81: 5980–5989.

Stojilovic N, Bender ET, Ramsier RD. Surface chemistry of zirconium. *Prog in Surf Sci.* 2005; 78: 101–184.

Strandman E, Landt H. Oxidation resistance of dental chromium-cobalt alloys. *Quintes of dental tech.* 1982; 6: 67–74.

Srivastava VK, Kapoor A, Azadmanjiri J, Wang J, Berndt CC. Effect of sliding brass pin over the adhere surfaces of magnetron sputtered monolayer composites. *Int J Mater Eng*. 2016; 6: 72–76.

Sukenik CN, Balachander N, Culp LA, Lewandowska K, Merritt K. Modulation of cell-adhesion by modification of titanium surfaces with covalently attached self-assembled monolayers. *J Biomed Mater Res*. 1990; 24: 1307–1323.

Sumita M, Hanawa T. Failure processes in biometallic materials. *Bioeng*. 2003; 9: 131–167.

Sumita M, Hanawa T, Teoh SH. Development of nitrogen-containing nickel-free austenitic stainless steels for metallic biomaterials – Review. *Mater Sci & Eng C-Biom and Supr Sys*. 2004; 24: 753–760.

Sumner DR, Galante JO. Determination of stress shielding-design versus materials versus interface. *Clin Ortho and Rela Res*. 1992: 202–212.

Sun X, Xie M, Wang G, Sun H, Cavanagh AS, Travis JJ, *et al*. Atomic layer deposition of TiO 2 on graphene for supercapacitors. *J Electrochem Soc*. 2012; 159: A364–A369.

Sury P, Semlitsch M. Corrosion behavior of cast and forged cobalt-based alloys for double-alloy joint endoprosthesis. *J Biomed Mater Res*. 1978; 12: 723–741.

Tan AW, Pingguan-Murphy B, Ahmad R, Akbar SA. Review of titania nanotubes: Fabrication and cellular response. *Ceram Int*. 2012; 38: 4421–4435.

Tang X, Ahmed T, Rack HJ. Phase transformations in Ti-Nb-Ta and Ti-Nb-Ta-Zr alloys. *J of Mater Sci*. 2000; 35: 1805–1811.

Tengvall P, Elwing H, Sjoqvist L, Lundstrom I, Bjursten LM. Interaction between hydrogen-peroxide and titanium – A possible role in the biocompatibility of titanium. *Biomater*. 1989; 10: 118–120.

Teo AJT, Mishra A, Park I, Kim YJ, Park WT, Yoon YJ. Polymeric biomaterials for medical implants and devices. *ACS Biomater Sci Eng*. 2016, 2, 454–472.

Thompson GJ, Puleo DA. Ti-6Al-4V ion solution inhibition of osteogenic cell phenotype as a function of differentiation timecourse in vitro. *Biomater*. 1996; 17: 1949–1954.

Tiwari JN, Tiwari RN, Kim KS. Zero-dimensional, one-dimensional, two-dimensional and three-dimensional nanostructured materials for advanced electrochemical energy devices. *Prog in Mater Sci*. 2012; 57: 724–803.

Tracana RB, Pereira ML, Abreu AM, Sousa JP, Carvalho GS. Stainless-steel corrosion products cause alterations on mouse spleen cellular-poulations. *J Mater Sci-Mater in Med*. 1995; 6: 56–61.

Vartiainen J, Shen YF, Kaljunen T, Malm T, Vaha-Nissi M, Putkonen M *et al*. Bio-based multilayer barrier films by extrusion, dispersion coating and atomic layer deposition. *J App Poly Sci*. 2016; 133: 6.

Vaughan TJ, McCarthy CT, McNamara LM. A three-scale finite element investigation into the effects of tissue mineralisation and lamellar organisation in human cortical and trabecular bone. *J Mech Behav of Biomed Mater*. 2012; 12: 50–62.

Vercaigne S, Wolke JGC, Naert I, Jansen JA. Histomorphometrical and mechanical evaluation of titanium plasma-spray-coated implants placed in the cortical bone of goats. *J Biomed Mater Res*. 1998; 41: 41–48.

Walczak J, Shahgaldi F, Heatley F. In vivo corrosion of 316L stainless-steel hip implants: Morphology and elemental compositions of corrosion products. *Biomater*. 1998; 19: 229–237.

Wan J, Yan X, Ding JJ, Wang M, Hu KC. Self-organized highly ordered TiO2 nanotubes in organic aqueous system. *Mater Charac*. 2009; 60: 1534–1540.

Wang K, Gustavson L, Dumbleton J. *Low Modulus, High-Stength, Biocompatible Titanium-Alloy for Medical Implants*. Published: Warrendale; 1993, Minerals, Metals & Materials Soc.

Wang KY, Liu GH, Hoivik N, Johannessen E, Jakobsen H. Electrochemical engineering of hollow nanoarchitectures: Pulse/step anodization (Si, Al, Ti) and their applications. *Chem Soc Rev*. 2014; 43: 1476–1500.

Wang LN, Shen C, Shinbine A, Luo JL. Variation on wettability of anodic zirconium oxide nanotube surface. *Thin Solid Films*. 2013; 531: 277–283.

Wang T, Weng Z, Liu X, Yeung KWK, Pan H, Wu Shuilin. Controlled release and biocompatibility of polymer/titania nanotube array system on titanium implants. *Bioact Mater*. 2017; 2: 44–50.

Wang Y, Wen CE, Hodgson P, Li YC. Biocompatibility of TiO2 nanotubes with different topographies. *J Biomed Mater Rese Part A*. 2014; 102: 743–751.

Wasserman SR, Tao YT, Whitesides GM. Structure and reactivity of alkylsiloxane monolayers formed by reaction of alkyltrichlorosilanes on silicon substrates. *Langmuir*. 1989; 5: 1074–1087.

Weiner S, Wagner HD. The material bone: Structure-mechanical function relations. *Ann Rev of Mater Sci*. 1998; 28: 271–298.

Wennerberg A, Albrektsson T, Johansson C, Andersson B. Experimental study of turned and grit-blasted screw-shaped implants with special emphasis on effects of blasting material and surface topography. *Biomater*. 1996; 17: 15–22.

Williams DF. General concepts of biocompatibilitty. *Handbook of Biomater Prop. Part III.* 1998: 482–489.

Witte F, Kaese V, Haferkamp H, Switzer E, Meyer-Lindenberg A, Wirth CJ *et al.* In vivo corrosion of four magnesium alloys and the associated bone response. *Biomater.* 2005; 26: 3557–3563.

Wu HC, Wang TW, Sun JS, Lee YH, Shen MH, Tsai ZR, Chen CY, Hsu HC. Development and characterization of a bioinspired bone matrix with aligned nanocrystalline hydroxyapatite on collagen nanofibers. *Materials.* 2016; 9: 198.

Xie YT, Ao HY, Xin SG, Zheng XB, Ding CX. Enhanced cellular responses to titanium coating with hierarchical hybrid structure. *Mater Scie & Eng C-Mater for Biolog App.* 2014; 38: 272–277.

Yavari SA, Wauthle R, Bottger AJ, Schrooten J, Weinans H, Zadpoor AA. Crystal structure and nanotopographical features on the surface of heat-treated and anodized porous titanium biomaterials produced using selective laser melting. *App Sur Sci.* 2014; 290: 287–294.

Yu SY, Scully JR, Vitus CM. Influence of niobium and zirconium alloying additions on the anodic dissolution behavior of activated titanium in HCl solutions. *J Electrochem Soc.* 2001; 148: B68–B78.

Zander D, Köster U. Corrosion of amorphous and nanocrystalline Zr-based alloys. *Mater Sci and Eng A.* 2004; 375–377: 53–59.

Zemła J, Lekka M, Raczkowska J, Bernasik A, Rysz J, Budkowski A. Selective protein adsorption on polymer patterns formed by self-organization and soft lithography. *Biomacromolecules.* 2009; 10: 2101–2109.

Zheng YB, Juluri BK, Mao X, Walker TR, Huang TJ. Systematic investigation of localized surface plasmon resonance of long-range ordered Au nanodisk arrays. *J App Phy.* 2008; 103.

3 Graphene-Polymer Nanocomposites for Biomedical Applications

*Duo Wu, Mohammad Ramezani, and Minna Hakkarainen**

CONTENTS

3.1 Introduction 51
3.2 Properties of Graphene Family 52
 3.2.1 Physical Properties and Surface Chemistry 52
 3.2.2 Toxicity and Antimicrobial Properties 53
3.3 Polymer/Graphene Nanocomposites 53
3.4 Polymer/Graphene Nanocomposite Scaffolds 54
 3.4.1 Biomedical Applications of Graphene Materials 54
 3.4.2 Polymer/Graphene Scaffolds 54
 3.4.3 State-of-the-Art Fabrication Techniques 55
  3.4.3.1 Graphene/Nanocomposite Scaffolds by Electrospinning 56
  3.4.3.2 Graphene/Nanocomposite Scaffolds by 3D Printing 57
 3.4.4 Polymer/Graphene Solid-State Scaffolds and Hydrogel Scaffolds 58
3.5 Future Perspectives 60
Acknowledgments 60
References 61

3.1 INTRODUCTION

Carbon-based nanomaterials like carbon nanotubes (CNT), graphene, graphene oxide (GO), reduced graphene oxide (rGO), graphene quantum dots (GQD), graphene oxide quantum dots (GOQD) and fullerenes, are quite novel classes of materials that can provide solutions to many problems in science and technology (Novoselov *et al.*, 2004; Hernandez *et al.*, 2008; Kim *et al.*, 2009) due to the interesting and, in many cases, outstanding physical, mechanical, chemical, electronic, and optoelectronic properties (Ramanathan *et al.* 2008; Pang *et al.* 2009). Graphene family materials especially are rapidly emerging and could revolutionize multiple fields from electronics to energy storage and biomedical applications. Chapter 3 presents some key properties of graphene materials and their applications in polymer nanocomposites concentrating especially on the fabrication, preparation, and properties of polymer/graphene scaffolds and hydrogels for biomedical applications and tissue engineering.

Graphene is a two-dimensional (2D) sp^2 carbon network with C-C bond length of 1.42 Å rearranged in a honeycomb lattice having a thickness of a single carbon atom. Graphene was first discovered by Novoselov *et al.* (2004). Its preparation from graphite and characterization led to the Nobel Prize in 2010 (Geim and Novoselov, 2007). GO is an extensively oxidized derivative of graphene consisting of more hydrophobic basal planes with some hydroxyl and epoxy groups and hydrophilic edges with plenty of carboxylic groups (Guo *et al.*, 2011). rGO is produced from GO under reducing conditions (high temperature and chemical reductants). It is usually formed through

* Corresponding author.

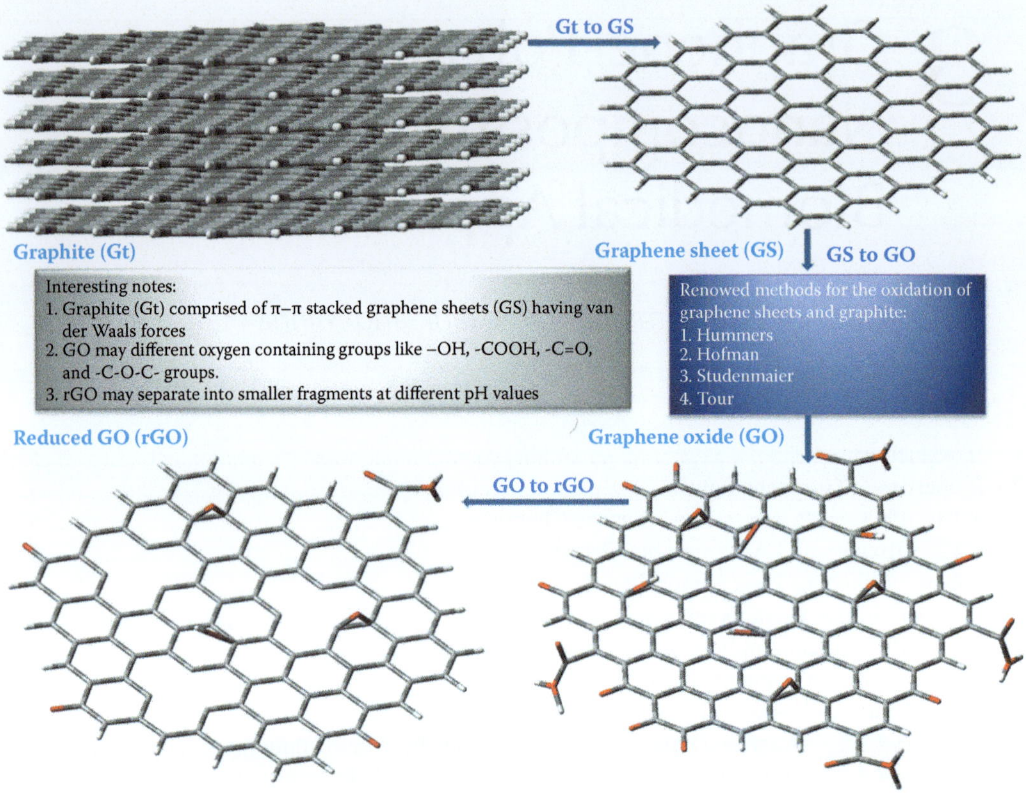

FIGURE 3.1 Examples of graphene family materials.

removal of CO/CO_2, which results in higher hydrophobicity, reduced dispersibility in water and reduced surface charge. At the same time, new holes and defects are introduced (Park *et al.*, 2009; Bagri *et al.*, 2010). Graphene and its derivatives like GO and rGO (Figure 3.1) have emerged as promising materials for broad variety of applications such as bioimaging, sensors and biosensors, supercapacitors, membranes, catalysis, drug delivery, therapy, nanobiomaterials, and many others (Stankovich *et al.*, 2006; Dikin *et al.*, 2007; Paek *et al.*, 2009; Zhang *et al.*, 2010; Xu *et al.*, 2017).

3.2 PROPERTIES OF GRAPHENE FAMILY

3.2.1 Physical Properties and Surface Chemistry

Monolayer graphene provides 2600 m^2/g available surface area against neighboring atoms and molecules from two sides (Nel *et al.*, 2009). Increasing the number of layers reduces the available surface area (Sanchez *et al.*, 2011) and is the key item in determining the specific available surface area and bending stiffness of graphene materials. Lee *et al.* (2008) represented the inverse proportion of specific area to number of layers. They also noted the increase in stiffness with the third power of thickness of the materials or the number of layers. The surface chemistry of graphene family materials varies from highly hydrophobic graphene forming mainly π–π stacking interactions to GO with hydrophilic groups having the capability of forming hydrogen bonding, ionic interactions, and metal complexes to rGO with intermediate hydrophilicity and basal reactivity due to its defects (Hasan *et al.*, 2010; Yang *et al.*, 2010; Hsieh and Chen, 2011). Although the graphene family has reasonable purity, as the production does not require metal catalysts, the materials may contain chemical additives applied to the layer separation, which were not completely removed by final washing,

TABLE 3.1
Summary of the Physical Properties of Graphene

Property	Fragment Strength	Young's Modulus	Thermal Conductivity	Electron Mobility at (300 K)	Fermi Velocity (ν_F)	Lattice Constant	Specific Capacitance	Intrinsic Mobility
Value	125	~1100	~5000	15,000	10^6	~2046	100	200,000
Unit	GPa	GPa	$Wm^{-1}K^{-1}$	$cm^2V^{-1}s^{-1}$	$m\ s^{-1}$	Å	$F\ g^{-1}$	$cm^2V^{-1}s^{-1}$
Reference	Booth *et al.* 2008	Booth *et al.* 2008	Balandin *et al.* 2008	Geim and Novoselov 2007	Rozhkov *et al.* 2011 Novoselov *et al.* 2005	Schwierz 2010 Kwon *et al.* 2009	Stoller *et al.* 2008	Du *et al.* 2008

some residual intercalants and non-covalently attached small oxidative fragments (Kim *et al.*, 2010). Some interesting physical properties of graphene are collected and presented in Table 3.1.

3.2.2 Toxicity and Antimicrobial Properties

Development of nanomaterials for applications in living systems and environments urges the assessment of toxicity of nanomaterials through their interactions with living systems and environment (Oberdorster *et al.*, 2005). The potential toxicity of graphene and its derivatives in biological systems depends on the concentration, functionalization, oxygen content, surface charges, size, and use of reducing and stabilizing agents (Gurunathan and Kim, 2016). Graphene, GO, and rGO have also been reported to have antimicrobial properties (Hu *et al.*, 2010; Szunerits and Boukherroub, 2016). These properties were found to be size dependent (Perreault *et al.*, 2015). In comparison to graphene, GO and rGO were more frequently studied *in vivo* and *in vitro* owing to their better solubility, dispersibility, and stability under physiological and aqueous test conditions (Ali-Boucetta *et al.*, 2013). Pristine graphene, GO, and rGO have all showed some cytotoxic and/or genotoxic effects, whereas functionalized materials and modifications through decrease in sheet size resulted in increased dispersibility and stability in physiological conditions offering less toxicity (Hu *et al.*, 2011; Akhavan *et al.*, 2012; Sham and Notley, 2013). Also, small-sized graphene/graphene oxide quantum dots showed low toxicity in HeLa cells (Chong *et al.*, 2014) and osteoblastic MG63 cells (Wu *et al.*, 2017a). The toxicity of graphene materials may also stem from the contaminants introduced to these materials during manufacturing and modified and/or well-purified materials which have demonstrated less toxic effects. *In vivo* experiments show varying results depending on the type of material and concentration studied. However, graphene oxide at a low-to-medium concentration often exhibited no toxic effects in *in vivo* studies, which is of great interest for drug delivery and other biomedical applications (Wang *et al.*, 2011).

3.3 POLYMER/GRAPHENE NANOCOMPOSITES

Discovery of the first polymer nanocomposite by Okada *et al.*, (1990) initiated a flourishing research field, which has led to the development of materials with greatly improved properties. These materials have been implemented in variety of applications like automotive, aerospace, construction, optoelectronic, packaging, nanobiomaterials, biomedical materials, and technologies (Alexandre and Dubois, 2000; Ray and Okamoto, 2003). Concerning carbon-based fillers, drawbacks like the agglomeration of CNT incorporated into polymer bulks and restricted amount of high quality CNT with high production costs motivated scientists to find an alternative with lower production cost, but still desirable mechanical and optoelectronical properties. Nanocomposites with graphene or graphene derivatives are among the most promising alternatives to large variety of applications

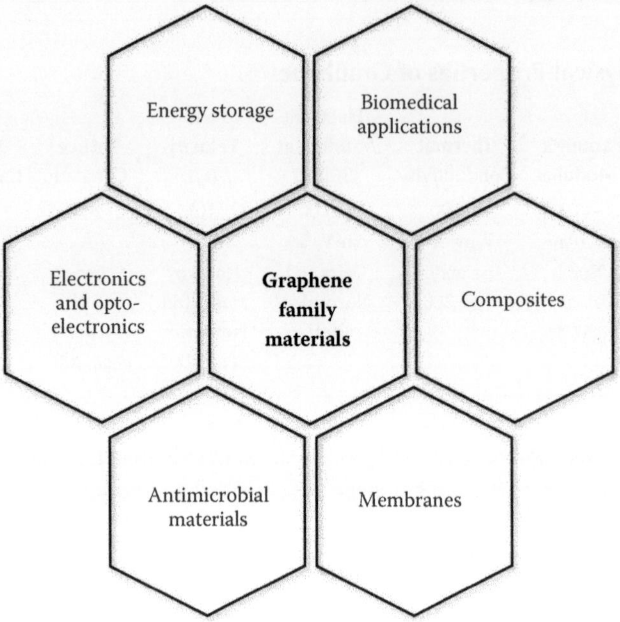

FIGURE 3.2 Anticipated applications for graphene family materials.

(Figure 3.2) (Verdejo *et al.,* 2011). A key factor directly affecting the overall properties of polymer nanocomposites is the dispersion of the filler in the polymer matrix which, to this time, has attracted significant scientific efforts. Homogenous and well-dispersed polymer/graphene systems have been prepared through covalent or non-covalent functionalization of graphene or graphene derivatives (Gatti *et al.,* 2016). The functionalization generally improved mechanical properties, transport and thermal properties and modified the morphology of the resulting nanocomposites (Santos *et al.,* 2011; Sreeprasad *et al.,* 2011; Yan *et al.,* 2011; Some *et al.,* 2012). Tailoring the size and degree of oxygenation, also offers an effective way to improve the interfacial interactions (Xu *et al.,* 2016).

3.4 POLYMER/GRAPHENE NANOCOMPOSITE SCAFFOLDS

3.4.1 Biomedical Applications of Graphene Materials

Graphene materials are highly interesting for numerous biomedical applications. Scaffolds, drug delivery, gene delivery, bioimaging, biosensors, cancer therapy, and artificial muscles are among some of the reported potential applications for the graphene nanomaterials. Some examples are collected in Table 3.2, while Chapter 3 mainly concentrates on polymer/graphene nanocomposite scaffolds.

3.4.2 Polymer/Graphene Scaffolds

Tissue/organ repair has been traditionally treated by two therapies: (i) tissue grafting and organ transplantation or (ii) alloplastic or synthetic material replacement. However, both of them have limitations. To date, tissue engineering has emerged to address those limitations. Tissue engineering is based on transplanting a biofactor, such as cells, genes, or proteins, into a porous bioresorbable scaffold. Material selection and the porous structure of a scaffold play vital roles in tissue regeneration. The prerequisites for functioning scaffold include (i) a three-dimensional structure with high porosity to enable cell growth and transport of nutrients and metabolic waste; (ii) biocompatibility;

TABLE 3.2
Examples of Graphene Materials in Biomedical Applications

Reference	Nanocomposite System	Objectives
Nishida *et al.* 2014	GO-coated materials	Study of scaffold materials' bioactivity
Li *et al.* 2013	G/GO-HA	Synthesis and biocompatibility studies
Wen *et al.* 2012	GO-PEG-anticancer drugs	Intracellular drug delivery
Sahoo *et al.* 2011	GO-PVA-camptothecin	Cancer cell apoptosis
Pan *et al.* 2011	GS-PNIPAM-camptothecin	Toxicity assay and drug delivery
Chen *et al.* 2011	GO-PEI-gene molecules	Gene delivery and pDNA transfection
Bao *et al.* 2011	GO-chitosan-drug and DNA	Delivery in human cancer cell line
Tian *et al.* 2011	GO-PEG-ce6	Photodynamic therapy of cancer cells
Kim *et al.* 2011	NGO-PEG	Bioimaging
Min *et al.* 2011	Graphene/PDMS	Artificial muscles
Yang *et al.* 2010	GS-PEG	Tumor ablation by photothermal therapy

(iii) controlled degradation and bioresorption rate to match the rate of cell/tissue growth; (iv) surface chemistry that promotes cell attachment, proliferation, and differentiation; and (v) mechanical properties that match the mechanical properties of the tissues at the implantation site.

The biodegradable polymers for scaffold fabrication can be classified into two groups: (i) natural biopolymers, such as polysaccharides (*e.g.* alginate, starch, chitin/chitosan, hyaluronic acid derivatives) and proteins (*e.g.* collagen, soy, fibrin gels, silk); (ii) synthetic polymers (*e.g.* polylactide [PLA], polyglycolide [PGA], poly ε-caprolactone [PCL] and poly[vinyl alcohol] [PVA]). However, conventional single-polymer materials are not able to satisfy all the requirements for biomedical scaffold. Incorporation of nanotechnology in the development of biomedical scaffolds has emerged to address problems and offer scaffolds with multitude of functions. Among all the nanomaterials, graphene has some outstanding advantages such as high specific surface area, conductivity and possibility to introduce plenty of functional groups, which enables various biofunctionalizations. One of the major factors influencing the properties of nanocomposites is the interfacial adhesion between the nanoadditive and polymer matrix. Due to the 2D structure of graphene, both sides of a single graphene sheet can contribute to the addition of functional groups and adsorption of molecules. Both covalent and non-covalent modifications can be applied to impart graphene with special biological activity, compatibility, and colloidal stability. The most common modification is oxidation by Hummers method to prepare GO (Hummers Jr. and Offeman, 1958), while noncovalent modifications are often accomplished through hydrophobic forces or π–π interactions (Dreyer *et al.*, 2010; Loh *et al.*, 2010). GO can be further modified by coupling polymers to its carboxylic groups or by reduction to rGO. Compared to neat polymer scaffolds, polymer/graphene nanocomposite scaffolds often present improved mechanical properties; improved dimensional stability; electrical/thermal conductivity; water retention; introduction of bioactivity; and improved cell adhesion, differentiation, and function (Ramanathan *et al.*, 2008; Liang *et al.*, 2009; Fan *et al.*, 2010; Xu *et al.*, 2010; Depan *et al.*, 2011; Lo *et al.*, 2011; Sanchez *et al.*, 2011; Zhang *et al.*, 2011; Lu *et al.*, 2012; Sayyar *et al.*, 2013, Goenka *et al.*, 2014; Hassanzadeh *et al.*, 2015b; Kang *et al.*, 2015; Wang *et al.*, 2015; Wu *et al.*, 2017a; Wu *et al.*, 2017b). Figure 3.3 presents examples of detected biological effects of graphene materials and the underlying physicochemical and mechanical backgrounds.

3.4.3 State-of-the-Art Fabrication Techniques

As already stated above, one key prerequisite for tissue engineering scaffolds is high porosity and interconnected pore networks to facilitate cell seeding, diffusion of cells and transport of nutrients and cellular waste products through the scaffold. This will also increase the specific surface area

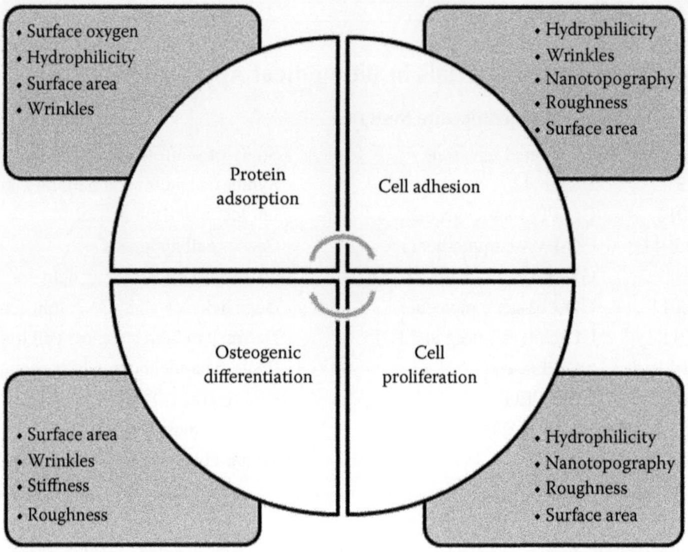

FIGURE 3.3 Potential biological effects of graphene materials and the underlying physicochemical and mechanical backgrounds, (Modified from Gu M *et al.*, *Tissue Eng Part B.* 2014; 20: 477–491.)

available for cell attachment and tissue ingrowth, which further assists the uniform distribution of cells. Besides high porosity, it is also crucial to have good regulation over the pore size and morphology. Bioresorbable three-dimensional (3D) nanocomposite scaffolds have been developed by using different zero-, one- and two-dimensional nanostructures and processing methods. These techniques mainly include solvent casting/salt leaching, gas foaming, emulsion freeze-drying, thermally induced phase separation, textile technologies (*e.g.* electrospinning), and rapid prototyping (*e.g.* 3D printing) (Langer, 1994; Ma and Langer, 1995; Zhang and Ma, 2001; Wei and Ma, 2004; Ma and Elisseeff, 2005; Rezwan *et al.*, 2006). Here, we will concentrate on the scaffolds prepared by electrospinning and rapid prototyping technologies.

3.4.3.1 Graphene/Nanocomposite Scaffolds by Electrospinning

Common technologies from the textile industry can be utilized and applied to fabricate highly porous scaffolds (Hutmacher *et al.*, 1999), *e.g.* non-woven mesh-like designs have been widely applied in tissue engineering (Puelacher *et al.*, 1996; Sittinger *et al.*, 1996; Rotter *et al.*, 1998). Freed *et al.* (1998) have reviewed in detail the electrospinning of bioresorbable PGA, PGA/PDLA, and PGA/PLLA nanofibers that have been successfully applied in tissue engineering applications. Electrospinning is a relatively simple processing method and the resulting product is a fibrous scaffold with a large surface-area-to-volume ratio and microscale porosity. The fiber thickness and morphology are controllable through several variables including the properties of the solution (*e.g.* viscosity, elasticity, conductivity, and surface tension), electric field strength, temperature, humidity, and the distance between the spinneret and the collecting plate. These variables and their influence were also summarized in a recent review (Huang *et al.*, 2003).

A few studies incorporated graphene materials in bioresorbable scaffolds by electrospinning. Wan *et al.* (2011) prepared PCL electrospun nanofibers which were reinforced by graphene oxide nanoplatelets. This study showed that the addition of only 0.3 wt.% GO resulted in PCL membranes with increased tensile strength, modulus, and energy at break. In addition, the bioactivity during biomineralization was improved, and over 94% porosity was retained. The improvement of the mechanical properties was explained by the good dispersion of GO with a high specific surface area, and strong interfacial adhesion between GO and PCL matrix. On the other hand, biomineralization tests also reflected that addition of GO could facilitate the nucleation of CaP on the scaffolds

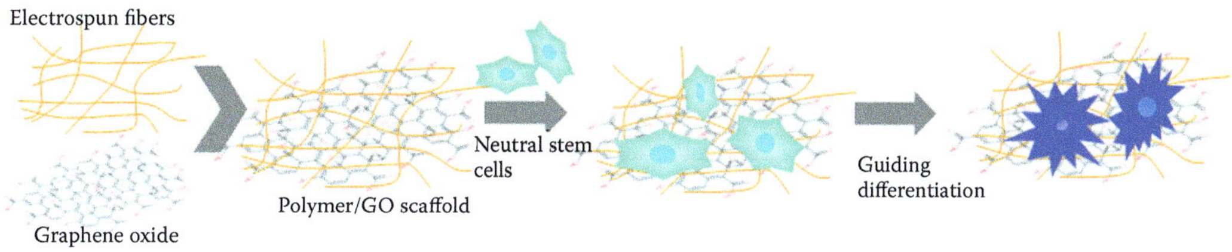

FIGURE 3.4 Electrospun polymer/GO fibrous scaffolds can promote differentiation of neural stem cells. (Redrawn and modified from Shah S *et al.*, Adv Mater. 2014; 26: 3673–3680.)

(Mohammadi *et al.,* 2017) as was also earlier shown for PCL/GO films (Hassanzadeh *et al.,* 2015b). The bioactive PCL/GO nanofibrous membranes with high porosity could, thus, have great potential acting as scaffolds for bone/cartilage tissue engineering. In other studies, a low amount of GO (0.3 and 0.5 wt%) could also significantly improve the adhesion and spreading of rat pheochromocytoma and mouse marrow mesenchymal stem cells (Song *et al.,* 2015). Another interesting PCL/GO nanofiber scaffold was able to selectively guide differentiation of neural stem cell (Shah *et al.*, 2014). As illustrated in Figure 3.4, PCL nanofibers were generated using electrospinning, followed by coating with GO and then seeding with neural stem cells. The cultured neural stem cell on PCL/GO scaffold showed enhanced differentiation into oligodendrocyte lineage cells.

Similar positive effects were obtained when GO was incorporated in PLA (Liu *et al.,* 2016) and PLGA (Luo *et al.,* 2015) electrospun fibers. For example, significant enhancement of storage and loss moduli were observed after chemical bonding of GO to PLGA (2 wt%) electrospun nanofibers (Yoon *et al.,* 2011). The enhanced hydrophilicity of GO–modified nanofibers enabled good compatibility with neuronal cells, resulting in enhanced cell proliferation and viability. Electrospinning technique allowed the fabrication of PVA/GO scaffolds with highly nanofibrous and porous structures. Strong hydrogen bonds between GO and PVA in PVA/GO composites led to greatly improved mechanical performance (Liang *et al.*, 2009; Bai *et al.*, 2010; Putz *et al.*, 2010) and the as-prepared PVA/GO scaffold could support the mouse osteoblastic cells attachment, growth and proliferation (Qi *et al.,* 2013). Recently, nanosized GO was shown to significantly improve the electrospinnability of starch scaffolds (Wu *et al.,* 2017b). GO could also introduce bioactivity and induce mineralization on the surface of the scaffolds simultaneously as good biocompatibility was retained.

3.4.3.2 Graphene/Nanocomposite Scaffolds by 3D Printing

Recently rapid prototyping techniques have been successfully applied for fabrication of multifunctional scaffolds with complex geometries, *e.g.* 3D printing enables the manufacture of porous scaffolds mimicking the microstructure of living tissues. 3D printing technology was firstly developed at the Massachusetts Institute of Technology (Sachs *et al.,* 1992) and is increasingly applied for fabrication of bioresorbable porous scaffolds for tissue engineering (Cima *et al.*, 1991; Park *et al.*, 1998). The scaffold is built in layers until the desired final form is achieved and complex 3D objects are obtainable by precise control of speed, flow rate, and drop position (Sittinger *et al.*, 1996). Biological agents, such as cells or growth factors, are relatively easily incorporated into the 3D printed porous scaffolds.

The increasing interest in patient-specific care and personalized medicine has further fueled the interest in 3D printing technologies to fabricate multicomponent scaffolds and to reach precise control over the arrangement of cells and bioactive nanomaterials in well-defined complex scaffold geometries. O'Brien *et al.* have explored the state of 3D printing in a recent review (O'Brien *et al.*, 2014). The review focuses on printing and function of nanomaterials in various complex tissue regeneration applications. 3D printing is able to create and control scaffolds at the microscale, which ultimately results in enhanced stem cell function and promotion of tissue formation.

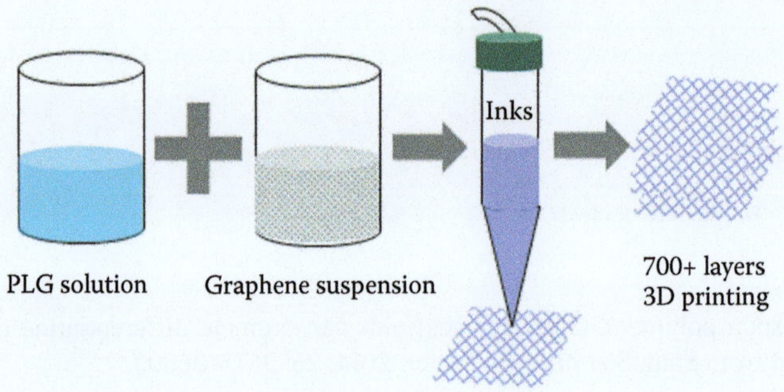

FIGURE 3.5 Illustration of 3D printing of graphene/PLGA scaffolds. (Redrawn and modified from Jakus AE *et al.*, *ACS Nano.* 2015; 9: 4636–4648.)

Introduction of 3D printing techniques to fabricate polymer/graphene scaffolds introduces a new dimension in the development of multifunctional scaffolds. Jakus *et al.* (2015) demonstrated the manufacture of 3D-printed polymer/graphene composites, consisting of graphene as the main component and polylactide-co-glycolide (PLGA) as the minor component. Figure 3.5, this graphene/PLGA scaffold was 3D printed from a liquid ink. The ink could be utilized under ambient conditions via extrusion-based 3D printing. The resulting scaffold was mechanically robust and flexible while retaining electrical conductivity. *In vitro* experiments in simple growth medium, without neurogenic stimuli, revealed that this scaffold supported human mesenchymal stem cell (hMSC) adhesion, viability, proliferation, and neurogenic differentiation. Furthermore, *in vivo* experiments indicated that the 3D graphene/PLGA scaffold had good biocompatibility over at least 30 days. In addition, surgical tests using a human cadaver nerve model illustrated that the scaffold was easy-to-handle and could be incorporated into surgical procedures. In another study, highly flexible scaffolds with good biocompatibility and improved mechanical robustness could be fabricated by incorporating GO into polyurethane/PLA 3D printed scaffolds (Chen *et al.*, 2017), while addition of graphene improved the cell attachments on the PCL scaffolds (Wang *et al.*, 2016). Electrically conductive scaffolds could be produced by adding only 3 wt% of graphene during 3D printing of poly(trimethylene carbonate) composite scaffolds (Sayyar *et al.*, 2016).

3.4.4 Polymer/Graphene Solid-State Scaffolds and Hydrogel Scaffolds

The mechanical properties of graphene make it highly suitable for the reinforcement of hydrogels and scaffolds for tissue engineering. Hydrogels are soft materials incorporating interconnected polymer chains and water that is retained in the network structure. Graphene containing hydrogels has been studied, as scaffolds for repairing tissues such as skin, bladder, cartilage, and bone (Drury and Mooney, 2003; Lu *et al.*, 2016). Hydrogels made of hydrophilic polymers such as polyvinyl alcohol (PVA) and poly(methyl methacrylate) (PMMA) can be utilized to repair cartilage and tendons, but the problem is the low mechanical strength and elasticity of these hydrogels (Drury and Mooney, 2003). Incorporation of GO to PVA (GO/PVA) increased both the tensile strength and elasticity modulus of the hydrogels (Liang *et al.*, 2009; Zhang *et al.*, 2011). Similar results were obtained when GO was incorporated into PMMA films (Ramanathan *et al.*, 2008).

A light-responsive poly(N-isopropylacrylamide) (PNIPAAm) hydrogel composite with functionalized GO was prepared by Lo *et al.*, (2011). GO was covalently bonded to PNIPAAm via esterification. The results showed that these hydrogels could undergo a volumetric change in response to infrared light illumination. Thus, an IR-responsive nanocomposite hydrogel was obtained. The obtained material also exhibited several times larger water uptake as compared to conventional PNIPAAm hydrogels. GO and CNT were also functionalized and covalently attached to poly(ethylene glycol fumarate) to form a crosslinked hydrogel for neural tissue engineering (Liu *et al.*, 2016).

Chitosan is a biopolymer that is obtained by alkaline deacetylation of chitin. Chitosan gels are osteoconductive and enhance bone formation. Structurally chitosan is similar to the major component of the extracellular matrix of bone, glycosaminoglycans. It can be used as biological adhesive, and it promotes wound-healing and cellular attachment and proliferation (Drury and Mooney, 2003; Di Martino *et al.*, 2005). Graphene can be utilized to reinforce chitosan films (chitosan/graphene) to improve their mechanical properties. Chitosan/graphene films were not toxic when used as a substrate for culture of L929 cells (Fan *et al.*, 2010). Morphological changes of chitosan/graphene films were dependent on the graphene concentration. GO-chitosan scaffolds synthesized by covalent attachment of GO with the amide groups of chitosan presented a promising approach to tune the degradation rate of scaffolds used for hard tissue regeneration. The hydrogels retained their size and shape under physiological pH conditions, holding a higher moisture content, lower degradation rate, and improved mechanical strength when compared to pure chitosan hydrogels. In addition, the cellular adhesion, proliferation, differentiation of MC3T3-E1 cells and calcium and phosphate deposition were significantly improved (Depan *et al.*, 2011). Collagen scaffolds with covalently attached GO were also developed (Kang *et al.*, 2015). The addition of GO improved the mechanical properties of the scaffolds and promoted the osteogenic differentiation of human MSC. The stiffness of the scaffolds increased threefold and no cytotoxicity was detected. Cell proliferation analyses exhibited clearly increased osteogenic differentiation after addition of GO making them of interest for bone tissue engineering.

Xu *et al.* (2010) illustrated the possibility to obtain hydrogels with high mechanical strength, stability, and absorption capacity by self-assemble of graphene oxide and DNA. This hydrogel also had a self-healing property. Wang *et al.*, (2015) created self-assembled peptide nanofibers (PNF) with specially-designed peptide molecules. GO was then modified with PNF to fabricate a new type of GO–PNF nanohybrid. After incubation in SBF for 2 weeks, the biomimetic mineralization of hydroxyapatite (HA) microspheres were investigated on the synthesized GO–PNF nanohybrid. The mineralization test indicated that GO regulated the growth of nanocrystals on the microspheres. Besides, the cell culture results indicated that the nanohybrid had good biocompatibility and enhanced the proliferation ability of the L-929 cells and MC3T3-E1 cells.

Lu *et al.* (2012) explored graphene modified hydrogels for wound healing by preparing chitosan-PVA-graphene scaffolds. The wounds treated with samples containing graphene healed completely and at a faster rate in both mice and rabbits as compared to the wounds treated with scaffolds without graphene. In another study chitosan/nano–GO type carbon dot hydrogels were successfully applied for wound dressing applications and showed great potential as antibacterial wound healing bandage (Omidi *et al.*, 2017). PCL/GO composites could be prepared by both blending route and covalent bonding route (Sayyar *et al.*, 2013). PCL/GO composites showed good compatibility for L929, C2C12 and PC12 cell lines. The covalently linked composite exhibited far better homogeneity and thus both Young's modulus and tensile strength improved more than twofold. GO also endowed PCL/GO composites electrical conductivities, showing potential use as conducting substrates for the electrically stimulated growth of cells.

Recently, a microwave-assisted route to transferring cellulose and starch to novel nanosized GO was developed (Hassanzadeh *et al.*, 2014a; Hassanzadeh *et al.*, 2014b; Wu and Hakkarainen, 2014, Adolfsson *et al.*, 2015; Hassanzadeh *et al.*, 2015a; Wu *et al.*, 2016). The production route included a microwave-assisted hydrothermal degradation of cellulose and a starch or waste paper feedstock and then oxidation of the carbon residue to reach GO, which is nanosized even at lateral dimensions (nGO, ~ 10–80 nm depending on the production process and raw material). It was shown that the nGO has an interesting self-assembling behavior (Hassanzadeh *et al.*, 2015a; Wu *et al.*, 2016). In blends with PCL, nGO acted as a multifunctional property-enhancing additive (Hassanzadeh *et al.*, 2015b). It could control the morphology and induce mineralization on the PCL films and in porous starch scaffold (Wu *et al.*, 2017a). nGO in PCL films reflected a concentration-dependent self-association and could thereby tune the nucleation, size, and amount of the formed CaP crystals. A highlight summary of polymer/graphene solid-state/hydrogel scaffolds is presented in Table 3.3.

TABLE 3.3
Examples of Polymer/Graphene Scaffolds

Scaffold	Highlights of the Study	Reference
Starch/GO	Improvement of mechanical properties and electrospinnability. Promotion of calcium phosphate deposition. Good biocompatibility.	Wu *et al.* 2017a, Wu *et al.* 2017b
PCL/GO	Covalent modification for better homogeneity, Young's improvement of modulus and strength. Good conductivity. Stimulation of cell growth. Promotion of calcium phosphate deposition.	Wang *et. al.* 2016; Song *et al.* 2015; Hassanzadeh *et al.* 2014; Sayyar *et al.* 2013
PLA/PLGA/GO	Enhanced biocompatibility, cell proliferation and differentiation. Improved mechanical properties.	Liu *et al.* 2016, Luo *et al.* 2015
DNA, peptide/GO	Self-assembled complex. High strength and self-healing properties. Promotion of calcium phosphate deposition and good biocompatibility. Enhanced cell adhesion and proliferation.	Wang *et al.* 2015; Xu *et al.* 2010
Collagen/GO	Improvement of stiffness and no toxicity. Promotion of osteogenic differentiation of human MSCs.	Kang *et al.* 2015
PVA/GO	Dispersion of GO, improvement of mechanical properties, antibacterial function, no toxicity.	Lu *et al.* 2012; Zhang *et al.* 2011; Liang *et al.* 2009
Chitosan/GO	Improvement of mechanical properties, lower degradation rate, support of cell growth and differentiation, promotion of calcium phosphate deposition.	Lu *et al.* 2012, Rotter *et al.* 1998, Di Martino *et al.* 2005
PNIPAAm/GO	Larger water uptake, IR responsive.	Lo *et al.* 2011
PMMA/GO	Superb interface interaction. Improvement of strength, modulus, and thermal stability.	Ramanathan *et al.* 2008

3.5 FUTURE PERSPECTIVES

Graphene nanocomposites have been recently explored in biomedical applications such as tissue engineering. The exceptional structure and properties of graphene and its derivatives impart the polymer scaffold highly improved mechanical properties and multi-functionalities such as electrical/thermal conductivity, bioactivity, the ability to direct cells adhesion, proliferation and differentiation, *etc*. Innovative techniques have been developed to assist fabricating of porous fibrous scaffolds and with proper selection of polymeric materials and adjusted fabrication pathways. Techniques like electrospinning and 3D printing are also applicable for preparation of polymer/graphene fibrous porous scaffolds. Although remarkable progress has already been made in the preparation and functionalization of polymer/graphene biomedical scaffolds, fewer literatures addressed the potential migration of graphene and its derivatives. Little is also known about the *in vivo* effect to the body and the clearance of these nanomaterials from body. The tailored graphene has different sizes, structures, properties, and functionalities, which may arouse interactions under different mechanisms. Therefore, it is highly desirable to perform further studies on the graphene–tissue interactions, toxicity and its metabolism. Graphene in a polymer/graphene scaffold imparts the ability of directing cell differentiation and inducing mineralization. The underlying mechanisms have not been fully disclosed. Further studies in understanding these mechanisms at cellular level will pace another step towards customized scaffolds for tissue engineering.

ACKNOWLEDGMENTS

D.W. gratefully appreciates the China Scholarship Council (CSC) for financial support.

REFERENCES

Adolfsson KH, Hassanzadeh S, Hakkarainen M. Valorization of cellulose and waste paper to graphene oxide quantum dots. *RSC Adv.* 2015; 5: 26550–26558.

Akhavan O, Ghaderi E, Akhavan A. Size-dependent genotoxicity of graphene nanoplatelets in human stem cells. *Biomaterials* 2012; 33: 8017–8025.

Alexandre M, Dubois P. Polymer-layered silicate nanocomposites: Preparation, properties and uses of a new class of materials. *Mater Sci Eng.* 2000; 28: 1–63.

Ali-Boucetta H, Bitounis D, Raveendran-Nair R, Servant A, Van den Bossche J, Kostarelos K. Purified graphene oxide dispersions lack *in vitro* cytotoxicity and *in vivo* pathogenicity. *Adv Healthcare Mater.* 2013; 2: 433–441.

Bagri A, Mattevi C, Acik M, Chabal YJ, Chhowalla M, Shenoy VB. Structural evolution during the reduction of chemically derived graphene oxide. *Nat Chem.* 2010; 2: 581–587.

Bai H, Li C, Wang X, Shi G. A pH-sensitive graphene oxide composite hydrogel. *Chem Commun.* 2010; 46: 2376–2378.

Balandin AA, Ghosh S, Bao W, Calizo I, Teweldebrhan D, Miao F. Superior thermal conductivity of single-layer graphene. *Nano Lett.* 2008; 8: 902–907.

Bao H, Pan Y, Ping Y, Sahoo NG, Wu T, Li L *et al.* Chitosan functionalized graphene oxide as a nanocarrier for drug and gene delivery. *Small* 2011; 7: 1569–1578.

Booth TJ, Blake P, Nair RR, Jiang D, Hill EW, Bangert U *et al.* Macroscopic graphene membranes and their extraordinary stiffness. *Nano Lett.* 2008; 8: 2442–2446.

Chen B, Liu M, Zhang L, Huang J, Yao J, Zhang Z. Polyethylenimine-functionalized graphene oxide as an efficient gene delivery vector. *J Mater Chem.* 2011; 21: 7736–7741.

Chen Q, Mangadlao JD, Wallat J, Leon AD, Pokorski JK, Advincula RC. 3D printing biocompatible polyurethane/poly(lactic acid)/graphene oxide nanocomposites: Anisotropic properties. *ACS Appl Mater Interfaces.* 2017; 9: 4015–4023.

Cima L, Vacanti J, Vacanti C, Ingber D, Mooney D, Langer R. Tissue engineering by cell transplantation using degradable polymer substrates. *J Biomech Eng.* 1991; 113: 143–151.

Chong Y, Ma YF, Shen H, Tu X, Zhou X, Xu J *et al.* The *in vitro* and *in vivo* toxicity of graphene quantum dots. *Biomaterials.* 2014; 35: 5041–5048.

Depan D, Girase B, Shah J, Misra R. Structure–process–property relationship of the polar graphene oxide-mediated cellular response and stimulated growth of osteoblasts on hybrid chitosan network structure nanocomposite scaffolds. *Acta Biomater.* 2011; 7: 3432–3445.

Dikin DA, Stankovich S, Zimney EJ, Piner RD, Dommett GH, Evmenenko G *et al.* Preparation and characterization of graphene oxide paper. *Nature.* 2007; 448: 457–460.

Di Martino A, Sittinger M, Risbud MV. Chitosan: A versatile biopolymer for orthopaedic tissue-engineering. *Biomaterials.* 2005; 26: 5983–5990.

Dreyer DR, Park S, Bielawski CW, Ruoff RS. The chemistry of graphene oxide. *Chem Soc Rev.* 2010; 39: 228–240.

Drury JL, Mooney DJ. Hydrogels for tissue engineering: Scaffold design variables and applications. *Biomaterials.* 2003; 24: 4337–4351.

Du X, Skachko I, Barker A, Andrei EY. Approaching ballistic transport in suspended graphene. *Nat Nanotechnol.* 2008; 3: 491–495.

Fan H, Wang L, Zhao K, Li N, Shi Z, Ge Z *et al.* Fabrication, mechanical properties, and biocompatibility of graphene-reinforced chitosan composites. *Biomacromolecules.* 2010; 11: 2345–2351.

Freed LE, Vunjak-Novakovic G. Culture of organized cell communities. *Adv Drug Deliv Rev.* 1998; 33: 15–30.

Gatti T, Vicentini N, Mba M, Menna E. Organic functionalized carbon nanostructures for functional polymer-based nanocomposites. *Eur J Org Chem.* 2016; 1071–1090.

Geim AK, Novoselov KS. The rise of graphene. *Nat Mater.* 2007; 6: 183–191.

Goenka S, Sant V, Sant S. Graphene-based nanomaterials for drug delivery and tissue engineering. *J Controlled Release.* 2014; 173: 75–88.

Gu M, Liu Y, Chen T, Du F, Zhao X, Xiong C *et al.* Is graphene a promising nano-material for promoting surface modification of implants or scaffold materials in bone tissue engineering? *Tissue Eng Part B.* 2014; 20: 477–491.

Guo F, Kim F, Han TH, Shenoy V, Huang J, Hurt RH. Hydration-responsive folding and unfolding transitions in graphene oxide liquid crystal phases. *ACS Nano.* 2011; 5: 8019–8025.

Gurunathan S, Kim JH. Synthesis, toxicity, biocompatibility, and biomedical applications of graphene and graphene-related materials. *Int J Nanomedicine.* 2016; 11: 1927–1945.

Hasan SA, Rigueur JL, Harl RR, Krejci AJ, Gonzalo-Juan I, Rogers BR *et al.* Transferable graphene oxide films with tunable microstructures. *ACS Nano.* 2010; 4: 7367–7372.

Hassanzadeh S, Adolfsson KH, Hakkarainen M. Controlling the cooperative self-assembly of graphene oxide quantum dots in aqueous solutions. *RSC Adv.* 2015a; 5: 57425–57432.

Hassanzadeh S, Adolfsson KH, Wu D, Hakkarainen M. Supramolecular assembly of biobased graphene oxide quantum dots controls the morphology of and induces mineralization on poly (ε-caprolactone) films. *Biomacromolecules.* 2015b; 17: 256–261.

Hassanzadeh S, Aminlashgari N, Hakkarainen M. Chemo-selective high yield microwave assisted reaction turns cellulose to green chemicals. *Carbohydr Polym.* 2014a; 112: 448–457.

Hassanzadeh S, Aminlashgari N, Hakkarainen M. Microwave-assisted recycling of waste paper to green platform chemicals and carbon nanospheres. *ACS Sustain Chem Eng.* 2014b; 3: 177–185.

Hernandez Y, Nicolosi V, Lotya M, Blighe FM, Sun Z, De S, *et al.* Nat. High-yield production of graphene by liquid-phase exfoliation of graphite. *Nanotechnology.* 2008; 3: 563–568.

Hsieh C-T, Chen W-Y. Water/oil repellency and work of adhesion of liquid droplets on graphene oxide and graphene surfaces. *Surf Coat Technol.* 2011; 205: 4554–4561.

Hu W, Peng C, Lv M, Li X, Zhang Y, Chen N, Fan C, Huang Q. Protein corona-mediated mitigation of cytotoxicity of graphene oxide. *ACS Nano.* 2011; 5: 3693–3700.

Hu YM, Shamaei-Tousi A, Liu YJ, Coates A. A new approach for discovery of antibiotics by targeting non-multiplying bacteria: A novel topical antibiotic for staphylococcal infections. *PLoS One.* 2010; 5: e11818.

Huang ZM, Zhang YZ, Kotaki M, Ramakrishna S. A review on polymer nanofibers by electrospinning and their applications in nanocomposites. *Compos Sci Technol.* 2003; 63: 2223–2253.

Hummers Jr. WS, Offeman RE. Preparation of graphitic oxide. *J Am Chem Soc.* 1958; 80: 1339–1339.

Hutmacher D, Teoh S, Zein I, Ranawake M, Lau S. Tissue engineering research: The engineer's role. *Med Device Technol.* 1999; 11: 33–39.

Jakus AE, Secor EB, Rutz AL, Jordan SW, Hersam MC and Shah RN. Three-dimensional printing of high-content graphene scaffolds for electronic and biomedical applications. *ACS Nano.* 2015; 9: 4636–4648.

Kang S, Park JB, Lee TJ, Ryu S, Bhang SH, La WG, Noh MK, Hong BH, and Kim BS. Covalent conjugation of mechanically stiff graphene oxide flakes to three-dimensional collagen scaffolds for osteogenic differentiation of human mesenchymal stem cells. *Carbon.* 2015; 83: 162–172.

Kim F, Cote LJ, Huang J. Graphene oxide: Surface activity and two-dimensional assembly. *Adv Mater.* 2010; 22: 1954–1958.

Kim H, Namgung R, Singha K, Oh, IK, Kim WJ. Graphene oxide-polyethylenimine nano construct as a gene delivery vector and bioimaging tool. *Bioconjug Chem.* 2011; 22: 2558–2567.

Kim KS, Zhao Y, Jang H, Lee SY, Kim JM, Kim KS, Ahn J-H, Kim P, Choi J-Y, Hong BH. Large-scale pattern growth of graphene films for stretchable transparent electrodes. *Nature.* 2009; 457: 706–710.

Kwon SY, Ciobanu CV, Petrova V, Shenoy VB, Bareno J, Gambin V, Petrov I, Kodambaka S. Growth of semi-conducting graphene on palladium. *Nano Lett.* 2009; 9: 3985–3990.

Langer R. Biodegradable polymer scaffolds for tissue engineering. *Bio/technology.* 1994; 12.

Lee C, Wei X, Kysar JW, Hone J. Measurement of the elastic properties and intrinsic strength of monolayer graphene. *Science.* 2008; 321: 385–388.

Li M, Wang Y, Liu Q, Li Q, Cheng Y, Zheng Y *et al. In situ* synthesis and biocompatibility of nanohydroxy-apatite on pristine and chitosan functionalized graphene oxide. *J Mater Chem B.* 2013; 1: 475–484.

Liang J, Huang Y, Zhang L, Wang Y, Ma Y, Guo T *et al.* Molecular-level dispersion of graphene into poly (vinyl alcohol) and effective reinforcement of their nanocomposites. *Adv Funct Mater.* 2009; 19: 2297–2302.

Liu C, Wong HM, Yeung KWK, TJong SC. Novel electrospun polylactic acid nanocomposite fiber mats with hybrid graphene oxide and nanohydroxyapatite reinforcements having enhanced biocompatibility. *Polymers.* 2016; 8: 287.

Liu X, Miller II AL, Park S, Waletzki BE, Terzic A, Yaszemski MJ *et al.* Covalent crosslinking of graphene oxide and carbon nanotube into hydrogels enhances nerve cell responses. *J Mater Chem B.* 2016; 4: 6930–6941.

Lo CW, Zhu D, Jiang H. An infrared-light responsive graphene-oxide incorporated poly (N-isopropylacrylamide) hydrogel nanocomposite. *Soft Matter.* 2011; 7: 5604–5609.

Loh KP, Bao Q, Ang PK, Yang J. The chemistry of graphene. *J Mater Chem.* 2010; 20: 2277–2289.

Lu B, Li T, Zhao H, Li X, Gao C, Zhang S *et al.* Graphene-based composite materials beneficial to wound healing. *Nanoscale.* 2012; 4: 2978–2982.

Lu J, Cheng C, He Y-S, Lyu C, Wang Y, Yu J *et al.* Multilayered graphene hydrogel membranes for guided bone regeneration. *Adv Mater.* 2016; 28: 4025–4031.

Luo Y, Shen H, Fang Y, Cao Y, Huang J, Zhang M *et al.* Enhanced proliferation and osteogenic differentiation of mesenchymal stem cells on graphene oxide-incorporated electrospun poly(lactic-co-glycolic acid) nanofibrous mats. *ACS Appl Mater Interfaces.* 2015; 7: 6331–6339.

Ma PX, Elisseeff J. *Scaffolding in Tissue Engineering*, CRC press. 2005.

Ma PX, Langer R. Degradation, structure and properties of fibrous nonwoven poly(glycolic acid) scaffolds for tissue engineering. *MRS Proceedings*, Cambridge Univ Press. 1995.

Min K, Jung JY, Han TH, Park Y, Jung C, Hong SM *et al.* Graphene electrodes for artificial muscles. Proceedings of the KJF- International Conference on Organic Materials for Electronics and Photonics. *Mol Cryst.* 2011; 539: 260–265.

Mohammadi S, Shafiei SS, Asadi-Eydivand M, Ardeshir M, Solati-Hashjin M. Graphene oxide-enriched poly(ε-caprolactone) electrospun nanocomposite scaffold for bone tissue engineering applications. *Bioact Compat Polym.* 2017; 32: 325–342.

Nel AE, Madler L, Velegol D, Xia T, Hoek EM, Somasundaran P *et al.* Understanding biophysicochemical interactions at the nano-bio interface. *Nat Mater.* 2009; 8: 543–557.

Nishida E, Miyaji H, Takita H, Kanayama I, Tsuji M, Akasaka T *et al.* Graphene oxide coating facilitates the bioactivity of scaffold material for tissue engineering, *Jpn J Appl Phys.* 2014; 53: 06JD04-1–7.

Novoselov K, Geim A, Morozov S, Jiang D, Grigorieva IV, Dubonos S *et al.* Two-dimensional gas of massless Dirac fermions in graphene. *Nature.* 2005; 438: 197–200.

Novoselov KS, Geim AK, Morozov SV, Jiang D, Zhang Y, Dubonos SV *et al.* Electric field effect in atomically thin carbon films. *Science.* 2004; 306: 666–669.

Oberdorster G, Oberdorster E, Oberdorster J. Nanotoxicology: An emerging discipline evolving from studies of ultrafine particles. *Environ Health Perspect.* 2005; 113: 823–839.

O'Brien CM, Holmes B, Faucett S, Zhang LG. Three-dimensional printing of nanomaterial scaffolds for complex tissue regeneration. *Tissue Eng Part B Rev.* 2014; 21: 103–114.

Okada A, Kawasumi M, Usuki A, Kojima Y, Kurauchi T, Kamigaito O. Synthesis and properties of nylon-6/clay hybrids. In: Schaefer DW, Mark JE, editors. *Polymer Based Molecular Composites.* Vol. 171. MRS symposium proceedings. Pittsburgh PA: Materials Research Society; 1990. pp. 45–50.

Omidi M. Yadegari A, Tayebi. Wound dressing application of pH sensitive carbon dots/chitosan hydrogel. *RSC Adv.* 2017; 7: 10638.

Paek SM, Yoo E, Honma I. Enhanced cyclic performance and lithium storage capacity of SnO_2/graphene nanoporous electrodes with three-dimensionally delaminated flexible structure. *Nano Lett.* 2009; 9: 72–75.

Pan Y, Bao H, Sahoo NG, Wu T, Li L. Water-soluble poly(N-isopropylacrylamide)-graphene sheets synthesized via click chemistry for drug delivery. *Adv Funct Mater.* 2011; 21: 2754–2763.

Pang S, Tsao HN, Feng X, Mullen K. Patterned graphene electrodes from solution-processed graphite oxide films for organic field-effect transistors. *Adv Mater.* 2009; 21: 3488–3491.

Park A, Wu B, Griffith LG. Integration of surface modification and 3D fabrication techniques to prepare patterned poly(L-lactide) substrates allowing regionally selective cell adhesion. *J Biomater Sci Polym Ed.* 1998; 9: 89–110.

Park S, An J, Jung I, Piner RD, An SJ, Li X, Velamakanni A, Ruoff RS. Colloidal suspensions of highly reduced graphene oxide in a wide variety of organic solvents. *Nano Lett.* 2009; 9: 1593–1597.

Perreault R, de Faria AF, Nejati S, Elimelech M. Antimicrobial properties of graphene oxide: Why size matters. *ACS Nano.* 2015; 9: 7226–7236.

Puelacher W, Vacanti J, Ferraro N, Schloo B, Vacanti C. Femoral shaft reconstruction using tissue-engineered growth of bone. *Int J Oral Maxillofac Surg.* 1996; 25: 223–228.

Putz KW, Compton OC, Palmeri MJ, Nguyen ST, Brinson LC. High-nanofiller-content graphene oxide–polymer nanocomposites via vacuum-assisted self-assembly. *Adv Funct Mater.* 2010; 20: 3322–3329.

Qi Y, Tai Z, Sun D, Chen J, Ma H, Yan X, Liu B, Xue Q. Fabrication and characterization of poly(vinyl alcohol)/graphene oxide nanofibrous biocomposite scaffolds. *J Appl Polym Sci.* 2013; 127: 1885–1894.

Ramanathan T, Abdala A, Stankovich S, Dikin D, Herrera-Alonso M, Piner R, Adamson D, Schniepp H, Chen X, Ruoff R. Functionalized graphene sheets for polymer nanocomposites. *Nat Nanotechnol.* 2008; 3: 327–331.

Ray SS, Okamoto M. Polymer/layered silicate nanocomposites: A review from preparation to processing. *Prog Polym Sci.* 2003; 28: 1539–641.

Rezwan K, Chen Q, Blaker J, Boccaccini AR. Biodegradable and bioactive porous polymer/inorganic composite scaffolds for bone tissue engineering. *Biomaterials.* 2006; 27: 3413–3431.

Rotter N, Aigner J, Naumann A, Planck H, Hammer C, Burmester G *et al.* Cartilage reconstruction in head and neck surgery: Comparison of resorbable polymer scaffolds for tissue engineering of human septal cartilage. *J Biomed Mater Res.* 1998; 42: 347–356.

Rozhkov AV, Giavaras G, Bliokh YP, Freilikher V, Nori F. Electronic properties of mesoscopic graphene structures: Charge confinement and control of spin and charge transport. *Phys Rep.* 2011; 503: 77–114.

Sachs E, Cima M, Williams P, Brancazio D, Cornie J. Three dimensional printing: Rapid tooling and prototypes directly from a CAD model. *J Eng Ind.* 1992; 114: 481–488.

Sahoo NG, Bao H, Pan Y, Pal M, Kakran M, Cheng HKF *et al.* Functionalized carbon nanomaterials as nanocarriers for loading and delivery of poorly water-soluble anticancer drug, a comparative study. *Chem Commun.* 2011; 47: 5235–5237.

Sanchez VC, Jachak A, Hurt RH, Kane AB. Biological interactions of graphene-family nanomaterials: An interdisciplinary review. *Chem Res Toxicol.* 2011; 25: 15–34.

Santos CM, Tria MCR, Vergara RAMV, Ahmed F, Advincula RC, Rodrigues DF. Antimicrobial graphene polymer (PVK-GO) nanocomposite films. *Chem Commun.* 2011; 47: 8892–8894.

Sayyar S, Bjorninen M, Haimi S, Miettinen S, Gilmore K, Grijpma D, Wallace G. UV Cross-linkable graphene/poly(trimethylene carbonate) composites for 3D printing of electronically conductive scaffolds. *ACS Appl Mater Interfaces.* 2016; 8: 31916–31925.

Sayyar S, Murray E, Thompson BC, Gambhir S, Officer DL, Wallace GG. Covalently linked biocompatible graphene/polycaprolactone composites for tissue engineering. *Carbon.* 2013; 52: 296–304.

Schwierz F. Graphene transistors. *Nat Nanotechnol.* 2010; 5: 487–496.

Shah S, Yin PT, Uehara TM, Chueng STD, Yang L, Lee KB. Guiding stem cell differentiation into oligodendrocytes using grapheme-nanofiber hybrid scaffolds. *Adv Mater.* 2014; 26: 3673–3680.

Sham AYW, Notley SM. A review of fundamental properties and applications of polymer-graphene hybrid materials. *Soft Matter.* 2013; 9: 6645–6653.

Sittinger M, Reitzel D, Dauner M, Hierlemann H, Hammer C, Kastenbauer E *et al.* Resorbable polyesters in cartilage engineering: Affinity and biocompatibility of polymer fiber structures to chondrocytes. *J Biomed Mater Res.* 1996; 33: 57–63.

Some S, Ho SM, Dua P, Hwang E, Shin YH, Yoo H *et al.* Dual functions of highly potent graphene derivative–poly-L-lysine composites to inhibit bacteria and support human cells. *ACS Nano.* 2012; 6: 7151–7161.

Song J, Gao H, Zhu G, Cao X, Shi X, Wang Y. The preparation and characterization of polycaprolactone/graphene oxide biocomposite nanofiber scaffolds and their application for directing cell behaviors. *Carbon.* 2015; 95: 1039–1050.

Sreeprasad TS, Shihabudheen MM, Deepti K, Chaudhuri K, Lourdu XP, Pradeep T. Transparent, luminescent, antibacterial and patternable film forming composites of graphene oxide/reduced graphene oxide. *ACS Appl Mater Interfaces.* 2011; 3: 2643–2654.

Stankovich S, Dikin DA, Dommett GH, Kohlhaas KM, Zimney EJ, Stach EA, Piner RD, Nguyen ST, Ruoff RS. Graphene-based composite materials. *Nature.* 2006; 442: 282–286.

Stoller MD, Park S, Zhu Y, An J, Ruoff RS. Graphene-based ultracapacitors. *Nano Lett.* 2008; 8: 3498–3502.

Szunerits, S, Boukherroub R. Antibacterial activity of graphene-based materials. *J Mater Chem B.* 2016; 4: 6892–6912.

Tian B, Wang C, Zhang S, Feng L, Liu Z. Photothermally enhanced photodynamic therapy delivered by nano-graphene oxide. *ACS Nano.* 2011; 5: 7000–7009.

Verdejo R, Mar Bernal M, Romasanta LJ, Lopez-Manchado MA. Graphene filled polymer nanocomposites. *J Mater Chem.* 2011; 21: 3301–3310.

Wan C, Chen B. Poly(ε-caprolactone)/graphene oxide biocomposites: Mechanical properties and bioactivity. *Biomed Mater.* 2011; 6: 055010.

Wang GM, Qian F, Saltikov C, Jiao YQ, Li Y. Microbial reduction of graphene oxide by Shewanella. *Nano Res.* 2011; 4: 563–570.

Wang J, Ouyang Z, Ren Z, Li J, Zhang P, Wei G, Su Z. Self-assembled peptide nanofibers on graphene oxide as a novel nanohybrid for biomimetic mineralization of hydroxyapatite. *Carbon.* 2015; 89: 20–30.

Wang W, Caetano G, Ambler WS, Blaker JJ, Frade MA, Mandal P *et al.* Enhancing the hydrophilicity and cell attachment of 3D printed PCL/graphene scaffolds for bone tissue engineering. *Materials.* 2016; 9: 992.

Wei G, Ma PX. Structure and properties of nano-hydroxyapatite/polymer composite scaffolds for bone tissue engineering. *Biomaterials.* 2004; 25: 4749–4757.

Wen H, Dong C, Dong H, Shen A, Xia W, Cai X, Song Y, Li X, Li Y, Shi D. Engineered redox-responsive PEG detachment mechanism in PEGylated nano-graphene oxide for intracellular drug delivery. *Small.* 2012; 8: 760–769.

Wu D, Bäckström E, Hakkarainen M. Starch derived nano-graphene oxide functionalized bioactive porous starch scaffolds. *Macromol Biosci.* 2017a; 17: 1600397.

Wu D, Hakkarainen M. A closed-loop process from microwave-assisted hydrothermal degradation of starch to utilization of the obtained degradation products as starch plasticizers. *ACS Sustain Chem Eng.* 2014; 2: 2172–2181.

Wu D, Samanta A, Srivastava RK, Hakkarainen M. Starch derived nano-graphene oxide paves the way for electrospinnable and bioactive starch scaffolds for bone tissue engineering. *Biomacromolecules.* 2017b; 18: 1582–1591.

Wu D, Xu H, Hakkarainen M. From starch to polylactide and nano-graphene oxide: Fully starch derived high performance composites. *RSC Adv.* 2016; 6: 54336–54345.

Xu H, Adolfsson KH, Xie L, Hassanzadeh S, Pettersson T, Hakkarainen M. Zero-dimensional and highly oxygenated graphene oxide for multifunctional poly(lactic acid) bionanocomposites. *ACS Sustainable Chem Eng.* 2016; 4: 5618–5631.

Xu H, Xie L, Hakkarainen M. Coffee ground-derived quantum dots for aqueous processable nanoporous graphene membranes. *ACS Sustainable Chem Eng.* 2017; 5: 5360–5367.

Xu Y, Wu Q, Sun Y, Bai H, Shi G. Three-dimensional self-assembly of graphene oxide and DNA into multifunctional hydrogels. *ACS Nano.* 2010; 4: 7358–7362.

Yan L, Zhao F, Li S, Hu Z, Zhao Y. Low-toxic and safe nanomaterials by surface-chemical design, carbon nanotubes, fullerenes, metallofullerenes, and graphenes. *Nanoscale.* 2011; 3: 362–382.

Yang K, Zhang S, Zhang G, Sun X, Lee ST, Liu Z. Graphene in mice, ultrahigh *in vivo* tumor uptake and efficient photothermal therapy. *Nano Lett.* 2010; 10: 3318–3323.

Yang ST, Chang Y, Wang H, Liu G, Chen S, Wang Y, Liu Y, Cao A. Folding/aggregation of graphene oxide and its application in Cu2+ removal. *J Colloid Interface Sci.* 2010; 351: 122–127.

Yoon OJ, Jung CY, Sohn IY, Kim HJ, Hong B, Jhon MS, Lee N-E. Nanocomposite nanofibers of poly (d, l-lactic-co-glycolic acid) and graphene oxide nanosheets. *Compos Part A Appl Sci Manuf.* 2011; 42: 1978–1984.

Zhang L, Wang Z, Xu C, Li Y, Gao J, Wang W, Liu Y. High strength graphene oxide/polyvinyl alcohol composite hydrogels. *J Mater Chem.* 2011; 21: 10399–10406.

Zhang L, Xia J, Zhao Q, Liu L, Zhang Z. Functional graphene oxide as a nanocarrier for controlled loading and targeted delivery of mixed anticancer drugs. *Small.* 2010; 6: 537–544.

Zhang R, Ma P. Processing of polymer scaffolds: Phase separation. *Methods Tissue Eng.* 2001; 7: 715–724.

4 Lipid-Based Nanocarriers in Lymphatic Transport of Drugs: Retrospect and Prospects

Vikas Rana, Sunil Kamboj, and Sheshank Sethi*

CONTENTS

4.1 Introduction ... 67
4.2 Lymphatic Drug Transport System: An Overview ... 68
4.2.1 Biological Transporters ... 69
4.2.1.1 ABC Transporter (ATP Binding Cassettes Transporter) ... 70
4.2.1.2 SLC Transporters (Solute Carrier Transporter) ... 71
4.2.1.3 Coupling Efflux Transporters with Metabolic Enzymes ... 71
4.3 Approaches to Target Drugs *via* the Lymphatic System ... 72
4.3.1 Lipid-Based Formulations ... 72
4.3.1.1 Emulsions/Microemulsions/SMEDDS/SNEDDS ... 73
4.3.1.2 Lipid Nanoparticles ... 81
4.3.1.3 Liposomes ... 85
4.4 Conclusions ... 92
References ... 92

4.1 INTRODUCTION

Nowadays, drug discovery is mainly aimed at the development of potent drugs, the potential outcome of which results in the launch of a number of lipophilic drugs (Bayliss *et al.*, 2016). This led to the origin of new areas of research associated with such drugs, including, bioavailability, drug design and food-related effects. To resolve these issues, a myriad of formulation techniques have been developed for the delivery of compounds with low aqueous solubility and high lipophilicity (Feeney *et al.*, 2016). The strategies that utilize the consistency of both lipids and drugs, along with potential of lymphatic transport, is the most acceptable route of drug transport for lipophilic drugs (Carrière *et al.*, 2016).

The intestinal lymphatic system is a pathway made for the transport of digested food-derived lipids, water-insoluble peptide and fat-soluble vitamins to the blood (Yáñez *et al.*, 2011). Interestingly, the drug transported across this system can bypass the liver and thus neglect hepatic first-pass metabolism. Further, the intestinal tract is richly supplied with lymph and blood vessels. Generally, the drugs or lipids given orally have to cross enterocytes (sequentially-transmitted intestinal epithelial cells). After being absorbed and crossing the enterocyte barrier, the drug molecule enters the portal circulation that transport the drug molecule to the liver *via* portal vein for the so-called hepatic first pass effect. However, the drugs with log P > 5 and with long-chain triglycerides (TG) with a solubility > 50 mg/g, are preferentially extracted into lymphatic circulation and there is movement into portal circulation (Figure 4.1) (Porter *et al.*, 2007; Jannin *et al.*, 2008; Porter *et al.*, 2008). Interestingly, the highly lipophilic drugs are associated with secretable enterocyte lipoproteins (chylomicrons), which easily enter into the mesenteric lymphatic circulation, instead of portal

* Corresponding author.

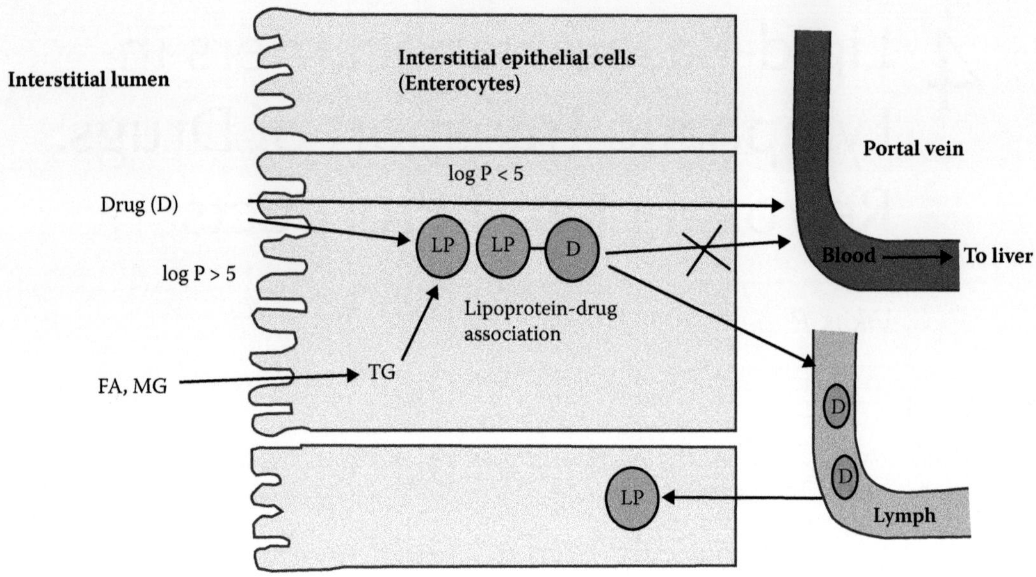

FIGURE 4.1 Process of drug absorption *via* the intestinal lymphatic system and portal vein (FA = fatty acid, MG = monoglyceride, TG = triglyceride, LP = lipoprotein).

circulation (Figure 4.1). Thus, the drugs that follow the lymphatic route do not undergo hepatic first pass metabolism. This also results in the enhancement of drug concentration in lymph nodes and lymph ducts and, therefore, can be the site of therapeutic drug action (Trevaskis *et al.*, 2009). Keeping in mind the therapeutic benefits or drugs that have suitable attributes for lymphatic transport, it is interesting and important to know the factors responsible for lymphatic drug transport. Chapter 4 will discuss the ideal carrier system and approaches to the lymphatic transport system.

4.2 LYMPHATIC DRUG TRANSPORT SYSTEM: AN OVERVIEW

The lymphatic system mainly exists in all parts of our body with the exception of the central nervous system. The main areas are the bone marrow, spleen, thymus gland, lymph nodes/nodules, tonsils, heart, liver and skin. This system is therefore mainly concerned with the transport of lymph (a watery clear fluid), that delivers lymphocytes and immune-related factors across the body. The major role of lymphocytes is to protect the body against the attack of antigens (such as bacteria and/or viruses). In addition, the lymphatic system is an important drainage fluid from tissues and cells, which drain the fluid back to the circulating blood (Khan *et al.*, 2013). Therefore, lymph is generally defined as a mixture of watery fluid and proteins that have been extracted out and filtered from blood plasma.

Other key functions of lymphatic system are absorption and transportation. The composition and flow of lymphatic fluid is summarized in Figure 4.2.

There are mainly three ways to deliver drugs through the intestinal lymphatic vessels:

1. The lymphatic capillaries are comprised of single-layered, non-fenestrated endothelial cells. The arrangement of these cells is such that they create a porous barrier in the lymphatic vasculature, which permits the macromolecules to the lymphatic system. Therefore, enhancement in the absorption of hydrophilic macromolecules and/or its conjugates is achieved by this porous barrier following the paracellular route using absorption enhancer (Porter *et al.*, 2007; Porter *et al.*, 2008; Pouton and Porter, 2008).
2. The aggregated lymphoid follicles that form Peyer's patches in the intestine lumen provide an entry point for drugs to lymphatic system (Dahan and Hoffman, 2008).

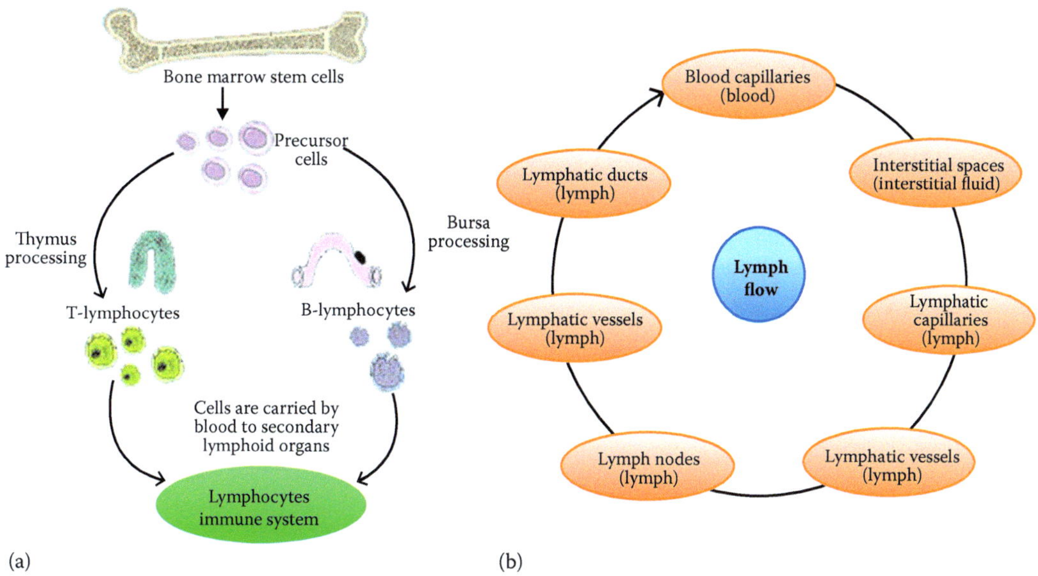

FIGURE 4.2 Schematic representation of (a) composition and (b) flow of lymphatic fluid.

3. The inhibition of P-glycoprotein and cytochrome P450 that leads to transcellular absorption or paracellular transport could be another entry point for drugs/lipids into lymphatic system (Pouton and Porter, 2008).

The efficiency of drug uptake by this route depends upon many factors:

a. *Particle size*: Particles in the size range of 50–3000 nm are capable of uptake by the Peyer's patches and subsequently transported into lymphatic system. However, particles of 3–10 μm are often retained within the Peyer's patches and in the lymph. Particles more than 10 μm are not taken up by the gastrointestinal (GI) tract (Liu *et al.*, 2008).
b. *Surface hydrophobicity*: The higher hydrophobic microparticles are generally uptake by GI tract.
c. *Specific targeting ligands*: The M cells or lectin-specific monoclonal antibodies which bind to specific carbohydrate residues on M cells are reported to enhance uptake of microparticulate delivery system.

4.2.1 Biological Transporters

Biological transporters are proteins attached with the membrane that mediate the movement of substrate across the biological membrane. These biological transporters are located throughout the body, more specifically in the epithelia of most of the organs with/without barrier attributes like the brain, testis, placenta, liver, intestine and kidney (Giacomini and Huang, 2013). The biological transporters mainly belong to two super families *i.e.* ATP-binding cassette (ABC) and solute-carrier (SLC) super families. The ABC transporters gather energy from ATP hydrolysis and work as an efflux transporter. However, SLC transporters are responsible for the uptake of small molecules into cells (Lin *et al.*, 2015). The interest to utilize both these super families for drug development and delivery is growing day by day particularly for those with broad substrate specificities like organic anion transporter I (OAT1; SLC22A6) or multi-drug resistance protein-I (MDR1; P-glycoproteins or ABCB1).

The lipids and drugs are transported into the intestinal lymphatic system via biological transporters. Thus, the intestinal lymphatic transport is utilized as a transport pathway for most of the lipophilic drug. In addition, these lipophilic drug transporters are also responsible for the separation of lipophilic drug transport into lymph versus portal circulation.

4.2.1.1 ABC Transporter (ATP Binding Cassettes Transporter)

These are lipid transport-proteins identified on the membrane of enterocytes (both the basolateral and apical membrane). They exist as intracellular lipid binding proteins and facilitate the absorption (entry) and intracellular transport (movement across) of endogenous lipids and lipids from diet (exogenous) (Stremmel *et al.*, 2001; Agellon *et al.*, 2002). Hence, an ABC transporter involves the intestinal lipid absorption and entry of drugs solubilized in these lipids (exogenous and endogenous), exploited in the lymphatic transport of drugs. Table 4.1 summarizes the example of some lipids/surfactants that carries solubilized drugs in lipids and transport it to lymphatic system.

Interestingly, the lipid uptake from intestinal lumen requires several proteins on the apical brush border membrane *i.e.* N1PC1 (Niemann Pick C1-like), SRB-2 (Scavenger receptor B-1) and CD36.

TABLE 4.1
Instances of Drugs (Inhibitors or Substrates) with or without Lipids and/or Surfactants that Utilizes Biological Transporters for their Uptake or Efflux in to the Lymphatic System

Examples of Lipids/Surfactants Transported *via* ABC Transporters

S. No.	Lipids/Surfactants	Metabolic Enzyme	Examples of Drugs Associated with Transporters
1.	Cremophor	CYP3A	Cyclosporine
2.	Tween® 80/Tween® 20	CYP3A	Lovastatin
3.	Solutol HS-15	CYP3A	Ketoconazole
4.	Pluronic block copolymer	CYP3A	Ritonavir
5.	Sucrose monolaurate	No report	Tamoxifen
6.	Labrasol	No report	Amiodarane

Example of Substrates and Inhibitors of ABC Transporters

S. No.	Transporters	Substrates	Inhibitors
1.	BCRP/MXR (*ABCG2*)	Methotrexate, Topotecan, Imatinib	Oestrone, 17β-oestradiol, Fumitre-morgin C
2.	BSEP/SPGP, cBAT, ABCB11 (*ABCB11*)	Taurocholic acid, pravastatin, Bile acids	Cyclosporin A, Rifampicin, Glibenclamide
3.	MRP2/ABCC2, cMOAT (*ABCC2*)	Methotrexate, Etoposide	Cyclosporine, Delaviridine, Efavirenz, Emtricitabine
4.	MRP3/ABCC3 (*ABCC3*)	Oestradiol-17β-glucuronide, Methotrexate	Delaviridine, Efavirenz, Emtricitabine
5.	MDR3/ABCB4 (*ABCB4*)	Paclitaxel, Digoxin, Vinblastine	Verapamil, Cyclosporine

Examples of Substrates and Inhibitors of SLC Transporters

S. No.	Transporters	Substrates	Inhibitors
1.	OCT2 (*SLC22A2*)	Pindolol, Ranitidine, Oxaliplatin	Cetirizine, Testosterone, Quinidine
2.	OAT3 (*SLC22A8*)	Cefaclor, Ceftizoxime	Probenecid, Novobiocin
3.	OAT1 (*SLC22A6*)	Adefovir, Cidofovir, Zidovudine	Probenecid, Novobiocin
4.	OATP1B3/OATP-8 (*SLCO1B3*)	Digoxin, Telmisartan,	Cyclosporine, Ritonavir, Lopinavir
5.	OATP1A2/OATP-A (*SLCO1A2*)	Bile salts, Ouabain, Digoxin, Levofloxacin	Naringin, Ritonavir, Lopinavir, Saquinavir, Rifampicin

However, two ABC half transporters located on the apical membrane of enterocytes (ABCG5 and ABCG8) are responsible for cholesterol efflux back to the intestinal lumen to limit the cholesterol absorption. Another ABC transporter present at basolateral membrane is ABCA1 that mediates the high-density lipoprotein secretion from the enterocytes (Hui *et al.*, 2008). Table 4.1 summarizes the examples of substrates and inhibitors of various ABC transporters.

4.2.1.1.1 P-glycoprotein (P-gp)

P-gp was the first active transporter protein and a member of a large diverse family of ATP-binding cassette (ABC) efflux proteins. The lipophilic and cationic agents are the substrate inhibitor for P-gp. For example, anticancer agents, antibiotics, antivirals, immunosuppressive agents, calcium channel blockers and plant chemicals are mainly found in normal diet. The nature of P-gp is to limit the permeation of certain compounds by effluxing compounds from the enterocytes back in to the lumen, thus limiting their bioavailability (Schinkel and Jonker, 2003; Urquhart and Kim, 2009). In addition, the disruption of P-gp activity leads to acute problems with regard to drug disposition. This is probably due to the reason that lipids are the substrate of P-gp that plays a role in the lymphatic transport of drugs.

4.2.1.1.2 Multidrug resistance associated proteins (MdPs)

The main function of *MdPs* is possibly associated with the elimination of compounds from the cells through efflux. This class of transporter can grant resistance to cytotoxic drugs like peptides, vincristine, heavy metal anions, as well as endogenous metabolites such as bilirubin glucuronides (Borst *et al.*, 2000; Urquhart and Kim, 2009; Yáñez *et al.*, 2011).

4.2.1.2 SLC Transporters (Solute Carrier Transporter)

The solute carrier transporters interact with a various array of substrates like organic and inorganic ions and amino acids, which are needed for protein synthesis. There are approximately 395 membrane-spanning SLC transporters that are organized into 52 families (Geier *et al.*, 2013; Hediger *et al.*, 2013; Schlessinger *et al.*, 2013). Many SLC transporters have been reported to be modulated by drugs and contribute to a disease phenotype (Fotiadis *et al.*, 2013). Table 4.1 summarizes more information. The different classes of drugs (like diuretics, neurotransmitter-reuptake inhibitors for neuropsychiatrics, uric acid transporter inhibitors, glucose transporter inhibitors for diabetes, etc.) are used to modulate drug delivery in diseased conditions (Lin *et al.*, 2015). The drugs can have inhibitory activity or stimulatory activity that leads to decrease or increase in transporter activity thereby potentially beneficial effect or adverse effects (Ulanovskaya *et al.*, 2011; Hu *et al.*, 2013). For example, the inhibitors are dopamine transporters (DAT, SLC6A3), serotonin transporters (SERT; SLC6A4), noradrenaline transporters (NET; SLC6A2) (MacDougall and Griffith 2008), etc. Riluzole (used to cure amyotrophic lateral sclerosis and inhibit glutamate release) was found to enhance the transport activity of EAAT1 (SLC1A3) by taking extracellular glutamate from the synapses and minimizing the glutamate level in the synaptic cleft (Fumagalli *et al.*, 2008; Carbone *et al.*, 2012).

4.2.1.3 Coupling Efflux Transporters with Metabolic Enzymes

The association between one metabolic enzyme, at least two proteins and single transporter is responsible for transportation of drugs or its metabolites. The 'double jeopardy' theory proposed by Benet *et al.* stated that the substrate is assumed to be absorbed at the apical membrane and is associated with CYPs (Benet *et al.*, 2004). This substrate undergo metabolism and hence labelled as 'prosecuted' for the first time by CYPs. The efflux transporter, P-gp is assumed to take the parent drug and efflux back it into the intestinal lumen and parent drug comes back into basolateral side of the membrane. Hence, this unmetabolized parent drug reaches either the lymphatic or the portal capillaries. However, if this intact drug returns to the intestinal lumen through efflux transportation, it can be absorbed once again at the apical side of the intestinal membrane. Thus, this leads to metabolism again. This repeated reabsorption gives the term 'double jeopardy' to this theory. However, the change in the activity of P-gp would result in less repeated chance of exposure time for the substrate in question at the apical membrane. This would result in enhancement of substrate inside the intestine (*i.e.* increase in absorption). Thus, the bioavailability of both the substrate and/or its metabolite in the systemic circulation is enhanced. Alternatively, the disruption in the CYPs activity, leads to decrease in amount of metabolite and thus allowing more intact substrate to reach systemic circulation.

4.3 APPROACHES TO TARGET DRUGS *VIA* THE LYMPHATIC SYSTEM

Targeted drug delivery is mainly achieved by using carries with specified affinity to the target tissues. The key approach to target drugs via lymphatic system is chemical modification of drugs (prodrug) or by pharmaceutical modification. In a chemical modification approach also known as prodrug approach, the drug substance is covalently attached with lipophilic carriers that overall enhanced the lipophilic behaviour of drug. This promotes lymphatic delivery of drugs. On the contrary, in a pharmaceutical formulation approach, drug substances are loaded in a carrier system like emulsion, liposomes, nanoparticles, etc. that promotes lymphatic delivery of drugs. Therefore, various techniques were developed to achieve lymphatic delivery of drugs. The approaches used are summarized in the form of flow diagram as Figure 4.3.

4.3.1 Lipid-Based Formulations

The lymphatic route of absorption provides an opportunity to increase the bioavailability of highly lipophilic drugs (Log P > 5) with enhanced solubility in triglycerides (Cs > 50 mg/mL) (Jannin *et al.*, 2008; Trevaskis *et al.*, 2009). Lipid systems consisting of alcohol esters of unsaturated LCFA (long-chain fatty acids) have been shown to increase the bioavailability of certain drugs by utilizing the lymphatic route, e.g., saquinavir with polyglyceryl oleate (Plurol® Oleique CC497) or ethoxylated castor oil (Cremophor® EL) (Griffin and O'Driscoll, 2006), ontazolast with glyceryl monooleate (Peceol™) (Hauss *et al.*, 1998), and halofantrin with various triglycerides and derivatives (Charman and Stella, 1991; O'Driscoll, 2002).

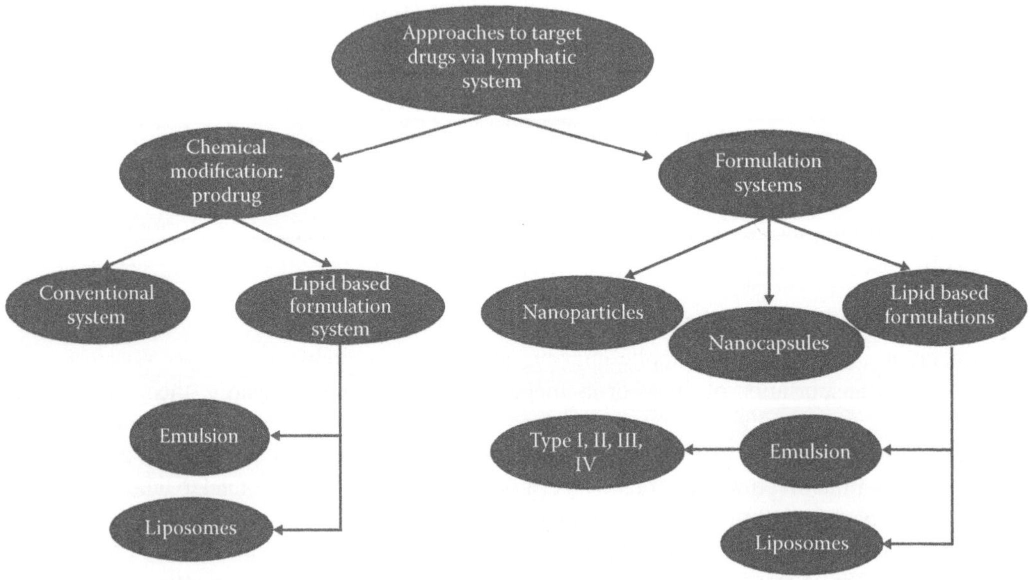

FIGURE 4.3 Flow diagram depicting approaches to deliver drugs *via* the lymphatic system.

The administration of formulation derived lipids loaded with drug is most beneficial approach to increase lymphatic drug transport. The lipid based formulations can be solutions, suspensions and self-emulsifying formulations containing liquid or semisolid triglycerides, mixed mono and diglycerides, surfactants, either alone or as mixture with hydrophobic cosolvent. Pouton and Porter (2008) developed a lipid formulation classification system (LFCS) and it is summarized in Table 4.2.

The LFCS system highlights the characteristics and the potential limitations of the various types of formulations in obtaining small particle sizes and the possibilities of drugs to precipitate with formulation derived lipid digestion and intestinal fluid dilution.

Structural features of lipids: Lymphatic drug transport is generally increased by lipid formulations containing long chain (C18) lipids as opposed to medium-chain (C8–10) and short-chain (C4) lipids. Fatty acids with chain lengths of 14 and above are primarily transported into intestinal lymph and shorter-chain fatty acids are primarily transported into portal blood (Yáñez *et al.*, 2011).

Excipient effects on metabolism and transporters: The excipients have effects on interactions with cellular transporters and drug metabolism. The inhibitory abilities of surfactants on the regulation, expression and functional capacity of various CYP isoforms were observed.

For instance, polyoxyethylated/pegylated surfactants such as PEG-15-hydroxystearate, cremophor, polysorbates (*e.g.* Tween 80, Tween 20) have all been identified as CYP3A isozymes inhibitors (Elnaggar *et al.*, 2009; Yáñez *et al.*, 2011). In this case, inhibition is apparent in expression (molecular level) and/or intrinsic metabolic activity (functional capacity). An example of a lipid excipient that modulates with specific effects on the expression of efflux transporters (*i.e.* P-gp) is seen with Gelucire® and Peceol®. Due to a decrease in expression, a significant decrease in P-gp efflux activity observed on Caco-2 cell monolayers. However, the decrease in P-gp efflux was solely attributed due to decrease in expression is still not clear. In another example, Triton X-100 (a non-ionic detergent) has been shown to prevent drug binding and thus modulate the functional capacity of P-gp.

4.3.1.1 Emulsions/Microemulsions/SMEDDS/SNEDDS

The major function of lipids in oral delivery is emulsification followed by solubilizing the drug in the gastrointestinal tract (Jannin *et al.*, 2008). Thus, screening for suitable lipids is based on solubility data that is treated as an indicator of inclusion/exclusion criteria for lipids, which often do not correlate with the physiological behaviour of the dosage form. A drug solubilized in hydrophilic cosolvents (PEG, propylene glycol, ethanol, etc.) is diluted with the surrounding physiological fluids when it enters the gastrointestinal tract, leading to a rapid loss in solubility and precipitation of the drug. However, an emulsified drug is less prone to precipitation or binding with other elements. Hence, the emulsification or micellar dispersion of the drug in the gastrointestinal milieu and not the solubility in the dosage form alone should be of primary concern. Some examples of drugs transported use microemulsions, SMEDDS, SEDDS, etc., which are shown in Table 4.3.

Once a list of suitable excipients is identified, a binary drug excipient screening using solubility, compatibility and stability should be examined so as to obtain the lipid system(s) most appropriate for the drug in question. For designing SMEDDS or SNEDDS systems, a blend of lipid, surfactant/cosurfactant is screened on the basis of solubility and affinity of drug for each component. Thus, the aim should be the overall solubilizing/dispersing power of SMEDDS or SNEDDS systems compared to the solubility of drugs in the individual components. Whether the aim is to achieve drug solubility that is sustainable *in vivo*; preventing the drug against binding to elements in the gastrointestinal tract; inhibition or saturation of efflux and metabolic barriers; and/or promotion of lymphatic transport, defining the main objectives will help to narrow the focus on the appropriate type of excipient(s) to address the set of barriers specific to the drug molecule. Successful lipid-based drug delivery systems require a holistic approach to formulation (Porter *et al.*, 2007; Chen *et al.*, 2008; Jannin *et al.*, 2008; Porter *et al.*, 2008).

An injectable emulsion formulation for the lymphatic transport of mitomycin C was developed that showed selective uptake after injection into regional lymphatics, occurring in the order of o/w > w/o > aqueous solution. In another report gelatin spheres in oil emulsion was developed for the purpose

TABLE 4.2
Formulation Classification System of the Four Essential Types of Lipid Formulations

Constituents/Attributes	Type I	Type II	Type IIIA	Type IIIB	Type IV
Definition	Oily formulation without surfactant and cosolvents	PCMs with low HLB surfactant	Self-emulsifying/PCMs with water-soluble components	PCMs with water-soluble components and low oil content	Oil-free formulation based on surfactants and cosolvents
Excipients	**% w/w of Oil/Surfactant or Cosurfactant of Different LFCS**				
Triglycerides/mixed Glycerides	100%	40–80%	40–80%	< 20%	–
Water insoluble surfactants	N/A	20–60%	–	–	0–20%
Water soluble surfactants	N/A	N/A	20–40%	20–50%	30–80%
Hydrophilic cosolvent	N/A	N/A	0–40%	20–50%	0–50%
Physical Characterization					
Particle size of dispersion (nm)	Coarse	100–250	100–250	50–100	N/A
Self-emulsifying behaviour	Poor	Solvent capacity unaffected	Some loss of solvent capacity	Significant phase changes and potential loss of solvent capacity	High
Significance of Digestibility	Crucial requirement	Not crucial, but is likely to occur	Not crucial, but may be inhibited	Not required, but is unlikely to occur	No
Advantages	GRAS status; simple; excellent capsule compatibility	Unlikely to lose solvent capacity on dispersion	Clear, dispersion; drug absorption without digestion	Clear dispersion; drug absorption without digestion	Good solvent, disperses to micellar solution
Disadvantages	Formulation has poor solvent capacity, unless drug is highly lipophilic	Turbid o/w dispersion	Possible loss of solvent capacity on dispersion; less easily digested	Likely loss of solvent capacity on dispersion	Loss of solvent capacity on dispersion; may not be digestible

TABLE 4.3
Lymphatic Transport of Drugs Using Emulsion, SMEDDS and SNEDDS as Formulation System

S.No.	Formulation	Drug	Components		Lymphatic Transport Model	Lymphatic Transport	Remarks	Reference
			Lipid	Surfactant/ Co-Surfactant				
1.	SNEDDS	Tamoxifen	Corn oil	Labrasol and Transcutol-P	Cannulation of the mesenteric lymph duct	3.04-fold improved uptake	Lymphatic transport enhanced as compared to suspension	Sandhu *et al.*, 2017
2.	SNEDDS	Bicalutamide (BCT) and Hesperetin (HSP)	Castor oil	Kolliphor EL and triton X-100	Chylomicron flow blocking model	The AUC of both BCT and HSP < in cycloheximide treated as compared to saline group	SNEDDS formulations showed higher lymphatic transport	Arya *et al.*, 2017
3.	SMEDDS	Huperzine A	Castor oil	Cremophor RH40 and propylene glycol	Lymph node extraction from rats	Lymphatic pathway transport of SMEDDS and Hup-A suspension were about 40% and 5%, respectively	Lymphatic transport enhanced as compared to suspension	Li *et al.*, 2017
4.	SNEDDS	Lopinavir	Maisine	Tween 80 and Transcutol HP	Chylomicron flow block SPIP studies	3.9-fold increase in AUC	Lymphatic transport enhanced as compared to suspension	Garg *et al.*, 2016
5.	SNEDDS	N-251	Capryol 90, Carbitol	Cremophor RH40, Cremophor EL	Lymph node extraction from rats	40 µg/g of lymph	Long chain SNEDDS showed higher lymphatic transport	Imada *et al.*, 2015
6.	SNEDDS	Diacerein	Maisine	Tween 80, Cremophor EL and Transcutol P	Chylomicron flow blocking model	–	Maisine support the lymphatic transport	El-Laithy *et al.*, 2015
7.	SNEDDS	Simvastatin	Captex 300 (MCT), Soybean oil (LCT),	Cremophor RH40/ Ethanol	*In vitro* lipolysis		LCT>MCT in lymphatic transport	Thomas *et al.*, 2012

(*Continued*)

TABLE 4.3 (CONTINUED)
Lymphatic Transport of Drugs Using Emulsion, SMEDDS and SNEDDS as Formulation System

S.No.	Formulation	Drug	Components		Lymphatic Transport Model	Lymphatic Transport	Remarks	Reference
			Lipid	Surfactant/ Co-Surfactant				
8.	SMEDDS	Sirolimus	MCT	Cremophor RH40, propylene glycol	Chylomicron flow blocking model	13.7% of fraction transported with maximum amount of oil	SMEDDS formulations containing ≥ 25% MCT showed higher lymphatic transport	Sun *et al.*, 2011
9.	Self-emulsifying pellets	Silymarin	Akoline MCM®, Miglyol®	Tween 80, soy lecithin and propylene glycol	Lymph cannulated rats	Silybin A and silybin B absorbed by the lymphatic route were 8 and 6 µg	Higher lymphatic transport achieved as compared to pure extract	Iosio *et al.*, 2011
10.	Microemulsion	Methylene blue	Crodamol GTCC	Span 80-Tween 80/ transcutol P	Mice, sentinel lymph node biopsy	3 µg/mg of lymph	Microemulsion produced good lymphatic tropism	Wang *et al.*, 2009
11.	Microemulsion	Halofantrine	Soybean oil	Polysorbate 60 or 80	Chylomicron flow blocking model	11.8% of the dose	Higher lymphatic transport achieved	Lind *et al.*, 2008
12.	Microemulsion and mixed micelles	Saquinavir	Oleic acid	Plurol Oleique, Labrasol	Lymph cannulated rats	0.025–0.05% of the dose administered	Microemulsion produced higher lymph concentrations	Griffin and O'Driscoll, 2006
13.	Microemulsion	Halofantrine	Captex 355, Capmul MCM	Cremophor EL, Maisine 35-1 and ethanol	Thoracic lymph duct and cephalic vein–cannulated fasted greyhound dogs	28.3% for LCT and 5% for SCT	Long-chain glycerides appear to be more effective with respect to lymphatic transport	Khoo *et al.*, 2003

(*Continued*)

TABLE 4.3 (CONTINUED)
Lymphatic Transport of Drugs Using Emulsion, SMEDDS and SNEDDS as Formulation System

S.No.	Formulation	Drug	Components		Lymphatic Transport Model	Lymphatic Transport	Remarks	Reference
			Lipid	Surfactant/ Co-Surfactant				
14.	SMEDDS	Halofantrine	1,3-dioctanoyl-2-linoleyl-sn-glycerol (MLM) and 1,3-dilinoyl-2-octanoyl-sn-glycerol (LML)	Cremophor EL, Maisine 35-1 and ethanol	Lymph cannulated canines,	17.9% in MLM and 27.4 in LML	LML showed higher lymphatic transport	Holm *et al.*, 2003
15.	SEDDS	Ontazolast	Soybean oil	Tween 80, Gelucire 44/14	Lymph cannulated rats	9% when administered as an oil-in-water emulsion	Lymphatic transport increased 9 times	Hauss *et al.*, 1998
16.	Emulsion	Radiolabeled iodohippuric acid (IH, water-soluble model compound)	Sesame oil	Hydrogenated castor oil (HCO-60) and sorbitan sesquioleate (SO-15)	Rats sacrificing at different time intervals, radioactivity assay	20% using w/o emulsion	w/o emulsion was the most favorable	Hashida *et al.*, 1977

of minimizing the instability of w/o emulsion. Interestingly, antifilarial drug in gelatine nanoparticles was loaded in an oil emulsion system. This colloidal system holds great potential for a lymphotropic carrier system. Further, an emulsion formulation loaded with anticancer drugs (Pirarubicin, and LIPIODOL®) was effective against gastric cancer and metastatic lymph nodes. The emulsion delivery of these drugs was retained over 7 days at the injection site and in the regional lymph node (Nakamoto *et al.*, 1975).

Hashida *et al.* (1977) used four different preparations to investigate the lymphatic transport. In the iliac lymph node, the radioactivity (expressed as percentage of dose per gram of tissue) was very high indicating 10-fold to more than a 100-fold accumulation when compared to the radioactivity detected in the blood. Among the four preparations, the W/O-G emulsion exhibited the highest concentration, reaching more than 20% of dose per gram of lymph nodes after 24 h; this was followed by W/O emulsion, and the lowest levels of concentration were obtained after injection of sesame oil. Results from the left iliac lymph nodes were similar. They postulated that the accumulation of radioactivity in the right iliac lymph nodes took place primarily during the absorption phase, when the concentration of the oil transported from the injection site directly through the peripheral lymph vessels would be at its highest levels. In another conformational study, the photomicrographs of frozen sections of the thigh muscle and iliac lymph nodes at 15 min after injection of W/O emulsion using lipophilic dye (Sudan blue). The oil droplets were seen 15 min after injection, in the marginal sinus of the regional lymph node and some of them have also intruded into the intermediary sinus adjacent to the trabeeula. These droplets were smaller (approximately 1–5 µm in diameter) than those observed at the injection site. The results indicated that a part of the injected sesame oil was delivered directly from the injection site to the regional lymph node as droplets have retained their shape and size. They also pointed out that if anticancer agents were loaded into these oily droplets, they would be transported to the lymphatic system with these carriers. The W/O emulsion made from sesame oil was the most suited dosage formulation due to its highest ability for transport of the drug to the lymphatics while the other W/O emulsion was somewhat less efficient.

In another attempt, Hauss *et al.* (1998) prepared SEDDS for the lymphatic delivery of ontazolast. The lymphatic concentration of ontazolast was attained 3-fold compared to its suspension. The results indicated that ontazolast was transported lymphatically by solubilization in the chylomicron core triglyceride. Interestingly, the total amount of ontazolast distributed via lymphatic route over 8 h was significantly higher for the lipid-based formulation ($p < 0.05$) as compared to the suspension. Further, the critical factor that had the most impact on the total amount of ontazolast delivered lymphatically was the duration of transport produced by the formulation. In addition, the highest lymphatic delivery of ontazolast was achieved with the emulsion and peceol formulations, as they provided the highest rates of triglyceride transport. It was also observed that there was a trend towards the minimum release (completion) of ontazolast from the peceol solution; the higher effective lymphatic triglyceride delivery was achieved. Thus, this pointed towards the occurrence of enhanced uptake of drug by lymph. However, the SEDDS formulations which lead to the faster Ontazolast absorption from the gastrointestinal tract resulted in enhanced concentrations of ontazolast in chylomicron triglyceride. The combined data pointed towards the fraction of absorbed drug delivered lymphatically depending on (1) the rate of drug absorption from the gastrointestinal tract and (2) the rate of continuous triglyceride delivery.

On oral administration of halofantarine and simivastatin in different types of triglycerides, *viz.* long-chain triglycerides (LCT), medium-chain triglycerides (MCT) and short-chain triglycerides (SCT) to the lymph-cannulated fasted rats, it was evident that content of LCT was 3–9-fold higher as compared to MCT or SCT content. This may be due to the stimulatory effect of exogenously administered LCT on the secretion and lymphatic transport of endogenous LCT. The effect is hypothesized to diminish with a decrease in the chain length of the exogenously administered triglyceride, *i.e.* LCT > MCT > SCT (Caliph *et al.*, 2000; Thomas *et al.*, 2012).

In a similar investigation by Holm *et al.* (2003), the benefit of lipidic self-microemulsifying drug delivery systems (SMEDDS) containing triglycerides with a defined structure is that it improves the lymphatic transport and the portal absorption of a poorly water-soluble drug, halofantrine. They studied two different structured triglycerides *i.e.*1,3-dioctanoyl-2-linoleyl-sn-glycerol (C8:0-C18:2-C8:0) (MLM) and 1,3-dilinoyl-2-octanoyl-sn-glycerol (C18:2-C8:0-C18:2) (LML). The main components of optimised formulation for halofantrine, was triglyceride (MLM/LML), Cremophor EL, Maisine 35-1 and ethanol. The findings pointed that the successful lymphatic transport *via* thoracic duct was 27.4% of the dose of animals dosed with LML SMEDDS and 17.9% for MCM.

The level of lymphatic transport of halofantarine after administration of the long-chain lipid formulation was significantly high (28.3% of dose) than that seen after administration of the medium-chain formulation (5.0% of dose). Hence, this was one of the first study to demonstrate intestinal lymphatic delivery associated with small amounts of lipid present within a single, lipid-based dosage form in the fasted state. Furthermore, microemulsions based on long-chain glycerides appear to be more effective with respect to lymphatic transport than the equivalent medium-chain formulation (Khoo *et al.,* 2003).

Saquinavir belongs to the category of HIV protease inhibitors. It is lipophilic in nature and prone to first-pass metabolism. This may be the reason for its low oral bioavailability. An attempt was made to redirect the absorption pathway from portal system to the lymphatic system. For this purpose, three formulations using oleic acid as main component was prepared. These formulations are oleic acid microemulsion, D-alpha tocopheryl polyethylene glycol 1000 succinate (TGPS) oleic acid mixed micelles and an oleic acid mixed micelles. The results suggested that 0.025–0.05% of the dose administered undergoes lymphatic transport. In addition, the findings pointed towards a higher level of microemulsion in mesenteric lymph as compared to that in micellar formulation (Griffin and O'Driscoll, 2006).

Further, the best suitable lymphatic delivery for methylene blue (MB) was examined through different formulations (solution (MB-S), MB water-in-oil microemulsion (MB-ME) and MB multiple microemulsion (MB-MME) to carry the MB target to regional lymph nodes and show lymphatic tropism after subcutaneous administration. The results suggested that multiple microemulsions resided in the lymph nodes for a long time after MB-ME administration compared with the methylene blue solution (MB-S) (Wang *et al.,* 2009).

Lind *et al.* (2008) explored the potential of polysorbate 60 and 80 in the intestinal lymphatic delivery of a poorly water-soluble compound, halofantrine, using a chylomicron flow blocking approach in rats. For this purpose, male Sprague–Dawley rats were initially administered intraperitoneally with 3.0 mg/kg cycloheximide or saline and were treated by dosing with 6.7 mg/kg halofantrine in 0.4 or 1.0 g/kg polysorbate 60 or 80, 0.33 g/kg soybean oil or 0.33 g/kg soybean oil + 1.0 g/kg polysorbate 80 by gavage. The results showed no significant differences in halofantrine transport to the systemic blood and lymphatic transport between the two polysorbates at the tested dosages. Furthermore, there was an acceptable correlation between lymphatic in the chylomicron flow blocking model and the halofantrine delivery to blood. The amount of halofantrine carried directly into the systemic circulation was found to be 4.9%, and the amount carried to the lymphatic system was 11.8% when dosed in soybean oil. A similar amount of halofrantine was obtained when the formulation containing long-chain triglyceride was administered using conscious, mesenteric lymph-cannulated rat models. The chylomicron flow-blocking model using cycloheximide was used for predicting the amount of halofantrine delivered by the intestinal lymphatic system when administered in a natural long-chain triglyceride (Lind *et al.,* 2008).

Negatively charged vinpocetine (VCP)-SMEDDS was compared with its putative form and marketed formulation, negatively charged VCP-SMEDDS showed 3.4- and 2.9-fold enhancement, respectively, in terms of percentage drug release at 3 h. Relative bioavailability of negatively charged VCP-SMEDDS and positively charged VCP-SMEDDS improved by 1.85- and 1.91-fold, respectively, with respect to VCP powdered drug suspension. The reasons for improved bioavailability of VCP would be contributed towards the enhancement in penetration ability of VCP-SMEDDS, improved drug release and enhanced lymphatic uptake (Chen *et al.,* 2008). The better performance of VCP-SMEDDS may be ascribed to the spontaneous formation of microemulsion in the GI tract where the drug was presents in solubilized form; the small droplet size of VCP-SMEDDS provides a large interfacial surface area for better drug absorption. Furthermore, the enhancement in bioavailability of vinpocetine from SMEDDS may be ascribed to its enhancement of lymphatic uptake via transcellular pathway due to the presence of long-chain triglyceride (Maisine™35-1). It was also reported that the long-chain triglycerides promoted the synthesis of lipoproteins, enhanced lymphatic uptake and thus minimized the first pass effect of vinpocetine.

Iosio *et al.* (2011) prepared solid, self-emulsifying pellets for delivering milk thistle extract (silymarin). These pellets were developed via an extrusion/spheroniation technique, using a self-emulsifying system (propylene glycol, Akoline MCM™, Tween 80, soy lecithin and Miglyol), lactose monohydrate and microcrystalline cellulose. When administered as pellets, silymarin was absorbed through the lymphatic route; however, it was not present in the lymph at the quantifiable limits when silymarin was administered as a dry extract. The total amounts of silybin B and silybin A absorbed by the lymphatic transport were 6 µg and 8 µg, respectively.

The raloxifene SMEDDS exhibited improved penetration ability, when it was compared with the drug suspension of raloxifene in the *in vitro* permeability studies. SMEDDS formulation was expected to enhance the oral bioavailability of raloxifene, as it was evidenced by the increased permeation ability of raloxifene SMEDDS. The mechanistic approach responsible for enhanced absorption might be the improvement in oral bioavailability, which would also be stimulated by enhanced lymphatic delivery due to the presence of the triglyceride core of chylomicrons (Thakkar *et al.*, 2011).

Sun *et al.* (2011) explored the effect of different concentrations of oil or surfactants present in the self-microemulsifying drug delivery systems on the lymphatic delivery of sirolimus by utilizing the chylomicron flow blocking approach and single-pass intestinal perfusion technique. The formulations having oil phases facilitating drug transport *via* the lymphatic system, whereas oil-free formulations showed almost no drug absorption through the lymphatic system. Furthermore, the preparations containing oil ranging from 25% to 35% (w/w) afforded more significantly enhanced lymphatic delivery than the formulation containing 15% oil. The formulations, which had a higher percentage of oil, facilitated the higher lymphatic delivery of drugs and lead to the enhancement of oral bioavailability. When the maximum amount of oil was used in the SMEDDS, then the maximum of 13.7% of lipid fraction was transported to the lymph. Thus, a lipid threshold seemed to be a likely explanation for the differences between the formulations having lower and higher concentrations of oil.

SNEDDS formulations were used to enhance the oral bioavailability of N-251 and a comparison was also made between SNEDDS made from MCT and LCT. The results revealed that lymphatic transport was probably being the possible mechanism for the enhancement of oral bioavailability by LCT-SNEDDS (Imada *et al.*, 2015). MCT-SNEDDS and LCT-SNEDDS exhibited higher value of area under curve (AUC) (MCT-SNEDDS, 56.9 µg/g·h; LCT-SNEDDS, 181.5 µg/g·h) and higher C_{max} (MCT-SNEDDS, 14.9 µg/g; LCT-SNEDDS, 42.3 µg/g). The results clearly signify that LCT-SNEDDS comprised of long-chain fatty acids, remarkably enhance the lymphatic delivery of N-251 after the uptake from the small intestine as compared to MCT-SNEDDS comprised of medium-chain fatty acids (Caliph *et al.*, 2000; Imada *et al.*, 2015).

In another endeavour, El-Laithy *et al.* (2015) developed self-nanoemulsifying self-nanosuspension-loaded systems (SNESNS) of diacerein with a high drug loading capacity. Drug loaded SNESNS were developed by homogenizing diacerein into maisine that gave the highest drug solubility. The notable increase in the *in vitro* dissolution, portal absorption and lymphatic transport of SNESNS-D was observed.

To improve the oral bioavailability and lymphatic uptake of lopinavir, Garg *et al.* (2016) took the advantage of solid self-nanoemulsifying oily formulations. Improved oral bioavailability assists in targeting the viral reservoir *i.e.* lymphatic system for the complete inhibition of HIV. The results suggested that solid self-nanoemulsifying oily formulations remarkably enhanced the lymphatic delivery of lopinavir.

4.3.1.2 Lipid Nanoparticles

Solid lipid nanoparticles (SLNs) are a novel drug delivery system which comprises of biodegradable physiological lipids or lipidic substances being solid at body and room temperature with an average particle size ranging from 50 to 1000 nm. Solid lipid nanoparticles are believed to be the most efficient lipid based colloidal system, commenced in the early 1990s. The production cost for formulating solid lipid nanoparticles is not much higher than those established for the production of parenteral emulsions, as the excipients required for the development of solid lipid nanoparticles are comparatively cheap. Solid lipid nanoparticles comprised of triglycerides which resemble chylomicrons may have the ability to transform the absorption characteristics of drugs, including avoidance of presystemic hepatic metabolism (Singh *et al.*, 2014). Solid lipid nanoparticles could be considered a good formulation strategy for encapsulating drugs with low intrinsic solubility in the gastrointestinal tract and poor oral bioavailability, permitting transportation into the blood circulation *via* intestinal lymphatic. Table 4.4 contains the updated list of various drugs that were used for lymphatic targeting using lipid nanoparticles.

The effect of the subcutaneous route on the biodistribution and tumour uptake of solid lipid nanoparticles loaded with drugs was assessed by Reddy *et al.* (2005). Etoposide-loaded tripalmitin (ETP-SLN) radiolabeled with ^{99m}Tc were developed, and the ETP-SLN were injected intravenously, intraperitoneally and subcutaneously to mice with lymphoma tumours. Radioactivity and gamma scintigraphy measurements were used to estimate the tumour uptake and biodistribution of the ETP-SLN in the blood and visceral organs. After 24 h of subcutaneous injection, the ETP-SLN revealed an apparently higher degree of tumour uptake (59- and 8-fold higher than that of the intravenous and intraperitoneal routes, respectively) and diminished the build-up in reticuloendothelial system (*i.e.* spleen, lung and liver) as compared to that of the other two routes. After 1 h of subcutaneous administration, the tumour concentration of ETP-SLN was 23.3% of that at 24 h post injection, depicting a lower extent of initial uptake as compared to intravenous and intraperitoneal routes, whose initial uptake ratios were 48.1% and 29.7%, respectively. The moderate improvement in tumour uptake of the drug from the subcutaneous injection site after the lower initial uptake stipulated that the solid lipid nanoparticles loaded with drug could effectively be used for sustained drug release. Hence, a proximal subcutaneous injection could be suggested as the favourable route for lymphatic delivery over that of intravenous and intraperitoneal routes.

Paliwal *et al.* (2009) developed solid lipid nanoparticles loaded with methotrexate (SLN-MX) that resembled chylomicrons. Solid lipid nanoparticles of methotrexate were formulated by using Compritol 888 ATO, glycerol monostearate, tristearin and stearic acid. The configuration of such solid lipid nanoparticles may modify the duration of drug absorption predominantly to and through the lymphatic route and regions, probably following a transcellular path of lipid absorption, especially by the enterocytes and polar epithelial cells of the intestine. Recurrent lymphatic concentration of methotrexate followed by the oral administration of SLN-MX was determined by mesenteric duct cannulation. Furthermore, the comparative study was also conducted on solid lipid nanoparticles loaded with methotrexate (SLN-MX) which affirmed that the formulation comprises of Compritol 888 ATO remarkably enhanced the oral bioavailability of MX, presumably following solid lipid nanoparticles constituting lipid digestion and coabsorption *via* lymphatic delivery. Solid lipid nanoparticle-based formulations showed a significant increase in the concentration of MX in the lymphatic system when compared with drug suspension of MX.

Negi *et al.* (2013) described that solid lipid nanoparticles loaded with lopinavir (SLN-LP) undergo lymphatic uptake from GI tract, which leads to the improvement of oral bioavailability. The maximum lymphatic transport was detected for LP solution and SLN-LP at 1 h and 2 h, respectively. The time required for achieving maximum plasma concentration (t_{max}) in case of LP solution and SLN-LP were 1 and 1.5 h, respectively. Optimized SLNs formulation of lopinavir was found to show remarkably higher lymphatic uptake for SLNs with respect to lopinavir solution. This delay in achieving peak plasma concentration was due to the intracellular processing of lipids during

TABLE 4.4
Lipid Nanoparticle as a Medium to Enhance the Lymphatic Transport of Drugs

S.No.	Drug	Components Lipid	Particle Size	Charge	Entrapment Efficiency	Lymphatic Transport Model	Lymphatic Transport	Reference
1.	Curcumin	Glyceryl monostearate and soya lecithin	245.1 nm	-10.4	78.5%	Lymph node extraction from rats	Enhanced lymphatic uptake of curcumin	Baek and Cho, 2017
2.	Efavirenz	Gelucire 44/14 and Compritol 888 ATO	168.92 nm	–	39.4%	Chylomicron flow blocking model	Long chain triglyceride showed enhanced bioavailability of the EFV along with targeted drug delivery of EFV to the reservoir sites of HIV virus, i.e lymphatics	Makwana *et al.*, 2015
3.	Tripterine	Sodium oleate, and soybean oil	150 nm	–	99.95%	Chylomicron flow blocking model	2-fold enhanced lymphatic transport	Zhang *et al.*, 2014
4.	Raloxifene	Glyceryl tribehenate	167 ± 3 nm	–	> 92%	Chylomicron flow blocking model	1.3-fold enhanced lymphatic transport as compared to Cyclohexamide blocked rats	Ravi *et al.*, 2014
5.	Lopinavir	Stearic acid,	180.6 ± 2.32	-13.4 ± 0.56	91.5 ± 1.3%	Mesenteric duct cannulation in Wistar rats	1.6–1.8 μg/ml	Negi *et al.*, 2013
6.	Lopinavir	Compritol 888 ATO	239 nm	-27	100%	Mesenteric duct cannulation in Wistar rats	9.68 ± 0.88 μg	Alex *et al.*, 2011
7.	Methotrexate	Stearic Acid (SA), Glycerol Monostearate (MS), Tristearin (TA), and Compritol 888 ATO (CA)	SA, 131.0 MS, 166.2 TA, 140.4 CA, 120.1	SA, -28.3 MS, -14.4, TA, -10.7, CA,-12.0	SA, 57.3 MS, 65.5, TA, 60.9, CA, 69.2	Mesenteric duct cannulation in albino rats	≈ 25% of drug using Compitrol as lipid	Paliwal *et al.*, 2009

(*Continued*)

TABLE 4.4 (CONTINUED)
Lipid Nanoparticle as a Medium to Enhance the Lymphatic Transport of Drugs

		Components						
S.No.	Drug	Lipid	Particle Size	Charge	Entrapment Efficiency	Lymphatic Transport Model	Lymphatic Transport	Reference
8.	Etoposide	Tripalmitin	358 nm	-46.6	98.96	Radiolabellig on tumour-bearing Balb/c mice	8- and 59-fold higher than that of the intraperitoneal and intravenous routes	Reddy *et al.*, 2005
9.	Tobramycin	Stearic acid	70–125 nm	–	34819 molecules of tobramycin/ particle of SLN	Lymph duct cannulation in Wistar rats	≈ 16 µg/g	Cavalli *et al.*, 2003
10.	Radiolabelled	Compritol 888 ATO	200 nm	-17.2	97 ± 2% labeling efficiency	Gamma-scintigraphy	7.0 ± 1.9% activity in periaortic lymph nodes	Videira *et al.*, 2002

lymphatic transport. The reason for enhanced oral bioavailability was due to higher lymphatic transport of SLN-LP. This avoids the CYP450 3A mediated first pass metabolism as well as P-gp efflux transport of LP.

In another study, radiolabelled compritol-SLNs with an average particle size of 200 nm were prepared and nebulized using an ultrasonic nebulizer (Videira *et al.*, 2002). By gamma scintigraphy, it was determined that the particles carried by aerosol droplets could reach the alveolar regions of the lung. Furthermore, the mechanism underlying the clearance of ^{99}Tc-HMPAO radiolabelled Compritol-SLN with respect to free ^{99}Tc-HMPAO was explored. Free ^{99}Tc-HMPAO was eliminated via blood circulation, while radiolabelled SLN were cleared from the lungs via the lymphatics to the periaortic, axillar and inguinal lymph nodes, probably after phagocytosis by macrophages. The elimination was quite rapid in the initial phase, as 50% of activity was cleared from the lungs after 13 min. Maximum values of activity in the lymph nodes were reached 40 min postnebulization. However, after 4 h of inhalation, 23.3% of activity was still visible in the lung, whereas the activity was 16% in the lymph nodes, which means that there was a significant retention of SLN in the lung.

Raloxifene HCl (RHCl) exhibits low bioavailability (< 2%) in humans due to extensive (> 90%) metabolism in the gut and poor intrinsic solubility. Solid lipid nanoparticles composed of glyceryl tribehenate were developed by Ravi *et al.* (2014) to improve the oral bioavailability of RHCl. Optimized RHCl-loaded SLN had a high encapsulation efficiency (> 92%) and mean particle size of 167 ± 3 nm. Upon oral administration, RHCl was loaded into glyceryl tribehenate, which underwent lymphatic delivery by two mechanisms: (1) stimulating chylomicron production and transport *via* triglyceride rich lipoproteins (2) through M-cells and gut-associated lymphoid tissue that consist of lymphoidal follicles forming Peyer's patches. To interpret the role of chylomicrons and lymphatic uptake in RHCl-SLN, rats were treated with cycloheximide (lymphatic transport inhibitor) and the pharmacokinetic parameters were evaluated. Cycloheximide (CHX) is known to obstruct the lymphatic uptake without disturbing other absorption pathways. Results obtained from lymphatic uptake study revealed that the plasma concentration of CHX-treated rats was remarkably lower than control group. In the treated rats group, AUC_{0-t} of RHCl was reduced by 29% and the maximum plasma concentration (C_{max}) was reduced by 34%. This could be elucidated by suppression of lymphatic uptake of SLN by CHX. This study exhibits the priority of lymphatic uptake in oral absorption of RHCl-SLN (Ravi *et al.*, 2014).

Zhang *et al.* (2012) prepared Candesartan cilexetil-encapsulated solid lipid nanoparticles (CC-SLN) with an aim to enhance their oral bioavailability. The fluorescence microscopy revealed that yellow fluorescence was detected in the caco-2 cells, which symbolizes that CC-SLN was scattered in the lysosomes and endoplasmic reticulum after their internalization. As endoplasmic reticulum is crucial in the lymphatic uptake of lipidic materials, the co-localization of CC-SLNs with the endoplasmic reticulum could be beneficial to its intestinal transport via the lymphatic pathway. To further explore the subsequent transport of CC-SLN after its uptake into the enterocytes, cycloheximide (CHX) was used to obstruct the lymphatic uptake without disturbing other absorption pathways. The experimental observation implied that the plasma concentration of CC in rats treated with cycloheximide was much lower when compared to rats treated with saline. When CC-SLN were injected intraduodenally to rats treated with cycloheximide (CHX), the AUC_{0-t} value of CC-SLNs decreased to about 30%, whereas the maximum plasma concentration (C_{max}) of CC from CC-SLN was notably reduced by 40%. This could be associated to the hinderance of intestinal lymphatic uptake by lymphatic transport inhibitors (CHX). Hence, it could be reasoned that the lymphatic delivery plays a significant role in the intestinal uptake of CC-SLN into blood circulation (Zhang *et al.*, 2012).

The poor orally bioavailable lopinavir was loaded in glyceryl behenate-based solid lipid nanoparticles (LP-SLN) for intestinal lymphatic transport, which is to be given in highly active antiretroviral therapy (HAART). Hot homogenization followed by ultrasonication techniques were used to develop lopinavir-encapsulated solid lipid nanoparticles having an average particle size of 230 nm, polydispersity index < 0.27 and zeta potential of approximately -27 mV. It was evident from the intestinal lymphatic uptake study that excipients of SLN enhanced the concentration of lopinavir in lymph, which was approximately 4.91-fold higher when compared to the lopinavir drug solution using 0.5% (w/v) methyl cellulose as a suspending agent (Alex *et al.*, 2011).

The effect of the intestinal efflux process and lipid vehicle containing paclitaxel (PC) was studied in mesenteric lymph-duct-cannulated anesthetized rat models. For this, the intraduodenal injection of PC-solution alone, PC-solution pretreated with a P-gp inhibitor (*i.e.* verapamil) and PC mixed with a 1:2 (w/w) mixture of glyceryl monooleate (GMO):linoleic acid (LA) was administered. It was suggested that the GMO:LA greatly increased the lymphatic uptake of PC, but did not enhance the oral bioavailability of PC. However, the prior administration of the P-gp inhibitor significantly increased the oral bioavailability by 1.8-fold, with a 3.5-fold enhanced lymphatic uptake. The enhancement was observed only with joint influence of P-gp and the coadministration of the GMO:LA mixture (Cai *et al.*, 2016).

Cavalli *et al.* (2003) confirmed the transmucosal uptake of tobramycin loaded solid lipid nanoparticles (Tb-SLN) to the lymphatic system. The lymphatic uptake of Tb-SLN was detected by the transmission electron microscopy, which reveals that Tb-SLNs were localized in plasma and lymph after 21 h of administration. Tobramycin-loaded SLNs developed were uniform in shape and size. After the administration of Tb-SLNs, they maintain their size and morphological characteristics for not less than 21 h in both biological fluids. The TEM micrographs revealed that concentration of Tb-SLNs was significantly higher in lymph, even if the blood volume is higher. Tb-SLN (2.5% of Tb) follows lymph behaviour and further confirms the appearance of Tb-SLNs in lymph and exhibits a higher plasma concentration of tobramycin. After *in vivo* administration, the lymphatic network (including nodes and capillaries) could be considered as a factor to elucidate the non-identical behaviour of three types of Tb-SLNs. The quantity of Tb-SLN accumulated in the lymph nodes could enhance from Tb-SLN (5% of Tb) to Tb-SLN (1.25% of Tb) followed by a steady release of the retained solid lipid nanoparticles. To further support this hypothesis, a concentration of Tb in lymph and lymph nodes was analyzed at different time intervals. After 6 h of intraduodenal injection of Tb-SLN, concentration of Tb in lymph nodes was 25 mg/g, whereas in lymph it was 7 mg/g. At 21 h of intraduodenal administration, a remarkable amount of Tb-SLN (2.5% of Tb) was still available in the mesenteric lymph nodes (Cavalli *et al.*, 2003).

The lipidic nanospheres of tripterine (an anticancer drug) was developed by Zhang *et al.* (2014) with the objective to enhance the intrinsic solubility, oral bioavailability and to find out the best possible mechanism responsible for oral absorption of tripterine. From the above findings, it was concluded that the ameliorated intestinal permeability and lymphatic uptake were responsible for the improved oral absorption of lipidic nanospheres. Lipidic formulations preferentially enter the blood circulation *via* the lymphatic route. An invasive *in vivo* technique (chylomicron flow blockage technique), was recently enforced to evaluate the intestinal lymphatic uptake in the absence/presence of chylomicron inhibitors such as cycloheximide, pluronic L-81 and colchicines. The results suggested that the absorption of tripterine loaded lipidic nanospheres was remarkably inhibited by cycloheximide, suggesting that the lymphatic transport had a benefaction to the absorption of tripterine loaded lipidic nanospheres.

4.3.1.3 Liposomes

Liposomes (particle size 20–10,000 nm) are biodegradable in nature that have a core loaded with one or more bilayers consisted of synthetic/natural lipids. This core, due to its lipid-loving nature, finds its place for the lipophilic drug delivery. The liposomes loaded lipophilic drugs delivered into the lymphatic tissues after their administration through intraperitoneal, intramuscular or subcutaneous injection (Nishioka and Yoshino, 2001). Liposomes larger in size were mostly retained by the lymph nodes through physical filtration. However, the liposomes that were smaller in size passed unretained by the lymph nodes. Further, the lymphatic uptake of small-sized liposomes is high. The route of administration and surface charge of liposomes and were proclaimed to be crucial for the lymphatic transport of compounds. Also, after lymphatic transport, liposomes passed by the lymphatic vessels and reached the lymph nodes. A fraction of liposomes were retained there. The phagocytosis by macrophages is considered to be the most important mechanism of transport of colloidal particles in lymph nodes. Further, different varieties of liposomes were developed to attain improved targeting ability, such as immunoliposomes (Kim *et al.*, 1993), PEGylated liposomes (Torchilin, 1994) and galactosylated liposomes (Kim and Jeong, 1997). The exclusive facet of the liposomes is that the lipidic drugs can be incorporated in the phospholipid bilayer, whereas the hydrophilic drugs can be loaded in the aqueous layer. Liposomes are generally processed from synthetic as well as natural phospholipids and cholesterol, and may also include other proteins and lipids. The mean particle size of the liposomes ranges from 20 to 30 nm to several microns. This particle size range is variable relying upon the type of method followed and the lipidic material used for formulating liposomes. On the basis of structure, liposomes are broadly classified as small, unilamellar vesicles (size ranging from 25 to 100 nm) and large unilamellar vesicles (size ranging from 200 to 1000 nm). The exclusive and constructive facet of the liposomes exists in their recognition as a foreign body by the mononuclear phagocytic system. In earlier reports, it was mentioned that HIV mainly congregates in the macrophages and mononuclear phagocytic systems of the body; this forms the basis of targeting these reservoirs by encapsulating antiretroviral drugs in liposomes with minimal side effects of putative forms of the drug (Ojewole *et al.*, 2008). These vesicular carriers do have certain disadvantages, like poor stability; high-scale up-cost; low encapsulation efficiency; the presence of solvent residues in the final formulation, which is unacceptable owing to their possible toxicity both in the bloodstream and poor stability (Gupta and Jain, 2010). A list of various drugs and antigens that were used in lymphatic targeting using liposomes are summarized in Table 4.5.

Moghimi (2006) proposed that by inclusion of mPEG-phospholipid, methoxypoly (ethylene glycol) into the liposomal bilayer, the lymphatic distribution and rate of drainage of subcutaneous injection can be controlled. The most remarkable response was observed when 15 mol% mPEG-lipid (Average mol mass = 350 Da) was used to formulate liposomes. These lipidic vesicles were drained rapidly from the site of injection into the initial lymphatics in contrast to unmodified liposomes, and they were reserved positively by the scavengers of the regional lymph node. Modifying the molecular layout of surface mPEG and IgG-PEG chains to a 'nearly overlapped mushroom' regime further improved the target identification of immuno-PEG2000 liposomes without negotiating their rate of drainage from the interstitium. The lipidic bilayer was enriched with 20 mol% phosphatidylserine to optimize the lymph node retention of these vesicles. Moghimi and Moghimi (2008b) postulated that the subcutaneous injection of small-sized immuno-poly(ethyleneglycol)-liposomes, constituting 1 mol% PEG2000–phospholipid and 10 mol% mPEG350–phospholipid in their lipid bilayer and where IgG1 is combined to the distal end of PEG2000, not only drains quickly from the interstitial spaces into the initial lymphatic system, but they also deposit effortlessly among the lymph nodes draining the region when it is compared with non-PEG-bearing immunoliposomes where IgG is directly linked to the phospholipid. In their other study, Moghimi and Moghimi (2008a) reported that when immuno-poly(ethyleneglycol)-liposomes were injected interstitially, they were conveyed to the regional lymph node by permeating through the thin-walled and fenestrated lymphatic vessel. The defined Fc region of coupled antibodies present on the liposome surface promotes vesicular identification and clearance by lymph node macrophages *via* the Fc receptor.

Kaur *et al.* (2008) prepared zidovudine (ZVD)-loaded surface modified liposomes (SM liposomes). The surface of liposomes was modified by introduction of site-specific ligand (mannose) and the incorporation of charges (negative or positive) in order to increase localization to lymphatic – specifically to spleen and lymph – node. Ligand-coated surface modified liposomes were developed using mannose-terminated stearylamine (mannose conjugate) while negatively and positively charged nanosized surface modified liposomes (120 ± 10 nm) were developed using dicetylphosphate and stearylamine. All the surface modified liposomal formulations exhibited a biphasic ZVD release, whereas liposomes coated with mannose greatly reduced the drug release of ZVD compared to conventional liposome. The organ distribution pattern of the surface modified liposomes

TABLE 4.5
Liposomes as a Nanocarrier to Enhance the Lymphatic Transport of Drugs

S.No.	Formulation	Drug	Components: Lipid	Particle Size	Entrapment Efficiency	Lymphatic Transport Model	Lymphatic Transport	Remarks	Reference
1.	DOTAP-PEG-mannose liposomes (LP-Man)	OVA antigen	Istearoyl-sn-glycero-3-phosphoethanolamine-N-[amino (poly-ethylene glycol)-2000] (DSPE-PEG-NH_2) and -α-d-manno pyranosylphenylisothiocyanate	–	Mannose coated liposome showed highest antigen entrapment	Dynamic lymphatic trafficking in mice	2.5% uptake by lymph of Mannose coated liposomes	Liposomes accumulate in lymph and enhance the uptake of antigen by lymph	Wang *et al.*, 2014
2.	IgG1-PEG2000-liposomes	IgG_1	(4′-(4″-maleimidophenyl) butyroyl)-phosphatidylethanolamine (MPB-PE), mPEG350-distearoylp hosphatidylethanolamine (mPEG350-DSPE) and MPB-PEG2000-DSPE	126 ± 30 nm with PDI 0f 0.06	69 ± 16 µg rat IgG1 bound per µ mol phospholipid	Animals were sacrificed at various time points post liposome injection and associated radioactivity was measured	The IgG1-PEG2000-liposomes accumulate efficiently among the lymph nodes	Liposome deposition among the draining lymph nodes was enhanced dramatically	Moghimi and Moghimi, 2008b
3.	Surface engineered liposomes	Zidovudine	Stearylamine (SA) and dicetyl phosphate (DCP)	Mannose-coated (127 nm, PDI 0.12); negatively charged (128nm, PDI 0.073); positively charged (126 nm, PDI 0.098); Liposome (122nm, PDI 0.063)	Mannose-coated (20.03); negatively charged (22.41); positively charged (24.22); Liposome (18.45)	Lymph nodes were isolated from sacrificed rats at different time interval	Highest uptake was the highest for mannose coated SE liposomes (≈ 15% of free ZDV)	Mannose coated > negatively charged > positively charged > lip. In lymph	Kaur *et al.*, 2008
4.	Liposomes	Cefotaxime	Pro-lipo cuo® (proliposomal mixture)	326.4 ± 2.3nm; PDI 0.38 ± 0.01	23.9 ± 1.0%	Cannulation of the mesenteric duct	2-fold enhancement in lymphatic transport	Liposomes enhance the lymphatic transport	Sheue Nee Ling *et al.*, 2006

(Continued)

TABLE 4.5 (CONTINUED)
Liposomes as a Nanocarrier to Enhance the Lymphatic Transport of Drugs

S.No.	Formulation	Drug	Components: Lipid	Particle Size	Entrapment Efficiency	Lymphatic Transport Model	Lymphatic Transport	Remarks	Reference
5.	Immuno-PEG liposomes	IgG	Distearoylphosphatidylethanolamine	110–120 nm	–	Scintigraphy	Lymph node uptake was 3–4-fold as compared to unmodified	Liposomes decorated with longer surface mPEG chains exhibit faster drainage rates	Moghimi, 2006
6.	Liposomes	Adriamycin	Egg phosphatidyl choline and cholesterol	120 nm	98%	Tissues collected at different time interval of rabbits	2.894-fold enhancement in the mean volumes of axillary lymph node	Lymphatic chemotherapy appears to be a promising method to induce apoptosis in lymph node metastases	Ling *et al.*, 2005
7.	Liposomes	2,5,7,8-tetramethyl-2R-(4R,8R,12-trimethyltridecyl) chroman-6-yloxyacetic acid (α-TEA), and 9-nitro-camptothecin	1,2-dilauroyl-*sn*-glycerol-3-phosphocholine (DLPC)	–	–	Sacrificing BALB/c mice at different time intervals	The number of microscopic metastases in the axillary and brachial lymph nodes were significantly reduced in comparison to the control group	α-TEA + 9-NC inhibit microscopic metastases in lymph nodes	Lawson *et al.*, 2004
8.	[^{111}In]-avidin and [^{99m}Tc]-biotin-liposomes	Avidin and biotin	N-Biotinoyl Distearoyl Phosphoethanolamine	134 nm (0.079 PI)	Labelling efficiency 80 to 90%	Gama Scintigraphy in rats	3.3% uptake in mediastinal nodes by pleural injection	Mediastinal node targeting using the avidin/ biotin-liposome was enhanced	Medina *et al.*, 2004

(Continued)

TABLE 4.5 (CONTINUED)
Liposomes as a Nanocarrier to Enhance the Lymphatic Transport of Drugs

S.No.	Formulation	Drug	Components: Lipid	Particle Size	Entrapment Efficiency	Lymphatic Transport Model	Lymphatic Transport	Remarks	Reference
9.	Anti-HLA-DR immunoliposomes	Indinavir	Dipalmitoylphosphatidylcholine (DPPC) and dipalmitoylphosphatidylglycerol (DPPG), Distearoylphosphatidylethanolamine-[poly-(ethyleneglycol) 2000] maleimide (DSPE-PEG-MAL)	100–120 nm	Fab- conjugated to liposomes (5–15 µg/µmol of lipids)	Tissues collected at different time interval of C3H mice	126-fold accumulation of Indinavir using anti-HLA-DR immunoliposomes	Anti-HLA-DR immunoliposomes enhances the accumulation of drug in lymphatic tissue	Gagné *et al.*, 2002
10.	Liposomes bearing anti-HLA-DR Fab′ fragments	Anti-HLA-DR Fab′ fragments	Dipalmitoylphosphatidylcholine (DPPC) and dipalmitoylphosphatidylglycerol (DPPG), N-(4-(p-maleimidophenyl)butyryl)-dipalmitoylphosphatidylethanolamine (MPB-DPPE)	85 ± 6 nm	15–20% coupling of Fab′ fragments to liposomes	Tissues collected at different time interval of C3H mice	2.89-fold in brachial lymph nodes and 1.57-fold in cervical lymph nodes	Higher accumulation in the cervical and brachial lymph nodes	Dufresne *et al.*, 1999

revealed a noteworthy reduction in the free quantity of ZVD in the serum, whereas a remarkably increased concentration of ZVD was observed in the lymph nodes and spleen. Fluorescent microscopy recommended the improved localization and uptake of the surface modified liposomes in the spleen and lymph nodes.

Wang *et al.* (2014) developed liposomal vaccines for lymphatic delivery by encapsulating OVA antigens with targeting encapsulated OVA antigens with DTAP (1, 2-Dioleoyl-3-trimethylammonium-propane) cationic liposomes or DTAP-PEG-mannose liposomes (DPM-LP). The result of *in vivo* imaging revealed that most of the cationic liposomes localized near the injection site, whereas DPM-LP exclusively localized in draining the spleen and the lymph nodes, and also reinforced the uptake by resident antigen-presenting cells. In contrast, lymphatic-targeted (DPM-LP) vaccines obtained sustained antibody production and robust recall responses 3 months after priming, recommending lymphatic targeting rather than antigen depot-stimulated long-term memory responses. The improved long-term immunological memory by DPM-LP was associated to vigorous germinal centre responses as well as increased Tfh cells and central memory $CD4^+$ T cells in the secondary lymphoid organs.

Harrington *et al.* (2000) proclaimed enhanced delivery of anti-neoplastic drugs to tumour site using PEGylated liposomes as vehicles for subcutaneous and intratumoral injection. The pharmacokinetic performance and tissue distribution of intratumorally or subcutaneously administered PEGylated liposome-encapsulated ^{111}In-labeled diethylenetriaminepentaacetic acid (IDA) were compared with the unencapsulated IDA. The PEGylated liposomes improve the carrier residence time and concentration in the head and neck tumours was also increased in a nude mouse xenograph model.

The major problem associated with conventional liposomal formulations is that they are poorly retained in the draining lymph nodes at less than 2% of injected dose. In order to improve the retention of liposomes to lymph nodes, avidin-activated, biotin-coated liposomes were developed, resulting in the extended lymph node localization and prolonged lymphatic retention in rabbit models (Phillips *et al.*, 2000). Liposomes coated with biotin were administered subcutaneously into the hind feet of rabbits, accompanied by an adjacent subcutaneous injection of avidin, which leads to the development of liposomal aggregates in the lymphatic vessels. The retention of the liposomes was increased by 6.7- and 7.1-fold in the iliac and popliteal nodes, after 24 h of injection.

The liposomal preparation of adriamycin, when administered intravenously, has considerably enhanced the pharmacokinetic performance of drug as compared to standard intravenous therapy of adriamycin, along with improved half-life and AUC. Ling *et al.* (2005) correlated the subcutaneous administration of liposomal formulation of adriamycin (LP-AD) with the standard intravenous injection of doxorubicin in a rabbit breast cancer model. Female New Zealand rabbits were immunized with VX2 carcinomas and they were treated three times with 1 mg/kg equivalent of intravenous adriamycin, subcutaneous liposomal preparation of adriamycin or saline, until the axillary lymph nodes became 5 mm in diameter. After treatment with the liposomal formulation of adriamycin, the nodal volume was measured 2 days after the completion of treatment and it was compared between different treated groups. The results suggested that the treatment with liposomal preparation of adriamycin decreased the nodal volume by 1.8-fold as compared to adriamycin drug suspension.

Trubetskoy *et al.* (1998) demonstrated that when liposome-encapsulated drugs were injected subcutaneously, they got absorbed into the systemic circulation by the aid of manual massage at the site of injection. A manual massage for 5 min could induce the release up to 40% of the subcutaneously injected 200 nm ^{111}In-angiotensin II incorporated PEG-liposomes. Additionally, when PEG-shielded liposomes were compared with the unmodified liposomes, localization of PEG-shielded liposomes in the blood was increased due to the lower uptake by the lymph node macrophages. The concentration of drug-incorporated PEGylated liposomes and radioisotope-labelled liposomes in blood remains constant during observation period of 45 min, whereas the amount of the plain liposomes decreased shortly post-massage. The results also recommended that the pharmacokinetic profile and drug release pattern of the formulation could be inflected through the alteration of the liposomal surface structure.

Two formulation concerns which are to be considered when liposomal formulation is administered orally are intestinal absorbability and stability. Hashida *et al.* (1984) studied the *in vivo* absorption characteristics of liposome loaded with carboxyfluorescein and the concentration of liposomal formulation released in the lymph and plasma was compared with the free dye. No difference was observed in the concentration in lymph and plasma between the two formulations, suggesting that drug loaded liposomes unveiled poor absorption across the intestinal mucosa. Although, the coadministration of lipid-surfactant mixed micelles significantly improved the penetration of the drug loaded liposomes. This inductive effect was observed due to the interaction between the lipid-surfactant micelle and the lipid bilayer of the intestinal cell membrane.

Sheue Nee Ling *et al.* (2006) estimated a liposomal formulation for the oral delivery of cefotaxime in a rat model by comparing four forms of the drug: an aqueous-free drug, the physical mixture of the drug, empty liposomes and a liposomal formulation. The data suggested that the liposomal formulation of the cefotaxime showed a 2.7-fold enhancement in its oral bioavailability as compared to the aqueous drug, and a 2.3-fold enhancement for the physical mixture of drug. In addition, they also described that the liposomal formulation lead to a remarkable improvement of the lymphatic transportation of the cefotaxime as compared to other two formulations. Hence, the liposomal systems may be useful delivery system for poorly bioavailable hydrophilic drugs, which show low oral bioavailability, encouraging their lymphatic uptake in the intestinal lymph as well as their systemic bioavailability.

Antineoplastic drugs, when transported *via* the pulmonary route, could be localized in the lungs and draining lymphatics, while circumventing important issues correlated with systemic chemotherapy including: poor drug penetration into the lymphatic vessels and surrounding lymph nodes, serious non-targeted toxicities and first pass clearance in the case of oral delivery. Latimer *et al.* (2009) explored the utility of delivering a liposomal formulation of a vitamin E analogue (α-TEA) and paclitaxel in an aerosol, for the treatment of metastases and murine mammary tumours (Latimer *et al.*, 2009). The pulmonary delivery of sequential combination treatments remarkably lowers the mammary tumour burden as compared to controls as well as the individually administered liposomal drugs. In addition, the aerosolized combination regimen reduced the total number of cells staining positive for cell markers CD31 and Ki67, suggesting an enhanced inhibition of lymphatic and pulmonary metastases.

In another study, Lawson *et al.* (2004) correlated the anti-proliferative potential of a dilauroylphosphatidylcholine-based liposome transported α-TEA, 9-nitro-camptothecin (9-NC) and a combination therapy of 9-NC and α-TEA in a metastatic murine mammary tumour model. Each individual treatment and the combination treatment was entrapped in liposomes, and transported via an aerosol to treat the lungs and the surrounding lymph node metastases. The results of Ki-67 immunostaining suggested that there were fewer proliferative cells present in the animals treated with the combination therapy as compared to that of the animals treated with 9-NC alone. However, statistically there was no significant difference observed between the 9-NC + α-TEA group and α-TEA group alone. All three treatments (9-NC + α-TEA, 9-NC alone and the α-TEA alone) exhibited a high number of apoptotic cells by a TUNNEL assay, when compared to the empty liposome control group. Furthermore, the *in vivo* anticancer efficacy studies exhibited that the combination treatment therapy greatly obstructs the tumour progression as compared to each treatment alone, which leads to the extended survival rate.

Desormeaux and Bergeron (2005) synthesized immunoliposomes conjugated with antibodies (anti-HCA-DR) that were stabilized with PEG. These conjugates were found to efficiently deliver indinavir to the lymphoid tissues of female C3H for the effective treatment of HIV-1. The results suggested 2.9-fold enhanced accumulation of anti-HCA-DA immunoliposomes in the cervical lymph nodes of C3H mice. Further, the stabilization of these immunoliposomes with PEG enhanced the transportation of indinavir into the lymphoid tissue for 15 d post injection. In addition, a subcutaneous injection in a dose of 2 μmol lipid/gm tissue of PEG-stabilized immunoliposomes exhibited a 0.9 to 1.9 fold higher AUC in the mesenteric, cervical, inguinal, brachial and popliteal lymph nodes of C3H mice. Overall, the immunoliposomes of indinavir is reported as the efficient formulation to inhibit HIV-1 replication in cultured cells (Gagné *et al.*, 2002).

Medina *et al.* (2004) evaluated the mediastinal node uptake and pharmacokinetics of [^{99m}Tc]-biotin and [^{111}In]-avidin-liposomes with the help of scintigraphic imaging by administering either intraperitoneal or intrapleural injection. Results suggested that the higher movement of ^{99m}Tc-blue-biotin-liposomes into the mediastinal nodes occured when intraperitoneal injection was administered. After 12 h of injection, the mediastinal node retention was approximately 2-fold higher when 5 mg dose was used. The results revealed that when liposomes were administered peritoneal, they were cleared rapidly from the cavity *via* the peritoneal lymphatic in the diaphragm. The results of biodistribution study of [^{111}In]-avidin at 44 h revealed 1.3% uptake in mediastinal nodes by intraperitoneal injection *vs.* 3.3% with intrapleural injection.

4.4 CONCLUSIONS

The increasing development of highly lipophilic drugs, along with the increasing interest in understanding the mechanisms by which drugs access the lymph, is the main cause for a development of intestinal lymphatic drug transport. It has lately been identified as a suitable alternative to the enhanced bioavailability and systemic exposure of extensive first-pass metabolized drugs. Further efforts need to be undertaken upfront to unravel novel and improved drug delivery strategies using lymphatic transport in order to circumvent this arduous challenge.

REFERENCES

Agellon LB, Toth MJ, Thomson AB. Intracellular lipid binding proteins of the small intestine. *Mol Cell Biochem.* 2002; 239: 79–82.

Alex MA, Chacko A, Jose S, Souto E. Lopinavir loaded solid lipid nanoparticles (SLN) for intestinal lymphatic targeting. *Eur J Pharm Sci.* 2011; 42: 11–18.

Arya A, Ahmad H, Tulsankar S, Agrawal S, Mittapelly N, Boda R, Bhatta RS, Mitra K, Dwivedi AK. Bioflavonoid hesperetin overcome bicalutamide induced toxicity by co-delivery in novel SNEDDS formulations: Optimization, in vivo evaluation and uptake mechanism. *Mater Sci Eng C.* 2017; 71: 954–64.

Baek JS, Cho CW. Surface modification of solid lipid nanoparticles for oral delivery of curcumin: Improvement of bioavailability through enhanced cellular uptake, and lymphatic uptake. *Eur J Pharm Biopharm.* 2017; 117: 132–40.

Bayliss MK, Butler J, Feldman PL, Green DV, Leeson PD, Palovich MR, Taylor AJ. Quality guidelines for oral drug candidates: Dose, solubility and lipophilicity. *Drug Discov Today.* 2016; 21: 1719–1727.

Benet LZ, Cummins CL, Wu CY. Unmasking the dynamic interplay between efflux transporters and metabolic enzymes. *Int J Pharm.* 2004; 277: 3–9.

Borst P, Evers R, Kool M, Wijnholds J. A family of drug transporters: The multidrug resistance-associated proteins. *J Natl Cancer Inst.* 2000; 92(16): 1295–1302.

Cai Q, Den X, Li Z, An D, Shen T, Zhong M. Effects of lipid vehicle and P-glycoprotein inhibition on the mesenteric lymphatic transport of paclitaxel in unconscious, lymph duct-cannulated rats. *Drug Deliv.* 2016; 23: 147–153.

Caliph SM, Charman WN, Porter CJ. Effect of short, medium, and long-chain fatty acid-based vehicles on the absolute oral bioavailability and intestinal lymphatic transport of halofantrine and assessment of mass balance in lymph-cannulated and non-cannulated rats. *J Pharm Sci.* 2000; 89: 1073–1084.

Carbone M, Duty S, Rattray M. Riluzole elevates GLT-1 activity and levels in striatal astrocytes. *Neurochem Int.* 2012: 60: 31–38.

Carrière F. Impact of gastrointestinal lipolysis on oral lipid-based formulations and bioavailability of lipophilic drugs. *Biochimie.* 2016; 125: 297–305.

Cavalli R, Bargoni A, Podio V, Muntoni E, Zara GP, Gasco MR. Duodenal administration of solid lipid nanoparticles loaded with different percentages of tobramycin. *J Pharm Sci.* 2003; 92: 1085–1094.

Charman W, Stella V. Transport of lipophilic molecules by the intestinal lymphatic system. *Adv Drug Deliv Rev.* 1991; 7: 1–14.

Chen Y, Li G, Wu X, Chen Z, Hang J, Qin B *et al.* Self-microemulsifying drug delivery system (SMEDDS) of vinpocetine: Formulation development and *in vivo* assessment. *Biol Pharm Bull.* 2008; 31: 118–125.

Dahan A, Hoffman A. Rationalizing the selection of oral lipid based drug delivery systems by an in vitro dynamic lipolysis model for improved oral bioavailability of poorly water soluble drugs. *J Control Release.* 2008; 129: 1–10.

Desormeaux A, Bergeron MG. Lymphoid tissue targeting of anti-HIV drugs using liposomes. *Methods Enzymol.* 2005; 391: 330–351.

Dufresne I, Désormeaux A, Bestman-Smith J, Gourde P, Tremblay MJ, Bergeron MG. Targeting lymph nodes with liposomes bearing anti-HLA-DR Fab′ fragments. *Biochim Biophys Acta Biomembr.* 1999; 1421: 284–94.

El-Laithy HM, Basalious EB, El-Hoseiny BM, Adel MM. Novel self-nanoemulsifying self-nanosuspension (SNESNS) for enhancing oral bioavailability of diacerein: Simultaneous portal blood absorption and lymphatic delivery. *Int J Pharm.* 2015; 490: 146–154.

Elnaggar YS, El-Massik MA, Abdallah OY. Self-nanoemulsifying drug delivery systems of tamoxifen citrate: Design and optimization. *Int J Pharm.* 2009; 380: 133–141.

Feeney OM, Crum MF, McEvoy CL, Trevaskis NL, Williams HD, Pouton CW *et al.* 50years of oral lipid-based formulations: Provenance, progress and future perspectives. *Adv Drug Deliv Rev.* 2016; 101: 167–194.

Fotiadis D, Kanai Y, Palacín M. The SLC3 and SLC7 families of amino acid transporters. *Mol Aspects Med.* 2013; 34: 139–158.

Fumagalli E, Funicello M, Rauen T, Gobbi M, Mennini T. Riluzole enhances the activity of glutamate transporters GLAST, GLT1 and EAAC1. *Eur J Pharmacol.* 2008; 578: 171–176.

Gagné JF, Désormeaux A, Perron S, Tremblay MJ, Bergeron MG. Targeted delivery of indinavir to HIV-1 primary reservoirs with immunoliposomes. *BBA-Biomembranes.* 2002; 1558: 198–210.

Garg B, Katare O, Beg S, Lohan S, Singh B. Systematic development of solid self-nanoemulsifying oily formulations (S-SNEOFs) for enhancing the oral bioavailability and intestinal lymphatic uptake of lopinavir. *Colloids Surf B.* 2016; 141: 611–622.

Geier EG, Schlessinger A, Fan H, Gable JE, Irwin JJ, Sali A, Giacomini KM. Structure-based ligand discovery for the Large-neutral Amino Acid Transporter 1, LAT-1. *Proc Natl Acad Sci USA.* 2013; 110: 5480–5485.

Giacomini KM, Huang SM. Transporters in drug development and clinical pharmacology. *Clin Pharmacol Ther.* 2013; 94: 3–9.

Griffin BT, O'Driscoll CM. A comparison of intestinal lymphatic transport and systemic bioavailability of saquinavir from three lipid-based formulations in the anaesthetised rat model. *J Pharm Pharmacol.* 2006; 58: 917–925.

Gupta U, Jain NK. Non-polymeric nano-carriers in HIV/AIDS drug delivery and targeting. *Adv Drug Deliv Rev.* 2010; 62: 478–490.

Harrington KJ, Rowlinson-Busza G, Syrigos KN, Uster PS, Vile RG, Stewart JSW. Pegylated liposomes have potential as vehicles for intratumoral and subcutaneous drug delivery. *Clin Cancer Res.* 2000; 6: 2528–2537.

Hashida M, Egawa M, Muranishi S, Sezaki H. Role of intramuscular administration of water-in-oil emulsions as a method for increasing the delivery of anticancer to regional lymphatics. *J Pharmacokinet Biopharm.* 1977; 5: 225–239.

Hashida N, Murakami M, Yoshikawa H, Takada K, Muranishi S. Intestinal absorption of carboxyfluorescein entrapped in liposomes in comparison with its administration with lipid-surfactant mixed micelles. *J Pharmacobiodyn.* 1984; 7: 195–203.

Hauss DJ, Fogal SE, Ficorilli JV, Price CA, Roy T, Jayaraj AA *et al.* Lipid-based delivery systems for improving the bioavailability and lymphatic transport of a poorly water-soluble LTB4 inhibitor. *J Pharm Sci.* 1998; 87: 164–169.

Hediger MA, Clémençon B, Burrier RE, Bruford EA. The ABCs of membrane transporters in health and disease (SLC series): Introduction. *Mol Aspects Med.* 2013; 34: 95–107.

Holm R, Porter CJ, Edwards GA, Müllertz A, Kristensen HG, Charman WN. Examination of oral absorption and lymphatic transport of halofantrine in a triple-cannulated canine model after administration in self-microemulsifying drug delivery systems (SMEDDS) containing structured triglycerides. *Eur J Pharm Sci.* 2003; 20: 91–97.

Hu G, Henke A, Karpowicz Jr RJ, Sonders MS, Farrimond F, Edwards R, Sulzer D, Sames D. New fluorescent substrate enables quantitative and high-throughput examination of vesicular monoamine transporter 2 (VMAT2). *ACS Chem Biol.* 2013; 8: 1947–1954.

Hui DY, Labonté ED, Howles PN. Development and physiological regulation of intestinal lipid absorption. III. Intestinal transporters and cholesterol absorption. *Am J Physiol Gastrointest Liver Physiol.* 2008; 294: 839–843.

Imada C, Takahashi T, Kuramoto M, Masuda K, Ogawara Kl, Sato A *et al.* Improvement of Oral Bioavailability of N-251, a Novel Antimalarial Drug, by Increasing Lymphatic Transport with Long-Chain Fatty Acid-Based Self-Nanoemulsifying Drug Delivery System. *Pharm Res.* 2015; 32: 2595–2608.

Iosio T, Voinovich D, Perissutti B, Serdoz F, Hasa D, Grabnar I *et al.* Oral bioavailability of silymarin phytocomplex formulated as self-emulsifying pellets. *Phytomedicine.* 2011; 18: 505–12.

Jannin V, Musakhanian J, Marchaud D. Approaches for the development of solid and semi-solid lipid-based formulations. *Adv Drug Deliv Rev.* 2008; 60: 734–746.

Kaur CD, Nahar M, Jain, NK. Lymphatic targeting of zidovudine using surface-engineered liposomes. *J Drug Target.* 2008; 16: 798–805.

Khan AA, Mudassir J, Mohtar N, Darwis Y. Advanced drug delivery to the lymphatic system: Lipid-based nanoformulations. *Int J Nanomedicine.* 2013; 8(1): 2733–2744.

Khoo SM, Shackleford DM, Porter CJ, Edwards GA, Charman WN. Intestinal lymphatic transport of halofantrine occurs after oral administration of a unit-dose lipid-based formulation to fasted dogs. *Pharm Res.* 2003; 20: 1460–1465.

Kim CK, Choi YJ, Soo-Jeong L, Lee MG, Lee SH, Hwang SJ. Lymph node targeting and pharmacokinetics of [^{3}H] methotrexate-encapsulated neutral large unilamellar vesicles and immunoliposomes. *Int J Pharm.* 1993; 98: 9–18.

Kim CK, Jeong EJ. Enhanced lymph node delivery and immunogenicity of hepatitis B surface antigen entrapped in galactosylated liposomes. *Int J Pharm.* 1997; 147: 143–151.

Latimer P, Menchaca M, Snyder RM, Yu W, Gilbert BE, Sanders BG *et al.* Aerosol delivery of liposomal formulated paclitaxel and vitamin E analog reduces murine mammary tumor burden and metastases. *Exp Biol Med.* 2009; 234: 1244–1252.

Lawson KA, Anderson K, Snyder RM, Simmons-Menchaca M, Atkinson J, Sun LZ *et al.* Novel vitamin E analogue and 9-nitro-camptothecin administered as liposome aerosols decrease syngeneic mouse mammary tumor burden and inhibit metastasis. *Cancer Chemother Pharmacol.* 2004; 54: 421–431.

Li F, Hu R, Wang B, Gui Y, Cheng G, Gao S *et al.* Self-microemulsifying drug delivery system for improving the bioavailability of huperzine A by lymphatic uptake. *Acta Pharm Sin B.* 2017; 7: 353–360.

Lin L, Yee SW, Kim RB, Giacomini KM. SLC transporters as therapeutic targets: Emerging opportunities. *Nat Rev Drug Discov.* 2015; 14: 543.

Lind ML, Jacobsen J, Holm R, Müllertz A. Intestinal lymphatic transport of halofantrine in rats assessed using a chylomicron flow blocking approach: The influence of polysorbate 60 and 80. *Eur J Pharm Sci.* 2008; 35: 211–218.

Ling R, Li Y, Yao Q, Chen T, Zhu D, Jun Y *et al.* Lymphatic chemotherapy induces apoptosis in lymph node metastases in a rabbit breast carcinoma model. *J Drug Target.* 2005; 13: 137–142.

Liu J, Scolland D, Reilly R, Wu X, Johnston M. Effect of particle size on the lymphatic distribution of ^{111}indium-aminopolystyrene through intrapleural administration. *Lymphology.* 2008; 41: 153–160.

MacDougall IJ, Griffith R. Pharmacophore design and database searching for selective monoamine neurotransmitter transporter ligands. *J Mol Graph Model.* 2008; 26: 1113–1124.

Makwana V, Jain R, Patel K, Nivsarkar M, Joshi A. Solid lipid nanoparticles (SLN) of Efavirenz as lymph targeting drug delivery system: Elucidation of mechanism of uptake using chylomicron flow blocking approach. *Int J Pharm.* 2015; 495(1): 439–446.

Medina LA, Klipper R, Phillips WT, Goins B. Pharmacokinetics and biodistribution of [^{111} In]-avidin and [^{99m}Tc]-biotin-liposomes injected in the pleural space for the targeting of mediastinal nodes. *Nucl Med Biol.* 2004; 31: 41–51.

Moghimi M, Moghimi SM. Lymphatic targeting of immuno-PEG-liposomes: Evaluation of antibody-coupling procedures on lymph node macrophage uptake. *J Drug Target.* 2008a; 16: 586–590.

Moghimi SM. The effect of methoxy-PEG chain length and molecular architecture on lymph node targeting of immuno-PEG liposomes. *Biomaterials.* 2006; 27: 136–144.

Moghimi SM, Moghimi M. Enhanced lymph node retention of subcutaneously injected IgG1-PEG 2000-liposomes through pentameric IgM antibody-mediated vesicular aggregation. *BBA-Biomembranes.* 2008b; 1778: 51–55.

Nakamoto Y, Masayo F, Tetsuo N, Kimura T, Hitoshi S. Studies on pharmaceutical modification of anticancer agents. I. Enhancement of lymphatic transport of mitomycin C by parenteral emulsions. *Chem Pharm Bull (Tokyo).* 1975; 23: 2232–2238.

Negi JS, Chattopadhyay P, Sharma AK, Ram V. Development of solid lipid nanoparticles (SLNs) of lopinavir using hot self nano-emulsification (SNE) technique. *Eur J Pharm Sci.* 2013; 48: 231–239.

Nishioka Y, Yoshino H. Lymphatic targeting with nanoparticulate system. *Adv Drug Deliv Rev.* 2001; 47: 55–64.

O'Driscoll CM. Lipid-based formulations for intestinal lymphatic delivery. *Eur J Pharm Sci.* 2002; 15: 405–415.

Ojewole E, Mackraj I, Naidoo P, Govender T. Exploring the use of novel drug delivery systems for antiretroviral drugs. *Eur J Pharm Biopharm.* 2008; 70: 697–710.

Paliwal R, Rai S, Vaidya B, Khatri K, Goyal AK, Mishra N *et al.* Effect of lipid core material on characteristics of solid lipid nanoparticles designed for oral lymphatic delivery. *Nanomedicine: NBM.* 2009; 5: 184–191.

Phillips WT, Klipper R, Goins B. Novel method of greatly enhanced delivery of liposomes to lymph nodes. *J Pharmacol Exp Ther.* 2000; 295: 309–313.

Porter CJ, Pouton CW, Cuine JF, Charman WN. Enhancing intestinal drug solubilisation using lipid-based delivery systems. *Adv Drug Deliv Rev.* 2008; 60: 673–691.

Porter CJ, Trevaskis NL, Charman WN. Lipids and lipid-based formulations: Optimizing the oral delivery of lipophilic drugs. *Nat Rev Drug Discov.* 2007; 6: 231–248.

Pouton CW, Porter CJ. Formulation of lipid-based delivery systems for oral administration: Materials, methods and strategies. *Adv Drug Deliv Rev.* 2008; 60: 625–637.

Ravi PR, Aditya N, Kathuria H, Malekar S, Vats R. Lipid nanoparticles for oral delivery of raloxifene: Optimization, stability, in vivo evaluation and uptake mechanism. *Eur J Pharm Biopharm.* 2014; 87: 114–124.

Reddy LH, Sharma R, Chuttani K, Mishra A, Murthy R. Influence of administration route on tumor uptake and biodistribution of etoposide loaded solid lipid nanoparticles in Dalton's lymphoma tumor bearing mice. *J Control Release.* 2005; 105: 185–198.

Sandhu PS, Kumar R, Beg S, Jain S, Kushwah V, Katare OP, Singh B. Natural lipids enriched self-nano-emulsifying systems for effective co-delivery of tamoxifen and naringenin: Systematic approach for improved breast cancer therapeutics. *Nanomed Nanotech Biol Med.* 2017; 13(5): 1703–1713.

Schinkel AH, Jonker JW. Mammalian drug efflux transporters of the ATP binding cassette (ABC) family: An overview. *Adv Drug Deliv Rev.* 2003; 55(1): 3–29.

Schlessinger A, Yee SW, Sali A, Giacomini KM. SLC classification: An update. *Clin Pharmacol Ther.* 2013; 94: 19–23.

Sheue Nee Ling S, Magosso E, Abdul Karim Khan N, Hay Yuen K, Anne Barker S. Enhanced oral bioavailability and intestinal lymphatic transport of a hydrophilic drug using liposomes. *Drug Dev Ind Pharm.* 2006; 32: 335–345.

Singh I, Swami R, Khan W, Sistla R. Lymphatic system: A prospective area for advanced targeting of particulate drug carriers. *Expert Opin Drug Deliv.* 2014; 11: 211–229.

Stremmel W, Pohl J, Ring A, Herrmann T. A new concept of cellular uptake and intracellular trafficking of long-chain fatty acids. *Lipids.* 2001; 36: 981–989.

Sun M, Zhai X, Xue K, Hu L, Yang X, Li G *et al.* Intestinal absorption and intestinal lymphatic transport of sirolimus from self-microemulsifying drug delivery systems assessed using the single-pass intestinal perfusion (SPIP) technique and a chylomicron flow blocking approach: Linear correlation with oral bioavailabilities in rats. *Eur J Pharm Sci.* 2011; 43: 132–140.

Thakkar H, Nangesh J, Parmar M, Patel D. Formulation and characterization of lipid-based drug delivery system of raloxifene-microemulsion and self-microemulsifying drug delivery system. *J Pharm Bioall Sci.* 2011; 3: 442.

Thomas N, Müllertz A, Graf A, Rades T. Influence of lipid composition and drug load on the *in vitro* performance of self-nanoemulsifying drug delivery systems. *J Pharm Sci.* 2012; 101: 1721–1731.

Torchilin VP. Immunoliposomes and PEGylated immunoliposomes: Possible use for targeted delivery of imaging agents. *Immunomethods.* 1994; 4: 244–258.

Trevaskis NL, Shackleford DM, Charman WN, Edwards GA, Gardin A, Appel-Dingemanse S *et al.* Intestinal lymphatic transport enhances the post-prandial oral bioavailability of a novel cannabinoid receptor agonist via avoidance of first-pass metabolism. *Pharm Res.* 2009; 26: 1486

Trubetskoy VS, Whiteman KR, Torchilin VP, Wolf GL. Massage-induced release of subcutaneously injected liposome-encapsulated drugs to the blood. *J Control Release.* 1998; 50: 13–19.

Ulanovskaya OA, Cui J, Kron SJ, Kozmin SA. A pairwise chemical genetic screen identifies new inhibitors of glucose transport. *Chem Biol.* 2011; 18: 222–230.

Urquhart BL, Kim RB. Blood-brain barrier transporters and response to CNS-active drugs. *Eur J Clin Pharmacol.* 2009; 65(11): 1063–1070.

Videira MA, Botelho M, Santos AC, Gouveia LF, De Lima JP, Almeida AJ. Lymphatic uptake of pulmonary delivered radiolabelled solid lipid nanoparticles. *J Drug Target.* 2002; 10(8): 607–613.

Wang C, Liu P, Zhuang Y, Li P, Jiang B, Pan H *et al.* Lymphatic-targeted cationic liposomes: A robust vaccine adjuvant for promoting long-term immunological memory. *Vaccine.* 2014; 32: 5475–5483.

Wang S, Yang R, Yao H, Zhou G, Zhang Y, Yang B *et al. In vivo* lymphatic targeting of methylene blue with microemulsion and multiple microemulsion. *Drug Deliv.* 2009; 16: 371–377.

Yáñez JA, Wang SW, Knemeyer IW, Wirth MA, Alton KB. Intestinal lymphatic transport for drug delivery. *Adv Drug Deliv Rev.* 2011; 63: 923–942.

Zhang X, Zhang T, Zhou X, Liu H, Sun H, Ma Z *et al.* Enhancement of oral bioavailability of tripterine through lipid nanospheres: Preparation, characterization, and absorption evaluation. *J Pharm Sci.* 2014; 103: 1711–1719.

Zhang Z, Gao F, Bu H, Xiao J, Li Y. Solid lipid nanoparticles loading candesartan cilexetil enhance oral bioavailability: *In vitro* characteristics and absorption mechanism in rats. *Nanomedicine: NBM.* 2012; 8: 740–747.

5 Nanocarriers in Early Diagnosis of Cancer

Gurpal Singh, Rajneet Kaur Khurana, Atul Jain, Taranvir Kaur, and Bhupinder Singh**

CONTENTS

5.1 Introduction 98
5.2 Cancer and Pathophysiology of Cancer 98
5.3 Present Conventional Diagnostic Methods 100
    5.3.1 Physical Examination 100
    5.3.2 Lab Test 100
        5.3.2.1 Blood Tests Used to Diagnose Cancer 100
        5.3.2.2 Urine Tests for Diagnosis 100
    5.3.3 Imaging Techniques 100
        5.3.3.1 Computerized Tomography 100
        5.3.3.2 Nuclear Scan (Radionuclide Scan) 101
        5.3.3.3 Positron Emission Tomography 101
        5.3.3.4 Magnetic Resonance Imaging 101
        5.3.3.5 Ultrasound Imaging 101
        5.3.3.6 Fluoroscopy 101
        5.3.3.7 Mammography 102
        5.3.3.8 Pathological Techniques 102
5.4 Nanotechnology 103
    5.4.1 Iron Oxide Nanoparticles 103
    5.4.2 Gold Nanoparticles 104
    5.4.3 Carbon Nanotubes 104
    5.4.4 Quantum Dots 104
        5.4.4.1 Quantum Dot-Doped Silica Nanoparticles 106
    5.4.5 Nanocantilevers 106
    5.4.6 Nanowires 107
5.5 Biomarkers of Cancer 107
5.6 Bioconjugation 107
    5.6.1 Direct Absorption 107
    5.6.2 Covalent Coupling 115
        5.6.2.1 Carboxylic Group Involved Coupling 115
        5.6.2.2 Amine Group-Involved Coupling 115
        5.6.2.3 Thiol Group-Involved Coupling 115
5.7 Conclusions 115
Acknowledgments 117
References 117

* Corresponding authors.

5.1 INTRODUCTION

Cancer still remains one of the world's most horrifying diseases, as the number of patients affected by it continues to increase over the years. The cancer cases reported in 2008 are expected to increase from 12.7 million to 20.3 million in 2030. About 7.6 million deaths are happening every year worldwide because of cancer. The disease is recognized by the condition wherein abnormal cells divide without any control, and are able to invade other tissues. Some of the most commonly diagnosed cancers around the world include lung cancer (1.8 million), liver cancer (0.8 million), colorectal (1.4 million), stomach cancer (0.7 million) and breast cancer (1.7 million) (Bray *et al.*, 2013; globocan.iarc.fr). However, with time, significant understanding has been gained in the management of cancer, though it still remains to be a terminal (i.e., fatal) disease.

Present conventional imaging techniques such as computer tomography, ultrasound imaging, magnetic resonance, nuclear scan, fluoroscopy and mammography are widely used in the detection of various cancers. These techniques are, however, not capable of quantifying the actual tumor volume in the specified area, but only help in locating and visualizing the tumor before treatment. Further, such techniques are also quite expensive to avail, can only be carried out by trained professionals and cannot be taken to field conditions. Additionally, the use of radiolabeled components is also in practice, which is toxic for the body. Moreover, these techniques are less sensitive and less accurate for early cancer detection (Kumar *et al.*, 2008; Choi *et al.*, 2010; Xing *et al.*, 2010). Although an invasive biopsy after imaging followed by a histopathological examination is a preferable method, it is quite painful. Being invasive, it is important to ensure that it is carried by skilled professionals, and is not very helpful in the early diagnosis of cancer (Daltrey and Kissin, 2000; Zagouri *et al.*, 2008; Gong, 2013). A noninvasive technique for the detection of cancer is in its early stages is of great interest since early cancer diagnosis, in combination with precise cancer therapies, and could significantly increase the survival rate of patients. Nanomedicine, an emerging multidisciplinary research area that integrates chemistry, engineering, biology and medicine with nanotechnology, has the potential to provide novel diagnostic tools for detection of primary cancers at their earliest stages, and to provide improved therapeutic protocols (Erickson *et al.,* 2008; Joshua *et al.*, 2011). Nanoscale imaging technology significantly improves the precision and accuracy of tumor diagnosis, and nanocarriers are able to significantly improve the accuracy of dose and targeted drug delivery and reduce the toxic side effects.

Nanotechnology has immense prospects of application in the diagnosis and treatment of cancers. The basic rationale is that nanotechnology has opened up the possibility of controlling and manipulating structures at the molecular level, leading to the creation of nanomaterials of novel surface architectures and properties. Attaching antibodies/targeting agents (such as receptor, ligands) to the surface of nanocarriers in order to achieve the specific targeting of cancerous cells is a promising modality for therapeutic and diagnostic oncology that will take the disease treatment to different dimension (Chapman *et al.*, 1999; Carter, 2006; Alyssa *et al.*, 2015). Nanotechnology is in a unique position to transform cancer diagnostics to produce a new generation of detection techniques with higher reliability, sensitivity and specificity. These are some of the basic reasons owing to which nanotechnology has been considered as a revolutionary step in the diagnostics and treatment of cancer (Sun *et al.*, 2015). The chapter endeavors to provide a brief review on cancer, its pathophysiology, use of conventional imaging techniques, different nanoparticles (NPs) used for diagnosis in cancer, biomarkers of cancer, as well as various bioconjugation techniques employed for the purpose.

5.2 CANCER AND PATHOPHYSIOLOGY OF CANCER

Cancer is a disease where abnormal cells divide aggressively without any control, and are able to invade other tissues. The disease can attack anywhere and anytime in the body. Normally, human cells grow and divide to form new cells, and the older cells are constantly replaced with new cells. However, as the deadly disease develops within the body, the orderly process seems to break down.

The cells thus become abnormal, as damaged cells still survive within the body with the growth of further new cells, when these are not actually needed. These extra cells divide rapidly without stopping, leading eventually to the growth of tumors.

Cancer is named on the organ or type where it happens within the body, e.g., if the cancer begins in colon, then it is usually called colon cancer. The tumor can either be benign or malignant. Benign tumors are not considered as cancerous, as the body cells appear to be close to normal in appearance. They grow slowly and also do not invade tissues or spread to other parts of the body. On the other hand, malignant tumors spread and invade into huge numbers, leading to the fast growth of tumors. These tumors invade into different parts of the body, i.e., through the blood, lymph system, and form new tumors (Stojan *et al.*, 2004).

The human body comprises of over a trillion cells which divide at a standard rate and pace. While cancer is developed in the body, normal cells get transformed into the cancer cells. Cancer cells have a different DNA than healthy cells, which can cause great harm to the body. The beginning of a cell getting transformed into cancer cell begins with change of proto-oncogenes to oncogenes. Proto-oncogenes are genes which are important to maintain normal cell growth within human body. However, on the contrary, oncogenes are genes which cause cell growth to change and get faster. While an individual is suffering from cancer, cells grows abnormally and also divide very quickly. The second step towards creating a cancer cell is turning off the tumor suppressor genes, which help in the stoppage of cancer from propagating in the healthy cells. These genes slow down cell growth, while turning these off leads to abnormal cell growth and the quick division of cells. The last step in the formation of a cancer cell is when DNA repair genes are turned off. DNA repair genes are quite helpful for ensuring healthy cells to know if something is wrong with its DNA and how to fix it (Figure 5.1). However, once the DNA repair genes are turned off, the cell is not able to detect when it is sick and is, therefore, unable to fix any problems with its DNA (Knudson, 2001; Carlo and Croce, 2008).

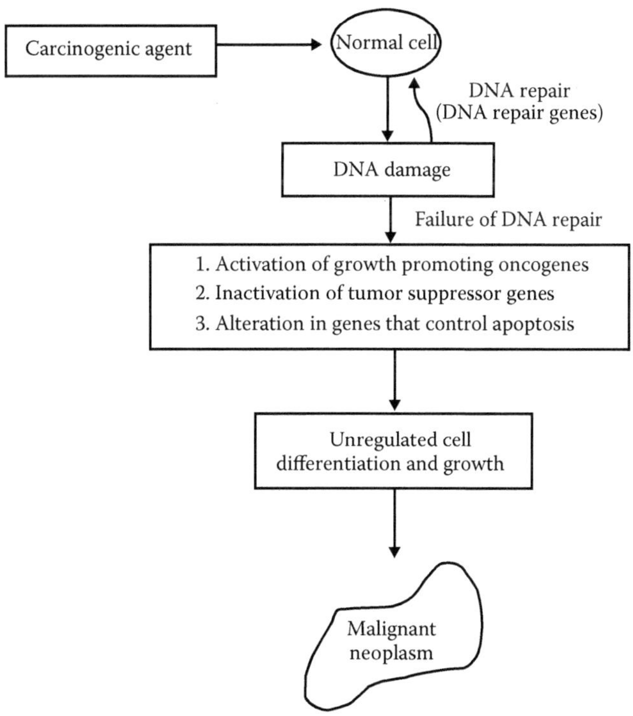

FIGURE 5.1 Various steps in the progression of a malignant neoplasm.

5.3 PRESENT CONVENTIONAL DIAGNOSTIC METHODS

The conventional methods of detecting cancer include the following:

5.3.1 Physical Examination

This is typically the first screening method, which is recommended for further testing. However, this method is not considered as accurate and sensitive. If the doctor suspects cancer, then the following conditions should exist:

a. Suspected mass of tissue or fluid accumulation developed in the abdominal area
b. Any sign of jaundice observed through skin and eye color
c. Suspected abnormal size of liver
d. Cyst formation in any part of body which is hard but not painful

5.3.2 Lab Test

High or low levels of certain substances within the body can act as indicators of cancer. Hence, it is important to conduct laboratory tests with blood, urine and other body fluids, which can measure these substances and help the physicians in arriving at a proper diagnosis. However, it should be noted that laboratory tests, though vital tools, cannot always be relied upon as a definitive indication for the detection of cancer (mayoclinic.org).

5.3.2.1 Blood Tests Used to Diagnose Cancer

a. *Complete blood count (CBC)*: This test measures the amount of different types of blood cells within a sample of blood, and can be used for detecting cancer, if a high rate of abnormal cells is detected.
b. *Blood protein testing*: This test is conducted to examine the various proteins within the blood and helps in detecting abnormal immune system proteins that at times get elevated in certain people with multiple myeloma.
c. *Circulating tumor cell tests*: The experimental blood test is conducted to find cells which have broken away from the original site and are floating in the bloodstream.

5.3.2.2 Urine Tests for Diagnosis

A cytology examination of urine is conducted to detect cancer and other diseases which might be a part of urinary tract. The urinary sample from the patient is studied under microscope to look for growth of abnormal cells. These abnormal cells in urine might lead to inflammation and also can be considered as signs of cancer in kidney, bladder or ureter (Smith *et al.*, 2014).

5.3.3 Imaging Techniques

Imaging is used for investigating the interior of the body noninvasively using a computer, which helps in locating the abnormal tissue masses in different locations in the body.

5.3.3.1 Computerized Tomography

This test is conducted after a contrast agent like iodine dye is administered through intravenous injection. The scanner sends X-rays across the designated parts of the body, which helps in studying the visible cancerous mass within the body through use of X-ray films. This computerized tomography (CT) scan can be used for studying different parts of the body, such as chest, belly, pelvis, leg and arm. The cross-sectional images captured at the time of CT scan, can be further reformatted in multiple planes and three-dimensional (3-D) images. These images can be viewed on computer monitor, printed on film, or even transferred onto the apt electronic media. This method is considered as one of the best methods for detecting cancers, since the images help the physicians to diagnose the tumor, determine its size and also its location within the body (Joshua *et al.*, 2011).

5.3.3.2 Nuclear Scan (Radionuclide Scan)

In order to conduct this scan, an injection of a radioactive material (tritium, technetium, promethium, etc.) is administered within the patient. The radioactive material flows through the blood stream of patient and gets collected in certain bones and organs. The scanner then detects and measures radioactivity, and pictures of bones or organs are captured on the computer screen for further diagnosis. It is quite beneficial for getting images of organs or areas of the body which cannot be captured using normal X-ray machines. The growth of abnormal tissues within the body are clearly visible using these kinds of tests.

5.3.3.3 Positron Emission Tomography

Herein, the patient is given an injection of radiolabeled glucose. This glucose is absorbed by cancerous cells at a much faster rate than the normal tissues present within the body. The scanner can take detailed 3-D pictures of organs and other body parts wherein radiolabeled glucose has been absorbed. This method is used in clinical oncology (i.e., the medical imaging of tumors) and for the clinical diagnosis of certain diffuse brain diseases. Positron emission tomography (PET) is also considered as one of the most important tools for neuroimaging, cardiology, bacterial infectious diseases and also musculoskeletal imaging (Khan *et al.*, 2003; clinicalkey.com).

5.3.3.4 Magnetic Resonance Imaging

Herein, the patient is administered an injection of the contrast agent (i.e., iron oxide and gadolinium) and then asked to lie on a movable bed. The subject is moved into a magnetic resonance imaging (MRI) scanner that is surrounded with giant circular agent. The magnet creates a strong magnetic field by aligning protons when exposed to the beams of radio waves. The radio waves are then absorbed by a contrast agent containing cancerous tissues, which produces a signal. It is used to be detected by a receiver portion of MRI scanner. The scanner can provide different details about tissues, like the one with very similar structural densities, grey and white brain matter. MRI is also considered as an important research tool for neurological, musculoskeletal, liver, gastrointestinal and cardiovascular imaging (Giussani *et al.*, 2010; Masselli and Gualdi, 2012; Khan and Henderson, 2013).

5.3.3.5 Ultrasound Imaging

Herein, a thin coating of jelly is applied over the area of body, wherein the patient has to be imaged. With the use of handheld device, called a transducer, sound waves are directed through the skin towards specific tissues. As the sound waves are reflected back by the tissue, a pattern is formed by a wave, which creates two-dimensional (2-D) image of tissue on the computer. This technique is quite a good noninvasive imaging technique for imaging soft tissue organs without X-rays. Ultrasound is considered an important tool for imaging different organs of the body like the heart, pelvis, kidney, liver, reproductive organs, fetus and blood vessels (Jensen, 2007; Yuan *et al.*, 2013).

5.3.3.6 Fluoroscopy

In this method, the injection of an iodine-based contrast agent is administered to the patient. The patient to be tested is located in between the radiation source and the fluorescent screen. The radiation source emits X-rays to capture the live image of the internal body organs. Fluoroscopy is a commonly used imaging technique for pediatric patients, through which different surgeries can be performed. (Mahadevappa, 2001).

5.3.3.7 Mammography

Mammography is a specialized type of medical imaging that uses a low-dose X-ray system to visualize the interiors of breasts. Imaging with X-rays involves exposing a part of the body to a small dose of ionizing radiation to produce pictures of the inside of the body (Mark, 2010; Magnus *et al.*, 2015). There are three main types of advancements in the field of mammography, including:

a. *Digital mammography*: In this type of mammography, X-ray film is replaced with electronics that convert X-rays into mammographic pictures of the breast. The technique involves a low radiation dose. The images are then transferred to the computer that can be reviewed later on by the radiologist and can also be used for long-term storage (Mark, 2010; Magnus *et al.*, 2015).
b. *Computer-aided detection*: In this system, the system searched digitized mammographic images for abnormal areas of mass that may indicate the presence of cancer (Mark, 2010).
c. *Breast tomosynthesis*: This is an advanced form of breast imaging, where multiple images of the breast from different angles are captured and reconstructed into a 3-D image set (Mark, 2010).

5.3.3.8 Pathological Techniques

5.3.3.8.1 Biopsy

In this procedure, sample of the tissue is taken from the tumor and then subjected for histopathological examination. The procedure allows the pathologists to establish the nature of cells and their surface morphology to determine whether they are cancerous or not (Prasant *et al.*, 2014). The sample can be removed as follows:

a. Use of a thin needle attached with syringe to withdraw tissue or fluid
b. Use of an endoscope, a thin lighted tube inserted through the mouth, followed by use of a special tool by the surgeon to remove tissue or cell through the tube
c. Use of surgery, to be done in one of the two ways, *viz.* excisional biopsy (i.e., the surgeon takes out entire tumor and is subjected to histopathological examination), or incisional biopsy (i.e., the surgeon takes out part of tumor for histopathological examination)

5.3.3.8.2 Histochemistry

This method can be used by pathologist for identification of chemical compounds within and between biological cells with the usage of histology stains, indicators and a light (optical) and electron microscope (Helen and Jane, 2003).

5.3.3.8.3 Immunohistochemistry

In this method, the cell components (i.e., antigens) are combined with the special antibodies on the formalin-fixed paraffin sections on slide. The complex of antigen antibodies can be subjected to identification of light microscopy with the help of pathologist, either by a fluorescent dye or by an enzyme system (Helen and Jane, 2003).

These conventional imaging techniques for cancer diagnosis usually work at the tissue level, which has a low efficiency, and specifically fails to detect cancer cells at very early stages of the disease (Mark, 2010). Also, such conventional strategies are highly invasive, nonspecific, often associated with traumatic side effects to both tumor and healthy cells. Cancer has a good chance to be treatable and the prognosis could be favorable, provided it is detected early. Therefore, nanotechnology tools are being developed to detect cancer at early stages with improved sensitivity and specificity at tumor site. Advances in nanotechnology have demonstrated the promise of NPs for noninvasive tumor imaging (Lee *et al.*, 2007). Different nanodevices have been applied in cancer diagnosis such as gold NP, quantum dots, magnetic NP, carbon nanotubes and cantilevers are explained below.

5.4 NANOTECHNOLOGY

The unprecedented expansion in the field of nanomedicine for the development of newer NPs with a smaller size and large surface-area-to-volume ratio has emerged as a potential tool for the diagnosis and treatment of cancer. NPs help in binding, absorption, and carrying of drugs, DNA, RNA, proteins and probes with higher efficiency. Tunable size, shape and surface characteristics also enable them to have high stability, high carrier capacity and the ability to incorporate both hydrophilic and hydrophobic substances and compatibility with different administration routes, thereby making them highly attractive in many aspects of oncology. These engineered NPs with fluorescent probes with integrated imaging and carrier functionalities have become excellent tools for molecular diagnostics and delivery of therapeutics molecules (Thakor and Gambhir, 2013). With a reproducible particle size, narrow size distribution and simpler synthetic routes generating large-scale, cost-effective products, they are the important prerequisites for clinical translation (Gobbo *et al.,* 2015). Figure 5.2 represents a schematic diagram for commonly employed NPs.

The following sub-sections tend to highlight the various engineered and functionalized NPs carrying a potential dual role in diagnosis and treatment of cancer.

5.4.1 Iron Oxide Nanoparticles

Iron oxide NPs are a class of biomedical materials that are particularly suited for biomedical applications of cancer treatment and diagnosis. Iron oxide NPs with nanocrystalline magnetite (Fe_3O_4) cores have great potential for use in oncological medicine due to their biocompatibility,

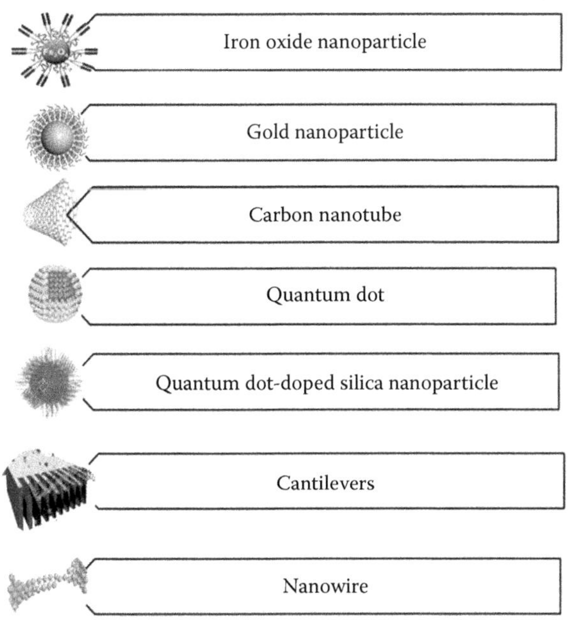

FIGURE 5.2 Schematic diagram for plausible engineered nanoparticles.

biodegradability, facile synthesis and ease with which they may be tuned and functionalized for specific applications. Additionally, spherical magnetite NPs with diameters less than approximately 20 nm will exhibit superparamagnetic behavior, a property that is exploited to enhance contrast in magnetic resonance imaging. Typically, superparamagnetic iron oxide nanoparticles (SPIONs) are comprised of a biocompatible coating that provides space for ample of functional groups for conjugation of additional tumor targeting and therapeutic moieties. Being an efficient contrast agents for MRI, iron oxide NPs hold significant potential both for imaging and image-guided delivery, as these can be detected with high sensitivity; the iron and the polymer components employed along with it are biocompatible and degradable (Pantic, 2010).

5.4.2 Gold Nanoparticles

Gold NPs are emerging as promising agents for cancer therapy and are being investigated as drug carriers, photothermal agents, contrast agents and radiosensitizers. Being powerful imaging labels and contrast agents, due to their unusual optical properties, biocompatibility, absence of cytotoxicity, resistance to photobleaching and absorbance at near-infrared light region (NIR), they have been successfully used in both cancer diagnosis and therapy (Fruscella *et al.*, 2016).

The important properties of gold NPs include:

- Small size
- Biocompatibility
- High atomic number (i.e., high-Z)
- Ability to bind with targeting agents

Being small, they can penetrate throughout the body, preferentially accumulating at tumor sites owing to the enhanced permeability retention effect. Contrast materials (such as gold) improve the definition of heavily vascularized tumors by increasing photoelectric photon absorption, increasing the accuracy of tumor diagnosis and staging and aiding volume and definition in radiotherapy planning (Jain *et al.*, 2012).

5.4.3 Carbon Nanotubes

Carbon nanotubes (CNTs) have attracted much attention by researchers worldwide in recent years owing to their small dimensions and unique architecture in diagnosis of diseases, particularly in nanodrug delivery (Singh *et al.*, 2015). Of the diverse applications of CNTs, recent findings have shown that nanotubes have been used as diagnostic tools for pharmaceutical applications (Beg *et al.*, 2011).

5.4.4 Quantum Dots

The flexible surface chemistry, unique optical properties, high sensitivity and multiplexing capabilities of quantum dots (QDs) certainly make them the most promising tool for diagnostic imaging. QDs not only act as a commendable alternative to magnetic and radioactive imaging, these engineered nanocrystals serve as excellent carriers for delivering chemotherapeutic drugs and genes for altering tumor cells. QDs are some of the extremely popular optical probes to label different components within a cell for detection and delivery purposes (Tripathi *et al.*, 2016). QDs hold massive capabilities for the detection of many cancer markers simultaneously and hold tremendous promise for sorting out the complex gene expression profiles of cancers, thereby providing accurate clinical diagnosis. Undoubtedly, QDs have become an indispensible tool for targeting, imaging, and performing specific therapeutic functions (Tripathi *et al.*, 2015) (Figure 5.3a and b).

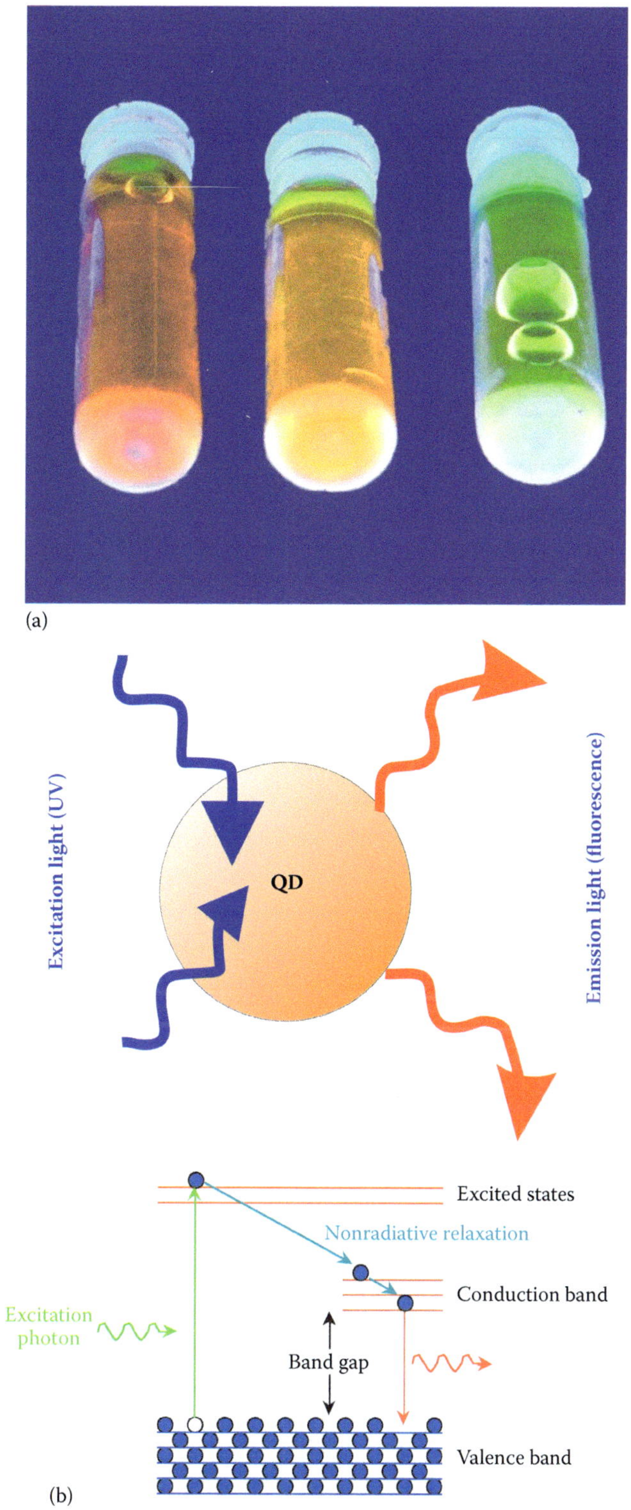

FIGURE 5.3 (a) CdTe QDs (i) 4.3 nm, orange; (ii) 3.8 nm, yellow; (iii) 3.7 nm, green; as seen under UV light (λ ~ 380 nm) (These QDs were synthesized in our lab; unreported). (b) Excitation and emission pattern of fluorescent QDs.

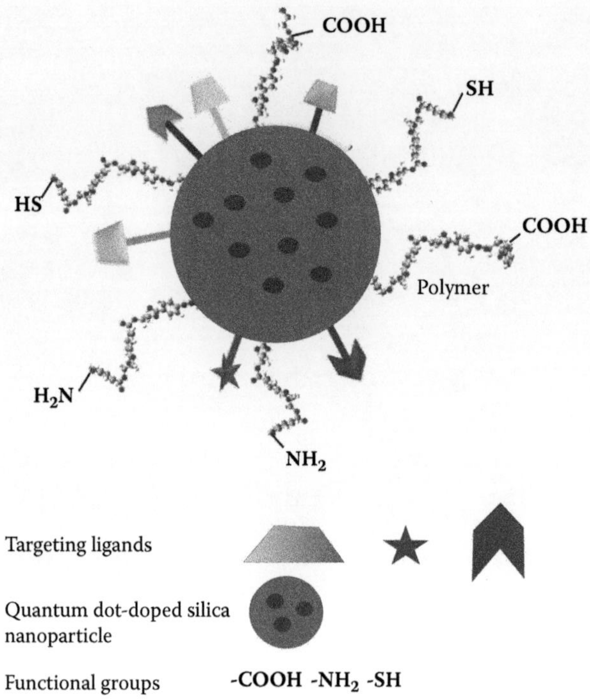

FIGURE 5.4 Fluorescent image of functionalized QD-doped silica nanoparticles.

5.4.4.1 Quantum Dot-Doped Silica Nanoparticles

Of late, fabricated silica NPs doped with QDs and superficially functionalized with amino, carboxylic, thiol and phosphonate groups have recently emerged as novel nanomaterials with a multivalent loading capacity for easy detection or effective drug delivery (Bottini *et al.*, 2007) (Figure 5.4). Nanocantilevers and nanowires are biomolecular sensors classified as label-free measurement technologies, meaning the binding of the target biomolecule to its surface-bound capture agent is directly detected (Heath and Davis, 2008).

5.4.5 Nanocantilevers

These are designed as part of a larger diagnostic device and can provide rapid and selective detection of cancer-related molecules (Datar *et al.*, 2009). Nanoscale cantilevers, constructed as part of a larger diagnostic device, can provide rapid and sensitive detection of cancer-related molecules. Nanoscale cantilevers are primarily built using semiconductor electron-beam lithography techniques. Such micron-sized devices, comprising many nanometer-sized cantilevers, can detect single molecules of DNA, RNA or protein (Bohunicky and Mousa, 2011). As a cancer cell secretes its molecular products, the antibodies coated on the cantilever fingers selectively bind to these secreted proteins. These antibodies have been designed to pick up one or more different, specific molecular expressions from a cancer cell. The physical properties of the cantilevers change as a result of the binding event. This change could be read in real time and provide not only information about the presence and the absence, but also the concentration of different molecular expressions (Hassanzadeh *et al.*, 2011). Biosensing technologies, based on cantilever arrays, have the potential of detecting multiple biomarkers exhibited during metastasis for multi-target detection with high sensitivity and selectivity using very small sample volumes. These arrays can be used as noninvasive and ultrasensitive bioassays, and provide an opportunity to increase the sensitivity of detection of tumor-associated antigens with a smaller sample size and at much earlier stages of disease progression *vis-à-vis* the current medical diagnostic technologies (van Vlerken and Amiji, 2006). These tiny levers, which are anchored at one end, can be engineered to bind to molecules that represent some of the changes associated with cancer. They may bind to altered DNA sequences or proteins that are present in certain types of cancer. By monitoring the bending of the cantilevers, the presence or absence of these molecules could be detected (Klein *et al.,* 2005).

5.4.6 Nanowires

Nanowires, having a high surface-to-volume ratio, are used as biosensors for disease detection. Various nanowires have also been applied to biomarker detection including silicon nanowires, Indium oxide (In_2O_3) nanowires, gold nanowires and conducting polymer nanowires (Liu *et al.,* 2016; Namdari *et al.,* 2016).

Semiconducting nanowires have exceptional physical, optical and electronic properties and excellent biocompatibility. Also, surface of the nanowires can be modified by with well-known methods. This advantage makes it a promising platform for sensitive detection of biomarkers (Choi *et al.,* 2010). Particularly, nanosized sensing wires are laid down across a microfluidic channel, as these have incredible properties of selectivity and specificity. As particles flow through the microfluidic channel, the nanowire sensors pick up the molecular signatures of these particles and can immediately relay this information through a connection of electrodes to the outside world. These nanodevices are man-made constructs made up with carbon, silicon and other materials, which have the capability to monitor the complexity of biological phenomenon and relay the information, as it is monitored to the medical care provider (Shamsudhin *et al.,* 2016). Table 5.1 presents some of the instances in the literature highlighting the utility of the nanotechnology in cancer diagnosis.

5.5 BIOMARKERS OF CANCER

Cancer markers help in detecting the presence of certain types of cancer in the body, and in monitoring the progress of cancer treatment. Cancer markers are substances found in the blood, body fluids or tissues that are produced by cancer cells. If a certain cancer marker is found in the body, it can indicate that the cancer is still present and ongoing treatment may still be recommended. Cancer biomarkers also helpful in establishing a specific diagnosis. This is particularly the case when there is a need to determine whether the tumors are of primary or metastatic origin. Some of the cancer biomarkers are enlisted in Table 5.2 (Tokumaru and Coyle, 1992; Salomon *et al.*, 1995; Alivisatos, 1996; Bruchez *et al.*, 1998; Chan and Nie, 1998; Hammarstrom, 1999; Greenlee *et al.*, 2000; Dubertret *et al.*, 2002; Sidransky, 2002; Suzuki *et al.*, 2008; Anajwala *et al.*, 2010; Xiaoxiao *et al.*, 2010; Jana *et al.*, 2014; labtestsonline.org).

5.6 BIOCONJUGATION

Colloidal inorganic NPs have been a key area of interest in biomedicine, especially when it comes to diagnostics, imaging and sensing. The surface of NPs needs to be modified in order to have enhanced biocompatibility and functionality for carrying out *in vitro* and *in vivo* applications. Surface functionalization can provide NPs with water-solubility, functionality and stability in a physiological environment. Later on, these NPs can be conjugated with biomolecules of interest. such as nucleic acid, proteins, peptide, folic acid and antibodies for its usage in biomedical applications (Figure 5.5). Various methods of conjugation have been carried out and some of the methods are explained below.

5.6.1 Direct Absorption

In this method, biomolecules of interest can interact directly with different kinds of NPs. Nonspecific staining of certain cellular structures may be achieved by injecting hydrophilic NPs into cytoplasm

TABLE 5.1
Some of the Literature Instances on the Utility of the Nanotechnology in Cancer Diagnosis

S. No	Nanosystem	Drug, Ligand, Coating Material	Imaging Modility	*In Vitro*/*In Vivo* Studies	Inference(s)	Reference
1	Thermally cross-linked SPION (TCL-SPION)	Anti-biofouling polymer Cy5.5 (dye) doxorubicin (Drug)	MRI	T2-weighted fast-spin echo and optical fluorescent images of major organs and antitumor efficacy of Dox@ TCL-SPION explored in LLC allograft animal model.	Tumorreduction was approximately 63% relative to control and 38% for the DOX group. Intense signal at target tumor site shows the highest uptake at tumor site and the lowest uptake in non-target liver site.	(Yu *et al.*, 2008)
2	Functionalized conjugated, and ^{64}Cu-labeled SPIONS	cRGD (Ligand) Doxrubicin (drug)	PET/MRI	*In vitro* cell uptake and cytotoxicity studies against U87MG cells were observed. *In vivo* studies conducted on female athymic nude tumor-bearing mice and their PET images were taken using a microPET/micro CT Inveon rodent model scanner.	*In vitro* studies revealed cRGD-conjugated SPIONs exhibited a higher cellular uptake than cRGD-free ones. Moreover, cRGD-conjugated SPIONs showed a much higher level of localization at tumor site.	(Yang *et al.*, 2011)
3	Surface-coated Bismuth-iron Oxide Nanocomposite (BIONs)	Dextran	CT and MRI successfully deployed	*In vitro* cell viability, cytotoxicity on Hep G2 and BJ5ta cells and biodegradation studies were investigated and *in vivo* CT/MRI and biodistribution were carried in wild-type C57BL/6J mice.	The developed BIONs were found to be biocompatible and biodegradable. Also possessed strong X-ray attenuation T2-weighted MR contrast properties.	(Naha *et al.*, 2014)
4	Polymer-conjugated SPIONs (HA-PEG10-SPIONs)	Methoxy PEG Hyaluronic acid Dopamine (coating on SPION)	MRI	Capability of both the HA-PEG10-SPIONs and HA-SPION (without PEG) for hyperthermic effect in head and neck cancer were studied. Cellular uptake and viability studies carried on SCC7 and NIH3T3 cell. *In vivo* MRI cancer diagnosis was also performed.	HA-SPION (without PEG) showed nearly 40% MR T2 contrast compared to the 20% MR T2 contrast of the HA-PEG10-SPION over a 3 h time period. Less than 40% cell viability of SCC7 cell line for both HA-SPIONs and HA-PEG10-SPIONs were observed. Specific targeting of HA to the CD44 was also observed.	(Thomas *et al.*, 2015)

(Continued)

TABLE 5.1 (CONTINUED)
Some of the Literature Instances on the Utility of the Nanotechnology in Cancer Diagnosis

S. No	Nanosystem	Drug, Ligand, Coating Material	Imaging Modility	*In Vitro*/*In Vivo* Studies	Inference(s)	Reference
5	Surface-functionalized (AuNP)	PEG (for stealth properties) Biotin (targeting agent) beta-cyclodextrin (drug pocket) Paclitaxel (drug)	Optic imaging	*In vitro* studies on cancer cells such as HeLa, A549, and MG63 were compared with NIH3T3 cells (normal cells).	CLSM measurements revealed Au nanoparticles with rhodamine B-linked beta-cyclodextrin has a higher affinity to cancer cells such as HeLa, A549, and MG63 compared to NIH3T3 cells. Significant anti-cancer effects were specifically observed againts HeLa cancer cells.	(Heo *et al.*, 2012)
6	Gold nanorods (GNR)	cRGD (Tumor-targeting ligands) 4,7-triazacyclononane-N, N′, N″-triacetic acid (NOTA), a macrocyclic ^{64}Cu chelator (for PET imaging) Doxorubicin (DOX)	PET	*In vitro* cell uptake, flow cytometric analysis of U87MG cancer cells and *in vivo* biodistribution studies using PET imaging performed.	*In vitro* studies showed cRGD-conjugated nanocarriers exhibited higher cellular uptake and cytotoxicity than non-targeted ones. However, similar *in vivo* biodistribution of GNR-DOX-cRGD and GNR-DOX (without ligand) observed through PET imaging.	(Xiao *et al.*, 2012)
7	Photosensitizer-loaded Gold Vesicles (GV-Ce6)	Ce6 (photosensitizer)	NIR fluorescence/thermal/photoacoustic imaging	Cellular uptake and localization studies of GV-Ce6 were done on human breast cancer cells (MDA-MB-435) using flow cytometer. *In vivo* therapeutic efficacy of GV-Ce6 was assessed by monitoring the growth rate of tumor in Athymic nude mice and NIR fluorescence/thermal/photoacoustic imaging were also performed.	High Ce6 loading efficiency in GVs, enhanced cellular uptake of Ce6, trimodality NIR fluorescence/thermal/PA imaging and synergistic PTT/PDT treatment with improved efficacy using a single wavelength continuous wave laser irradiation are the special feature of developed vesicles.	(Lin *et al.*, 2013)

(*Continued*)

TABLE 5.1 (CONTINUED)
Some of the Literature Instances on the Utility of the Nanotechnology in Cancer Diagnosis

S. No	Nanosystem	Drug, Ligand, Coating Material	Imaging Modility	*In Vitro*/*In Vivo* Studies	Inference(s)	Reference
8	Gold-coated SPIONS	MUC-1 aptamer	MRI	*In vitro* cell uptake, cytotoxicity, MRI, photothermal therapy were carried out on HT-29 human colon cancer cell line.	Aptamer-Au@SPIONs shown to have higher uptake in MUC-1 positive cells compared to MUC-1negative cells.	(Azhdarzadeh *et al.*, 2016)
9	Dendrimer-entrapped α-tocopheryl succinate AuNPs	α-tocopheryl succinate (TOS), PEGylated folic acid (FA)	CT	Flow cytometry and confocal analysis were done to check the specific binding and cell uptake of Au-TOS-FA DENPs to U87MG (a human glioma cell line) cells expressing high-level FA receptors. *In vitro* as well as *in vivo* targeted CT imaging of cells and BALB/c nude tumor-bearing mice were carried out.	Multifunctional Au DENPs exert the specific targeting as well as therapeutic action against FAR-over-expressing cancer cells.	(Zhu *et al.*, 2014)
10	Surface modified Fe_3O_4@Au core/shell nanostars (Fe_3O_4@Au-HA NSs)	Hyaluronic acid	MRI, CT, and thermal imaging	*In vitro* studies including cytocompatibility, cellular uptake of Fe_3O_4@Au-HA NSs by HeLa cells over-expressing CD44 receptors was investigated. *In vivo* photothermal ablation and MR and CT imaging of HeLa tumors were carried out. Photothermal ablation of tumors characterized by TUNEL staining and further confirmed by histological examination of tumor sections.	Fe_3O_4@Au-HA NSs found to be multifunctional nanoplatform for MR/CT imaging of cancer cells. Also, these can be used as a platform for photothermal therapy of cancer cells and as a competent theranostics of different types of CD44 over-expressing receptor cancer.	(Li *et al.*, 2015)
11	IONPs decorated fullerene (C_{60})	Polyethylene glycol (PEG2000) FA (ligand)	MRI	*In vitro* targeting potential against MCF-7 cells was explored and *in vivo* S180 tumor model was used for examining MRI, PDT and RTT.	C_{60}-IONP-PEG-FA showed no toxicity; possessed a powerful tumor diagnostic tool as well as selective photothermal ablation characteristics, along with RTT.	(Shi *et al.*, 2014)

(*Continued*)

TABLE 5.1 (CONTINUED)
Some of the Literature Instances on the Utility of the Nanotechnology in Cancer Diagnosis

S. No	Nanosystem	Drug, Ligand, Coating Material	Imaging Modility	*In Vitro/In Vivo* Studies	Inference(s)	Reference
12	PEG- oxidized multi-walled carbon nanotubes modified with ligand (O-MWNTs-PEG-ANG)	Angiopep-2 (ligand) for LRP receptor overexpressed on the BBB and glioma cells Doxorubicin	Fluorescence imaging	*In vitro* studies including cytotoxicity and intracellular tracking of FITC-O-MWNTs-PEG-ANG were evaluated. *In vivo* targeting potential and anti-glioma activity evaluated in BCEC and C6 cells Balb/c mice, bearing intracranial C6 glioma. Biological safety of DOX-O-MWNTs-PEG-ANG was evaluated by histopathological analysis.	Developed CNTs shown better anti-glioma effect than DOX. Histopathological analyses, showed lower cardiac toxicity of developed system w.r.t-free DOX.	(Ren *et al.*, 2012)
13	Graphene-based magnetic plasmonic nanocomposite	PEG	MR and X-ray imaging	*In vivo* imaging and cancer photothermal therapy carried on female BALB/c mice bearing 4T1 murine breast tumor.	Excellent tumor ablation and imaging demonstrated in 4T1 breast tumor-bearing mice.	(Shi *et al.*, 2013b)
14	IONPs decorated surface of fullerene (C_{60}) (C_{60}-IONP-PEG/ HMME)	PEG (2000) Hematoporphyrin monomethyl ether (HMME)	MRI	Assessment of cell viability and cell uptake of C_{60}-IONP-PEG/HMME in B16-F10 cancer cells were examined. Also, ROS productions were by using DCFH-DA fluorescent probe. *In vitro* and *in vivo* PDT, MRI and biodistribution were carried out.	Developed system showed an excellent magnetic targeting ability in cancer therapy. Significant difference in cell killing effect was observed between free HMME and C60-IONP-PEG/ HMME.	(Shi *et al.*, 2013a)
15	PLGA nanoparticles Functionalized with aptamer associated QDs	QD 605 (semiconductor) EpCAM aptamer (ligand) Nutlin-3a (drug)	Confocal bioimaging	Various *in vitro* studies including cellular uptake, cytotoxicity, apoptosis, immunoblotting and the imaging of cancer cells in 2-D monolayer and 3-D spheroid models were investigated.	Superior therapeutic potential and localization of targeted QD in comparison to unconjugated QD in HEK-293 and ZR751 cells.	(Das *et al.*, 2015)

(Continued)

TABLE 5.1 (CONTINUED)
Some of the Literature Instances on the Utility of the Nanotechnology in Cancer Diagnosis

S. No	Nanosystem	Drug, Ligand, Coating Material	Imaging Modility	*In Vitro*/*In Vivo* Studies	Inference(s)	Reference
16	QD with Fe_3O_4-filled CNTs (Fe_3O_4@ CNT-HQDs)	CdTe (semiconductor) SiO_2-coated quantum dots (HQDs) Transferrin (ligand) Doxorubicin hydrochloride (DOX)	Optical imaging	*In vitro* cell viability and uptake assays in Hela cells and HEK293 human kidney cells were carried out.	Improved stability, selectivity, fluorescence and high drug-loading efficiency facilitated the simultaneous imaging/mapping of cancer cells. Only a minute amount Fe3O4@CNT-HQDs is required for effective diagnostic and therapeutic applications.	(Chen *et al.*, 2012)
17	Surface-coated QDs	Carboxymethyl chitosan (coating material and as a Zn^{2+} sensor) CdTe (semiconductor)	Photoluminescence	*In vitro* studies for intracellular Zn^{2+} sensing using PC-3M cell lines investigated.	CMC–CdTe QDs found to be biocompatible, sensitive and selective for intracellular Zn^{2+} sensing in prostate cancer cells.	(Ma *et al.*, 2014)
18	Functionalized quantum dots probe	Daunorubicin (drug) $CuInS_2$ (semiconductor)	Photoluminescence NIR	*In vitro* cell imaging and MTT assay were carried out.	MUC1–QD conjugate delivered the drug to the targeted site and sensed the prostate cancer cells through photoluminescent intensity of QDs.	(Lin *et al.*, 2014)
19	Surface-coated and ligand-conjugated liposome containing drug and QD	Docetaxel TPGS (emulsifier, stealth property) Folic acid (ligan)	CLSM	*In vitro* Cell uptake, cytotoxicity assays on folate receptor expressing MCF-7 cells.	Significantly high cell uptake and cytotoxicity of folate receptor targeted liposomes in comparison to non-targeted liposomes.	(Muthu *et al.*, 2012)
20	Quantum dot embedded si nanosphere	ZnO QDs	Fluorescence microscopy	*In vitro* evaluation of viability, fluorescence and cytoskeletal organization.	The embedding of ZnO QDs on silica nanospheres led to significant increase in photoluminescence intensity also promoted the internalization in HeLa cells. The fluorescence emission from QDs was stable for 3 days, indicating excellent stability toward photobleaching.	(Depan and Misra, 2014)

TABLE 5.2
Cancer Biomarkers of Different Cancers

Cancer	Markers	Characteristics	Sample
Prostate	PSA (Prostate specific antigen)	High sensitivity in all stages; also elevated from some non-cancer causes	Blood
	PSMA (Prostate specific membrane antigen	Level tends to increase with age	Blood
Breast	CA 15-3, 27, 29 (Cancer antigen 15-3, 27, 29)	Elevated in benign breast conditions. Either CA 15-3 or CA 27, 29 could be used as marker	Blood
	Estrogen & Progesterone receptors	Overexpressed in hormone-dependent cancer	Tissue
	Her-2/neu	Only 20–30% of patients are positive to Her-2 oncogene that is present in multiple copies	Tissue
Lung (non-small cell)	CEA (Carcinoembryonic antigen)	Used in colon cancer detection and in combination with NSA increase specificity	Blood
Lung (small cell)	NSE (Neuron-specific enolase)	Better sensitivity towards specific types of lung cancer	Blood
Bladder	NMP22 (Matritech's nuclear matrix protein), BTA (Bladder tumor antigen)	NMP-22 assays tend to have greater sensitivity than BTA assays	Urine
Pancreatic	BTA	Composed of basement membrane complexes	Urine
	CA 19-9 (Carbohydrate antigen 19-9)	Elevated also in inflammatory bowel disease, sometimes used as colorectal cancer biomarkers	Blood
Germ cell cancer of ovaries	CA 72-4 (Cancer antigen 72-4)	No evidence that this biomarker is better than CA-125 but may be useful in combination	Blood
	AFP (Alpha-fetoprotein)	Also elevated during pregnancy and liver cancer	Blood
Multiple myeloma and lymphomas	B2M (Beta-2 microglobulin)	Present in many other conditions, including prostate cancer and renal cell carcinoma	Blood
	Monoclonal Immunoglobulins	Overproduction of an immunoglobulin, usually detected by protein electrophoresis	Blood, Urine
Metastatic melanoma	S100B	Subunit of the S100 protein family	Serum
	TA-90 (Tumor-associated glycoprotein Antigen)	Could be used to monitor patients with high risks of developing the disease	Serum
Thyroid medullary Carcinoma	Calcitonin	Secreted mainly by parafollicular C cell	Blood, Serum
Lymphomas	B2M	Present in many other conditions, including prostate cancer and renal cell carcinoma	Serum
Lung (non-small cell), epithelial, colorectal, head and neck	EGFR (Her-1)	Binding of the protein to a ligand induces receptor dimerization and tyrosine auto-phosphorylation and leads to cell proliferation	Tissue
Colorectal, lung, breast, pancreatic and bladder	CEA (Carcinoembryonic antigen)	Subtle posttranslational modifications might create differences between tumor and normal CEA	Serum

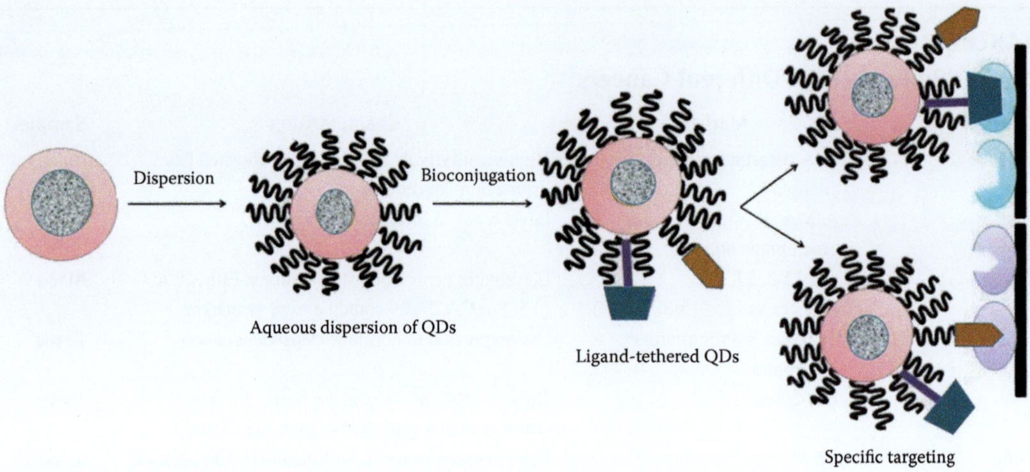

FIGURE 5.5 Schematic representation of bioconjugated aqueous dispersible QDs.

(Hermanson, 2007). The interaction is usually weak and nonspecific. Biomolecules with particular functional groups can be directly linked to NPs, e.g., thiol groups from cytosine amino residue can be used to coordinate peptides or proteins to noble metal NPs (Akerman *et al.*, 2002) and QDs (Tan *et al.*, 2007). Another approach is to absorb biomolecules to surface of NPs through electrostatic interactions, e.g., a positively charged surface of NPs can be attached to the negatively charged

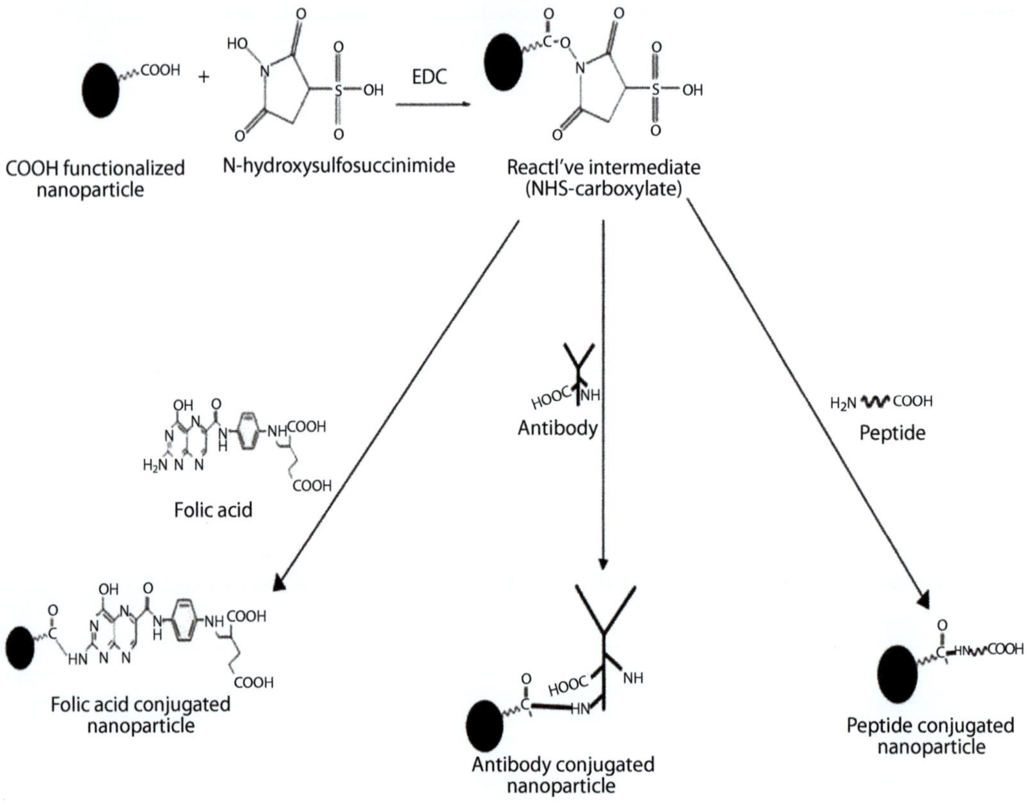

FIGURE 5.6 Schematic representation of carboxylic group involved coupling with folic acid, antibodies and peptide.

nucleic acid (Kell *et al.*, 2008; Jiang and Zhang, 2010). Under the physiochemical environment, the proteins with natural positive (or negative) charge can adhere to NPs with negative (or positive) charges. There are some drawbacks with this approach, such as sensitivity to pH and salt concentration, and a high degree of nonspecific binding.

5.6.2 Covalent Coupling

This method is used to provide the specific and stable conjugation of biomolecules with NPs. Usually, various coupling strategies are carried out on functional groups on the surface of NPs, including the carboxylic group, amide group and thiol group, which can be used to crosslink the biomolecules.

5.6.2.1 Carboxylic Group Involved Coupling

One of the most common coupling method for bioconjugation of NPs involve carboxylic group, which can react with primary amines. The most common coupling agent used here is 1-ethyl-3-(dimethylaminopropyl) carbodiimide (EDC). The EDC conjugation strategy has been applied to various proteins such as enzymes, antibodies, amine-terminated nucleic acids and small molecules with amine groups etc to carboxylic group functionalized NPs (Hermanson, 2008). Moreover, the coupling efficiency is increased by addition of stabilizing agents such as n-hydroxysuccinimide (NHS) or sulfo-NHS by formation of a succinimide ester intermediate (Figure 5.6).

5.6.2.2 Amine Group-Involved Coupling

In this coupling method, an amine group can react with carboxylic group to form an amide bond. The most common coupling reagent used for conjugation is glutaraldehyde. NH_2-functionalized NPs were activated with glutaraldehyde. The free aldehyde group generated during activation can conjugate to amino groups of biomolecules such as folic acid, antibodies, proteins, peptides, etc. (Figure 5.7).

5.6.2.3 Thiol Group-Involved Coupling

In this coupling method, the thiol group can selectively conjugate with primary amine groups. The reaction starts rapidly under the mediation of reagents, such as maleimides and iodo-acetamides. The most common coupling reagent used for conjugation is sulfosuccinimidyl-4-(maleimidomethyl) cyclohexane-1-carboxylate (sulfo-SMCC). Amine-functionalized NPs can be conjugated with various thiolated biomolecules, including thiol-modified peptide, residue thiol-teminated DNA, or peptides and proteins with free or reduced cysteine (Cai *et al.*, 2006, Hermanson, 2007) (Figure 5.8).

5.7 CONCLUSIONS

Cancer nanotechnology has shown great promise in providing novel methods for detecting cancer in its early stages, potentially leading to the improved degree of success in diagnosis and treatment of the disease. Traditional imaging strategies are highly invasive, nonspecific, and often associated with toxicity to both tumor and healthy cells. Using nanotechnology, clinicians are being able to be more proficient in delineating tumor margins. The development of various novel nanomaterials has enabled the detection of cancer biomarkers with greater precision and sensitivity that could not be achieved before. Antibodies, proteins or peptides can be conjugated to the surface of NPs by direct absorption; while covalent coupling can promote target imaging too. Consistent and concerted research efforts further need to be undertaken to harvest enormous potential of nanotechnology to detect the cancer in the early stages, and monitor the disease with much greater treatment accuracy.

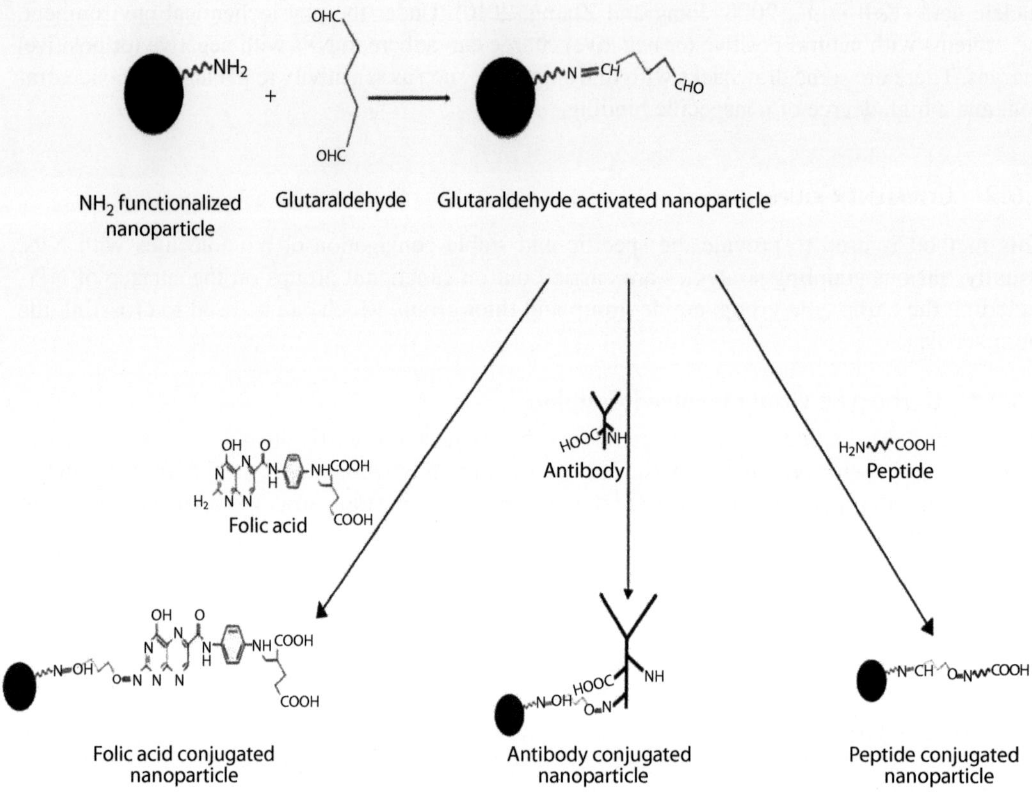

FIGURE 5.7 Schematic representation of amine group involved coupling with folic acid, antibodies and peptide.

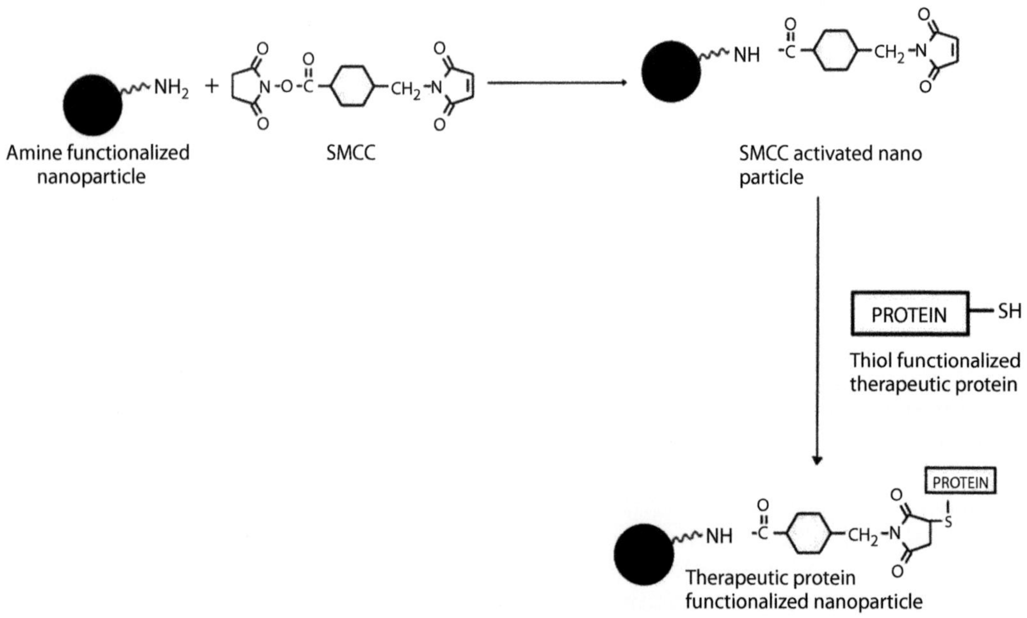

FIGURE 5.8 Schematic representation of an SMCC-activated amine functionalized nanoparticle coupling with thiol functionalized protein.

ACKNOWLEDGMENTS

Dr. Gurpal Singh deeply acknowledges the financial assistance provided by UGC, New Delhi.

REFERENCES

Akerman ME, Chan WCW, Laakkonen P, Bhatia SN, Ruoslahti E. Nanocrystal targeting *in vivo*. *Proc Natl Acad Sci USA*. 2002; 99(20): 12617–21.

Alivisatos AP. Semiconductor Clusters, Nanocrystals, and Quantum Dots. *Science*. 1996; 271: 933–7.

Alyssa BC, Chenxia MG, Jennifer RF, Stacey NB, Timothy JM, Chad AM. Nanoparticle probes for the detection of cancer biomarkers, cells, and tissues by fluorescence. *Chem Rev*. 2015; 115: 10530–74.

Anajwala CC, Jani GK, Vijayendra Swamy SM. Current trends of nanotechnology for cancer therapy. *Int J Pharm Sci Nanotechnol*. 2010; 3(3): 1–14.

Anon., *Henry's Clinical Diagnosis and Management by Laboratory Methods*. 22nd ed. Philadelphia, Pa. Elsevier Saunders; 2011. Available online at: http://www.clinicalkey.com. (Accessed 18 April, 2017).

Anon., Protein electrophoresis Lab Tests. Available online at: http://labtestsonline.org/understanding/analytes/electrophoresis/tab/test. (Accessed 18 March, 2017).

Azhdarzadeh M, Atyabi F, Saei AA, Varnamkhasti BS, Omidi Y, Fateh M *et al*. Theranostic MUC-1 aptamer targeted gold coated superparamagnetic iron oxide nanoparticles for magnetic resonance imaging and photothermal therapy of colon cancer. *Colloids Surf B Biointerfaces*. 2016; 143: 224–32.

Beg S, Rizwan M, Sheikh AM, Hasnain MS, Anwer K, Kohli K. Advancement in carbon nanotubes: Basics, biomedical applications and toxicity. *J Pharm Pharmacol*. 2011; 63: 141–63.

Bohunicky B, Mousa SA. Biosensors: The new wave in cancer diagnosis. *Nanotechnol Sci Appl*. 2011; 4: 1–10.

Bottini M, D'Annibale F, Magrini A, Cerignoli F, Arimura Y, Dawson MI *et al*. Quantum dot-doped silica nanoparticles as probes for targeting of T-lymphocytes. *Int J Nanomedicine*. 2007; 2: 227–33.

Bray F, Ren JS, Masuyer E, Ferlay J. Global estimates of cancer prevalence for 27 sites in the adult population in 2008. *Int J Cancer*. 2013; 132: 1133–45.

Bruchez M Jr, Moronne M, Gin P, Weiss S, Alivisatos AP. Semiconductor nanocrystals as fluorescent biological labels. *Science*. 1998; 281: 2013–6.

Cai W, Shin DW, Chen K, Gheysens O, Cao Q, Wang SX. Peptide-labeled near-infrared quantum dots for imaging tumor vasculature in living subjects. *Nano Lett*. 2006; 6: 669–76.

Carlo M, Croce CM. Oncogenes and cancer. *N Engl J Med*. 2008; 358: 502–11.

Carter PJ. Potent antibody therapeutics by design. *Nat Rev Immunol*. 2006; 6: 343–57.

Chan WC, Nie S. Quantum dot bioconjugates for ultrasensitive nonisotopic detection. *Science*. 1998; 281: 2016–8.

Chapman AP, Antoniw P, Spitali M, West S, Stephens S, King DJ. Therapeutic antibody fragments with prolonged in vivo half-lives. *Nat Biotechnol*. 1999; 17: 780–3.

Chen ML, He YJ, Chen XW, Wang JH. Quantum dots conjugated with Fe_3O_4-filled carbon nanotubes for cancer-targeted imaging and magnetically guided drug delivery. *Langmuir*. 2012; 28: 16469–76.

Choi YE, Kwak JW, Park JW. Nanotechnology for early cancer detection. *Sensors*. 2010; 10: 428–55.

Daltrey IR, Kissin MW. Randomized clinical trial of the effect of needle gauge and local anaesthetic on the pain of breast fine-needle aspiration cytology. *Br J Surg*. 2000; 87: 777–9.

Das M, Duan W, Sahoo SK. Multifunctional nanoparticle-EpCAM aptamer bioconjugates: A paradigm for targeted drug delivery and imaging in cancer therapy. *Nanomedicine*. 2015; 11: 379–89.

Datar R, Kim S, Jeon S, Hesketh P, Manalis S, Boisen A *et al*. Cantilever Sensors: Nanomechanical tools for diagnostics. *Mrs Bull*. 2009; 34(06): 449–4.

Depan D, Misra RD. Structural and physicochemical aspects of silica encapsulated ZnO quantum dots with high quantum yield and their natural uptake in HeLa cells. *J Biomed Mater Res A*. 2014; 102(9): 2934–41.

Dubertret B, Skourides P, Norris DJ, Noireaux V, Brivanlou AH, Libchaber A. *In vivo* imaging of quantum dots encapsulated in phospholipid micelles. *Science*. 2002; 298: 1759–62.

Erickson D, Mandal S, Yang AHJ, Cordovez B. Nanobiosensors: Optofluidic, electrical and mechanical approaches to biomolecular detection at the nanoscale. *Microfluid Nanofluidics*. 2008; 4: 33–52.

Fruscella M, Ponzetto A, Crema A, Carlonil G. The extraordinary progress in very early cancer diagnosis and personalized therapy: the role of oncomarkers and nanotechnology. *J Nanotech*. 2016; 2016: 1–18.

Giussani C, Roux FE, Ojemann J, Sganzerla EP, Pirillo D, Papagno C. Is preoperative functional magnetic resonance imaging reliable for language areas mapping in brain tumor surgery? Review of language functional magnetic resonance imaging and direct cortical stimulation correlation studies. *Neurosurgery*. 2010; 66: 113–20.

Globocan (2012) v1.0, cancer incidence and mortality worldwide: IARC cancer base no. 11 [Internet]. Lyon, France: International agency for research on cancer. Available online at: http://globocan.iarc.fr. (Accessed on 9 March, 2017).

Gobbo OL, Sjaastad K, Radomski MW, Volkov Y, Prina-Mello A. Magnetic nanoparticles in cancer theranostics. *Theranostics*. 2015; 5: 1249–63.

Gong Y (2013). Breast cancer: Pathology, cytology, and core needle biopsy methods for diagnosis. *Breast and Gynecological Cancers: An Integrated Approach for Screening and Early Diagnosis in Developing Countries*, Springer Science and Business Media New York, pp. 19–37.

Greenlee RT, Murray T, Bolden S, Wingo PA. Cancer statistics. *CA Cancer J Clin*. 2000; 50: 7–33.

Hammarstrom S. The carcinoembryonic antigen (CEA) family: Structures, suggested functions and expression in normal and malignant tissues. *Semin Cancer Biol*. 1999; 9(2): 67–81.

Hassanzadeh P, Fullwood I, Sothi S, Aldulaimi D. Cancer nanotechnology. *Gastroenterol Hepatol Bed Bench*. 2011; 4: 63–9.

Heath JR, Davis ME. Nanotechnology and Cancer. *Annu Rev Med*. 2008; 59: 251–65.

Helen EG, Jane AI (2003). Basic staining and histochemical techniques and immunohistochemical localizations using bone sections. *Handbook of Histology Methods for Bone and Cartilage* (ISBN97:8-1-59259-417-7). Springer Science, pp. 281–286.

Heo DN, Yang DH, Moon HJ, Lee JB, Bae MS, Lee SC *et al*. Gold nanoparticles surface-functionalized with paclitaxel drug and biotin receptor as theranostic agents for cancer therapy. *Biomaterials*. 2012; 33(3): 856–66.

Hermanson GT. Preparation of colloidal gold-labeled proteins. In *Bioconjugate Techniques* (Second Edition), Academic Press: New York; 2007; pp. 924–35.

Hermanson GT. Zero-length crosslinkers. In *Bioconjugate Techniques* (Second Edition), Academic Press: New York; 2008; pp. 213–33.

Jain S, Hirst DG, O'Sullivan JM. Gold nanoparticles as novel agents for cancer therapy. *Br J Radiol*. 2012; 85(1010): 101–13.

Jana S, Klara B, Pavel A, Vera L. Beta 2-microglobulin as a diagnostic marker in cerebrospinal fluid: A follow-up study. *Dis Markers*. 2014; 495402: 1–6.

Jensen JA. Medical ultrasound imaging. *Prog Biophys Mol Biol*. 2007; 93: 153–65.

Jiang S, Zhang Y. Upconversion nanoparticle-based FRET system for study of siRNA in live cells. *Langmuir*. 2010; 26(90): 6689–94.

Joshua ER, Serge Y, Ameena M, Mohit V, Frank XG. Nanotechnology and diagnostic imaging: New advances in contrast agent technology. *Nanomed Nanotech*. 2011; 2: 1–12.

Kell AJ, Stewart G, Ryan S, Peytavi R, Boissinot M, Huletsky A *et al*. Vancomycin-modified nanoparticles for efficient targeting and preconcentration of gram-positive and gram-negative bacteria. *ACS Nano*. 2008; 2: 1777–88.

Khan FR, Henderson JM (2013). Deep Brain Stimulation Surgical Techniques. In: Lozano AM, Hallet M. Brain Stimulation: Handbook of Clinical Neurology 116. Amsterdam: Elsevier, pp. 28–30.

Khan TS, Sundin A, Juhlin C, Langstrom B, Bergström M, Eriksson B. 11C-metomidate PET imaging of adrenocortical cancer. *Eur J Nucl Med Mol Imaging*. 2003; 30: 403–10.

Klein KM, Zheng J, Gewirtz A, Sarma DSR, Rajalakshmi S, Sitaraman SK (2005). Array of Nano-Cantilevers as a Bio-Assay for Cancer Diagnosis. Proceedings of 55th Electronic Components and Technology Conference, IEEE-CPMT and EIA, May 31–June 3, 2005, Orlando, FL, pp. 583–587.

Knudson AG. Two genetic hits (more or less) to cancer. *Nat Rev Cancer*. 2001; 1: 157–62.

Kumar R, Basu S, Torigian, Anand V, Zhuang H, Alavi A. Role of modern imaging techniques for diagnosis of infection in the era of 18F-fluorodeoxyglucose positron emission tomography. *Clin Microbio Rev*. 2008; 209–24.

Lee JH, Huh YM, Jun YW, Seo JW, Jang JT, Song HT *et al*. Artificially engineered magnetic nanoparticles for ultra-sensitive molecular imaging. *Nat Med*. 2007; 13(1): 95–9.

Li J, Hu Y, Yang J, Wei P, Sun W, Shen M *et al*. Hyaluronic acid-modified Fe_3O_4-Au core/shell nanostars for multimodal imaging and photothermal therapy of tumors. *Biomaterials*. 2015; 38: 10–21.

Lin J, Wang S, Huang P, Wang Z, Chen S, Niu G *et al*. Photosensitizer-loaded gold vesicles with strong plasmonic coupling effect for imaging-guided photothermal/photodynamic therapy. *ACS Nano*. 2013; 7: 5320–9.

Lin Z, Ma Q, Fei X, Zhang H, Su X. A novel aptamer functionalized $CuInS_2$ quantum dots probe for daunorubicin sensing and near infrared imaging of prostate cancer cells. *Anal Chim Acta*. 2014; 818: 54–60.

Liu Q, Aroonyadet N, Song Y, Wang X, Cao X, Liu Y *et al*. Highly sensitive and quick detection of acute myocardial infarction biomarkers using In_2O_3 nanoribbon biosensors fabricated using shadow masks. *ACS Nano*. 2016; 10: 10117–25.

Love C, Din AS, Tomas MB, Kalapparambath TP, Palestro CJ. Radionuclide bone imaging: An illustrative review. *RadioGraphics*. 2003; 23: 341–58.
Ma Q, Lin ZH, Yang N, Li Y, Su XG. A novel carboxymethyl chitosan-quantum dot-based intracellular probe for Zn^{2+} ion sensing in prostate cancer cells. *Acta Biomater.* 2014; 10: 868–74.
Magnus L, Mette LL, Michael B, Mette K. Benefits and harms of mammography screening. *Breast Cancer Res*. 2015; 17: 1–12.
Mahadevappa M. Fluoroscopy: Patient radiation exposure issues. *RadioGraphics*. 2001; 21: 1033–45
Mark AH. Digital mammography imaging: Breast tomosynthesis and advanced applications. *Radiol Clin North Am*. 2010; 48: 917–29.
Masselli G, Gualdi G. MR imaging of the small bowel. *Radiology*. 2012; 264: 333–48.
Mayo Clinic Rochester Minn. Available online at: http://www.mayoclinic.org/diseases-conditions/cancer /news/con-20032378. (Accessed on 1 February, 2017).
Muthu MS, Kulkarni SA, Raju A, Feng SS. Theranostic liposomes of TPGS coating for targeted co-delivery of docetaxel and quantum dots. *Biomaterials*. 2012; 33: 3494–501.
Naha PC, Zaki AA, Hecht E, Chorny M, Chhour P, Blankemeyer E *et al.* Dextran coated bismuth-iron oxide nanohybrid contrast agents for computed tomography and magnetic resonance imaging. *J Mater Chem B*. 2014; 2: 8239–48.
Namdari P, Daraee H, Eatemadi A. Recent advances in silicon nanowire biosensors: synthesis methods, properties, and applications. *Nanoscale Res Lett*. 2016; 11: 406–12.
Pantic I. Magnetic nanoparticles in cancer diagnosis and treatment: Novel approaches. *Rev Adv Mater Sci*. 2010; 26: 67–73.
Prasant MC, Singhai A, Pathak AD, Ali FM. Focus on basic principles of biopsy: A review. MRJMMS. 2014; 2: 001–6.
Ren J, Shen S, Wang D, Xi Z, Guo L, Pang Z *et al.* The targeted delivery of anticancer drugs to brain glioma by PEGylated oxidized multi-walled carbon nanotubes modified with angiopep-2. *Biomaterials*. 2012; 33: 3324–33.
Salomon D, Brandt R, Ciardiello F, Normanno N. Epidermal growth factor-related peptides and their receptors in human malignancies. *Crit Rev Oncol Hematol*. 1995; 19: 183–232.
Shamsudhin N, Tao Y, Sort J, Jang B, Degen CL, Nelson BJ *et al.* Magnetometry of Individual Polycrystalline Ferromagnetic Nanowires. *Small*. 2016; 12: 6363–9.
Shi J, Wang L, Gao J, Liu Y, Zhang J, Ma R *et al.* A fullerene-based multi-functional nanoplatform for cancer theranostic applications. *Biomaterials*. 2014; 35: 5771–84.
Shi J, Yu X, Wang L, Liu Y, Gao J, Zhang J *et al.* PEGylated fullerene/iron oxide nanocomposites for photodynamic therapy, targeted drug delivery and MR imaging. *Biomaterials*. 2013a; 34: 9666–77.
Shi X, Gong H, Li Y, Wang C, Cheng L, Liu Z. Graphene-based magnetic plasmonic nanocomposite for dual bioimaging and photothermal therapy. *Biomaterials*. 2013b; 34: 4786–93.
Sidransky D. Emerging molecular markers of cancer. *Nat Rev Canc*. 2002; 2: 210–9.
Singh B, Beg S, Lohan S, Raza K (2015). Carbon Nanotubes (CNTs) and their Applications in Drug Delivery Systems Vol. 6, Novel Drug Delivery. In: B Singh, OP Katare and JN Govil, eds., *Recent Trends in Biotechnology* (6 Volume Series; Series ISBN: 1-62699-050-6), Studium Press, Houston LLC, USA, pp. 269–293.
Smith A, Balar AV, Milowsky MI, Chen RC (2014). Bladder cancer. In: Niederhuber JE, Armitage JE, Doroshow JH, Kastan MB and Tepper JE, *eds., Abeloff's Clinical Oncology* (5th ed. Philadelphia), PA: Elsevier Churchill Livingstone, pp. 2224–2234.
Stojan R, Zorica S, Boris D. The pathogenesis of neoplasia. *Arch Oncol*. 2004; 12: 35–7.
Sun B, Fang Y, Li Z, Chen Z, Xiang J. Advances in the application of nanotechnology in the diagnosis and treatment of gastrointestinal tumors. *Mol Clin Oncol*. 2015; 3: 274–280.
Suzuki H, Lee IYS, Maeda N. Laser-induced emission from dye doped nanoparticle aggregates of poly (DL-lactide-co-glycolide). *Int J Phys Sci*. 2008; 3: 42–4.
Tan WB, Jiang S, Zhang Y. Quantum-dot based nanoparticles for targeted silencing of HER2/neu gene via RNA interference. *Biomaterials*. 2007; 28: 1565–71.
Thakor AS, Gambhir SS. Nanooncology: The future of cancer diagnosis and therapy. *CA Cancer J Clin*. 2013; 63: 395–418.
Thomas RG, Moon MJ, Lee H, Sasikala AR, Kim CS, Park IK *et al.* Hyaluronic acid conjugated superparamagnetic iron oxide nanoparticle for cancer diagnosis and hyperthermia therapy. *Carbohydr Polym*. 2015; 131: 439–46.
Tokumaru K, Coyle JD. A collection of experiments for teaching photochemistry. *Pure & Appl Chem*. 1992; 64: 1343–82.

Tripathi SK, Khurana RK, Kaur G, Teenu, Singh B (2016). Quantum dots: Dynamic tools in cancer nanomedicine. Nanobiomaterials in Medical Imaging: Applications of Nanobiomaterials, Elsevier; 2016; pp. 71–100.

Tripathi SK, Kaur G, Khurana RK, Kapoor S, Singh B. Quantum dots and their potential role in cancer theranostics. *Crit Rev Ther Drug Carrier Syst.* 2015; 32: 461–502.

Van Vlerken LE, Amiji MM. Multi-functional polymeric nanoparticles for tumor-targeted drug delivery. *Expert Opin Drug Deliv.* 2006; 3: 205–16.

Xiao Y, Hong H, Matson VZ, Javadi A, Xu W, Yang Y *et al.* Gold nanorods conjugated with doxorubicin and cRGD for combined anticancer drug delivery and PET imaging. *Theranostics.* 2012; 2: 757–68.

Xiaoxiao H, Wang K, Cheng Z. *In vivo* near-infrared fluorescence imaging of cancer with nanoparticles based probes. *Nanomed Nanobiotechnol.* 2010; 2: 349–66.

Xing L, Todd NW, Yu L, Fang H, Jiang F. Early detection of squamous cell lung cancer in sputum by a panel of micro RNA markers. *Mod Pathol.* 2010; 23: 1157–64.

Yang X, Hong H, Grailer JJ, Rowland IJ, Javadi A, Hurley SA *et al.* cRGD-functionalized, DOX-conjugated, and (6)(4)Cu-labeled superparamagnetic iron oxide nanoparticles for targeted anticancer drug delivery and PET/MR imaging. *Biomaterials.* 2011; 32: 4151–60.

Yu MK, Jeong YY, Park J, Park S, Kim JW, Min JJ *et al.* Drug-loaded superparamagnetic iron oxide nanoparticles for combined cancer imaging and therapy *in vivo. Angew Chem Int Ed Engl.* 2008; 47: 5362–5.

Yuan Z, Quan J, Yunxiao Z, Jian C, Zhu H, Liping G. Diagnostic value of contrast-enhanced ultrasound parametric imaging in breast tumors. *J Breast Cancer.* 2013; 16: 208–13.

Zagouri F, Sergentanis TN, Gounaris A, Koulocheri D, Nonni A, Domeyer P *et al.* Pain in different methods of breast biopsy: Emphasis on vacuum-assisted breast biopsy. Breast. 2008; 17(1): 71–5.

Zhu J, Zheng L, Wen S, Tang Y, Shen M, Zhang G *et al.* Targeted cancer theranostics using alpha-tocopheryl succinate-conjugated multifunctional dendrimer-entrapped gold nanoparticles. *Biomaterials.* 2014; 35: 7635–46.

6 Dendrimers: Emerging Anti-Infective Nanomedicines

*Keerti Jain**

CONTENTS

6.1 Introduction ... 121
6.2 Mechanisms of Action ... 122
6.2.1 Electrostatic Interaction with the Membrane ... 122
6.2.2 Effect of Surface Groups on Dendrimers–Membrane Interactions ... 123
6.2.3 Role of Membrane Fluidity on Dendrimers–Membrane Interactions ... 123
6.2.4 Inhibition of Pathozymes ... 125
6.3 Engineering of Dendrimers for Antimicrobial Activity ... 125
6.3.1 Surface Engineering of Dendrimers ... 125
6.3.2 Formation of Nanohybrids ... 129
6.4 Designing Novel Dendrimers ... 131
6.4.1 Peptide Dendrimers ... 131
6.4.2 Carbosilane Dendrimers ... 132
6.4.3 PAMAM Dendrimers ... 132
6.4.4 Triazine Dendrimers ... 132
6.4.5 Miscellaneous ... 134
6.5 Dendrimers for the Delivery of Antimicrobial Agents ... 135
6.6 Conclusions ... 135
References ... 136

6.1 INTRODUCTION

The development of anti-infective agents brought the perception of conquering infectious diseases, but the ability of microorganisms to develop resistance rendered many of the currently available anti-infective drugs ineffective. This emergence of microbial resistance is presenting a serious threat in the treatment of infectious diseases. In the light of this fact, current research is undertaking the development of novel antimicrobial agents (Engler *et al.*, 2012; Muszanska *et al.*, 2014; Lind *et al.*, 2015).

In the modern era of nanotechnology, dendrimers are emerging as promising macromolecules with diverse biomedical applications. Dendrimers also represent a new class of polymers that have perfectly-branched architecture with a well-defined size, shape and molecular weight. Further, the presence of tailor-made functional groups allows for the smart engineering of these nano-architectures, rendering them perfect candidates to be explored for various pharmaceutical and medicinal applications. Many researchers are working on the development of novel dendrimer molecules and engineering previously-designed dendrimer molecules for antimicrobial activity, as well as for delivery of antimicrobial agents (Jain *et al.*, 2015a,b,c; Grabchev *et al.*, 2015; Jain and Jain, 2015; Rai *et al.*, 2016).

Dendrimers are classical examples of branched antimicrobial agents attributed to their unique architecture, which offers the presentation of the required number of functional groups at a high local concentration (Figure 6.1). Although, a large number of different dendrimeric molecules have been developed

* Corresponding author.

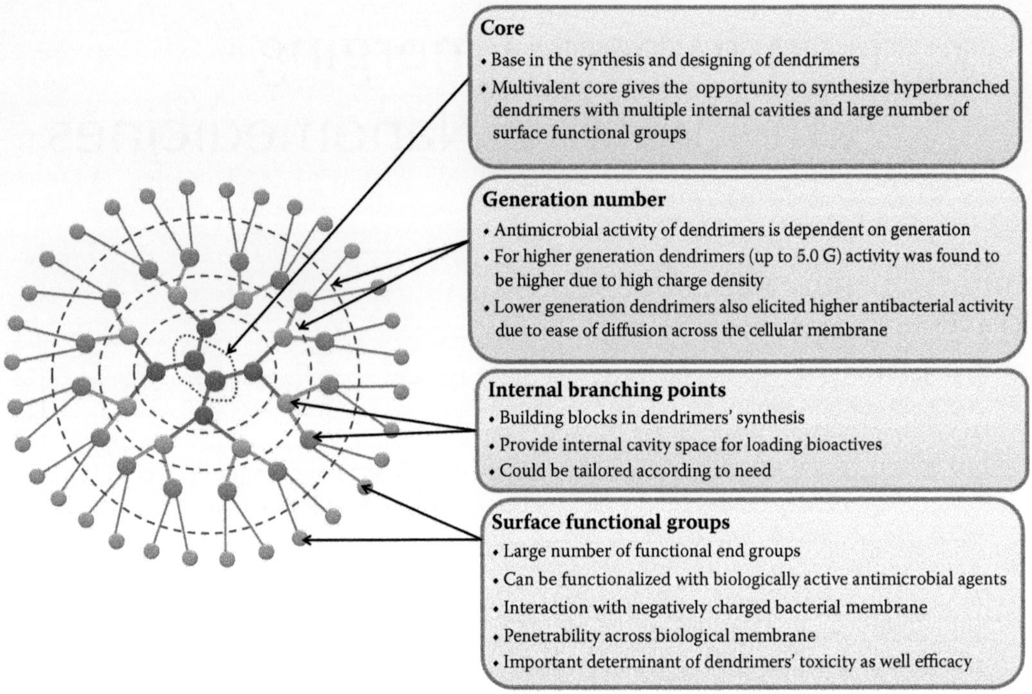

FIGURE 6.1 Schematic presentation of the typical structure of dendrimers with the role of each component of the dendrimeric architecture.

using different core groups, branching units and surface functional groups; poly(propylenamine) (PPI), poly(amidoamine) (PAMAM), carbosilane, phosphate, branched peptide and L-lysine-based dendrimers represent the most common dendrimers investigated for biomedical applications, as well as for antimicrobial activity (Engler *et al.*, 2012; Jain and Jain, 2013; Kaur *et al.*, 2016b; Singh *et al.*, 2016). Scientists have been investigating many strategies to develop dendrimer-based antimicrobial agents, including engineering previously-synthesized dendrimers, designing novel dendrimers and nanohybrids based on dendrimers will be the topic of discussion in the following sections of Chapter 6, after the discussion on the mechanisms of the actions of dendrimers speculated or investigated for their antimicrobial activity.

6.2 MECHANISMS OF ACTION

Many researchers have previously found that dendrimers with cationic surface charges interact with biological membranes (Chen *et al.*, 2000; Chen and Cooper, 2002; Jain *et al.*, 2010; Soni *et al.*, 2015; Kaur *et al.*, 2015; Lind *et al.*, 2015). The cationic charge of some of the dendrimers causes them to interact electrostatically with negatively charged biological membranes and, as a result, at a particular concentration, dendrimers have been observed to elicit bactericidal activity as a result of the disruption of the bacterial cell membrane (Grabchev *et al.*, 2015). Although, many scientists suggested that the antimicrobial activity of dendrimers is attributed to the interactions of dendrimers with a microbial cell membrane, other mechanisms (including the inhibition of pathozymes, chelation of ions present in membrane, formations of hydrogen bonds etc.) have also been proposed as mechanisms of the antimicrobial action of dendrimers.

6.2.1 Electrostatic Interaction with the Membrane

The antimicrobial activity of dendrimers is dependent on two major factors related to dendrimeric architecture; one, the charge density on the surface of molecule, and two, the ability of dendrimers to penetrate the bacterial cell membrane (Chen *et al.*, 2000). In an investigation conducted by Chen and Cooper, the membrane disruption was observed to be the mechanism of the antimicrobial action of dendrimers, which was further found to be affected by the presence of calcium ions. The presence of calcium ions limited the efficacy of dendrimers as antimicrobial agents increased the required concentration to elicit antimicrobial activity. Further, the investigation on the interaction of PPI dendrimers functionalized with quaternary ammonium compounds with bacterial membranes showed that these macromolecules are more potent than their monomer counterparts (Chen and Cooper, 2002).

6.2.2 Effect of Surface Groups on Dendrimers–Membrane Interactions

Wang *et al.* (2010) investigated the interaction mechanisms of hydroxyl, amine and carboxyl-terminated PAMAM dendrimers with bacterial cell membranes to elucidate its mechanism of antimicrobial activity as well as toxicity. In this study, it was observed that although the antimicrobial activity of amine-terminated dendrimers (PAMAM-NH_2) with surface cationic charge were the highest among all three dendrimers subjected to this investigation (PAMAM-NH_2, PAMAM-COOH and PAMAM-OH), its cytotoxicity towards human cervical epithelial cells lines were also high. PAMAM-NH_2 was found cytotoxic at concentrations above 10 μg/mL. In contrast to amine-terminated dendrimers, neutral PAMAM dendrimers with surface hydroxyl groups (PAMAM-OH) were found to elicit no cytotoxicity at concentrations as high as 1 mg/mL. Further, the interaction mechanism showed that these neutral PAMAM-OH dendrimers interact with the membrane *via* changing the permeability of the outer membrane, whereas 4.0G PAMAM dendrimers with amino groups and 3.5G PAMAM dendrimers with surface carboxyl groups (PAMAM-COOH) damaged the outer membrane and inner membrane leading to bacterial cell lysis. The possible mechanism of antibacterial action of G4-PAMAM-NH_2, G4-PAMAM-OH and G3.5-PAMAM-COOH was potentially due to the binding of polycationic PAMAM-NH_2 dendrimers to polyanionic liposaccharides present in *Escherichia coli*, the formation of hydrogen bonds between PAMAM-OH and membrane-located hydrophilic O-antigen of *E. coli* and chelation of divalent ions present in the outer membrane of *E. coli* with PAMAM-COOH, respectively.

6.2.3 Role of Membrane Fluidity on Dendrimers–Membrane Interactions

Apart from surface charge of dendrimers and membrane fluidity, dendrimeric architecture also plays an important role in dendrimer–membrane interactions. Klajnert *et al.* (2006a) examined the interaction of peptide dendrimers with model lipid membranes and observed that interactions of dendrimers with lipid membranes occur in two steps without affecting the integrity of membrane. These two steps are; electrostatic interactions between the charged lipid membrane and surface groups of dendrimers followed by hydrophobic interactions between arms of dendrimers and acyl chains of lipid membranes.

In a recent study, Lind *et al.* (2015) examined the interaction mechanism of an antimicrobial peptide dendrimer, BALY, with model lipid bilayers. BALY is a low molecular weight flexible and amphiphilic dendrimer containing four long arms with lipophilic 2-chlorocabobenzoxy terminal groups. It has been synthesized from lysine amino acids and contains an indole moiety at C-terminus (Figure 6.2). At physiological pH, BALY carries six positive charges and has been found active against gram-positive bacteria, *Staphyloccocus aureus*, and gram-negative bacteria, *E. coli*. The interaction mechanism of this peptide dendrimers with biological membrane was elucidated using model lipid bilayers 1,2-dipalmitoyl-sn-glycero-3-phos-phocholine (DPPC) and 1-Palmitoyl-2-oleoyl-sn-glycero-3-phosphocholine (POPC) by applying the techniques of molecular dynamics simulation and neutron reflection. The interaction of BALY with these fluid membranes showed membrane thinning. Followed by membrane thinning in fluid state, expansion was observed in the gel phase, followed by the insertion in the hydrophobic region of membrane at the lipid phase transition temperature. These observations suggested that the interaction of this peptide dendrimer with the lipid membrane is dependent on the membrane's fluidity.

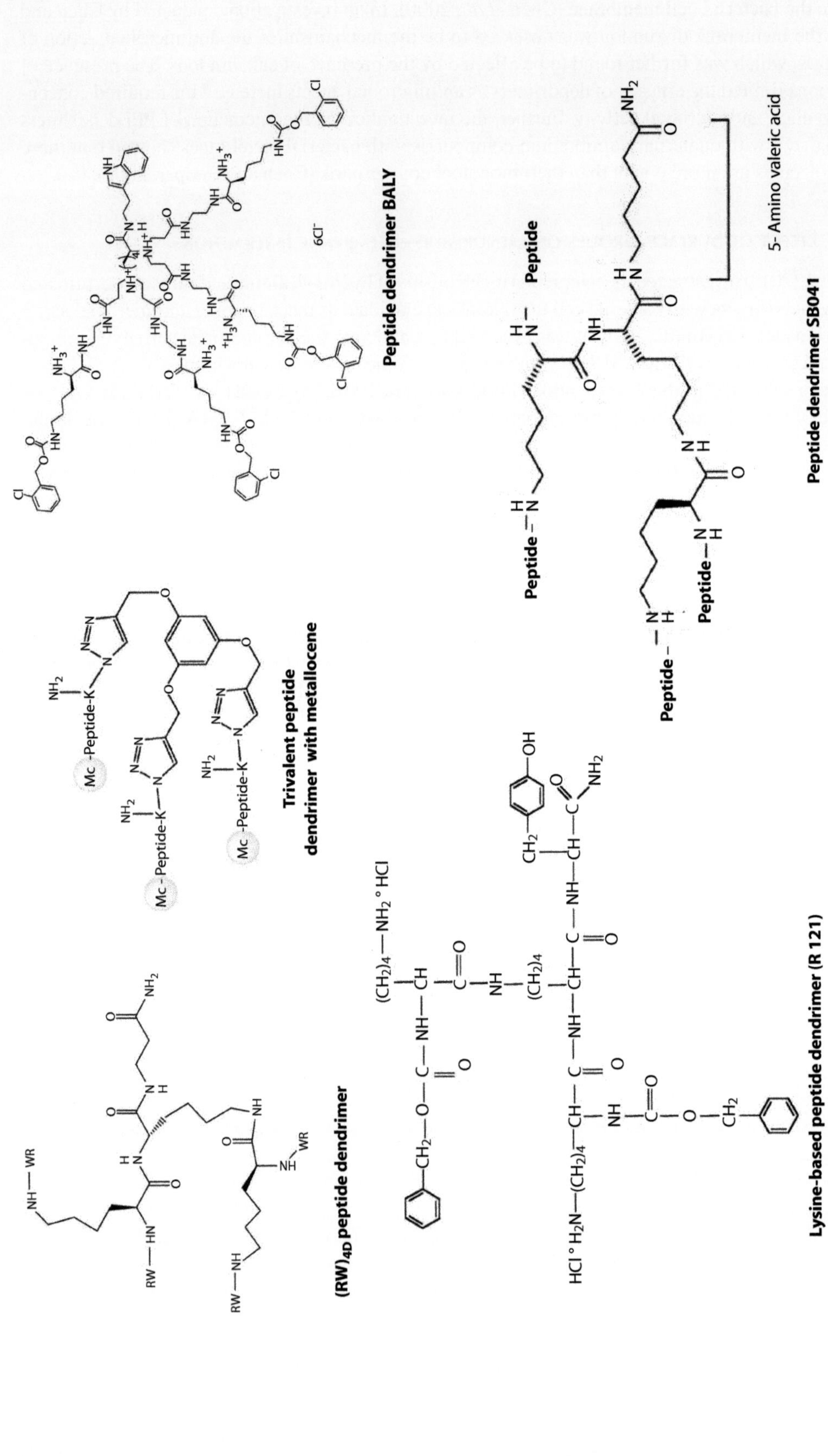

FIGURE 6.2 Chemical structure of some antimicrobial peptide dendrimers.

6.2.4 Inhibition of Pathozymes

Carta *et al.* (2015) observed that the antimicrobial activity of benzene sulfonamide–modified dendrimers was found to be associated with inhibition of carbonic anhydrase enzymes of pathogenic bacteria, fungi and protozoa. Further, this association was also found to be dependent on the generation number (Table 6.1).

The antimicrobial action of dendrimers depends on their architecture and surface charges, as well as the membrane's condition. Apart from that, dendrimers can also inhibit target microorganisms *via* their interaction with its cellular component, enzymes, ions, etc. The variation in the antimicrobial action of dendrimers could be anticipated due to the unique architecture of dendrimers which allowed the inclusion of different functionalities as part of the three architectural components of dendrimers, namely its core, branching units and surface groups (Figure 6.1).

6.3 ENGINEERING OF DENDRIMERS FOR ANTIMICROBIAL ACTIVITY

Dendrimers, as 21st-century polymers, have been investigated for various biomedical applications including the delivery of anticancer, antimicrobial and anti-inflammatory drugs, proteins, peptides, hormones, nucleic acids, diagnostic agents, contrast agents, etc. They have also shown promising results as catalysts for industrial applications, as well as in photodermal therapy, boron neutron capture therapy, etc. Apart from the delivery of bioactives, dendrimers have also shown therapeutic activity including anticancer, anti-angiogenic, antiviral, antibacterial, etc. (Jain *et al.*, 2010; Engler *et al.*, 2012; Li *et al.*, 2015; Singh *et al.*, 2015; Soni *et al.*, 2015; Kaur *et al.*, 2016a; Starpharma, 2016; Jain *et al.*, 2017).

6.3.1 Surface Engineering of Dendrimers

For antimicrobial activity, dendrimers have been engineered sequentially as well as for surface topology. Recently, 1.0G and 3.0G PPI dendrimers have been modified by Grabchev *et al.* (2015) to develop metallodendrimer complexes with promising antimicrobial activity. In this research work, dendrimers were first modified with 4-amino-1, 8-naphthalimide units followed by formation of complexes with Cu (II) and Zn (II) metal ions. *In vitro* evaluation against the six strains of microorganisms (*i.e. Bacillus subtilis, Micrococcus luteus, Pseudomonas aeruginosa, Acinetobacter johnsonii, E. coli, Saccharomyces cerevisiae* and *Candida lipolytica*) showed that 3.0G metallodendrimer complexes showed comparatively higher antimicrobial activity compared to 1.0G metallodendrimer complexes. The schematic presentation of the chemical structure of 4-amino-1, 8-naphthalimide modified 3.0G poly (propylenamine) dendrimers, as well as 3.0G metallodendrimer complex is given in Figure 6.3 (Grabchev *et al.*, 2015).

Carta *et al.* (2015) examined 4.0G PAMAM dendrimers by incorporating benzene sulfonamide groups as inhibitors of carbonic anhydrase enzymes present in pathogenic bacteria, fungi and protozoa including α-, β-, γ-, η-carbonic anhydrases. These PAMAM dendrimers with benzene sulfonamide showed the inhibition of eight pathogenic carbonic anhydrase enzymes present in the following eight pathogens *Vibrio cholerae, Trypanosoma cruzi, Leishmania donovani chagasi, Porphyromonas gingivalis, Cryptococcus neoformans, Candida glabrata* and *Plasmodium falciparum*. The inhibition efficiency was found to vary with generation of dendrimers for different pathozymes depending on the valency (Table 6.1).

The antibacterial activity of PAMAM dendrimers with surface amino group is limited by its considerable toxicity towards mammalian cells. Hence, to explore the antibacterial properties of PAMAM dendrimers without affecting normal cells, Xue *et al.* (2015) designed LED209-conjugated PAMAM dendrimers to develop a multifunctional antibacterial agent to combat the challenge of multidrug-resistant (MDR), gram-negative bacteria. LED209 was used in this study to increase specificity of PAMAM dendrimers towards gram-negative pathogens. It is a specific

TABLE 6.1
Summary of Dendrimers Based Prospective Anti-Infective Nanomedicines

Dendrimer	Antimicrobial Activity	References
	Dendrimers with Antimicrobial Activity	
Dendrimer peptides	• Dendrimer peptides containing lysine as core and branching units showed antimicrobial activity towards *S. aureus* and *E. coli*	Janiszewska *et al.*, 2003
Peptide dendrimers	• Lysine-based low molecular weight peptide dendrimers containing different proportions of aromatic and cationic groups were designed and evaluated for antimicrobial activity and toxicity profile. • Six dendrimers (designated as R 121, R 124, R 131 R 132, R 155, R 169 by author) showed antimicrobial activity which was found to be associated with cytotoxicity. • Among all the tested dendrimers, R 121 was found to be optimum, since it elucidated low cytotoxicity even at higher concentration (Figure 6.2).	Klajnert *et al.*, 2006b
PAMAM-OH dendrimers	• Showed promising activity towards *E. coli* without cytotoxicity at concentration up to 1 mg/mL	Wang *et al.*, 2010
SB041 peptide dendrimers	• Selective high activity towards gram-negative bacteria like *E. coli* and *P. aeruginosa* was observed. • Antimicrobial activity towards *E. coli* was unaffected by presence of high concentration of LPS.	Bruschi *et al.*, 2010
Dendrimeric Lipopeptides	• Consist of (Lys)Lys(Lys) dendron and C-8 or C-12 chain at C-end; antimicrobial activity towards *Candida* strains was found to be higher for dendrimeric lipopeptides with C12 residues. • The mechanisms of this antifungal activity included: (i) induction of potassium leakage from fungal cells, (ii) morphological alterations in fungal cells and (iii) inhibition of candidal β(1,3)-glucan synthase.	Janiszewska *et al.*, 2012
Organometallic Dendrimers	• MIC values were obtained at low micromolar concentration. • Found to be noncytotoxic to epidermal cell lines and mammalian red blood cells	Abd-El-Aziz *et al.*, 2015
Ultrashort Arg-Trp-Based Antimicrobial Peptides	• 15 antimicrobial peptides were synthesized by click chemistry (copper (I)-catalyzed alkyne-azide cycloaddition) containing exclusively arginine and tryptophan. • Potent antibacterial activity towards gram-positive multidrug-resistant *S. aureus* strain was observed with ultrashort peptide containing 2–3 amino acid residues whereas another ultrashort dipeptide on the trivalent scaffold showed MIC of 7.5 μM against gram-negative bacteria *A. baumannii*. • Among these 15 peptides, the 3 most active was subjected to hemolytic toxicity studies towards red blood cells and revealed <1% hemolysis at a concentration 10-fold higher than MIC.	Hoffknecht *et al.*, 2015
VivaGel®	• Active against HIV-1 and HSV • Currently under phase III clinical trial for treatment and prevention of bacterial vaginosis • Effective in preventing sexually transmitted infections • Available as condoms in few selected countries, including Australia and Japan	Starpharma, 2016; Li *et al.*, 2015; Antimisiaris and Mourtas, 2015

(*Continued*)

TABLE 6.1 (CONTINUED)
Summary of Dendrimers Based Prospective Anti-Infective Nanomedicines

Dendrimer	Antimicrobial Activity	References
Polyanionic carbosilane dendrimers	• Water soluble, safe and biocompatible • Incorporating either sulphated or naphthyl-sulphonated end groups • Inhibits HIV-1 and HSV-2 infection as an entry inhibitor • Showed promising efficiency as topical microbicides to prevent HIV-1 or HIV-2 vaginal/rectal transmission in humans	Antimisiaris and Mourtas, 2015; Ceña-Diez *et al.*, 2016a,b; Vacas-Córdoba *et al.*, 2016; Briz *et al.*, 2015
Peptide dendrimer, SB056	• SB056 is a semi-synthetic peptide with a dimeric dendrimer scaffold found active against gram-positive and gram-negative bacteria.	Batoni *et al.*, 2016
Glycopeptide	• Bactericidal activity was observed, which was peptide dendrimers.	Michaud *et al.*, 2016
Hyperbranched PAMAM dendrimers (with a trivalent core, diethylene triamine)	• Broad-spectrum antimicrobial and anti-biofilms activity towards *A. niger*, *C. albicans*, *S. aureus*, *B. subtilis*, *E. coli*, *P. aeruginosa*, and Methicillin-resistant *S. aureus*.	Labena *et al.*, 2016
SURFACE-ENGINEERED DENDRIMERS WITH ANTIMICROBIAL ACTIVITY		
4.0G PAMAM dendrimers with benzenesulfonamide groups	• Inhibited the cabonic anhydrase (CA) pathozymes of different pathogens with varying efficiency, depending on generation and multivalency. • 0.0G dendrimers with 4 sulfonamide groups and 3.0G dendrimers with 32 sulfonamide moieties showed highest inhibition of *Vibrio* enzyme VchCA and *Trypanosoma* enzyme TcCA, respectively. Whereas 1.0G (8 sulfonamides) and 2.0G (8 sulfonamides) dendrimers showed optimum inhibition of remaining pathozymes.	Carta *et al.*, 2015
Silver salts of 1.0G poly (propyl ether imine) (PETIM) dendron and dendrimers	• Presentation of multiple silver ions on single molecule of dendrimers to achieve higher antimicrobial efficacy at a reduced dose. • Increased antimicrobial activities were observed against sensitive strain of *S. aureus*, as well as methicllin resistant *S. aureus*.	Suleman *et al.*, 2015
Ciprofloxacin conjugated triazine dendrimers	• Antibacterial action was observed towards gram-positive and gram-negative bacteria, depending on hydrophobic and hydrophilic balance per repeat unit of dendrimers.	Vembu *et al.*, 2015
Mesoporous dendrimer-coated nanosized titanium dioxide	• Formed by association of ammonium-terminated and phosphonate-terminated phosphorus dendrimers with titanium dioxide by sol-gel process • Showed antimicrobial activity • Exhibited hemolytic toxicity and cytotoxicity depending on interface composition where ammonium-terminated dendrimer-coated titanium dioxide showed highest hemolytic toxicity and cytotoxicity.	Milowska *et al.*, 2015
Quaternary ammonium functionalized PAMAM dendrimer	• Potent bactericidal activity against gram-positive and gram-negative bacteria with negligible cytotoxicity	Zainul Abid *et al.*, 2016
PAMAM-ethyleneglycol dimethacrylate (EGDMA)		

(Continued)

TABLE 6.1 (CONTINUED)
Summary of Dendrimers Based Prospective Anti-Infective Nanomedicines

Dendrimer	Antimicrobial Activity	References
	DENDRIMER-BASED ANTIMICROBIAL NANOHYBRIDS	
PAMAM conjugated chitosan grafted on cotton fabric	• Increase in antimicrobial activity against *S. aureus* was observed in comparison to unmodified chitosan at a weight ratio of 1% of PAMAM chitosan.	Klaykruayat *et al.*, 2010
PPI conjugated chitosan grafted on cotton fabric	• 100% reduction of *S. aureus* and *E. coli* • Antimicrobial activity was retained even after washing for 12 times	Sadeghi-Kiakhani *et al.*, 2015
CdS and Ag_2S QDs immobilized on PAMAM-grafted carbon nanotubes	• The antibacterial activity of this nanohybrid was observed to be superior to carboxylated MWCNTs, PAMAM-grafted MWCNTs, PAMAM dendrimer, and CdS and Ag_2S QDs against *E. coli*, *P. aeruginosa* and *S. aureus*.	Neelgund *et al.*, 2012
PPI dendrimer-stabilized silver nanoparticles noncovalently functionalized on MWCNTs	• Showed catalytic and antibacterial activity	Murugan *et al.*, 2016
Chitosan-PAMAM Dendrimer/Ag Nano-emulsion treated cotton fabrics	• Improved antibacterial activity was observed	Aryabadie *et al.*, 2015
Hyaluronic acid (HA)/ PAMAM multilayers on a poly(3-hydroxybutyrate-co-4-hydroxybutyrate) [P(3HB-4HB)] substrate	• Prepared by layer-by-layer self-assembly method • Both, HA and PAMAM outer layer showed protein-resistant activity to bovine serum albumin with antiadhesion activity to *E. coli* • PAMAM outer layer inhibited *E. coli* while HA outer layer showed no inhibitory activity for *E. coli* • HA layer was speculated to improve biocompatibility as no cytotoxicity was observed towards L929 cells with developed multilayers.	Zhan *et al.*, 2015
Quaternary ammonium dendrimer in poly(methyl methcarylate) bone cement	• Showed promising antimicrobial activity with retention of mechanical properties of bone cements suggesting possible clinical application	Abid *et al.*, 2015

inhibitor of QseC (conserved histidine sensor kinase found in minimum of 21 MDR gram-negative bacteria infecting humans) of MDR gram-negative bacteria and nontoxic to human cells. It does not inhibit growth of bacteria and it is anticipated that it will exert least survival pressure on bacteria to develop resistance that can inhibit the virulence of these pathogens. The results according to the anticipated hypothesis revealed low cytotoxicity towards human cells and selectivity towards gram-negative bacteria, with promising antibacterial activity against many pathogenic gram-negative bacteria including *E. coli*, *P. aeruginosa*, *Klebsiella pneumoniae*, *Salmonella paratyphi*, *Salmonella typhimurium*, MDR-*Acinetobacter baumannii*, MDR-*E. coli*, MDR-*Shigella flexneri*, Extended Spectrum Beta-Lactamases (ESBLs)-*E. coli*.

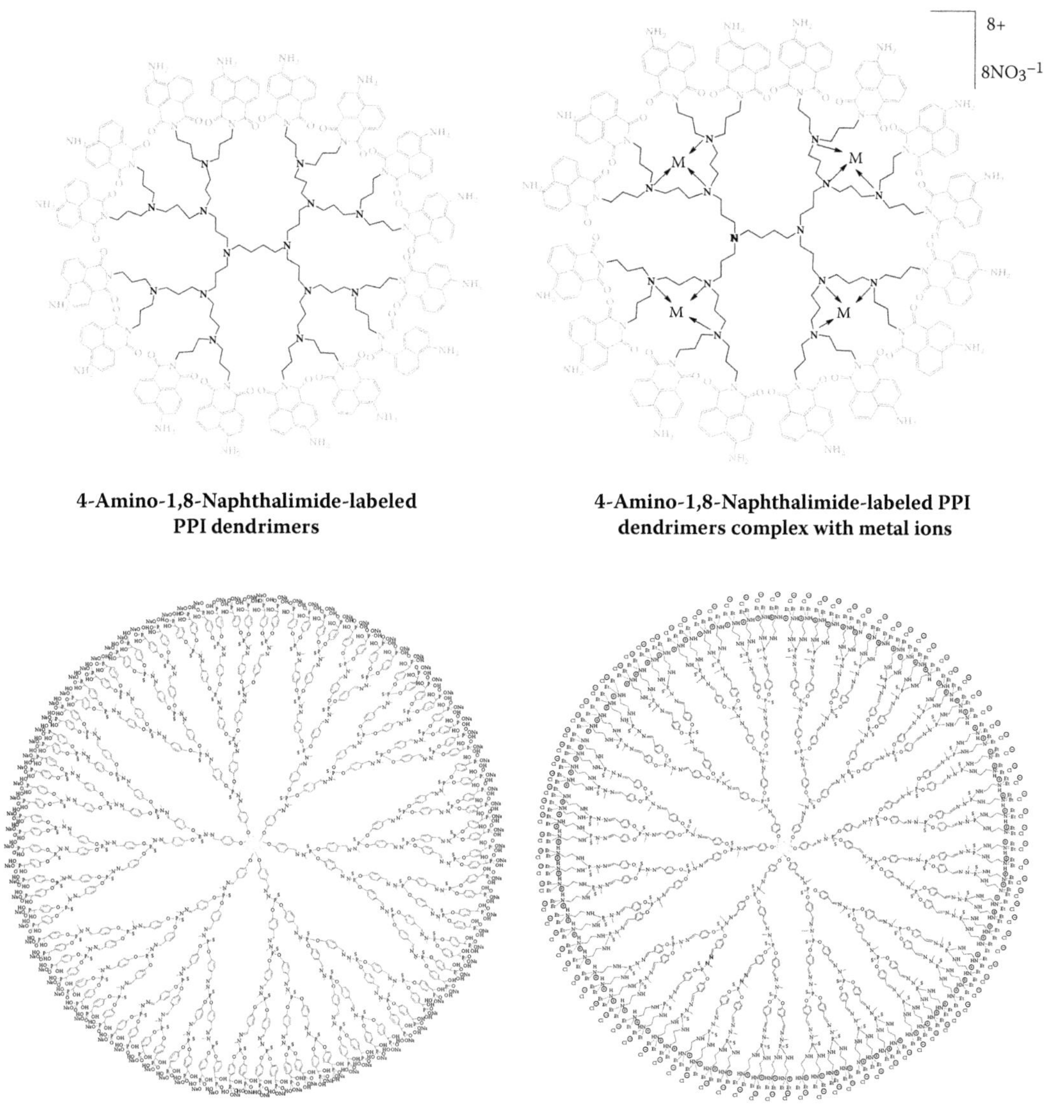

FIGURE 6.3 Surface-engineered dendrimers reported to exhibit antimicrobial activity.

6.3.2 Formation of Nanohybrids

Currently, nanohybrids are extensively being investigated for various biomedical applications and development of antimicrobial agent is not an exception to this strategy. Evaluation of nanohybrids formed using two or more nanomaterials for therapeutic activity as well as for delivery of bioactives including genes, drugs, peptides, hormones, diagnostic and contrast agents, etc. is currently under investigation (Neelgund *et al.*, 2012; Jain *et al.*, 2013; Sadeghi-Kiakhani *et al.*, 2015; Bajwa *et al.*, 2016).

Murugan and Vimala developed two nanohybrids using multiwalled carbon nanotubes (MWCNTs) with carboxyl functional groups (MWCNTs-COOH) and PPI dendrimers; first nano-hybrid was PPI dendrimers conjugated MWCNTs-COOH, and second nanohybrid was obtained by deposition of silver nanoparticles into these PPI functionalized MWCNTs-COOH. *In vitro* evaluation showed promising antimicrobial activity for these nanohybrids against *S. aureus*, *E. coli* and *B. Subtilis*. Further the formation of nanohybrids of MWCNTs-COOH with amphiphilic PPI dendrimers also aided in improving the dispersibility of CNTs as well as stability for more than six months. Among the two nanohybrids and MWCNTs-COOH, silver nanoparticles deposited PPI functionalized MWCNTs-COOH showed greatest antimicrobial activity followed by PPI functionalized MWCNTs-COOH and least activity was shown by MWCNTs-COOH (Murugan and Vimala, 2011) (Figure 6.4). Neelgund *et al.* (2012) developed a nanohybrid by grafting PAMAM dendrimers on MWCNTs followed by conjugation of Ag_2S and CdS quantum dots (QDs). The antibacterial activity of this nanohybrid was found to be superior than a single nanomaterial including carboxylated MWCNTs, PAMAM dendrimer, CdS and Ag_2S QDs as well as more than combination of two nanomaterials for example PAMAM-grafted MWCNTs.

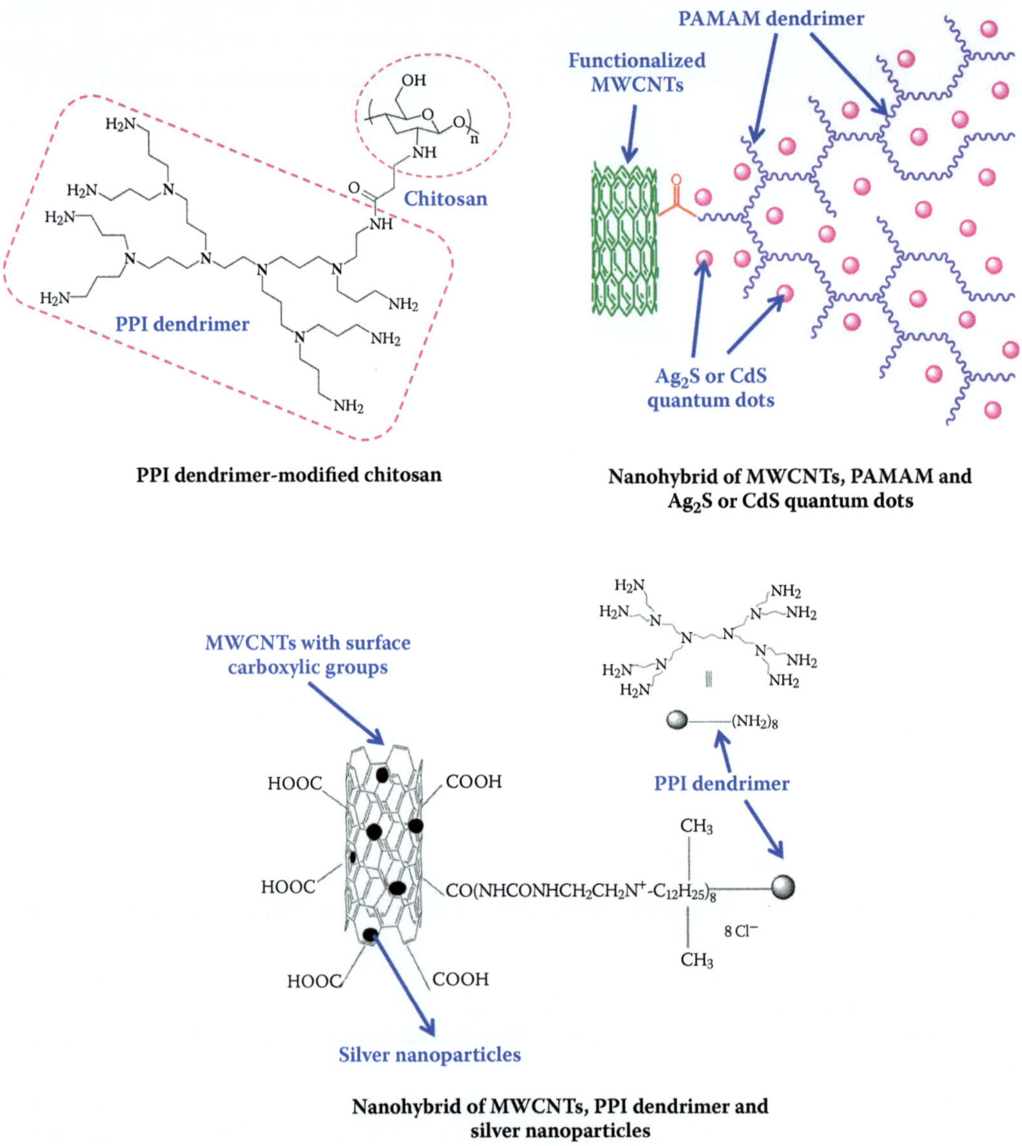

FIGURE 6.4 Dendrimer-based nanohybrids with antimicrobial activity.

Dendrimer-modified chitosan grafted on cotton represents another example of engineered dendrimers for antimicrobial activity (Figure 6.4) (Klaykruayat *et al.*, 2010; Sadeghi-Kiakhani *et al.*, 2015). In antimicrobial activity studies conducted by Sadeghi-Kiakhani *et al.* (2015), cotton fabric grafted with PPI dendrimers showed 100% reduction of bacterial burden at a 20% concentration of a chitosan-PPI dendrimer hybrid. Further the antimicrobial activity was found to be retained even after twelve washing periods. Similarly, Klaykruayat *et al.* (2010) also observed similar results for PAMAM modified chitosan grafted on cotton fabric. Summary of more studies exploring antimicrobial potential of nanohybrids based on dendrimers is given in Table 6.1.

Although only few studies are available on dendrimers based nanohybrids for antimicrobial activity but the results are promising and hence suggesting a strong possibility for the development of such nanohybrids to combat with the global challenge of developing microbial resistance to anti-infective agents.

6.4 DESIGNING NOVEL DENDRIMERS

To fight with the antimicrobial agents, resistant bacteria is becoming global problem. The need to develop potent and strong antimicrobial agents which can work against MDR opportunistic pathogens is pressing. Several strategies are being investigated by scientists to develop novel antimicrobial agents including dendrimers (Hou *et al.*, 2009; Engler *et al.*, 2012; Hoffknecht *et al.*, 2016).

6.4.1 Peptide Dendrimers

Peptide dendrimers are one common example of antimicrobial dendrimers. The combination of sequence modification and antimicrobial topology could also be a one promising strategy in the development of promising antimicrobial agents (Klajnert *et al.*, 2006a,b; Hou et al., 2009; Stach *et al.*, 2014; Hoffknecht *et al.*, 2015).

Klajnert *et al.* (2006b) designed lysine-based low molecular weight peptide dendrimers containing different proportions of aromatic and cationic groups in their architecture and evaluated for antimicrobial activity against gram-positive bacteria, gram-negative bacteria and fungal strains (namely *S. aureus*, *E. coli* and *Candida albicans*). Along with the evaluation of antimicrobial activity, the researchers also subjected these peptide dendrimers for toxicity evaluation, including genotoxicity, cytotoxicity and hematological toxicity. The results of these studies showed that the toxicity, as well as antimicrobial activity, of developed peptide dendrimers depended on the architecture of dendrimers, including the degree of branching, steric orientation, proportion and type of aromatic and cationic groups. It has also been speculated from this study that the three-dimensional architecture of developed dendrimers is similar to naturally occurring antimicrobial peptides, suggesting a strong possibility for the development of promising antimicrobial dendrimeric peptides in the near future.

Taking advantage of the unique properties of peptide dendrimers over their monomeric counterparts, Bruschi *et al.* (2010) developed a novel antimicrobial dendrimeric peptide, named SB041, by linking four units of pyr EKKIRVRLSA peptide to lysine core using 5-amino valeric acid chain (Figure 6.2). This peptide showed antimicrobial activity towards gram-negative (as well as some activity towards gram-positive) bacteria. But its activity was observed to prefer gram-negative bacteria equivalent to commercially available peptide antibiotics, polymyxin B and colistin. *In vitro* activity showed a high affinity of SB041 towards lipolysaccharides (LPS) of *E. coli* (Bruschi *et al.*, 2010).

Hoffknecht *et al.* (2016) combined peptide dendrimers with metallocenes and developed ferrocenoyl or a ruthenocenoyl group containing twelve new ultra short peptides. In this study, scientists evaluated the antibacterial activity of these peptide conjugates with metallocenes and without metallocenes by determining minimal inhibitory concentration (MIC) using five bacterial strains containing two gram-negative bacterial strains *i.e. E. coli*, *A. baumannii*, and three gram-positive strains *i.e. B. subtilis*, *S. aureus* and S. aureus (MRSA). The result of these antibacterial activity studies showed a higher activity for peptide conjugates containing metallocenoyl groups (Figure 6.2).

The product of Star Pharma, VivaGel® (SPL7013) is an excellent example of dendrimers based antiviral agent. It has been investigated successfully in phase I and II clinical trials for antiviral action. This 4.0G naphthalene di-sulfonic acid containing poly-L-lysine dendrimer has shown promising inhibitory action against HSV and HIV-1. Although the desired end point could not be achieved in phase III clinical trials but it has given a hope that in near or far future we can be able to develop dendrimers based antiviral drug. VivaGel®, a topical microbicide available in gel form, is chemically constituted of 4.0G poly-L-lysine dendrimer with 32 anionic naphthalene sulfonic acid groups. The antiviral action of this macromolecule with 16 kDa molecular weight is speculated to interactions of its negative surface groups with gp120, a glycoprotein with positive charge, present in viral envelop and hence interferes with viral entry to act as entry inhibitors (Li *et al.*, 2015; Antimisiaris and Mourtas, 2015). Instead of its failure in phase III clinical trial against HIV, VivaGel® is still being investigated for treatment and prevention of bacterial vaginosis and currently under phase III clinical trial. It is also being examined for prevention of sexually transmitted infections as well as available commercially as condoms (VivaGel® condom) in market of Australia under Ansell's LifeStyles® Dual Protect™ brand (Li *et al.*, 2015; Starpharma 2016).

6.4.2 Carbosilane Dendrimers

Carbosilane dendrimers have been also found to exert antiviral activity. Water-soluble polyanionic carbosilane dendrimers incorporating either sulphated or naphthyl-sulphonated end groups have also shown nonspecific microbicidal activity, including HIV-1, by blocking entry into epithelial cells without causing any irritation to vaginal epithelial tissue in mice models (Antimisiaris and Mourtas, 2015; Ceña-Diez *et al.*, 2016a,b). The chemical structure of a G3-S16-sulphated carbosilane dendrimer is shown in Figure 6.5. The binding affinity of these dendrimers with HSV-2 surface protein gB depends on generation and number of anionic surface groups. G1-S4 showed a greater binding affinity towards gB surface proteins in comparison to G2-S16. G1-S4 and G3-S16 inactivated HSV-2 by directly binding to it and G2-S16 bound to host cell-surface proteins. Muñoz-Fernández *et al.* also observed synergistic activity against HSV-2 for polyanionic carbosilane dendrimers with commercial antiviral drugs acyclovir and tenofovir. It has been also observed that polyanionic carbosilane dendrimers can also inhibit the propagation of HIV-1 infection *via* inhibiting cell-to-cell HIV transmission, even it is also active in the case of mock- and semen-exposed HIV-1 (Ceña-Diez *et al.*, 2016a,b; Vacas-Córdoba *et al.*, 2016).

6.4.3 PAMAM Dendrimers

Wang *et al.* (2010), investigated the potential of PAMAM dendrimers as topical microbicides and evaluated its efficacy to treat vaginal infections using the guinea pig model of chorioamnionitis as well as the mechanisms of action. This study showed that 4.0G PAMAM dendrimers with surface hydroxyl groups inhibited the *E. coli* uterine infection in guinea pig models after the topical cervical application of 500 μg of dendrimers. Recently Labena *et al.* (2016) has designed hyperbranched PAMAM dendrimers using a trivalent core molecule, diethyl triamine, and correspondingly using different molar ratios of branching units, methyl methacrylate (Figure 6.5). Two different types of these hyperbranched PAMAM dendrimers were synthesized with two different surface groups, esters and amines. The antimicrobial activity of these dendrimers was evaluated towards *Aspergillus niger*, *C. albicans*, *S. aureus*, *B. subtilis*, *E. coli*, *P. aeruginosa*, and methicillin-resistant *S. aureus*, with promising results suggesting strong future possibilities to develop clinically applicable antimicrobial agents.

6.4.4 Triazine Dendrimers

Triazine dendrimers are synthesized from 1,3,5-triazine represents a biodegradable type of dendrimers. Triazine dendrimers with surface hydroxyl groups have been found biocompatible with

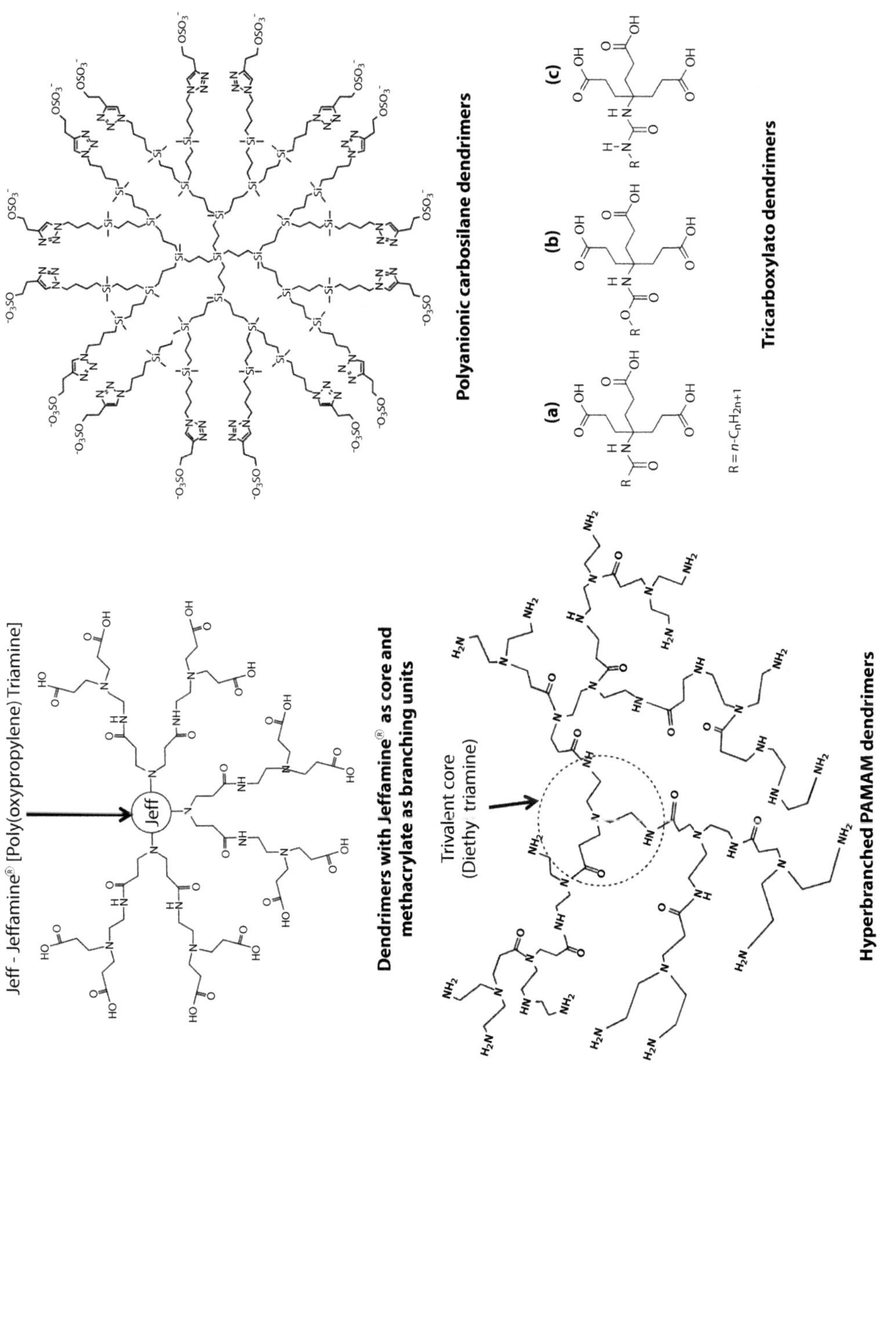

FIGURE 6.5 Chemical structures of dendrimers specifically designed for antimicrobial activity.

different speculated pharmaceutical and biomedical applications. Although these dendrimers are currently under investigation, their biocompatibility and biodegradability could be a plus point in prospective clinical applications (Jain *et al.*, 2010; Mehra *et al.*, 2016; Sreeperumbuduru *et al.*, 2016). Surface-engineered triazine dendrimers have been observed to elicit some promising antibacterial activity. Water-soluble triazine dendrimers with 1,4-diazabicyclo[2.2.2]octane (DABCO) end groups have shown antibacterial activity towards *S. aureus*, methicillin-resistant *S. aureus* and *E. coli* depending on two factors, including the type of alkyl substituent in the DABCO group and the valency of these dendrimeric constructs. The order of activity for alkyl groups on DABCO functionalities was as followed in the increasing order methyl < benzyl < dodecyl. The valency of constructs affected the antimicrobial activity in the following order: divalent/trivalent > tetravalent/hexavalent construct (Sreeperumbuduru *et al.*, 2016).

6.4.5 Miscellaneous

Scientists have also designed some other types of dendrimers using different core or branching units like tricarboxylato dendrimers, and Jeffamine®-based dendrimers. Tulu *et al.* (2009) designed water-soluble dendrimers using poly(oxypropylene) triamine (Jeffamine®) as the core and methacrylate as the branching units (Figure 6.5). The antimicrobial activity of these dendrimers was determined using disk diffusion method which was compared with references including vancomycin, tetracycline, penicillin-G, ampicillin, ofloxacin and cefotaxime. MIC was determined using dilution method and compared with gentamycin and nystatin as antibacterial and antifungal reference antibiotics, respectively. The microorganisms used for this study were five gram-positive bacteria strains (*S. aureus*, *Bacillus cereus*, *Mycobacterium smegmatis*, *Listeria monocytogenes*, *Micrococcus luteus*), five gram-negative bacteria (*E. coli*, *K. pneumoniae*, *P. aeruginosa*, *Proteus vulgaris*, *Enterobacter aerogenes*) and five yeast cultures (including *C. albicans*, *Kluyveromyces fragilis*, *Rhodotorula rubra*, *Debaryomyces hansenii* and *Hanseniaspora guilliermondii*). The antifungal activity of dendrimers was referenced with the antifungal antibiotics including clotrimazole, ketaconazole, and nystatin. The results of these studies suggested that the developed dendrimeric constructs showed broad spectrum activity *i.e.* found active against gram-positive, gram-negative and yeast cultures.

Sugandhi *et al.* synthesized water-soluble tricarboxylato dendrimers and investigated them for antimicrobial activity against *C. albicans* and *Cryptococcus neoformans* by broth microdilution assay as well as determined the effect of hydrophobicity or tail topology, *i.e.* one or two tail and alkyl chain length on antimicrobial activity. The MIC value for these dendrimers showed higher activity of amphiphilic dendrimers with one tail as compared to two-tailed dendrimers. Although the scientists did not determine the mechanism of antimicrobial action of these dendrimers, they hypothesized that the activity may be due to the one or more of the following mechanisms: a change in fluidity and porosity of membranes, alteration in membrane associate pathways like cell signaling, a change in membrane permeability for nutrients and other important constituents, the inhibition of enzymes associate with membrane or involvement in fatty acid synthesis (Sugandhi *et al.*, 2007). Later on, the antibacterial, antifungal activity and cytotoxicity of these tricarboxylato dendrimers was also compared and results showed that these dendrimers could emerge as promising antifungal agents with possible clinical application in the treatment of vaginal infections (Macri *et al.*, 2009). A summary about some research based on applications of dendrimers as antimicrobial agents is given in Table 6.1.

The availability of few reports on antimicrobial activity of a specific dendrimers is a big stumbling block in the development of dendrimers based antimicrobial agents instead of some preclinical promising results. So, it is quite possible that a complete and thorough investigation of dendrimers may lead to a promising antiviral agent that can be explored for the fatal infectious diseases, including AIDS.

6.5 DENDRIMERS FOR THE DELIVERY OF ANTIMICROBIAL AGENTS

Dendrimers have also been explored as promising carrier systems for the delivery of antimicrobial drugs to improve bioavailability, increase the solubility of poorly aqueous soluble drugs, and alter the pharmacokinetic profiles for controlled delivery to improve the therapeutic index. The topical application of two potent quinolone antimicrobial agents (nadifloxacin and prulifloxacin) is limited by their poor aqueous solubility. Cheng *et al.* (2007) explored 5.0G PAMAM dendrimers as delivery vehicle to improve aqueous solubility of these quinolone antibiotics. In the *in vitro* antibacterial activity studies against *E. coli*, these dendrimers based formulation of nadifloxacin and prulifloxacin revealed strong antibacterial activity. Even in case of prulifloxacin, a twofold increase in activity was observed with dendrimers as vehicle in comparison to plain prulifloxacin. This increase in activity was speculated to be attributed to ability of dendrimers to penetrate biological membrane which facilitated the interaction of antibiotic with the target site *i.e.* DNA gyrase and topoisomerase IV, the two bacterial enzymes which are the target site for the mechanism of action of these quinolone antibiotics.

Dendrimers have also been investigated as synergist to antimicrobial agents as well as to deliver various antimicrobial drugs. Wrońska *et al.* (2015) investigated amoxicillin with maltose-conjugated PPI dendrimers for inhibition of gram-negative bacteria. At a concentration of PPI non-toxic to eukaryotic cells, an increase in antibacterial efficacy of amoxicillin was observed at the reduced dose of amoxicillin delivered with maltose-conjugated PPI dendrimers. The delivery of antimicrobial drugs with dendrimers has also resulted in improved pharmacokinetic profiles with targeted delivery and reduction in toxicity associated with the drug. The delivery of an antibiotic, Amphotericin B, with surface-engineered dendrimers, *i.e.* mannose-conjugated and muramyl dipeptide (MDP)–conjugated dendrimers resulted in improved therapeutic index with targeted delivery to macrophages which serve as safe shelter for many parasitic microorganisms during infection (Jain *et al.*, 2015b,c).

The incorporation of 2.0G and 3.0G PAMAM dendrimers into the hydrogel of a lipophilic antibacterial drug, erythromycin, resulted in shear-thinning non-Newtonian thixotropic systems with faster release of erythromycin associated with insignificant improvement in the antibacterial action of erythromycin. It was observed that the presence of PAMAM dendrimers caused the hydrogel to be less compact with more porous structure leading to increase in release rate of drug depending on generation number with sufficient stability at 40 ± 2 °C and 75% ± 5% RH for a period of three months (Wróblewska and Winnicka, 2015).

All these studies demonstrated that dendrimers could emerge as promising agent in the treatment of various infectious diseases either as anti-infective therapeutic agents or as a nanocarrier for delivery of anti-infective drugs. Although few constrains like toxicity of cationic-charged dendrimers are there but these could be alleviated by designing/engineering dendrimers with biocompatible surface groups.

6.6 CONCLUSIONS

Synthetic macromolecule dendrimers could provide an excellent platform to develop novel therapeutic agents. The unique architectures of dendrimers provide an opportunity to combine different core moieties, branching units and surface functional groups to develop a novel macromolecule with interesting therapeutic properties. As we have seen in Chapter 6, peptide dendrimers have been synthesized to combine antimicrobial properties of peptides with dendrimeric architecture for the multimeric presentation of peptide molecules. Hence, it was also noted that these dendrimers were more potent than their corresponding monomer counterparts. Dendrimers like PAMAM, PPI and PLL, which have been investigated extensively for drug and gene delivery, have also shown antimicrobial activity on formation of nanohybrids or surface engineering. Further carbosilane, tricaroxylato, triazine dendrimers have also shown promising antimicrobial activity which should be investigated systematically and thoroughly for *in vitro* and *in vivo* activity, so that they could be finally transformed into clinically applicable formulations with efficacy, safety and sufficient stability. Further, the size and surface charge also play an important role in the functioning of dendrimers as antimicrobial agents. The positive charge is also associated with toxicity which could be alleviated by smartly engineering the dendrimers to elicit biocompatibility. VivaGel® is also expected to be developed as a promising agent to prevent bacterial vaginosis and sexually transmitted diseases, which is currently under phase III clinical trial. PAMAM dendrimer has also shown antimicrobial activity instead of its cytotoxicity, which could be lessened by designing dendrimers with neutral surface groups. So finally, we conclude that dendrimers (or, more appropriately, dendrimeric architecture) encompasses a broad possibility to be developed as a broad-spectrum antimicrobial agent, provided a comprehensive investigation is carried out, since majorly available reports comprises only preclinical investigations at present.

REFERENCES

Abd-El-Aziz AS, Agatemor C, Etkin N, Overy DP, Lanteigne M, McQuillan K *et al.* Antimicrobial organometallic dendrimers with tunable activity against multidrug-resistant bacteria. *Biomacromolecules.* 2015; 16 (11): 3694–703.

Abid CK, Jain S, Jackeray R, Chattopadhyay S, Singh H. Formulation and characterization of antimicrobial quaternary ammonium dendrimer in poly (methyl methcarylate) bone cement. *J Biomed Mater Res B Appl Biomater.* 2015; 105: 3: 521–530.

Antimisiaris SG, Mourtas S. Recent advances on anti-HIV vaginal delivery systems development. *Adv Drug Deliv Rev.* 2015; 92:123–45.

Aryabadie S, Sadeghi-Kiakhani M, Arami M. Antimicrobial and dyeing studies of treated cotton fabrics by prepared chitosan-PAMAM dendrimer/Ag nano-emulsion. *Fibers and Polymers.* 2015; 16: 2529–37.

Bajwa N, Mehra NK, Jain K, Jain NK. Pharmaceutical and biomedical applications of quantum dots. *Artif Cells Nanomed Biotechnol.* 2016; 44(3): 758–68.

Batoni G, Casu M, Giuliani A, Luca V, Maisetta G, Mangoni ML *et al.* Rational modification of a dendrimeric peptide with antimicrobial activity: Consequences on membrane-binding and biological properties. *Amino Acids.* 2016; 48(3): 887–900.

Briz V, Sepúlveda-Crespo D, Diniz AR, Borrego P, Rodes B, de la Mata FJ *et al.* Development of water-soluble polyanionic carbosilane dendrimers as novel and highly potent topical anti-HIV-2 microbicides. *Nanoscale.* 2015; 7(35): 14669–83.

Bruschi M, Pirri G, Giuliani A, Nicoletto SF, Baster I, Scorciapino MA *et al.* Synthesis, characterization, antimicrobial activity and LPS-interaction properties of SB041, a novel dendrimeric peptide with antimicrobial properties. *Peptides.* 2010; 31(8): 1459–67.

Carta F, Osman SM, Vullo D, AlOthman Z, Del Prete S, Capasso C *et al.* Poly(amidoamine) dendrimers show carbonic anhydrase inhibitory activity against α-, β-, γ- and η-class enzymes. *Bioorg Med Chem.* 2015; 23(21): 6794–8.

Ceña-Diez R, García-Broncano P, de la Mata FJ, Gómez R, Muñoz-Fernández MÁ. Efficacy of HIV antiviral polyanionic carbosilane dendrimer G2-S16 in the presence of semen. *Int J Nanomedicine.* 2016b; 11: 2443–50.

Ceña-Diez R, Vacas-Córdoba E, García-Broncano P, de la Mata FJ, Gómez R, Maly M *et al.* Prevention of vaginal and rectal herpes simplex virus type 2 transmission in mice: Mechanism of antiviral action. *Int J Nanomedicine.* 2016a; 11: 2147–62.

Chen CZ, Beck-Tan NC, Dhurjati P, van Dyk TK, LaRossa RA, Cooper SL. Quaternary ammonium functionalized poly(propylene imine) dendrimers as effective antimicrobials: Structure-activity studies. *Biomacromolecules.* 2000; 1(3): 473–80.

Chen CZ, Cooper SL. Interactions between dendrimer biocides and bacterial membranes. *Biomaterials.* 2002; 23(16): 3359–68.

Cheng Y, Qu H, Ma M, Xu Z, Xu P, Fang Y *et al.* Polyamidoamine (PAMAM) dendrimers as biocompatible carriers of quinolone antimicrobials: An *in vitro* study. *Eur J Med Chem.* 2007; 42(7): 1032–8.

Engler AC, Wiradharma N, Ong ZY, Coady DJ, Hedrick JL, Yang YY. Emerging trends in macromolecular antimicrobials to fight multi-drug-resistant infections. *Nano Today.* 2012; 7: 201–22.

Grabchev I, Yordanova S, Vasileva-Tonkova E, Bosch P, Stoyanov S. Poly(propylenamine) dendrimers modified with 4-amino-1,8-naphthalimide: Synthesis, characterization and *in vitro* microbiological tests of their Cu(II) and Zn(II) complexes. *Inorganica Chimica Acta* 2015; 438: 179–88.

Hoffknecht BC, Prochnow P, Bandow JE, Metzler-Nolte N. Influence of metallocene substitution on the antibacterial activity of multivalent peptide conjugates. *J Inorg Biochem.* 2016; 160: 246–9.

Hoffknecht BC, Worm DJ, Bobersky S, Prochnow P, Bandow JE, Metzler-Nolte N. Influence of the multivalency of ultrashort arg-trp-based antimicrobial peptides (amp) on their antibacterial activity. *Chem Med Chem.* 2015; 10(9): 1564–9.

Hou S, Zhou C, Liu Z, Young AW, Shi Z, Ren D *et al.* Antimicrobial dendrimer active against *Escherichia coli* biofilms. *Bioorg Med Chem Lett.* 2009; 19(18): 5478–81.

Jain A, Jain K, Mehra NK, Jain NK. Lipoproteins tethered dendrimeric nanoconstructs for effective targeting to cancer cells. *J Nanopart Res.* 2013; 15(10): 1–18.

Jain K, Jain NK. Novel therapeutic strategies for treatment of visceral leishmaniasis. *Drug Discov Today.* 2013; 18(23–24): 1272–81.

Jain K, Jain NK. Vaccines for visceral leishmaniasis: A review. *J Immunol Methods.* 2015; 422: 1–12.

Jain K, Kesharwani P, Gupta U, Jain NK. Dendrimer toxicity: Let's meet the challenge. *Int J Pharm.* 2010; 394(1-2): 122–42.

Jain K, Mehra NK, Jain NK. Nanotechnology in drug delivery: Safety and toxicity issues. *Curr Pharm Des.* 2015a; 21(29): 4252–61.

Jain K, Verma AK, Mishra PR, Jain NK. Characterization and evaluation of amphotericin B loaded MDP conjugated poly (propylene imine) dendrimers. *Nanomedicine.* 2015b; 11(3): 705–13.

Jain K, Verma AK, Mishra PR, Jain NK. Surface-engineered dendrimeric nanoconjugates for macrophage-targeted delivery of Amphotericin B: Formulation development and *in vitro* and *in vivo* evaluation. *Antimicrob Agents Chemother.* 2015c; 59(5): 2479–87.

Janiszewska J, Sowińska M, Rajnisz A, Solecka J, Łącka I, Milewski S *et al.* Novel dendrimeric lipopeptides with antifungal activity. *Bioorg Med Chem Lett.* 2012; 22(3): 1388–93.

Janiszewska J, Swieton J, Lipkowskia AW, Urbanczyk-Lipkowska Z. Low molecular mass peptide dendrimers that express antimicrobial properties. *Bioorg Med Chem Lett.* 2003; 13: 3711–13.

Kaur A, Jain K, Mehra NK, Jain NK. Dendrimer Internalization: A Systematic Review. *J Colloid Sci Biotech.* 2015; 4(2): 99–109.

Kaur A, Jain K, Mehra NK, Jain NK. Development and characterization of surface engineered PPI dendrimers for targeted drug delivery. *Artif Cells Nanomed Biotechnol.* 2017; 45(3): 414–425.

Kaur D, Jain K, Mehra NK, Kesharwani P, Jain NK. A review on comparative study of PPI and PAMAM dendrimers. *J Nanopart Res.* 2016; 18(6): 1–14

Klajnert B, Janiszewska J, Urbanczyk-Lipkowska Z, Bryszewska M, Epand RM. DSC studies on interactions between low molecular mass peptide dendrimers and model lipid membranes. *Int J Pharm.* 2006a; 327(1–2): 145–52.

Klajnert B, Janiszewska J, Urbanczyk-Lipkowska Z, Bryszewska M, Shcharbin D, Labieniec M. Biological properties of low molecular mass peptide dendrimers. *Int J Pharm.* 2006b; 309(1–2): 208–17.

Klaykruayat B, Siralertmukul K, Srikulkit K. Chemical modification of chitosan with cationic hyperbranched dendritic polyamidoamine and its antimicrobial activity on cotton fabric. *Carbohydr Polym.* 2010; 80: 197–207.

Labena A, Kabel KI, Farag RK. One-pot synthesize of dendritic hyperbranched PAMAM and assessment as a broad spectrum antimicrobial agent and anti-biofilm. *Mater Sci Eng C Mater Biol Appl.* 2016; 58: 1150–9.

Li J, Yu F, Chen Y, Oupický D. Polymeric drugs: Advances in the development of pharmacologically active polymers. *J Control Release.* 2015; 219: 369–82.

Lind TK, Darré L, Domene C, Urbanczyk-Lipkowska Z, Cárdenas M, Wacklin HP. Antimicrobial peptide dendrimer interacts with phosphocholine membranes in a fluidity dependent manner: A neutron reflection study combined with molecular dynamics simulations. *Biochim Biophys Acta.* 2015; 1848: 2075–84.

Macri RV, Karlovská J, Doncel GF, Du X, Maisuria BB, Williams AA *et al.* Comparing anti-HIV, antibacterial, antifungal, micellar, and cytotoxic properties of tricarboxylato dendritic amphiphiles. *Bioorg Med Chem.* 2009; 17: 3162–8.

Mehra NK, Jain K, Jain NK. Triazine dendrimers: Biomedical application. *In:* Encyclopedia of Biomedical Polymers and Polymeric Biomaterials. Taylor and Francis: New York; 2016; pp. 8062–73.

Michaud G, Visini R, Bergmann M, Salerno G, Bosco R, Gillon E *et al.* Overcoming antibiotic resistance in *Pseudomonas aeruginosa* biofilms using glycopeptide dendrimers. *Chem Sci.* 2016; 7: 166–82.

Milowska K, Rybczyńska A, Mosiolek J, Durdyn J, Szewczyk EM, Katir N *et al.* Biological activity of mesoporous dendrimer-coated titanium dioxide: Insight on the role of the surface–interface composition and the framework crystallinity. *ACS Appl Mater Interfaces.* 2015: 7(36): 19994–20003.

Murugan E, Arumugam S, Panneerselvam P. New nanohybrids from poly (propylene imine) dendrimer stabilized silver nanoparticles on multiwalled carbon nanotubes for effective catalytic and antimicrobial applications. *Int J Polymeric Mat Polymeric Biomat.* 2016; 65: 111–24.

Murugan E, Vimala G. Effective functionalization of multiwalled carbon nanotube with amphiphilic poly(propyleneimine) dendrimer carrying silver nanoparticles for better dispersability and antimicrobial activity. *J Colloid Interface Sci.* 2011; 357(2): 354–65.

Muszanska AK, Rochford ET, Gruszka A, Bastian AA, Busscher HJ, Norde W *et al.* Antiadhesive polymer brush coating functionalized with antimicrobial and RGD peptides to reduce biofilm formation and enhance tissue integration. *Biomacromolecules.* 2014; 15(6): 2019–26.

Neelgund GM, Oki A, Luo Z. Antimicrobial activity of CdS and Ag_2S quantum dots immobilized on poly (amidoamine) grafted carbon nanotubes. *Colloids Surf B.* 2012; 100: 215–21.

Rai M, Ingle AP, Gaikwad S, Gupta I, Gade A, Silvério da Silva S. Nanotechnology based anti-infectives to fight microbial intrusions. *J Appl Microbiol.* 2016; 120(3): 527–42.

Sadeghi-Kiakhani M, Gharanjig K, Arami M. Grafting of prepared chitosan–poly(propylene) imines dendrimer hybrid as a biopolymer onto cotton and its antimicrobial property. *Ind Eng Chem Res.* 2015; 28: 78–85.

Singh J, Jain K, Mehra NK, Jain NK. Effect of anticancer drug on delivery potential of poly (propylene imine) dendrimers. *J Colloid Sci Biotech.* 2015; 4(2): 133–40.

Singh J, Jain K, Mehra NK, Jain NK. Dendrimers in anticancer drug delivery: Mechanism of interaction of drug and dendrimers. *Artif Cells Nanomed Biotechnol.* 2016; 4(7): 1626–34.

Soni N, Jain K, Gupta U, Jain NK. Controlled delivery of gemcitabine hydrochloride using mannosylated poly (propyleneimine) dendrimers. *J Nanopart Res.* 2015; 17(11): 1–17.

Sreeperumbuduru RS, Abid ZM, Claunch KM, Chen HH, McGillivray SM, Simanek EE. Synthesis and antimicrobial activity of triazine dendrimers with DABCO groups. *RSC Adv.* 2016; 6: 8806–10.

Stach M, Siriwardena TN, Köhler T, van Delden C, Darbre T, Reymond JL. Combining topology and sequence design for the discovery of potent antimicrobial peptide dendrimers against multidrug-resistant Pseudomonas aeruginosa. *Angew Chem Int Ed Engl.* 2014; 53(47): 12827–31.

Starpharma. Available online at: www.starpharma.com/vivagel. (Accessed on 19 June, 2016).

Sugandhi EW, Falkinham JO 3rd, Gandour RD. Synthesis and antimicrobial activity of symmetrical two-tailed dendritic tricarboxylato amphiphiles. *Bioorg Med Chem.* 2007; 15(11): 3842–53.

Suleman N, Kalhapure RS, Mocktar C, Rambharose S, Singh M, Govender T. Silver salts of carboxylic acid terminated generation 1 poly (propyl ether imine) (PETIM) dendron and dendrimers as antimicrobial agents against *S. aureus* and MRSA. *RSC Adv.* 2015; 5: 34967–78.

Tulu M, Aghatabay NM, Senel M, Dizman C, Parali T, Dulger B. Synthesis, characterization and antimicrobial activity of water soluble dendritic macromolecules. *Euro J Med Chem.* 2009; 44: 1093–99.

Vacas-Córdoba E, Maly M, De la Mata FJ, Gómez R, Pion M, Muñoz-Fernández MÁ. Antiviral mechanism of polyanionic carbosilane dendrimers against HIV-1. *Int J Nanomedicine.* 2016; 11: 1281–94.

Vembu S, Pazhamalai S, Gopalakrishnan M. Potential antibacterial activity of triazine dendrimer: Synthesis and controllable drug release properties. *Bioorg Med Chem.* 2015; 23(15): 4561–6.

Wang B, Navath RS, Menjoge AR, Balakrishnan B, Bellair R, Dai H *et al.* Inhibition of bacterial growth and intramniotic infection in a guinea pig model of chorioamnionitis using PAMAM dendrimers. *Int J Pharm.* 2010; 395(1–2): 298–308.

Wróblewska M, Winnicka K. The effect of cationic polyamidoamine dendrimers on physicochemical characteristics of hydrogels with erythromycin. *Int J Mol Sci.* 2015; 16(9): 20277–89.

Wrońska N, Felczak A, Zawadzka K, Poszepczyńska M, Różalska S, Bryszewska M *et al.* Poly(propylene imine) dendrimers and amoxicillin as dual-action antibacterial agents. *Molecules.* 2015; 20(10): 19330–42.

Xue XY, Mao XG, Li Z, Chen Z, Zhou Y, Hou Z *et al.* A potent and selective antimicrobial poly (amidoamine) dendrimer conjugate with LED209 targeting QseC receptor to inhibit the virulence genes of Gram-negative bacteria. *Nanomedicine.* 2015; 11(2): 329–39.

Zainul Abid CKV, Jackeray R, Jain S, Chattopadhyay S, Asif S, Singh H. Antimicrobial efficacy of synthesized quaternary ammonium polyamidoamine dendrimers and dendritic polymer network. *J Nanosci Nanotech.* 2016; 16: 998–1007.

Zhan J, Wang L, Liu S, Chen J, Ren L, Wang Y. Antimicrobial hyaluronic acid/poly (amidoamine) dendrimer multilayer on poly(3-hydroxybutyrate-co-4-hydroxybutyrate) prepared by a layer-by-layer self-assembly method. *ACS Appl Mater Interfaces.* 2015; 7(25): 13876–81.

7 Production and Utilization of Nanofibers as Promising Biomaterials

Yaser Dahman, Valdir Mota Junior, Yan Xuan, Simon Nagy, Sumant Saini, Jasleen Kaur, and Bilal Khan*

CONTENTS

7.1 Nanotechnology 139
7.2 Nanofibers 140
7.3 Synthesis Techniques 143
7.3.1 Polymer Nanofibers Assembled by Electrospinning 143
7.3.2 Production of Self-Assembling Biomaterials for Tissue Engineering 144
7.3.3 Phase Separation, Pore Structure, and Properties of Nanofibrous Gelatin Scaffolds 146
7.3.4 Template Synthesis of Polypyrrole Nanofibers Insulated within One-Dimensional Silicate Channels: Hexagonal versus Lamellar for Recombination of Polarons into Bipolaons 146
7.3.5 Poly(Ethylene Terephthalate) Nanofiber Made by Sea-Island–Type Conjugated Melt Spinning and Laser-Heated Flow Drawing 147
7.3.6 Effect of Experimental Parameters on the Synthesis of Carbon Nanotube/Nanofiber-Supported Platinum by Polyol Processing Techniques 147
7.3.7 Nanostructured Zinc Oxide Cotton Fibers: Synthesis, Characterization, and Applications 148
7.3.8 Preparation of Cellulose Fibers with Antibacterial Ag-Loading Nano SiO_2 149
7.3.9 How Carbon Nanofibers Attach to Ni Foam 151
7.3.10 Study of Nanofiber Cellulose Production by *Glucanacetobacter xylinum* ATCC 10245 152
7.4 Applications of Nanofibers 154
7.4.1 Drug Delivery of Nanofibers 154
7.4.2 Nonwoven Thermal Insulation 154
7.4.3 3 Tissue Engineering 155
7.4.4 Protective Clothes 159
7.4.5 Mercury Detection 160
7.4.6 Use of Nanofibers as Noise Absorbers for Aircrafts 161
7.5 Conclusion 162
References 162

7.1 NANOTECHNOLOGY

In recent years, scientists around the world have been increasing their interest in developing materials with small dimensions. Nowadays, nanotechnology is the focus of a great number of studies, due

* Corresponding author.

to the expectation of the impact that nanostructured materials are able to provide in terms of the improvement of quality of life and environmental protection.

According to the National Nanotechnology Initiative (U.S. National Nanotechnology. 2017), a U.S. Government program that coordinates several agencies in the nano field, nanotechnology is "science, engineering, and technology conducted at the nanoscale, which is about 1 to 100 nanometers". One nanometer stands for 10^{-9} m, in other words, each nanometer is equivalent to a billionth of a meter, or a million times smaller than the diameter of a head of a pin, or 80,000 times smaller than the thickness of a hair.

The term "nanotechnology" was first introduced by the Japanese engineer, Norio Tanigushi, in reference to a new technology that extended beyond the control of materials and engineering in the microscale. However, the meaning of the term utilized nowadays is more similar to the concept originated by Eric Drexler, which corresponds to the methodology of fabrication involving the manipulation atom-by-atom (Regis 2004).

Research and developments in nanotechnology basically use two main approaches:

- Bottom-up: Manipulates structures in the nanoscale and combines them to form bigger components and systems.
- Top-down: Starts from a structure in the larger scale (e.g. micro, macro) in order to achieve materials in the nanoscale.

There are several sectors in the nanotechnology area, including nanocatalysts, nanotubes, nanomagnetics, and nanofibers, out of which the latter is the focus of this work.

7.2 NANOFIBERS

Nanofibers can be defined as slender, elongated, thread-like objects or structures in the nanoscale. Even though part of the academic community has somewhat agreed that just materials with at least one of the dimensions smaller than 100 nm are classified as nanomaterials, the commercial sector has adopted a more flexible criterion, according to which fibers with diameters under 1000 nm are in the category of nanofibers as well. Nanofibers are characterized by TG-DSC, XRD, FT-IR, SEM, TEM, and spectrophotometer. For calcined nanofibers (ex. $BiPO_4$:Tb^{3+}/Ce^{3+} nanofibers), they have a monoclinic phase with diameters randing from 80 to 200 nm as shown in Figure 7.1, adapted from Instituto Madrileño (Yang *et al.*, 2016).

Investments in nanoscience and nanotechnology are provided by funding agencies around the world are on the order of billions of dollars. According to BCC study entitled "Global Market and Technologies for Nanofibers", the global nanotechnology market was estimated to be worth over

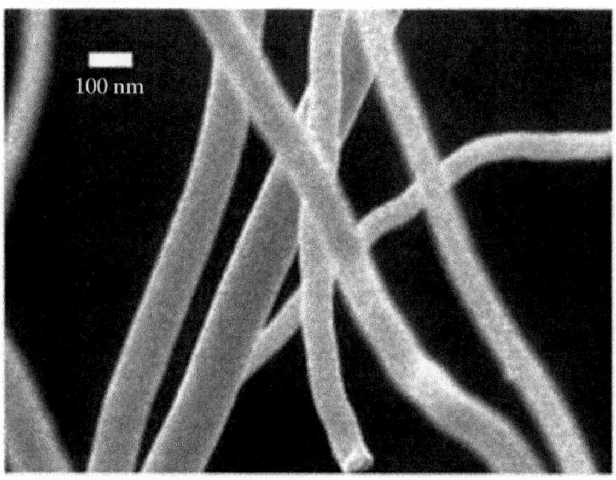

FIGURE 7.1 SEM image of carbon nanofibers.

$20 billion in 2020. Even though nanofibers represent just 1% of the nanotechnology market, the nanofibers sector is expanding at a faster rate than other segments of nanotechnology sector, indicating that their participation is expected to be greater in the future (Gagliardi 2013).

The effects of surface become more important in the nanoscale due to the increase of the ratio between the area and the volume of nanomaterials, including nanofibers. The behavior of nanofibers cannot be explained only by the theories of classical physics (quantum-like phenomena can be observed on the nanoscale), so a new theory linked to classical mechanics and quantum mechanics is formed. The El-Naschie's E-infinity theory is an example of theory that was created to explain the behavior of nanofibers and nanomaterials in general (Agop *et al.,* 2008).

Outstanding properties can be found in nanofibers due to its small dimensions:

- High surface energy
- Great surface reactivity
- High thermal and electrical conductivity
- High strength

Nanofibers are made of several types of materials, including polymers, composites, metals, and ceramics. In terms of materials, polymer nanofibers constitute the largest share of the total nanofibers market (Centre for Knowledge). Polymers mostly used include: polyvinyl chloride (PVC), polystyrene, polyvinylpyrrolidone (PVP), polyvinyl alcohol (PVA), polyvinyl butyral (PVB) and polyethylene glycol (PEO or POE). When dealing with inorganic nanomaterials, tungsten have been one of the most popular material to use due to their unique structure, luminescent behavior, and promising application (Jung *et al.,* 2016).

The applications of nanofibers include: reinforced fibers, support for enzymes, drug delivery system, fuel cells, photonics, pharmacy, wound dressing, medicine, filtration, tissue engineering, conducting polymers and composites, fiber templates for the preparation of functional nanotubes, and others.

There are some methods that are used to synthesize nanofibers, including drawing, template synthesis, phase separation, self-assembly, and electrospinning. The most widely used method is electrospinning. Figure 7.2 illustrates the increase of the number of publications in the electrospinning area.

Three countries are responsible for most of the research in the electrospinning nanofibers area: South Korea, China, and the United States. The most important institutes in the world in electrospinning nanofibers are found in He *et al.* (2008). The most important institutes in the world

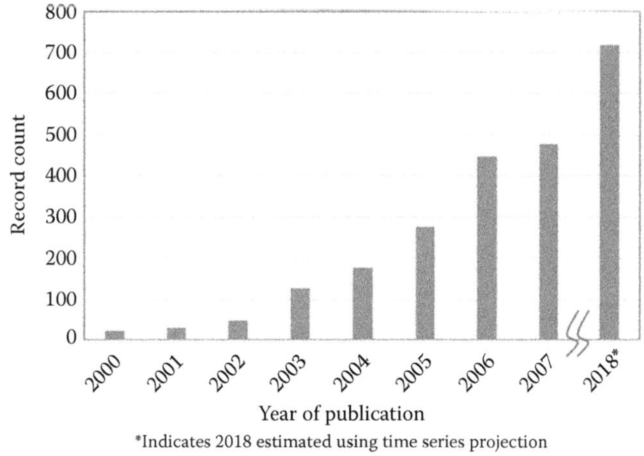

FIGURE 7.2 Increasing number of publications in electrospinning.

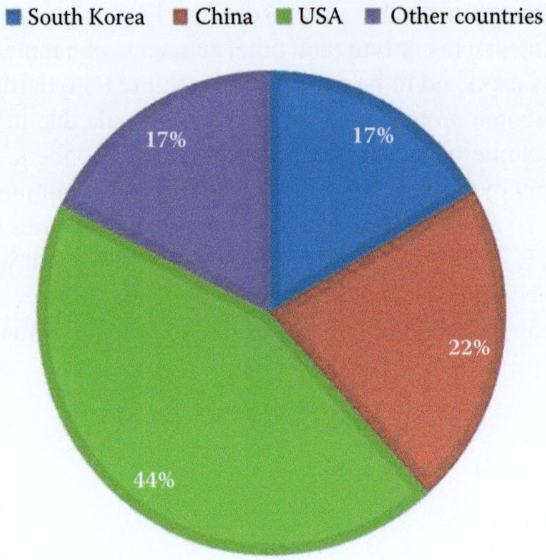

FIGURE 7.3 Global interest in electrospinning nanofibers.

engaged in electrospinning nanofibers till 2008 are listed as under, while their respective contribution towards quantum of publications is illustrated in the Figure 7.3.

1. National University of Singapore – Singapore
2. University of Akron – United States
3. Chonbuk National University – South Korea
4. Chinese Academy of Sciences – China
5. Drexel University – United States
6. Jilin University – China
7. Northeast Normal University – China
8. University of Massachusetts – United States
9. Chungnam National University – South Korea
10. Virginia Commonwealth University – United States

Besides, as on today, there are several other institutes across the world too that have been making tabglible mark in the domain, a few names among those are enumerated as follows:

1. Georgia Institute of Technology and Emory University – United States
2. University of British Columbia – Canada
3. University of Cincinnati – United States
4. University of Nebraska Medical Center – United States
5. National Physical laboratory – India
6. National University Corporation Shinshu University – Japan
7. Donghua University – China
8. Indian Institute of Technology – India
9. Universiti Malaysia Pahang – Malaysia
10. National Institute of Technology – India

An evidence that the nanofibers are becoming more important every year is the increase of the investment in the United States between 2000 and 2012, though slightly declining thereafter, as illustrated in Table 7.1 (U.S. National Nanotechnology Initiative; 2000, 2004, 2008, 2012, 2016, and 2017).

TABLE 7.1
Increase of Investment in Electrospinning in the United States, for Financial Years 2000–2017

	Actual ($ million)					
Federal Department or Agency	**2000**	**2004**	**2008**	**2012**	**2016**	**2018**
National Science Foundation	97	249	408	455.9	416.4	415
Department of Defense	70	222	460	368.2	130.4	131
Department of Energy	58	197	244.7	610.6	342.2	361
National Institutes of Health	32	70	304.5	464.8	423	382
National Institutes of Standards and Technology (NIST)	8	62	85.6	115.7	86.3	82
National Aeronautics and Space Administration (NASA)	5	31	17.4	32.3	15.8	6
Environmental Protection Agency	0	5	12.1	19.8	15.3	15
Total	**270**	**836**	**1532.3**	**2067.3**	**1429.4**	**1392**

Most of the applications require nanofibers with specific characteristics, which are possible to obtain just under certain experimental conditions (e.g. temperature, humidity, and solution concentration). The control of variables during the process of achievement of nanofibers is crucial to obtain the desired type of material.

7.3 TECHNIQUES OF SYNTHESIS

As in many industries, there are several techniques for the manufacturing of nanomaterials. The main areas include drawing, template synthesis, phase separation, self-assembly, and electrospinning. Each of these has their own advantages and disadvantages, as well as their specializations in specific uses and fields. The desired use of nanomaterials is most important when selecting a method of fabrication. Each has its own variety of techniques, where the solvents, polymers, scaffolds, etc. must be taken into account in order to assure the appropriate material properties are reached. Those reviewed in this report are just one of the many options in processes with many variables, and thus many alternatives for their synthesis are available.

7.3.1 Polymer Nanofibers Assembled by Electrospinning

This section covers the basic manufacturing methods used by several different groups of nanofibers via electrospinning. The system and process parameters, as well as their effects on the characteristics of the fibers are also considered. The average electrospun fibers' size and uniformity are important for process economics and the product's performance. In many applications, the average electrospun fibers' size and uniformity are important for the product's performance and process economics. Thus, it is desirable for electrospinning to achieve consistency (Moazeni *et al.*, 2017). A few possible applications of these nanofibers are also described in this report.

Fibers are created by passing a polymer fluid charged by a high voltage electrode through a capillary, where it coalesces into a cone before being collected on a plate. The major factors in the quality of the resulting fibers come from the orientation of the capillary (along with the charging electrode), the orientation of the collecting plate as well as any other adaptations made to it, and most importantly the properties of the polymer being used (Frenot and Chronakis 2003).

Capillaries can be placed singly or in groups at a multitude of angles, such as vertically, to allow gravity feed, with angles being used to control flow, along with a pump when required [such as for a feed from the bottom of the unit (Frenot and Chronakis 2003)]. The charging electrode is used either within the fluid or at the tip of the capillary, where the field created must reach the point where the repulsive electrostatic force overcomes the surface tension of the polymer fluid, ejecting the fluid

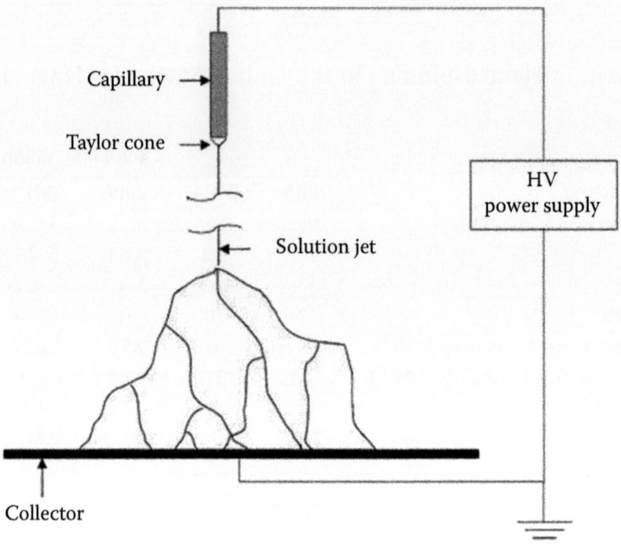

FIGURE 7.4 Scheme of an electrospinning equipment.

from the tip of the cone created. A schematic is shown in Figure 7.4 below (Frenot and Chronakis 2003). The distance between the capillary and the plate can affect the fibers as well, with bead structures appearing at low distances, increasing fiber diameters.

The collecting plate is generally either a metal sheet or a grid that is covered with a sheet of fabric, although rotating cylinders and other methods have been used. As the formation of the fibers is created by an electric field, parallel plates (as well as an evacuated apparatus) have been used to help control the curvature and increase the field strength used (Frenot and Chronakis 2003). The addition of ions into the air of the apparatus can also help control the properties of fibers created from particularly conductive polymer.

Polymer choice greatly influences the applications of the fibers created, as well as their synthesis. The concentration of the polymer must be high enough for entanglements while still allowing the polymer to move, enabling sufficient charge density for the fibers to be collected on the plate. The voltage capacity of the polymer used will affect the formation of the beads and fiber size, making voltage an important parameter with respect to the polymer used.

These methods could be applied to fibers used for membranes, biomedical structural elements (for structures used in artificial organs and the repair of physical wounds, as well as drug delivery), membranes, and other filters, reinforcement in composite and other materials, and structures for nanomachines (Frenot and Chronakis 2003).

7.3.2 Production of Self-Assembling Biomaterials for Tissue Engineering

The materials described within are all biomaterials, for use as 3D scaffolds for tissue engineering as well as for drug-release applications. As there are several alternatives for production, this article focuses on collagen, elastin, and *de novo* (from a small base unit, such as amino acid) production, creating protein and peptide chains. This self-assembly of molecules is possible via the diffusion of components which associate to specific molecules through noncovalent interactions, as well as strong (hydrogen and ionic bonds) and weak (van der Waals) interactions, with large amounts of these interactions overcoming their weakness for assembly (Kyle *et al.*, 2009). Several types of self-assembling systems are shown in Figure 7.5 below (Kyle *et al.* 2009).

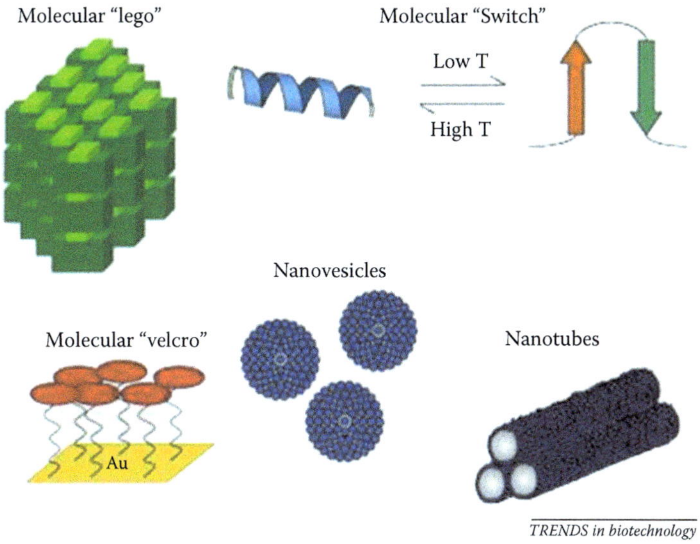

FIGURE 7.5 Various self-assembling peptide systems.

For peptide production, systems using chemical synthesis and recombinant production by transgenic organisms were described. Chemical synthesis is cost effective for small-sequence lengths, performing rapidly for small amino acid sequences, but becomes ineffective during the production of long chains (over 35 amino acids) due to production problems (Kyle *et al.* 2009). The use of transgenic organisms can be very cost effective, and is common in the industry, due to the simplicity of the systems employed, robustness of the microbes used, and lower overall costs. This method is superior than chemical synthesis for long peptides (over 35 amino acids – inferior for short chains), but is hampered by long lead times as well as dangers posed by potential transfer of harmful pathogens from the organism.

There is a natural source common in mammalian body proteins, with at least 27 different types identified, many of which have historically been used in engineering of many components such as valves, ligaments, nerves, and cartilage. They use a transgenic approach, with the common problems of pathogen transmission as well as excessive degradation, as having the collagen produced mimic its natural cousins difficult (Kyle *et al.*, 2009).

Elastin is likewise a natural source providing elasticity in tissues, etc. requiring the ability to extend and recoil through normal use. It is a crosslinked polymer that is highly insoluble, with the ability to form large complex arrays. The use of elastin from a natural source is less than ideal due to the difficulty in purification, as well as its high propensity to calcify upon implantation. For the production of elastin, a good response is typically received from physicochemical triggers, making it a great option for nanofiber production with further development (Foraida *et al.*, 2017).

The *de novo* systems developed include the production of building blocks, the use of "switches", a hook and velcro method, as well as molecular capsules and molecular cavities. They are easy to synthesize chemically and growth can be easily triggered in water by an increase of ionic strength or a change in pH, several examples of which are explained (Roberts 2017).

The authors also explain the new process they have developed using a tandem-repeat strategy for high-yield peptide production which uses a peptide protein within a microbe which is cleaved with a particular molecule on each monomer. This method has the highest reported level of recovery observed from a recombinant system, with emphasis on growth conditions to further maximize production levels.

7.3.3 Phase Separation, Pore Structure, and Properties of Nanofibrous Gelatin Scaffolds

Although natural collagen is an option for scaffolding, pathogen transmission is a concern. Gelatin nanofiber matrices are an option that have been traditionally created using a thermally induced phase separation (TIPS) technique, with macroporous scaffolds being created with the addition of a porogen-leaching process (Guo *et al.*, 2017). This gelatin is made from collagen by acid/base hydrolysis, which eliminates pathogens but lacks the proper architecture.

This paper presents a technique where gelatin is dissolved in an ethanol/water mixture, with the result being phase separated and immersed in cold ethanol, before being immersed in an exchange solvent (1,4-dioxane). The resulting gel was removed and frozen, where it was lyophilized (freeze-dried) for 1 week. Paraffin spheres were added to teflon molds which were preheated before the gelatin (dissolved in water and ethanol) was added (Li *et al.*, 2017). This composite was immersed in cold ethanol and freeze-dried. Finally, the paraffin spheres were dissolved in hexane. Chemical crosslinking of gelatin was performed with 1-Ethyl-3-(3-dimethylaminopropyl)carbodiimide (EDC) and N-Hydroxysuccinimide (NHS) chemicals, with solvent mixtures being used to prevent any matrix swelling. The resulting nanofiber had a diameter similar to that of natural collagen.

The solvent selection for the dissolution step can adjust the phase separation conditions during gelation of the gelatin, with solvent–gelatin interactions being observed as a critical component to the matrix creation. The solvent exchange step was found to be an important step in ensuring the nanofiber architecture was maintained, wherein conventional freeze-drying methods could lead to shrinkage. The gelatin concentration had little effect on the fibers' diameter during the scaffolding phase, with higher concentrations decreasing fiber length and porosity (although the structures were highly porous) (Chong *et al.*, 2007). This technique created scaffolds with very high surface areas compared to other techniques, which can help to enhance protein adsorption and cell adhesion. Pore size and pore interconnectivity could be further controlled using this technique. The crosslinking method used was also found to be less toxic than some others being developed, where the chemicals used (EDC) do not remain within structure. A major component in the structure of the nanofibers was the gelatin concentration. Increasing the available materials affected the fibers as well as the overall structure, as shown in Table 7.2 (Liu *et al.*, 2010).

This technique has great potential for control over several key parameters of nanofiber scaffolds, with new techniques enhancing the safety and biocompatibility of the structures.

7.3.4 Template Synthesis of Polypyrrole Nanofibers Insulated within One-Dimensional Silicate Channels: Hexagonal versus Lamellar for Recombination of Polarons into Bipolaons

Conjugated polymers oriented on a nanoscale have their benefits for oproeletronic (light detection and control) and have been fabricated using templates of unidirectionally-aligned channels and pores, with some of the current methods able to create polymers with electronic conductivity much

TABLE 7.2
Structural Parameters of 3D NF-Gelatin Scaffolds Prepared with Different Gelatin Concentrations

Gelatin Concentration (%)	Diameter of Fibers (nm)	Fiber Length (nm)	Porosity (%)	Surface Area (m^2/g)
5.0	177 ± 62	1181 ± 413	98.06 ± 0.07	33.63
7.5	161 ± 45	666 ± 156	97.51 ± 0.03	34.76
10.0	157 ± 49	497 ± 62	96.45 ± 0.19	36.92

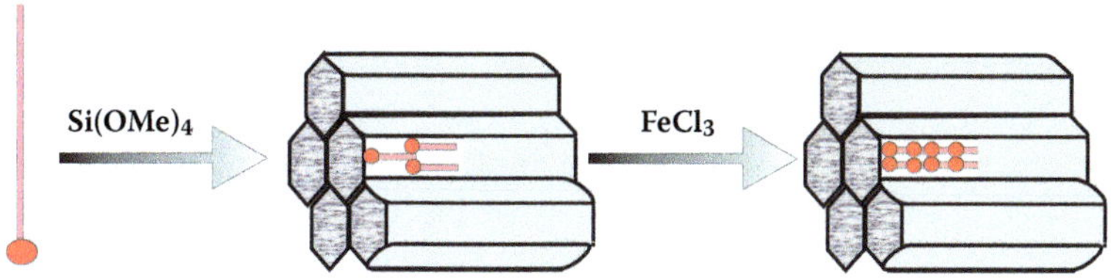

FIGURE 7.6 Template synthesis of a polypyrrole/silica nanocomposite material.

higher than traditional polymers (Ikegame *et al.*, 2003). This article outlines a method by which porous silica channels are filled using a gel-based approach, with the new method ensuring a dense filling of the channels.

The silica film was first synthesized to the desired shape by the addition of tetramethyl orthosilicate stirred with HCl to a specific hyrdocarbon, with the subsequent mixture being condensed and spin coated at 300 rpm onto a glass plate. After drying, the silica framework was completed, an example of which is shown in Figure 7.6 (Ikegame *et al.* 2003). This method showed that a high composition of the specific hydrocarbon used, with this high density being advantageous for the scaffold. Dipping the film in a solution of $FeCl_3$ in dethyl ether stained the film brown, exhibiting an X-ray diffraction absorption spectrum at 1200 nm, ideal for detecting electronic transition from a valence band of doped polypyrrole. The trials ran included the use of odecyltrimethylammonium bromide (DTAB), wherein varying degrees of adsorption could be exhibited by adjusting several factors, the key being the molar ratio of the specific hydrocarbon used by this method to DTAB (Ikegame *et al.* 2003).

These methods are potentially advantageous in that they would allow aligned conductive cables with an insulating coat, ideal for photon detection and other uses of conductive polymers (Peshoria and Narula 2017).

7.3.5 Poly(Ethylene Terephthalate) Nanofiber Made by Sea-Island–Type Conjugated Melt Spinning and Laser-Heated Flow Drawing

This article explores the construction of poly(ethylene terephthalate) (PET) nanofibers created using the technology and methods developed for ultrafine fibers (diameter of 10 micrometers). In particular, the sea-island method is used, wherein "island" fibers are created and then arranged in a "sea" component (with the sea component being extracted at a later time). A high draw ratio is used, where a flow drawing process by orientation induced crystallization allows for potentially limitless draw ratio. This drawn state tends to be unstable as temperature is increased with increasing draw ratio, with the viscosity profile along the spin-line having the greatest effect on the stability of the fiber (Nakata *et al.*, 2007). In this report, the PET nanofibers are created by laser-heated flow drawing using PET "islands" with a "sea" of nylon6, at a mass ratio of 50/50.

For the study performed by the authors, the fiber was heated by CO_2 laser perpendicular to the fiber direction, with mirrors used to heat the fibers uniformly. The resulting fibers were spun together and the nylon component removed by immersing the fiber in formic acid and applying ultrasonic waves (Nakata *et al.* 2007). The study was done with both individual fibers as well as fiber bundles, with both exhibiting similar attributes to regular PET fibers, with the added advantage of high draw rates as well as high surface area (Hwang *et al.*, 2002).

7.3.6 Effect of Experimental Parameters on the Synthesis of Carbon Nanotube/Nanofiber-Supported Platinum by Polyol Processing Techniques

Carbon nanotubes and nanofibers are becoming increasingly important and useful in a variety of implementations for both engineering and technology. Nanotubes are already in use for proton exchange membrane fuel cells, including as a support for Pt catalysts used within. Chapter 7 reviewed the synthesis techniques as well as the properties inherited by the catalyst materials, through several different detection techniques (Knupp *et al.,* 2008).

The procedures used within aimed to reduce catalyst loading (which is traditionally more challenging at the cathode/Pt catalyst of the fuel cell) as well as to reduce corrosion and deposition of catalyst particles. As the synthesis technique has been shown to have a major impact on the size of the nanoparticles, an important factor in catalyst performance. The nanotubes were exposed to a concentrated acid bath and sonicated (sound agitated), and treated at various times before dilution with a mixture of water and platinum based compound, which is finally neutralized via a centrifuge before being dried (Knupp *et al.* 2008). The heating methods used were both conventional (via an oil bath) as well as with microwave irradiation.

The effects of the multiple techniques and compound concentrations used on catalyst performance and specifications were analyzed with several detection methods, such as surface area measurements, FTIR spectroscopy, X-ray diffraction, among others. The microwave heating reduced synthesis time, with a slight increase in particle size observed, while the water content could be used to control particle size. The nanofibers used played an important role as well, with larger diameter tubes causing an increase in particle size. These methods can potentially be applicable in the creation of materials for specific situations (catalysts) as well as in alternate uses where a nanotube needs to be coated in an insulating or potentially conductive material (Lin *et al.,* 2010).

7.3.7 Nanostructured Zinc Oxide Cotton Fibers: Synthesis, Characterization, and Applications

Metal oxide nanoparticles such as zinc oxide (ZnO) are very important in adding antibacterial effects to cotton fibers. These metal oxide nanoparticles are tested as antibacterial agent against bacteria such *E. coli* and *S. aureus.* ZnO particles also have advantage of being nontoxic to humans. Although the exact mechanism is still in the research and development phase, the suggested mechanism could be due to the role of oxygen produced on the surface of zinc particles. The zinc ion release and formation of H_2O_2 might also be a possibility (Kim *et al.,* 2017).

Here are two ways to load ZnO nanoparticles on the cotton fibers. The first way is to simply coat the ZnO nanoparticles on the cotton fiber by cure process. The second way to load the ZnO nanoparticles on the cotton fiber is by ultrasonic irradiation. The filling of ZnO nanoparticles by ultrasonic irradiation process has been used in the research. The purpose of this research was to produce the ZnO nanoparticles coated cotton fibers to check its antibacterial properties against *E. coli* and *S. aureus.* ZnO particles act as disinfectant to remove any bacteria while cotton was used as a substrate which can be used to produce antibacterial clothing. Scanning electron microscope and X-ray diffraction analysis were used to study the sample of cotton filled with ZnO nanoparticles. Fourier transform infrared spectroscopy (FTIR) and UV-visible (UV-Vis) spectroscopy was used to analyze the ZnO nanoparticles on the cotton-coated fiber (Shahidi *et al.,* 2017).

Zinc sulphate, sodium hydroxide, ethanol, and cotton were used in the experiment. A UV-Vis spectrophotometer in the range of 200–800 nm was used to record the coated ZnO spectra. FTIR spectrophotometer in the range of 4000–400 cm^{-1} was used to record the infrared spectra of ZnO nanoparticles coated on the cotton. Energy-dispersive X–ray spectroscopy was used for the chemical analysis.

Cotton fibers were first washed with water and 5% sodium dodecyl sulfate for an hour. Fibers were again rinsed with distilled water and dried with a 10 ml of distilled water solution containing 0.12 g of $ZnSO_4$; $5H_20$ was socked with 0.05 g of cotton. An ultrasonic senator was used to irradiate the solution. A mass of 0.06 g of NaOH was added to the solution while being stirred. The mixture was again resonicated at 35–40 °C for an hour. The product was washed with distilled water to remove NaOH, dried in vacuum overnight at 60 °C. Titration was used to determine the

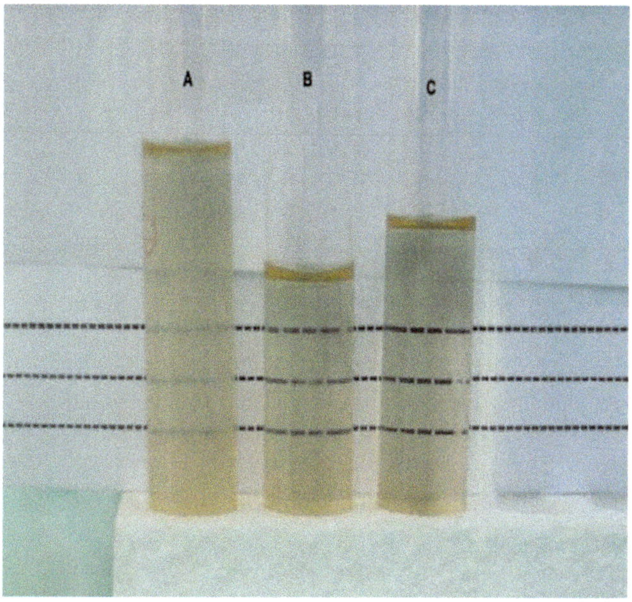

FIGURE 7.7 Photos of antimicrobial activity of ZnO-coated cotton (left positive control; middle sample; right negative control).

concentration of zinc in the fiber. The antimicrobial properties of ZnO particles coated in a cotton fiber were analyzed against the *E. coli* and *S. aureus*. A piece of the product was added to the tube containing the broth of bacteria and plated on nutrient agar plates.

X-ray diffraction analysis provided the presence of crystalline metal oxide particles on the surface of cotton. The sample of cotton containing nanofibers was also analyzed under SEM and EDX and nanosized ZnO particles were discovered on the surface of the cotton. The cotton fibers modified with antibacterial nanoparticles were tested against *E. coli* and *S. aureus*, as shown in the Figure 7.7.

The middle test tube containing the antibacterial sample showed the smallest growth of bacteria as compared with other test tubes.

The growth activity of the bacteria was analyzed for the sample and no growth was discovered in the tube containing the cotton-coated ZnO nanoparticles. Hence ZnO nanoparticles can be added to cotton fibers for antibacterial purposes. Antibacterial cotton can be used in clothing industry.

7.3.8 Preparation of Cellulose Fibers with Antibacterial Ag-Loading Nano SiO_2

Cellulose is an important naturally occurring fiber which can be used in composite materials (Nayigiziki 2016). The functional groups present in the cellulose can bond to other materials, modified and activated, to produce other composites. Various techniques are used to modify the properties of cellulose. For example, the adsorption properties of cellulose were enhanced by grafting acrylic acid on the surface of cellulose (Carlmark *et al.*, 2012). Silver nanoparticles are well known for their antimicrobial effect (Rai *et al.*, 2009). Silver nanoparticles loaded on nano-SiO_2 have excellent antibacterial properties. Antibacterial cellulose fibers can be produced by grafting silver-loaded SiO_2 nanoparticles on the surface of cellulose fibers. In this report, antibacterial cellulose was prepared by grafting silver-loaded SiO_2 nanoparticles on cellulose (Gorup *et al.*, 2017).

Natural cellulose fibers, potassium permanganate ($KMnO_4$), acrylamide, sodium hydroxide (NaOH), *E. coli* (ATCC 8739), and *S. aureus* (ATCC 6538) were used in the experiment. The silver nanoparticles and nano-SiO_2 was produced by adsorption methodology (Wang *et al.*, 2011).

Three steps were performed in the modification of cellulose fibers. In the first step the fibers were placed in $KMnO_4$ at 50 °C for 15 min, and then washed with distilled water. In the second step, the fibers were dipped in acrylamide solution for 90 min at 70 °C at the adjusted pH of 4–5. The fibers were again with water and ethanol mixture to remove any impurities such as unfixed polyacrylamide. The antibacterial nanoparticles were prepared by adding 5 g of silver loaded nano SiO_2 and 5 ml of KH–550 in 200 ml of distilled water. The mixture was stirred for 15 min at 60 °C, and the pH was adjusted to 6. The cellulose fibers were placed in the nanoparticle mixture for 1 h, and washed with distilled water, and dried for a few hours. The antibacterial nanoparticles of silver-loaded nano-SiO_2 were loaded onto the cellulose fiber in two steps. In the first step, the antibacterial nanoparticles adsorbed to the cellulose fibers by forming hydrogen bonds. In the second step, the strong bonds were formed between the amino acids of cellulose fibers and hydroxyl groups of antibacterial nanoparticles. The cellulose fibers were observed under SEM as shown in Figure 7.8.

It was clearly shown that antibacterial nano-SiO_2 silver-loaded nanoparticles were found on the surface of cellulose fibers, which was well dispersed. The layer of polymer was formed with the thickness of 50 to 110 nm. The antibacterial nanoparticles of silver-loaded nano-SiO_2 was around 200 nm. The cellulose fiber modified with silver-loaded nano-SiO_2 was tested against *E. coli* and *S. aureus* in the shake flasks. The growth of the bacteria was reduced significantly to 98% due to the antimicrobial properties of silver-loaded nano-SiO_2. The increase in silver content results in the decrease in growth of bacteria as shown in Table 7.3, where the bacteria growth is reduced to 98% (Wang *et al.* 2011).

It is concluded that the silver-loaded nano-SiO_2 composite antibacterial nanomaterials were grafted on the surface of cellulose fibers to increase the antimicrobial properties of cellulose fibers.

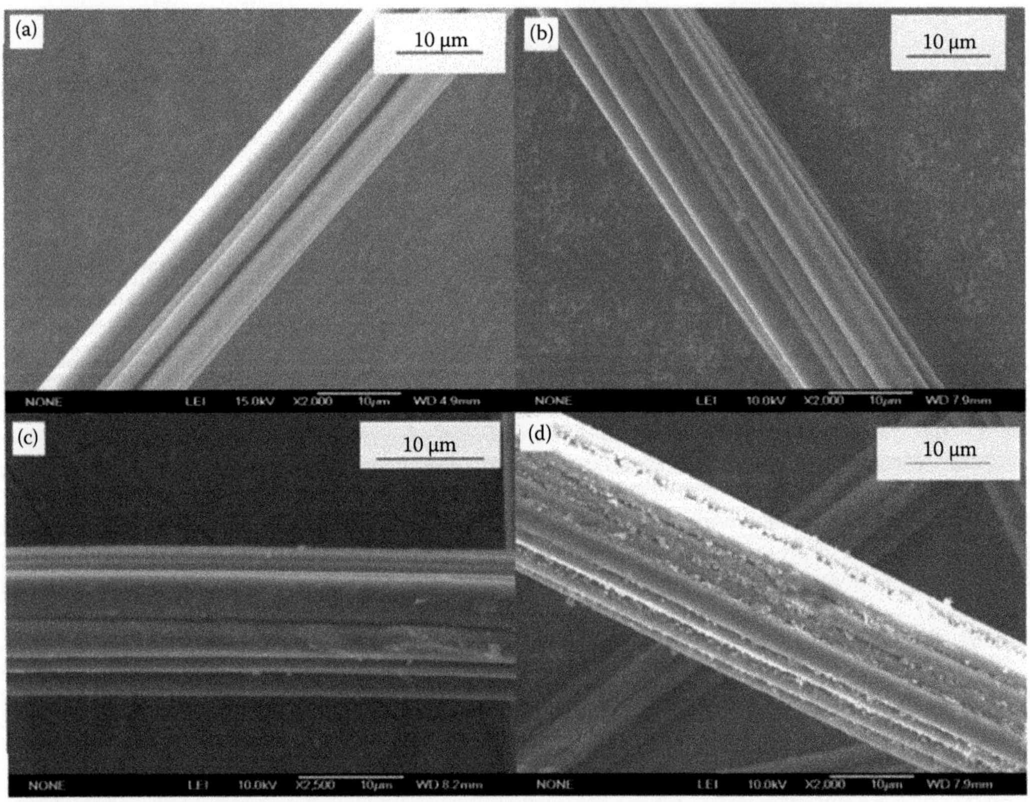

FIGURE 7.8 SEM images of cellulose fibers: (a) parent; (b) pre-modified; (c) polymerization grafted; and (d) grafted by Ag-loading SiO_2 nano-antibacterial materials.

TABLE 7.3
Particle Silver Content

Silver Constant	Mean Value (*E. coli*)	Mean Value (*S. aureus*)
0.0	0	0
0.011	54	41
0.019	74	74
0.024	80	76
0.034	98	97
0.036	98	98

The nano-SiO_2 composite has a layer thickness of 200 nm. The antibacterial cellulose showed excelled antibacterial properties against *E. coli* and *S. aureus*.

7.3.9 How Carbon Nanofibers Attach to Ni Foam

Carbon nanofibers have numerous numbers of applications in many fields ranging from biomedical, filtration, drug delivery and construction of strong materials with high tensile strength. A carbon nanofiber is composed of a material called fullerene. A carbon nanotube could be single-walled, multi-walled, and crystals. Carbon nanofibers can be produced cheaply by using nanoparticles such as Co, Ni, and Fe, as a catalyst to decompose hydrocarbon gasses. Carbon nanofibers can be used to produce a material with a high tensile strength. In this study, the carbon nanolayer is coated on the nickel metal foam. Two types of layers are formed on the surface of Ni metal foam to strengthen the material.

Ni foam with 99% purity was used in the form of a 20 × 10 cm^2 sheet, 5 mm thickness, and a geometric surface area of 1 m^2. Hydrogen, nitrogen, and ethylene were used with 99.9% purity, and ultra-pure water was used.

The mass of 0.2 g of Ni foam sample was oxidized in static air for 1 h by increasing the temperature up to 700 °C with the increment of 6 °C/min. In the next step the sample was reduced in the presence of 20% H_2/N_2 for 2 h at 700 °C. The temperature was reduced to 440 °C and carbon nanofibers were synthesized using gaseous mixture of 25% C_2H_4/N_2 with flow rate of 107 mL/min. Ethylene gas was decomposed to carbon and hydrogen. The carbon was deposited in the form of nanofibers on the surface of Ni foam. The amount of carbon was calculated using stoichiometry, by knowing the amount of hydrogen. The sample was cooled in the presence of nitrogen. The amount of carbon was also calculated experimentally by measuring the increase in the weight of Ni foam caused by carbon nanofibers. The sample was blown with 100 L/min of air for 1 min to remove any impurities, such as loose fibers. The physical features were studied using the SEM microscope. The cross-sectional area was calculated physically and the BET surface area was calculated by N_2 absorption isotherms. The chemical properties of the carbon layer deposited on the Ni foam were studied using gasification of hydrogen, water, and oxygen. The mechanical strength of the carbon nanofiber layer was determined by flowing the water over the carbon nanotube layer with velocity of 1m/s. The layer of carbon nanofibers was studied by SEM microscope, as shown in Figure 7.9.

From Figure 7.9, it was revealed that there were two layers of carbon deposited on the Ni foam surface. There was a dense carbon layer of 5 μm at the interface of Ni foam. The other layer composed of pure carbon nanofibers with a thickness of 30 μm was discovered on the top of the carbon layer. Experiments were also performed with a reduced concentration of ethylene (4%) and no layer of carbon nanofiber was discovered, only pure carbon was found to be deposited on the surface of Ni foam. It was found that the physical properties of the carbon layer deposited on the interface were identical to Ni, and the carbon nanofibers had grown out of the carbon layer.

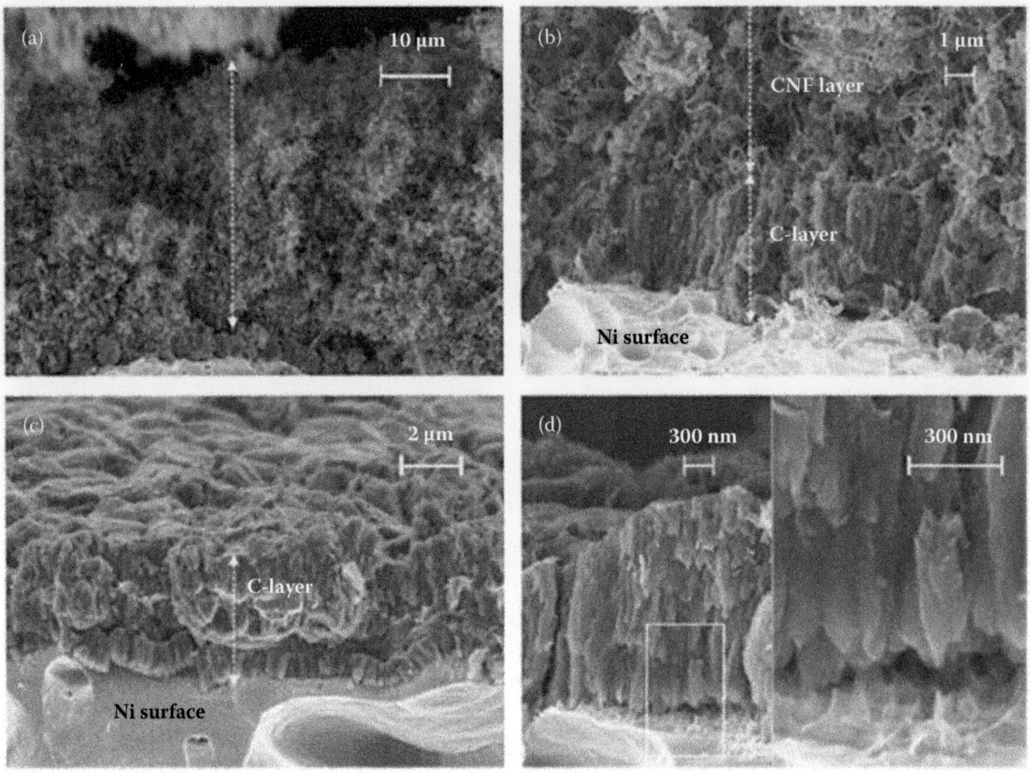

FIGURE 7.9 Cross-sectional images of foam: (a) CNFs on Ni foam deposited at 440 °C with 25% C_2H_4/N_2 for 1 h (CNF-C/Ni); (b) C-layer and CNF layer in CNF-C/Ni sample; (c) C-layer on Ni foam deposited at 440 °C with 4% C_2H_4/N_2 for 1 h (C/Ni); (d) left side of image shows the interface of C-layer and Ni surface in C/Ni sample and the right side of the image is the zoomedin image showing the rough interface. The arrows indicate the thickness of the layer.

It was concluded that two carbon layers were formed on the Ni foam surface. A pure carbon layer was followed by carbon nanofiber layer on the Ni foam surface. The presence of carbon nanofiber resulted in mechanical stability of the Ni foam.

7.3.10 Study of Nanofiber Cellulose Production by *Glucanacetobacter xylinum* ATCC 10245

Cellulose is the most common fiber extracted from plants. It has a molecular formula of $(C_6H_{10}O_5)_n$. Cellulose derivates can be formed by combining the hydroxyl groups with other materials. Cellulose can be produced by four methods. Cellulose is produced by the traditional process of extracting cellulose from lignin and hemicellulose. The other three processes in the production of cellulose are chemical, enzymatic, and biological synthesis. In this study, the biological synthesis of cellulose is carried out in the presence of bacteria named *Gluconobacter xylinum*, which is formerly known as *Acetobacter xylinus* or *Acetobacter xylinum. Gluconobacter* is a gram-negative bacterium with a rod-like shape, and belongs to the family of *Acetobacteracea.* The specialty of this bacterium is its ability to produce extracellular cellulose with a higher purity, crystallinity, better water absorption, and higher mechanical strength. The network of cellulose produced by this bacterium is also called pellicle. The conditions for its production include providing oxygen at the surface of cultivation medium, maintaining humidity of the cultivation medium, and protecting bacteria against UV rays. Nanocellulose produced by bacteria is gaining popularity over normal cellulose extracted from plants due to the higher purity, tensile strength, and higher water absorption capability compared to the naturally extracted cellulose. Static and agitated medium can be used to produce nanocellulose of different morphologies. Research is being done to use nanocellulose fibers in stent coating, drug delivery, tissue engineering, and wound healing. In this study, different cultivation media vessels were used to produce nanocellulose fibers.

Glucanacetobacter xylinum ATCC 10245 was used. According to the supplier instructions, the ampoule containing the bacteria was opened aseptically at the laminar air flow cabinet. Different cultivation media bottles were produced and inoculated with the *Glucanacetobacter* to produce bacterial cellulose. The cultivation media are given below:

1. A 500-mL flask was used to obtain 100 mL Lauria Bertani Broth, peptone 10 g/L, yeast 5 g/L, NaCl 5 g/L, and the pH was adjusted to 7.
2. A 500 mL flash containing 100 mL of Brain Heart Infusion Broth with controlled pH of 7.
3. A 500 mL rough bottle containing 100 mL of Hestrin – Schramm Broth, peptone 5 g, yeast 5 g, citric acid 1.15 g, disodium hydrogen phosphate 2.7 g, glucose 20 g. The pH was adjusted to 6.
4. Test tube and petri dish containing brain heart infusion agar
5. Petri dish containing LB medium and 2% agar
6. Solid medium containing glucono bacter oxidans, yeast 10 g, calcium carbonate 20 g, glucose 100 g, agar 15 g. The pH of the medium was adjusted to 6.8.

The cultivation media flasks were incubated around 30 °C for 2 to 7 days at static conditions. The gel formed at the liquid surface was removed after the cultivation period. The system was washed with deionized water at 50 °C. The cellulose membrane formed was purified by removing impurities by using 0.1 M NaOH at 90 °C for 30 minutes. The membrane was again washed with deionized water and a neutral pH of 7 was achieved. The system was then dehydrated to –40 °C by using a freeze dryer.

The bacterial cellulose nanofiber was studied under scanning electron microscopy (SEM), Fourier transform infrared spectroscopy (FTIR), and X–ray diffraction.

Different cultivation media vessels were used in the presence of *G. xylinum* to produce nanocellulose fibers in different morphologies. The results of different cultivation media are given in the Table 7.4 (Norouzian *et al.* 2011).

From Table 7.4, it is clear that the surface area is very important in the production of nanocellulose fiber and cellulose production occurs at the interface of air/liquid of solid/liquid. Under the observation of SEM, it proved that fibers' diameters were in the range of 30 nm–100 nm.

It can be concluded that bacterial nanocellulose fibers were produced under different cultivation media vessels. The nanocellulose produced had different morphologies in different media. The growth of the nanocellulose occurred at the interface of air/liquid and solid/liquid. Under the

TABLE 7.4
Production of Cellulose in Different Mediums

Medium	Celluose Production	Vessel
Brain Heart Agar	Solid medium/air interface cellulose production	Petri plate
Brain Heart Agar	No production of cellulose	Test tube
Luria Bertani Agar	No production of cellulose	Test tube
Luria Bertani Broth	Air/liquid interface cellulose production	Erlennmeyer
Brain Heart Infusion	-do-	-do-
Hestrain – Schramm	-do-	Rough bottle
Medium No. 125	Solid medium/air interface cellulose production	Petri plate

study of SEM, the fibers produced had the diameter range of 30–100 nm. Nanocellulose nanofibers produced from bacteria is gaining popularity due to their higher purity, tensile strength, and a higher liquid absorption compared to the normal cellulose. Nanocellulose fibers can be used in the applications of tissue engineering, drug delivery, stent coating, and wound healing (Norouzian *et al.*, 2011).

7.4 APPLICATIONS OF NANOFIBERS

7.4.1 Drug Delivery of Nanofibers

Electrospining can be used to produce nanofibers. Electrospun nanofibers can be used in a variety of medical fields ranging from tissue engineering, drug delivery, wound healing, etc. In this report, the drug delivery methods are explained by using polymeric nanofibers. The drug is loaded or encapsulated in a nanofiber and delivered to the target organ by blending and electrospining, and core shell coaxial electrospinning.

There are two techniques to deliver the drugs using nanofibers:

1. Blending and electrospinning
2. Core shell coaxial electrospinning

In the blending and electrospinning technique, the drug molecules are mixed with the polymer molecules and electrospinning is performed to load the drug molecules in the nanofiber. The nanofibers loaded with drug molecules are also called encapsulated fibers (Dave *et al.*, 2013). When the drug is delivered to target organs using nanofibers, it was found that the release of the drug from the nanofibers was constant, which means that drug release from the nanofibers follow the zero order rate law (Hu *et al.*, 2014).

In core shell coaxial electrospinning technique, the core shell structure is created in the polymer by using a coaxial spinneret. The drug is loaded in the hollow core. The drug molecules are usually linked to the fiber by physical or chemical bonds. For example, the drug methylene blue was bonded to nanofibers by ionic bonds. The nanofiber PDLLA was used and lysozome protein was used as a model drug to study the structure. The fibers had a core shell structure, as seen through scanning electron microscope (Ahire *et al.*, 2016).

Controlling the drug release in nanofibers is still in its research and development phase. The release of drug from the polymer depends upon the many factors, such as polymer degradation limit, polymer structure, pH, and temperature. The drug can also be controlled by electrospinning parameters, such as mesh sizes and fiber diameter.

The nanofibers have high surface area, which means that a higher amount of drugs can be delivered using nanofibers, compared to normal encapsulation techniques. The surface area can be further increased by using porous nanofibers. For example the porous nanofiber iPMMA provides more binding sites, hence higher amounts of drug can be loaded and delivered by using iPMMA (Li *et al.*, 2017). The properties of the drug, such as its bioreactivity, remain unchanged. The fiber molecules also protect the drug molecules from the other molecules which may react and affect the quality of drug molecules. For example, polymer nanofiber molecules prevents the reaction of drug molecules with enzymes and growth factors.

7.4.2 Nonwoven Thermal Insulation

Nanofibers can be used in the military industry for the manufacturing of lightweight and cold-resistant clothing, hand wear, sleeping bags, etc. The fibers can be used as insulation to stop or reduce the flow of heat transfer. The heat transfer of fiber-based insulation decreases when the fiber size or diameter decreases. However, the radiation heat transfer through fibers with diameter that is

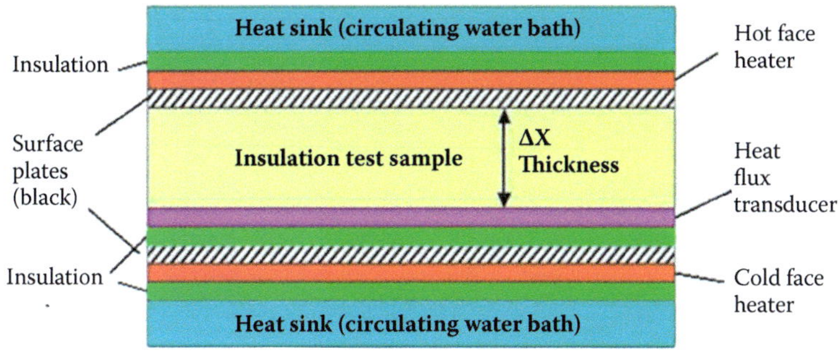

FIGURE 7.10 Heat flow meter schematic for thermal insulation testing. Thermocouples for temperature measurement at plate and heater surfaces not shown.

less than the infrared wavelength cannot be detected. Accordingly, the great challenge to reconcile radiation models to study the interaction between the fibers and the radiation. In general, heat transfer through nanofibers can be categorized as conduction, convection and radiation.

Research has been focused to produce lightweight, lower density, and compressible insulation fibers. When the fibers have a lower density, lightweight and compressible insulator, the heat transfer through conduction is negligible. However, the conduction through the air trapped within the insulation cannot be neglected (Rwawiire *et al.,* 2017).

To model the radiation process through nanofibers, it is assumed that fibers absorb, emit, and scatter thermal radiation. The optical properties of fibers are very important in radiation heat transfer. The scattering and absorption parameters of fiber insulation depend upon the optical properties, size, and shape of the polymer. For example, the coating on the silica fibers reduces the radiation heat transfer. It is also proven from the studies that nanofibers with a diameter in the range of 1 to 3 μm increases the thermal resistance, hence it is beneficial in stopping the flow of heat and making clothes warm (Feng *et al.,* 2017).

Heat is also transferred through nanofiber insulation through convection. Convection is diluted by the nanofiber insulation because the high surface area of nanofibers blocks the flow of air. Hence, by minimizing the convection heat transfer, warm clothing can be developed from nanofibers.

The experiment was performed to check the thermal resistance of various sample fibers. The fibers used were natural protein fibers (wool), polyester, meltblown pitch carbon nanofibers, electrospun nanopolyacrylonitrile, and silica aerogel fibers. The thermal resistance of the sample materials was calculated from the standard test method with the help of a heat flow meter, as shown in the Figure 7.10.

The temperature was set to 50 °C on the upper plate and 10 °C on the lower plate. The sample was placed in the middle with known thickness and thermal resistance was calculated through standard test methods.

It is clear from Figure 7.11 that the meltblown carbon nanofibers and electrospun nanopolyacrylonitrile had higher thermal resistance and higher densities compared to the normal fibers. It can be concluded that nanofibers show higher thermal resistance and can be used as thermal insulation to provide warm clothing for military use. The flowrate of heat in nanofibers can be greatly reduced via conduction, convection, and radiation to provide warm clothing.

7.4.3 Tissue Engineering

In the past ten years, nanofibers have been developed as fabricating scaffolds that can possibly imitate the architecture of natural human tissue at the nanometer scale (Figure 7.12). They have the potential to provide enhanced cell adhesion due to their large surface area and porosity. Moreover, due to the likeness of their 3D architecture to natural ECM, they provide an excellent microenvironment/

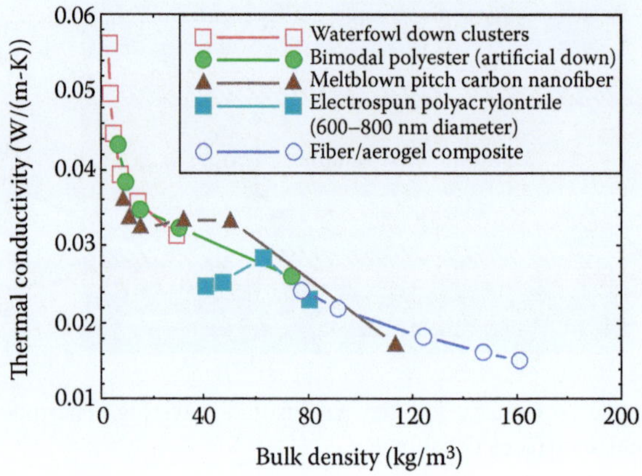

FIGURE 7.11 Effect of synthesis on fiber properties, thermal conductivity measurements as a function of density for battings of various fiber types and sizes.

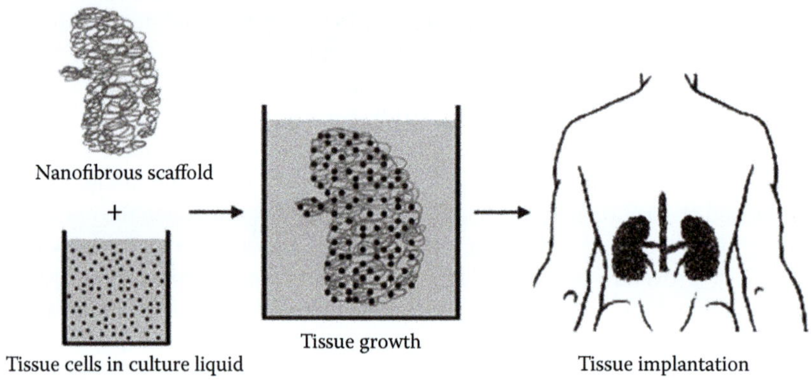

FIGURE 7.12 Illustration of tissue engineering.

nanoenvironment for cells to grow and perform their regular functions. In the meantime, nanofibers act as a vehicle to deliver drugs, proteins, and DNA. Therefore, they are used as scaffolds for musculoskeletal tissue engineering, skin tissue engineering, etc. Electrospun fibers with a high porosity and large surface-to-volume ratio have been used in many biomedical applications including antibacterial meshes, mats for hemostasis, wound dressing, and drug delivery. The nanofibers have been specially helpful in tissue engineering because the nanofiber membranes resemble the natural extracellular matrix, and promote proliferation and migration of cells (Liu *et al.*, 2017). Hydroxyapatite, chitosan, carbon and aluminum nanofibers are the materials that have been identified as nanofibrous scaffolding materials for musculoskeletal tissue engineering (including bone, cartilage, ligament, and skeletal muscle tissue engineering). Even though nanofibers could have been used as scaffolds for a variety of tissue types, musculoskeletal tissue is the most thoroughly studied.

The physical characteristics of bone tissue, such as pore size, mechanical strength, hardness, and overall 3D structure, determine the design of tissue engineering. Satisfying those properties, studies demonstrated that polycaprolactone (PCL) based nanofibrous scaffolds are the suitable materials for this area. Also the fetal bovine chondrocytes (FBC)–seeded PCL nanofibers has shown a strong ability to retain their chondrocytic phenotype by expressing cartilage-specific extracellular matrix genes (Rana *et al.*, 2017). Thus, PCL nanofibers are a potential material for cartilage tissue engineering.

Some scientists also discovered that HA with β-tricalcium phosphate (β-TCP) could be used to develop biodegradable nanocomposite porous scaffolds. Integrated with β-TCP, B-TCP/HA scaffolds worked together as a matrix to fabricate the porous scaffolds by a gel casting technique with the polymer sponge. This study introduced nanocomposites with HA nanofibers as a good scaffolding system for supporting applications such as bone tissue engineering.

Moreover, aligned nanofibers can be applied tomusculoskeletal tissue engineering due to their excellent adaptability in the use of ligament tissue engineering and PEU nanofiber scaffold systems.

Nanowoven silk fibroin nanofibers have been developed for skin tissue engineering. Using the technique of electrospinning and with the properties of high porosity and high surface-area-to-volume ratio, fibroin nanofibers coated with type I collagen were able to enhance the adhesion and spreading of keratinocytes and fibroblasts. Therefore, the silk fibroin nanofibers appeared to be the most promising candidate as a scaffold for skin tissue engineering (Rajabi *et al.,* 2017).

Wound healing is a way to repair damaged tissues. When the tissue is in the process of recovery, the dressings will protect the wound from invasion by the bacteria, and therefore accelerate wound healing. The best materials for wound dressing should have the following properties: being able to maintain a moist environment, infection resistant, good permeability, and ability to absorb wound exudate, etc. (Chen *et al.,* 2017). With the increasing awareness of environmental protection in recent years, the first priority in the development of the dressing materials is to be environment friendly. In addition to aforementioned advantages, the materials should be easily recyclable or biodegradable. Currently a variety of materials are widely used for wound dressings, however the most notable one is nanofibers, as they are affordable, creatable and biocompatible. In addition, they can be prepared into an equal-dimension nanounit; therefore, they are widely used in the biomedical area (Li *et al.,* 2017).

Currently, the most widely used nanomaterials are PLLA and PLGA. These materials can be made into a variety of shapes; such as fibers, granular, mesh, etc. (Liu *et al.,* 2010). These nanomaterials have been extensively used in tissue engineering in the area of tissue repair, cellular support, transplant, and others.

These nanomaterials not only close the wound and provide excellent oxygen permeability, but they also to stimulate the regeneration of the dermis and control water evaporation. Further, the ultra-fine porosity of polyurethane (PU) nanofiber membranes disallows invasion by exogenous microorganisms (Wright *et al.,* 2017). PU has good mechanical strength and biocompatibility. Moreover, it is a low-cost material and is easy to process and mold. Electrospun PU has a high pore density and is ultra-hydrophobic, which would make it good for waterproofing and permeability. Wound dressings made from electrospun PU are able to simulate the microenvironment of tissue growing, which will promote the injured tissues to regenerate the new and healthy tissues, therefore, they are used as a material for wound dressings. Electrospun PU has good market prospects.

In the preparation of the dressing material, if the film was added into antibacterial and anti-inflammatory additives, they could help reduce the scar formation and accelerate the wound healing process. Ag ion is a known antibacterial and anti-inflammatory material. When the Ag ions are added to the electrospun film, the dressing will become antimicrobial and, in return, it will speed up the process of wound recovery. In the prospect of sales and marketing, this electrospun film is affordable.

The method of electrospinning is used to add the Ag ions in the dressing film. First, with a parallel ground electrode as electrospinning receiver and under the feasible condition, the polymer spinning solution was prepared into the film formed by the parallel polymer nanofibers; then Ag ions were introduced into the film, thus the Ag ions were adsorbed evenly on the surface of the electrospun polymer nanofiber. This preparation method is simple with high efficiency, and it requires low energy consumption (Lagaron *et al.,* 2017).

Polymer nanofibers are arranged parallel in the dressing film, which is good for cell movement in the direction of the polymer nanofibers and the healing of the wound (Figure 7.13) (Soletti *et al.,* 2010). The Ag ions on the nanofibers are antibacterial, which helps reduce the chance of wound infection. Composed by the polymer electrospun nanofibers, the film has a large quantity of sub-micron poreson pores, so they are very good at air permeability. After the film is applied to the wound, the cells will move along the perpendicular direction of the wound, which could accelerate

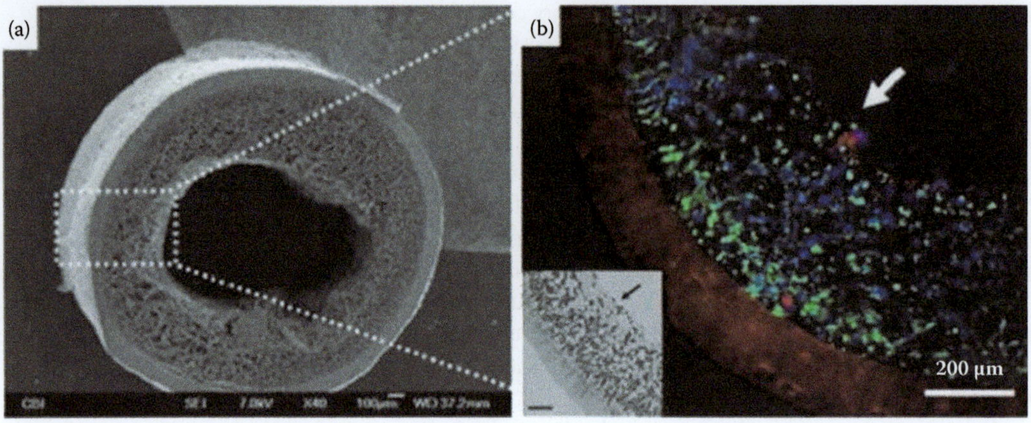

FIGURE 7.13 Cross-sectional images of a double-layered small diameter polyurethane vascular graft, before (a) and after (b) cell culturing (blue, nuclei; green, F-action; red, scaffold).

the speed of the fusion of cells on both sides of the wound. Therefore, the wound healing process is shortened.

For blood vessel tissue engineering, the research proved that electrospun collagen and elastin nanofibers were able to mimic the complex architecture required of vascular constructs. They were able to provide good mechanical properties that are desired in the environment of the blood stream, and microfibrous and nanofibrous scaffolds synthesized from natural polymers such as collagen and elastin could be useful in the engineering of artificial blood vessels. Poly(glycerol sebacate) (PGS) is a biodegradable and elastic polymer which has been widely used as a tissue engineering scaffolding biomaterial. But, PGS cannot be used while it is uncured because it is difficult to electrospin into nanofibers, therefore it is used in addition to the electrospinning agents (Gao *et al.*, 2017).

Moreover, researchers processed a conventional polymer, PET, into a nonwoven nanofibrous mat by electrospinning, and modified its surface by grafting gelatin. Their study demonstrated enhanced spreading and proliferation of endothelial cells on modified PET nanofiber mats, while their phenotype is preserved. Based on this study, gelatin-modified PET nanofibers could be potential candidates for the engineering of vascular grafts.

As a carrier for catalysts in chemistry and biology, preserving high catalytic activity, increasing the stability, and simplifying the reaction process are very important qualities for nanofibers to have. Some chemical reactions using enzymes as catalysts have high selectivity and require mild reaction conditions. These basic properties would make them very suitable candidates for catalyst carrying. Due to their size, they have a high permeability to reactants and a very large reacting surface, both of which enhance the catalytic capability.

The catalyst could give away the nanofibers during the catalysis reaction if a molecular catalyst was integrated into the nanofibers (Stasiak *et al.*, 2007). The type of shell polymer and coating thickness could significantly impact the catalysis performance. Also, the surface of catalyst-loaded nanofiber, with a thin layer of polymer, remarkably retained more of the catalyst in the nanofibers, and hence the catalyst efficiency was improved simultaneously.

Here are two widely used catalytic materials on the commercial market: The first one is Pd-loaded PAN-AA nanofibers. First of all, this material is used because it has high activity and good recycling properties for hydrogenation of a-olefin at room temperature. Second, proved by numerous testing, the yield of the hexene to hexane catalyzed by the Pd/PAN-AA nanofibers was 4.7 times higher than that of Pd/γ-Al_2O_3 (Liu *et al.*, 2008). Last, PAN nanofibers are a good alternative to freeze lipase in enzyme area. The second most commonly used catalytic material are electrospun PVA-platinum (Pt)/TiO_2 nanofibers. They were used to prepare photocatalytic degradation of solid-state PVA (He and Gong 2003). Moreover, in comparison to TiO_2 nanoparticles, electrospun TiO_2

nanofibers demonstrated higher photocatalytic activities in terms of photocurrent generation by a factor of 3 because of the mesoporosity and nanoparticle alignment, which caused efficient charge separation through interparticle charge transfer along the nanofiber framework (Choi *et al.*, 2010). Other than TiO_2 nanofibers, many other inorganic nanofibers (such as ZnO and $LaCoO_3$) have also been explored as catalysts and enzymes.

The biggest challenge of tissue engineering is to imitate the architecture of ECM. The method of using nanofibers has the best result in preparing ECM synthetically. Electrospinning, self-assembly, and phase separation are commonly used to prepare nanofibers. However, electrospinning is the most common technique and has also shown the most promising results. Hence, by using the electrospinning technology and with the abundance of natural and synthetic biomaterials, the area of nanofiber synthesis has been fueled.

Regardless of their methods of synthesis, nanofibers have a high surface area and enhanced porosity for the scaffold system. With these properties, a significant effect was testified on cell adhesion, proliferation, and differentiation. Therefore, nanofibers are considered as the most suitable scaffold for musculoskeletal tissue engineering, skin tissue engineering, vascular tissue engineering, and the controlled delivery of drugs, proteins, and DNA. As a result, the nanofiber-based scaffolds are very promising in the future development of tissue engineering applications.

Even though scientists showed that the nanofibrous scaffold could be a good candidate material for controlled DNA and enzyme/protein delivery, it's still facing some limitations as a new technology, but nanofibers will be the super material in the tissue engineering area of the future.

7.4.4 Protective Clothes

Protective materials are required for different types of jobs, including agricultural workers. Some characteristics of the material, such as the pore size of the textile, are responsible for provide to the workers protection and comfort while executing their functions. Conventional protective materials are mostly composed of microfibers. Due to their enhanced properties, nanofibers were tested as an alternative material in the composition of protective clothes (Lee *et al.*, 2007).

Electrospinning solutions, with concentration ranging from 10% to 15% wt. polymer, were obtained by dissolving PU pellets in N-dimethylformamide (DMF). A nonwoven commercial material for protective clothing was used as substrate.

The nanofibers were achieved using the electrospinning technique, through which fibers are drawn by applying an electrical field that overcomes the surface tension of a polymeric solution, in this case the PU. Preliminary experiments indicated that optimum spinning condition to produce uniform nanoscale fibers was the concentration of 13% wt. PU solution with a 0.16 mm i.d. needle (from which jets of the solution are ejected) and an applied voltage of 20 KV. The nanofibers were then electrospun directly onto a polypropylene nonwoven substrate in order to form a layered fabric system.

By varying the direct spray time, it is possible to obtain materials with different web area densities (levels of electrospun fibers web layers). Two levels of web area density were chosen for this experiment: 1.0 and 2.0 g/cm^2. Some tests were realized comparing samples of these materials to samples of the original nonwoven substrate.

First, liquid penetration was tested. Three types of pesticide mixtures composed of atrazine (2-chloro-4-ethylamino-6-isopropylamino-1,3,5-triazine) and pendimethalin (N-(1-ethylpropyl)-3,4-dimethyl-2,6-dinitrobenzenamine) with distinct properties (e.g., surface tension and viscosity) were used in this step. The surface tension of the pesticide mixture 1 was found to be 38 dynes/cm which reduced to 20.57 dynes/cm for mixture 3. However, the viscosity for the three mixtures showed an escalating trend with highest viscosity of mixture 3, i.e., 20.80 mPa-s.

Second, air transport (according to ASTM D 737-96, Standard Test Method for Air Permeability of Textile Fabrics) and water vapor transport (according to ASTM E 96-00, Standard Test Method for Water Vapor Transmission of Materials) through the samples were evaluated as a measure of the level of comfort of a worker wearing the clothes.

Third, the pore size distribution of the materials were analyzed by using a capillary flow porometer.

There was significant reduction in the percentage of liquid penetration through both of the layered fabric systems compared to the substrate alone. For challenge mixtures (liquids with low surface tension and high viscosity), the percentage of penetration reduced from 85% (nonwoven substrate) to around 25% (nonwoven substrate with layers of PU nanofibers). Electrospun PU nanofiber layers filled the space among the original microfibers of the substrate, creating a new structure with smaller pores that acted as a more efficient barrier to liquid penetration.

The smaller pores in the structure of the layered fabric systems create a decrease of 33% to 51% in air permeability. However, even at the higher level of web density, the nanomaterials presented better levels of air permeability than the best conventional protective clothes available. The behavior of the materials in terms of water vapor was different. The water vapor transmission rate kept at a constant value for the three materials analyzed. This is due to the fact that the pores in the nanofibers were still large enough to allow water vapor transportation, which is an important characteristic that provides more comfort to workers, due to easier transpiration.

As expected, the pores of the layered fabric system were significantly smaller than pores of the nonwoven substrate. This can be attributed to the fill of micropores of the substrate with nanofibers of PU, generating a structure with billions of smaller pores.

Nanofibers sprayed on a surface of a microstructural protective material were able to change their properties, resulting in more comfortable clothes with protection performance enhanced. By controlling the parameters of manufacture, it is possible to create distinct materials with the appropriate characteristics for different applications.

7.4.5 Mercury Detection

Mercury ion, Hg(II), is one of the most hazardous pollutants for aquatic environments. Several techniques are used to detect and/or quantify Hg(II), such as atomic absorption spectroscopy (AAS), inductively coupled plasma atomic emission spectroscopy (ICP-AES), and inductively coupled plasma mass spectrometry (ICP-MS). Although conventional techniques exhibit great results in terms of accuracy and repeatability, they are generally expensive and require analytical skills. Danwittayakul *et al.* proposed the utilization of cellulose ester membranes coated by dithizone nanofibers in order to create a simpler, equally efficient, and less expensive method of detection of Hg(II) (Danwittayakul *et al.*, 2008). Nanofibers were selected as adequate materials due to their high surface area and rapid response.

The membranes coated by nanofibers were fabricated by filtering nanofiber dispersion through a cellulose membrane under a vacuum. Various growth periods were evaluated in order to obtain the optimum condition of dithizone nanofiber achievement. Then, the morphology and uniformity of nanofibers were analyzed using SEM and a TLC scanning densitometer, respectively. Three acids were used to improve the stability of membranes: hydrochloric, citric, and ascorbic acid. A colorimetric method was applied using a TLC scanning densitometer. Calibration curves were constructed by plotting the color intensity versus the Hg(II) concentration. The Hg(II) was determined by filtration of water samples of river, artificial waste, and sea water using different sample volumes and filtration rate of 5–6 mL/min. The results were compared with the ICP-MS technique to verify the efficiency of the method.

SEM images indicated that the size of dithizone nanofibers obtained was approximately 250 nm and the thickness of dithizone layer was less than 500 μm. Tests of stability indicated that dithizone membranes obtained using ascorbic acid were more stable than those obtained using hydrochloric and citric acids. It was found that the sample volume affects the detection limit. The detection limit of the technique at a sample volume of 100 mL was 0.057 ppb. The detection of 10 ppb Hg(II) did not interfere with the presence of 10,000 ppm Na, 5000 ppm K, 5000 ppm Ca, 6.4 ppm Cu(II), 100 ppm Fe(III), 100 ppm Zn(II), 100 ppm Pb(II) and 10 ppm Cd(II) cations when 2.5×10^{-4} M EDTA was added to the samples, indicating EDTA as a great masking agent for cations. The presence of 0.6 M Cl, 0.01 M Br, 0.1 M SO_4, 0.01 M PO_4, 0.001 M CO_3, and 0.01 M NO_3 also did not interfere the detection of 10 ppb Hg(II) (Danwittayakul *et al.*, 2008).

The filtration technique using cellulose ester membranes coated by dithizone nanofibers presented satisfactory results when compared to the ICP-MS. This characteristic indicates that these enhanced membranes can be used as a less expensive alternative for conventional techniques.

7.4.6 Use of Nanofibers as Noise Absorbers for Aircrafts

Interior noise is one of the most important issues related to aircraft. The noise is generated by aircraft during the steps of run-up, taxiing, take-off, overflying, and landing. The noise level varies depending on the models of aircraft and engines used. Reduction of interior noise is important in order to provide more comfort to airlines' customers and crew members, especially during long flights.

Several types of noise absorbers have been used on interior surfaces of aircrafts, such as membrane and porous absorbers. They are all based on the principle of the reduction of sound by transferring it into heat energy, due to the friction of sound energy moving through the noise absorber. Ticker absorbers provide, in general, better absorption of sound (and its subsequent conversion into heat). However, conventional tick absorbers reduce the fuel efficiency of the aircraft due to the increase of weight associated. Nanofibers were tested as an alternative material for aircraft interior noise reduction (Khan 2012).

Nanofibers were obtained from three different polymeric solutions: PVP, PVC, and polystyrene.

The process utilized to create the nanofibers from polymeric solutions was electrospinning. In this process, the fibers were drawn by applying an electrical field that overcomes the surface tension of the polymeric solution. The solution was transferred to a 10-mL syringe with a platinum electrode inside which was connected to a supply of high-voltage direct current (DC). Jets of the solution were ejected from the tip of the capillary tip and begin to stretch and whip around as they travel to a grounded collector placed at the distance of 20–30 cm to the tip. The electrospun nanofibers were then dried in an oven at 60 °C for 8 h to remove the residual solvent. A scanning electron microscope (SEM) was used to determine the morphology of the fibers achieved and their diameter.

Sound tests were conducted in the nanofibers by using an impedance measurement tube linked to a two-microphone system. A 100-mm diameter tube was used for tests at frequencies ranging from 150 to 1600 Hz, while a 29-mm diameter tube was used for frequency range of 1200 to 6400 Hz.

The sound tests evaluated the sound absorption coefficients (ratio of absorbed energy to the incident energy) at different sound frequencies of nanofibers with varied characteristics (weight, thickness, and composition). In general, heavier and thicker fibers increased the absorption coefficients values of the samples, due to the fact that the sound molecules interact more with higher surface area materials. However, for samples A and C of PVP, a distinct behavior was observed: even though sample A was lighter than sample C, its sound absorption coefficient was greater, indicating that at the frequency of 6000 Hz, nanofibers with specifications of sample A can be used as enhanced noise absorbers, creating the best reduction of sound while decreasing the overall weight of the aircraft.

The study indicated that some characteristics of nanofibers, such as greater porosity, less overall weight, high surface area, and the potential for bulk production through the electrospinning process, allow the use of this material in future noise control applications. The challenge for scientists is to find materials with the same behavior as samples A and C of PVP.

7.5 CONCLUSION

Lately, with the emergence of evidence definitive of the utility of nanomaterials like nanofibers, the emphasis has been on devising and developing newer efficacious and cost-efficacious techniques of their synthesis, eventually leading to improved comprehension of the nuances of their manufacturing process. Electrospinning is currently the best choice for large-scale manufacturing, but does not have the breadth of applications that other systems do. Chapter 7 provides a bird's eye view account on the potential uses of nanofibers. Verily, the future of nanotechnology looks promising, as new uses of nanomaterials like nanofibers and improvements in their properties and synthetic techniques are constantly being worked upon and put into use.

REFERENCES

Agop M, Paun V, Harabagiu A. El Naschie's ε (∞) theory and effects of nanoparticle clustering on the heat transport in nanofluids. *Chaos Solitons Fractals*. 2008; 37(5): 1269–78.

Ahire JJ, Robertson D, Neveling DP, van Reenen AJ, Dicks LMT. Hyaluronic acid-coated poly(d,l-lactide) (PDLLA) nanofibers prepared by electrospinning and coating. *RSC Adv*. 2016; 6(41): 34791–6.

Carlmark A, Larsson E, Malmström E. Grafting of cellulose by ring-opening polymerisation – A review. *Eur Polym J*. 2012; 48(10): 1646–59.

Chen S, Liu B, Carlson MA, Gombart AF, Reilly DA, Xie J. Recent advances in electrospun nanofibers for wound healing. *Nanomedicine (Lond)*. 2017; 12(11): 1335–52.

Choi SK, Kim S, Lim SK, Park H. Photocatalytic comparison of TiO_2 nanoparticles and electrospun TiO_2 nanofibers: Effects of mesoporosity and interparticle charge transfer. *J Phys Chem C*. 2010; 114(39): 16475–80.

Chong EJ, Phan TT, Lim IJ, Zhang YZ, Bay BH, Ramakrishna S et al. Evaluation of electrospun PCL/gelatin nanofibrous scaffold for wound healing and layered dermal reconstitution. *Acta Biomater*. 2007; 3(3): 321–30.

Danwittayakul S, Takahashi Y, Suzuki T, Thanaboonsombut A. Simple detection of mercury ion using dithizone nanoloaded membrane. *J Metal Mater Miner*. 2008; 18(2): 37–40.

Dave R, Jayaraj P, Ajikumar PK, Joshi H, Mathews T, Venugopalan VP. Endogenously triggered electrospun fibres for tailored and controlled antibiotic release. *J Biomater Sci Polym Ed*. 2013; 24(11): 1305–19.

Feng J, Mo W, Ma S, Wang Y, Yang J, Su X et al. Thermal shrinkage inhibition mechanism of fumed silica based thermal insulating composite. *Appl Therm Eng*. 2017; 113: 749–55.

Foraida ZI, Kamaldinov T, Nelson DA, Larsen M, Castracane J. Elastin-PLGA hybrid electrospun nanofiber scaffolds for salivary epithelial cell self-organization and polarization. *Acta Biomater*. 2017; 62: 116–27.

Frenot A, Chronakis IS. Polymer nanofibers assembled by electrospinning. *Curr Opin Colloid Interface Sci*. 2003; 8(1): 64–75.

Gagliardi M. Global markets and technologies for nanofibers. BCC Research; 2013 [cited 2017 Nov, 30]; Available from: https://www.bccresearch.com/market-research/nanotechnology/nanofibers-market-nan043c.html.

Gao H, Yang Y, Akampumuza O, Hou J, Zhang H, Qin X. A low filtration resistance three-dimensional composite membrane fabricated via free surface electrospinning for effective PM 2.5 capture. *Environ Sci Nano*. 2017; 4(4): 864–75.

Gibson P, Lee C. Application of nanofiber technology to nonwoven thermal insulation. *J Eng Fiber Fabr*. 2007; 2(2): 32–40.

Gorup LF, Neto FNS, Kubo AM, Souza JAS, Fernandes RA, Fernandes GL *et al*. Nanostructured Functional Materials: Silver Nanoparticles in Polymer for the Generation of Antimicrobial Characteristics. *Recent Advances in Complex Functional Materials*: Springer; 2017. pp. 271–92.

Guo J, Liu X, Lee Miller A, Waletzki BE, Yaszemski MJ, Lu L. Novel porous poly (propylene fumarate-co-caprolactone) scaffolds fabricated by thermally induced phase separation. *J Biomed Mater Res A* 2017; 105(1): 226–35.

He C-H, Gong J. The preparation of PVA–Pt/TiO_2 composite nanofiber aggregate and the photocatalytic degradation of solid-phase polyvinyl alcohol. *Polym Degrad Stab*. 2003; 81(1): 117–24.

He J-H, Liu Y, Mo L-F, Xu L. What is Electrospinning? In: He J-H, Liu Y, Mo L-F, Xu L, editors. Electrospun nanofibres and their applications. UK: iSmithers; 2008. pp. 6–13.

Hu X, Liu S, Zhou G, Huang Y, Xie Z, Jing X. Electrospinning of polymeric nanofibers for drug delivery applications. *J Control Release.* 2014; 185(Suppl. C): 12–21.

Hwang YN, Yoon JY, Choi YB. A sea-island typed composite fiber used in warp knitting, and a process of preparing for the same. Google Patents; 2002.

Ikegame M, Tajima K, Aida T. Template synthesis of polypyrrole nanofibers insulated within one-dimensional silicate channels: Hexagonal versus lamellar for recombination of polarons into bipolarons. *Angew Chem Int Ed.* 2003; 42: 2154–7.

Jung BO, Bae S-Y, Lee S, Kim SY, Lee JY, Honda Y et al. Emission characteristics of InGaN/GaN core-shell nanorods embedded in a 3D light-emitting diode. *Nanoscale Res Lett.* 2016; 11:215.

Khan N. Applications of electrospun nanofibers in biomedical field. *Drug Deliv.* 2012; 5: 63–73.

Kim JH, Joshi MK, Lee J, Park CH, Kim CS. Polydopamine-assisted immobilization of hierarchical zinc oxide nanostructures on electrospun nanofibrous membrane for photocatalysis and antimicrobial activity. *J Colloid Interface Sci.* 2017; 513: 566–74.

Knupp SL, Li W, Paschos O, Murray TM, Snyder J, Haldar P. The effect of experimental parameters on the synthesis of carbon nanotube/nanofiber supported platinum by polyol processing techniques. *Carbon.* 2008; 46: 1276–84.

Kyle S, Aggeli A, Ingham E, McPherson MJ. Production of self-assembling biomaterials for tissue engineering. *Trends Biotechnol.* 2009; 27(7): 423–33.

Lagaron JM, Solouk A, Castro S, Echegoyen Y. Biomedical applications of electrospinning, innovations, and products. *Electrospun Materials for Tissue Engineering and Biomedical Applications*: Elsevier; 2017. pp. 57–72.

Lee SJ, Yoo JJ, Lim GJ, Atala A, Stitzel J. In vitro evaluation of electrospun nanofiber scaffolds for vascular graft application. *J Biomed Mater Res A.* 2007; 83(4): 999–1008.

Li H, Williams GR, Wu J, Lv Y, Sun X, Wu H et al. Thermosensitive nanofibers loaded with ciprofloxacin as antibacterial wound dressing materials. *Int J Pharm.* 2017a; 517(1): 135–47.

Li Y, Chen B, Pun EYB, Lin H. Nanofiber electrospinning in samarium complex-doped PMMA. *New J Chem.* 2017b; 41: 15195–203.

Li Y, Yu S, Chen P, Rojas R, Hajian A, Berglund L. Cellulose nanofibers enable paraffin encapsulation and the formation of stable thermal regulation nanocomposites. *Nano Energy.* 2017c; 34: 541–8.

Lin Z, Ji L, Woodroof MD, Yao Y, Krause W, Zhang X. Synthesis and electrocatalysis of carbon nanofiber-supported platinum by 1-AP functionalization and polyol processing technique. *J Phys Chem C.* 2010; 114(9): 3791–7.

Liu M, Duan X-P, Li Y-M, Yang D-P, Long Y-Z. Electrospun nanofibers for wound healing. *Mater Sci Eng C Mater Biol.* 2017; 76: 1413–142.

Liu SJ, Kau YC, Chou CY, Chen JK, Wu RC, Yeh WL. Electrospun PLGA/collagen nanofibrous membrane as early-stage wound dressing. *J Membr Sci.* 2010; 355: 53–9.

Liu Y, He JH, Yu Jy, Zeng Hm. Controlling numbers and sizes of beads in electrospun nanofibers. *Polym Int.* 2008; 57(4): 632–6.

Moazeni N, Merati AA, Latifi M, Sadrjahani M, Rouhani S. Fabrication and characterization of polydiacetylene supramolecules in electrospun polyvinylidene fluoride nanofibers with dual colorimetric and piezoelectric responses. *Polymer.* 2017: 1–27.

Nakata K, Fujii K, Ohkoshi Y, Gotoh Y, Nagura M, Numata M et al. Poly (ethylene terephthalate) nanofibers made by sea-island-type conjugated melt spinning and laser-heated flow drawing. *Macromol Rapid Commun.* 2007; 28(6): 792–5.

Nayigiziki FX. *Physical characterization and antimicrobial properties of PVA-cellulose nanofiber based films*: University of Missouri, Columbia; 2016.

Norouzian D, Farhangi A, Tolooei S, Saffari Z, Mehrabi MR, Chiani M et al. Study of nano-fiber cellulose production by Glucanacetobacter xylinum ATCC 10245. *Pak J Biol Sci.* 2011; 14(15): 780–4.

Peshoria S, Narula AK. Structural, morphological and electrochemical properties of a polypyrrole nanohybrid produced by template-assisted fabrication. *J Mater Sci.* 2017: 1–13.

Rai M, Yadav A, Gade A. Silver nanoparticles as a new generation of antimicrobials. *Biotechnol Adv.* 2009; 27(1): 76–83.

Rajabi M, Firouzi M, Hassannejad Z, Haririan I, Zahedi P. Fabrication and characterization of electrospun laminin-functionalized silk fibroin/poly (ethylene oxide) nanofibrous scaffolds for peripheral nerve regeneration. *J Biomed Mater Res B Appl Biomater.* 2017: 1–10.

Rana D, Ratheesh G, Ramakrishna S, Ramalingam M. Nanofiber composites in cartilage tissue engineering. *Nanofiber Composites for Biomedical Applications.* 2017. pp. 325–44.

Regis E. The Incredible Shrinking Man. Wired; 2004 [cited 2017 Nov, 30]; Available from: https://www.nanotechnologist.com/misc/.

Roberts EK. *Solid state NMR structural evaluation of the SAF-p1/p2a co-assembling peptide nanofiber system*: Georgia Institute of Technology; 2017.

Rwawiire S, Tomkova B, Militky J, Hes L, Kale BM. Acoustic and thermal properties of a cellulose nonwoven natural fabric (barkcloth). *Appld Acoustics*. 2017; 116: 177–83.

Shahidi S, Rezaee H, Rashidi A, Ghoranneviss M. In situ synthesis of ZnO Nanoparticles on plasma treated cotton fabric utilizing durable antibacterial activity. *J Nat Fibers*. 2017: 1–9.

Soletti L, Hong Y, Guan J, Stankus JJ, El-Kurdi MS, Wagner WR *et al.* A bilayered elastomeric scaffold for tissue engineering of small diameter vascular grafts. *Acta Biomater.* 2010; 6(1): 110–22.

Stasiak M, Röben C, Rosenberger N, Schleth F, Studer A, Greiner A *et al.* Design of polymer nanofiber systems for the immobilization of homogeneous catalysts–Preparation and leaching studies. *Polymer.* 2007; 48(18): 5208–18.

The National Nanotechnology Initiative: Research and development leading to a Revolution in Technology and Industry. Supplement to the President's FY 2017 Budget. https://www.nano.gov/sites/default/files/pub_resource/nni_fy17_budget_supplement.pdf. Accessed on Dec, 3, 2017.

The National Nanotechnology Initiative: Research and development leading to a Revolution in Technology and Industry. Supplement to the President's FY 2016 Budget. https://www.nano.gov/sites/default/files/pub_resource/nni_fy16_budget_supplement.pdf. Accessed on Dec, 3, 2017.

The National Nanotechnology Initiative: Research and development leading to a Revolution in Technology and Industry. Supplement to the President's FY 2012 Budget. https://www.nano.gov/sites/default/files/pub_resource/nni_fy12_budget_supplement.pdf. Accessed on Dec, 3, 2017.

The National Nanotechnology Initiative: Research and development leading to a Revolution in Technology and Industry. Supplement to the President's FY 2008 Budget https://www.nano.gov/sites/default/files/pub_resource/nni_fy08_budget_supplement.pdf. Accessed on Dec, 3, 2017.

The National Nanotechnology Initiative: Research and development leading to a Revolution in Technology and Industry. Supplement to the President's FY 2004 Budget https://www.nano.gov/sites/default/files/pub_resource/nni_fy04_budget_supplement.pdf. Accessed on Dec, 3, 2017.

The National Nanotechnology Initiative: Research and development leading to a Revolution in Technology and Industry. Supplement to the President's FY 2000 Budget https://www.nano.gov/sites/default/files/pub_resource/nni_fy00_budget_supplement.pdf. Accessed on Dec, 3, 2017.

Wang S, Niu R, Jia H, Wei L, Dai J, Liu X et al. Preparation of cellulose fibres with antibacterial Ag-loading nano-SiO_2. *Bull Mater Sci*. 2011; 34: 629–34.

Wright MEE, Parrag IC, Yang M, Santerre JP. Electrospun polyurethane nanofiber scaffolds with ciprofloxacin oligomer versus free ciprofloxacin: Effect on drug release and cell attachment. *J Control Release*. 2017; 250: 107–15.

Yang Y, Liu B, Zhang Y, Lv X, Wei L, Wang X. Fabrication and luminescence of $BiPO_4$:Tb^{3+}/Ce^{3+} nanofibers by electrospinning. *Superlattices Microstruct*. 2016;90(Supplement C): 227–35.

8 Fibro-Porous Composite Nanobiomaterials for Enhanced Biointegration

*Andrew T. Wood, Dominique T. Everett, Katelynne J. Herron, and Vinoy Thomas**

CONTENTS

8.1 Introduction 165
8.2 Material Must-Haves for Efficient Response 166
  8.2.1 Biocompatibility 167
  8.2.2 Biomimicry 168
  8.2.3 Bioactivity 168
  8.2.4 Biodegradability 169
8.3 Electrospinning Methods of Fibrous Scaffold Production 169
  8.3.1 Core-Shell Fibers 171
  8.3.2 Janus Fibers 172
8.4 Cellular Response and the Importance of Nanoarchitecture 174
  8.4.1 Fiber Diameter 174
  8.4.2 Fiber Composition 175
  8.4.3 Organic–Bioinorganic Hybrid Fibers 176
  8.4.4 Mechanical Properties 177
  8.4.5 Fiber Alignment 180
8.5 Future Directions 181
Acknowledgments 181
References 181

8.1 INTRODUCTION

From the biblical story of creation where Eve was said to have been created from Adam's rib, to the Fra Angelico painting "The Healing of Justinian" depicting Saint Cosmas and Saint Damian transplanting a limb onto an injured soldier, to the prosthetic nose worn by Tycho Brahe after losing part of his in a sword fight over a debate on mathematical proofs, we have, as the human race, been infatuated with the idea of self-healing, self-creation, and self-improvement. All of this fervor has seen its culmination to recent technological advancements in the development of the field of tissue engineering. The birth of the field is often attributed to Dr. W.T. Green, M.D. who attempted to generate new cartilage by seeding chondrocytes onto spicules of bone and, subsequently, implanting them into mice (Vacanti, 2006). Since these beginnings, the field of tissue engineering has called upon multiple fields of research and expertise, including biology, chemistry, physics, and engineering in order to achieve its goal of bioremediation, bioregeneration, and bioenhancement.

Recent advances in this field have shown the ability to regenerate tissue systems and even fabricate entire organs in laboratory settings with tissue-engineered bones, blood vessels, liver, skeletal

* Corresponding author.

muscles, nerve conduits, and lungs, among other biological systems. These advances have not only highlighted the necessity of various disciplines working together, but have also revealed a prominent variable in the efficacy of the applications that is highly dependent on material selection. The selection of the appropriate material lies not only with the produced material itself, but also with the techniques used to synthesize, derive, and fabricate the scaffolds. These scaffolds are chosen as substrates that allow for adequate cellular attachment and resultant cellular migration, proliferation, propagation, differentiation, and, ultimately, morphogenesis, resulting in a fully regenerated tissue system. In recent research endeavors, the most favorable of the scaffold geometries has been found to be that of highly porous, fibrous substrates (Bhattarai *et al.*, 2004).

Due to the inherent geometry of fibrous substrates, these types of scaffolds are able to achieve high surface area to volume ratios, enhanced mechanical properties, and high porosity, with pore size being directly related to the diameter and volume fraction of the fibers. These fibrous scaffolds have shown the ability to produce rapid cellular attachment and enhanced cellular proliferation and differentiation that can be tuned for specific applications by varying both the material selection and the fabrication processes. The optimal scaffold for tissue engineering and regenerative medicine would mimic the native environment of tissue systems, known as the extracellular matrix (ECM), as closely as possible. This ECM is composed of a majority of water with a fibrillary network made up of collagens, elastins, laminins, and fibronectins throughout (Frantz *et al.*, 2010). Proteoglycans, such as hyaluronic acid, decorin, aggrecan, and perlecan, are also dispersed throughout the ECM. These compositional components provide both structural rigidity, support to organs and tissue systems, as well as biochemical and biomechanical cues to influence and direct cellular fates. The remaining volume of the ECM is composed of covalently crosslinked glycosaminoglycan (GAGs) chain networks, further improving the environment structural rigidity and biochemical/biomechanical cues for the speciation of cells.

With the knowledge of fibrous networks and their *in vivo* role in the naturally occurring generation of tissue systems, recent efforts in tissue engineering have centered on making scaffolds that mimic this environment as closely as possible. A majority of the work to this end has used a fabrication method known as electrospinning. Electrospinning is a technique which uses electrostatic forces between a charged polymer jet and a grounded collecting substrate to form fibrous membranes with individual fibers capable of reaching tens of nanometers in diameter. This technique is capable of accommodating the production of fibers from both synthetic and natural origins, as well as a blend of the two into multiple types of geometries, which will be outlined in details in subsequent sections. These nanofibrous substrates not only offer the morphology required to mirror the native ECM, this technique also avails cellular response through biochemical cues. While these responses are outlined throughout the chapter, it is worth briefly stating that the substrates can be functionalized or made into biocomposites capable of eliciting positive biological responses not available to some virgin synthetic constructs. While naturally derived materials such as collagen, casein, and silk can be made into such nanofibrous scaffolds, they typically have inferior mechanical properties when compared to synthetic materials, while their inherent biological response and typical biodegradability offer prominent advantages (He and Lu, 2016).

In Chapter 8, we will cover the basic requirements of all materials used in tissue engineering, as well as their responses upon implementation. Further, we look at the types of fibrous networks typically used in tissue engineering and their effects on cellular response. While it is well known that the fibrous substrate provides enhanced responses *in vivo*, we also address the effects of porosity on cellular response, biodegradation, mechanical properties, and the overall efficacy of the fibroporous substrates, in an effort to better outline how future directions can attempt to enhance biointegration of tissue engineering scaffolds.

8.2 MATERIAL MUST-HAVES FOR EFFICIENT RESPONSE

Fibrous materials have seen increased growth in the fields of tissue engineering, regenerative medicine, and drug delivery in recent years. High mechanical properties, ease of modification, and a striking saliency to the native ECM are all factors that have attributed to this growth in interest. The ECM found within living tissue systems is an environment rich in water, and contains a fibrillary network of various proteins including collagens, elastins, laminins, and fibronectins, as well as proteoglycans such as hyaluronic acid, decorin, aggrecan, and perlecan (Holloway *et al.*, 2014). Although majorly comprised of water, proteins, and polysaccharides, the ECM found in natural biological systems varies in its composition, depending on the tissue system it is found in, but a similar morphological environment is found among them all. The exact morphological and topological characteristics for each ECM are developed during morphogenesis of the tissue that initiates, through biophysical and biochemical cues, the production of the resultant type of environment. Since cells are incapable of being self-supporting during growth, this ECM offers a structural support and proliferation medium for cellular propagation, growth, and differentiation by providing scaffolding for cells to attach and migrate throughout, along with biochemical and biomechanical cues which prompt and direct cellular behavior (Levorson *et al.*, 2013). We now will highlight and expound upon the four major requirements for an effective scaffold as mentioned at the onset of Chapter 8, namely, biocompatibility, biomimicry, bioactivity, and biodegradability.

8.2.1 Biocompatibility

Biocompatibility is the foremost requirement necessary for a material to initiate an efficient response upon implementation *in vivo*. A definition of biocompatibility is difficult to produce since, for some applications, the desired compatibility may be one that elicits a response, whereas for other applications a response is undesired. In Chapter 8, we will use a working definition of biocompatibility stating that it is the capability of a material to provoke a desired response upon implementation *in vivo* without causing undue negative effects, either locally or systemwide, upon implantation and/or degradation. Using this definition, we will cover the major factors that influence *in vivo* responses and ways that seemingly harmful materials have been successfully implemented with exceptional results (Leipzig and Shoichet, 2009).

When determining if a material is truly biocompatible, researchers must understand not only the immediate impacts the material will have on the biological system, but also the long-term effects the material may elicit upon degradation *via* by-products produced as the material breaks down. These types of studies are commonly carried out and the cytotoxicity of the material in question begins to determine its ultimate biocompatibility. Commonly, cytotoxic effects are found at some time point after implementation when the material begins to be broken down, thus releasing molecules which may react unfavorably *in vivo*. As Chapter 8 focuses on fibrous material for applications in tissue engineering and regenerative medicine, it is worth mentioning that the rate at which fibrous scaffolds degrade is typically higher than that of the same material found in its bulk form. This is attributed to the increased surface area of the fiber which exposes more of the material to enzymatic or hydrolytic degradation than that of a bulk piece of the same material. With this in mind, it is thus crucial to know the by-products of such degradation pathways and to understand their potential for acute and chronic affects *in vivo*.

When attempting to control this rapid rate of degradation found among fibrous scaffolds, some researchers have coated these fibers in a covalently crosslinked gelatinous matrix thus forming a fiber–hydrogel composite. In this area, cytotoxicity can now arise from a secondary source the crosslinking agent used. A commonly used crosslinking agent in biomedical applications is glutaraldehyde, which creates covalently bonded bridges between the polymeric chains that comprise the gelatinous phase. The resultant issue with such methods is the cytotoxicity found among this and most other crosslinking agents if these reagents are not thoroughly removed, with even trace amounts resulting in decreased cellular activity and viability (Holloway *et al.*, 2014).

While there are multiple ways to determine cytotoxicity, two more methods (other than a viability live/dead count) worth mentioning for future researchers to reference are through an MTT assay and analysis of cellular proliferation and penetration through the scaffold cross-section (Norouzi *et al.*, 2015; Xie *et al.*, 2014). The MTT assay allows a researcher to quantify the metabolic activity of the cells within a scaffold, which reflects the number of viable, living cells still present. Through staining the cells and analyzing the cross-section of a scaffold, it is possible not only to quantify the viability of the seeded cells, but also the effects of cytotoxic reactions on the proliferation and morphogenesis of the cells.

8.2.2 Biomimicry

After the issue of biocompatibility has been addressed, the next step to take in order to choose the most effective material for the fabrication of a scaffold is the matter of biomimicry. This subject can be visualized easily by understanding that, in order to achieve optimal results, one would not choose a ceramic scaffold for soft tissue applications due to the striking differences between the two material types and their respective responses to applied loads. A better-known example, using biological inspiration to achieve superior results in materials, can be seen in our adaptation of the gecko's foot pads to fabricate high-strength adhesives. Optimal results for tissue engineering and regenerative medicine are accomplished when the material being implemented mimics the native tissue system in terms of its mechanical, morphological, and geometrical properties. For instance, human skin is known to have an elongation-to-fracture of up to 115% through fabricating a PLGA–collagen composite scaffold (Peh *et al.*, 2015). This ability to match the elasticity of the native tissue system is important, not only for the mechanical response that keeps the material from ultimate failure under loading, but also for the directing of stem cell fate. It is well known, and will be addressed in detail in a later section, that the stiffness of the scaffold is a factor that can be used to direct stem cells to particular lineages therefore, when it is matched, the seeded cells will be directed from these mechanical cues towards morphogenesis for the tissue system desired (Engler *et al.*, 2006; Leipzig and Shoichet, 2009; Macri-Pellizzeri *et al.*, 2015).

8.2.3 Bioactivity

Tying the two previous concepts together, we will briefly mention the requirement of bioactivity of tissue engineering scaffolds. While this is a seemingly new term in the area of regenerative medicine, we will have the working definition that a bioactive material is capable of producing a response, be it positive or negative, on a tissue system upon implantation. While bioactivity is chiefly a response of the material with biochemical cues within the natural environment, the morphological and mechanical properties of the scaffold also play a response. For instance, by increasing the porosity and density of the scaffold, it is possible to adjust this bioactivity to respond at an accelerated rate or, conversely, a more controlled rate, by decreasing the same variables. Also, seemingly bioinactive materials, typically those of a synthetic origin, can be modified to promote a beneficial bioresponse by implementing a bioactive material. Recently, by introducing bovine serum albumin (BSA) into an electrospun fibrous substrate, cells were given adequate nutrition, allowing the material to give increased rates of cellular proliferation and viability for smooth muscle cells (Ye *et al.*, 2015). Further, for enhanced bone-regeneration, doping polyurethane scaffolds with hydroxyapatite particulates, cell viability, and collagen deposition were increased compared to the pure polymer scaffolds (Tetteh *et al.*, 2014). These same effects can be seen by enhancing already bioactive materials such as poly(caprolactone) by coating them with a bioactive component such as poly(sodium styrene sulfonate) (pNaSS), resulting in an increased cellular response and resultant expression of collagen, fibronectin, and integrin $\alpha 1$ (Huot *et al.*, 2013). These results are typical for bio-doped materials, which appear to show that the materials chosen for successful biointegration can be adjusted in order that the mechanical properties can be matched with the ideal material and, if necessary, the scaffold material subsequently doped with bioactive materials to increase the cellular response.

8.2.4 Biodegradability

While the aforementioned requirements of biocompatibility, biomimicry, and bioactivity are non-negotiable for implementation into the body, the issue of biodegradability is one that is typically desired but not always necessary. For instance, an arterial stent would likely not be made of a material that would degrade *in vivo*, while a scaffold used to promote tissue system regeneration would need to degrade naturally without producing cytotoxic by-products. With this in mind, we will only cover materials that are biodegradable, as we are interested chiefly in the regeneration of those tissues. As such, biodegradation can be defined as the ability for a material to be broken down by naturally occurring phenomena in the body, typically through enzymatic or hydrolytic pathways.

Ideally, a scaffold for tissue engineering would have a degradation profile inversely proportional to the regeneration profile of the tissue system of interest. If the scaffold undergoes degradation at an accelerated rate, it is not capable of supporting the cells throughout their attachment, proliferation, differentiation, and ultimate remodeling. If the degradation rate is too slow, the regenerated tissue system retains implanted matter and is thus not entirely salient to natural tissue. Both of these defects in regards to the degradation rate, in part, may result in sub-ideal properties and cellular response in long-term applications. A second factor that is worth noting in regards to the biodegradation of fibrous scaffolds is that the material substrate should not lose its mechanical characteristics to a large degree upon degradation. Since cellular attachment, proliferation, and differentiation are dependent, in part, on the mechanical values of the substrate, if these properties change substantially upon the onset of degradation, the cellular response will also change. Rapidly degrading scaffolds have previously been shown to result in loss of cellular proliferation, population, and viability, and a decrease in angiogenesis (Sung *et al.*, 2004). As mentioned above, the inefficiency of the scaffolds to thoroughly support the regeneration of tissue during regeneration as they undergo degradation appears, at its onset, to be an unavoidable aspect of these fibrous substrates. In studying the effects of degradation on the mechanical response of the scaffold, the introduction to hydrolytic degradation of polyester-urethane scaffolds *in vitro* resulted in a decrease in the elastic modulus, ultimate tensile strength, and maximum elongation reducing from $E_{max} = 1.14 \pm 0.23$ MPa, $\sigma_{max} = 0.52 \pm 0.12$ MPa, and $\varepsilon_{max} = 176.8 \pm 21.9\%$ to $E_{deg.} = 0.43 \pm 0.26$ MPa, $\sigma_{deg.} = 0.033 \pm 0.028$ MPa, and $\varepsilon_{deg.} = 24.6 \pm 3.0\%$, respectively (Krynauw *et al.*, 2011). One method which has seen advancements recently is the incorporation of a fibrous phase within a biocompatible hydrogel matrix. These composites can be pictured by imagining a bowl of Jell-O® that was poured into a bowl of cooked spaghetti before setting the gelatin. By creating this biocomposite, scaffolds are able to maintain their biological favorability while preserving structural integrity. The gelatinous matrix further enhances the saliency of the scaffold towards the ECM-enhancing cellular response. This matrix phase can also be loaded with growth factors to further increase the cellular response, as well as direct the fate of the cells. While allowing for greater cellular activity, the gel matrix phase also shields the fibrous substrate from enzymatic and hydrolytic attack, prolonging its rate of degradation. This facet of the biocomposite allows the scaffold to maintain its dimensional stability, as well as its ability to direct cellular fate through mechanical cues throughout long-term applications that are not achievable with bare-fiber substrates.

8.3 ELECTROSPINNING METHODS OF FIBROUS SCAFFOLD PRODUCTION

Recently, an increase in interest for the process of electrospinning, a method which was introduced at the beginning of this chapter, has been spurred on by an increasing number of researchers recognizing its remarkable saliency to the ECM and the positive response *in vivo* of substrates fabricated in this fashion (Liao *et al.*, 2017). With this growing interest, the technologies to fabricate nanoscale fibrous structures have also developed. Among them, electrospinning has consistently shown itself to be a reproducible and adjustable method by which nanofibrous substrates can be formed to be applied to a variety of environments. The process of nanofiber production through electrospinning, depicted in Figure 8.1, is dependent on the electrostatic attraction of a charged polymeric jet to a grounded collecting apparatus which serves both as a depositing substrate as well as a drawing mechanism for the extruded polymer jet. Briefly, a syringe with a metallic needle is filled with a polymer solution and loaded into an automated syringe pump. A positive electrical charge is then placed on the needle, with the ground of the circuit placed onto the collection substrate. The applied potential difference between the needle and the ground coupled with the controlled extrusion of the polymer serves to provide a continual jet of polymer solution into the electric field that is then drawn and deposited onto the collector. Various morphologies, diameters, porosity, and geometries are attainable by adjusting the multiple processing parameters. Electrospinning offers a method by which nanofibers can be quickly made for subsequent application, but also is a technique that allows for the fabrication of blended polymers, composites, and cell-laden nanoarchitectures. In the following sections, we will cover various electrospinning methods and the subsequent properties of these nanofiber scaffolds and their effects on both cellular and mechanical responses.

Recently, research has been focused on various electrospinning techniques for developing scaffolds for soft-tissue cell transplant and hard-tissue regeneration (Liao *et al.*, 2017). The electrospinning of a nonwoven nanofibrous networks has been ideal for gene/cell delivery and the imitation of biological substrates for many aspects of tissue engineering such as tissue regeneration, immobilizing enzymes, catalyst systems, and even artificial blood vessels (Lee *et al.*, 2005). However, there are potential challenges with alternative tissue scaffolds as it relates to the diffusional limitations of mass transport, due to the high level of vascularity and cellularity requirements (Leong *et al.*, 2009; Zhu *et al.*, 2008).

The most notable advantage of the electrospinning technique as it relates to tissue engineering is the aforementioned ability to produce scaffolds that can closely mimic the fibrous-porous structure of ECM of human biological environments. There are physical similarities that are vital, such as the capability of cell interaction and proliferation due to the three-dimensional structure. There has been research that has established the way of functionalizing the surfaces of these spun membranes with ECM proteins, such as collagen, to induce cell attachment and interaction within the matrix (Bhattarai *et al.*, 2004). These electrospun scaffolds should maintain both their porosity and their structural stability under applied stresses within wet conditions which would be relevant for both *in vitro* and *in vivo* applications. For instance, there is a requirement of large pores throughout the membrane and structural stability when there are external stresses applied during cell growth and biodegradation of the scaffold. The degree of porosity and pore size should be controlled to satisfy

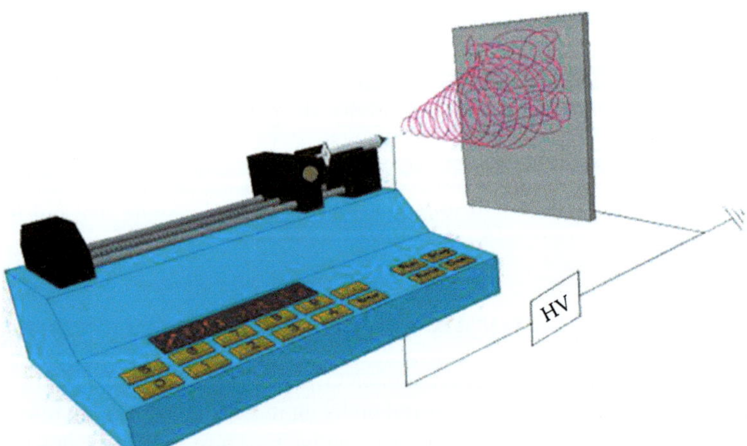

FIGURE 8.1 Typical setup for the electrospinning process where the polymeric jet is extruded at a controlled rate and has a positive charge on its surface. The charged jet is drawn via electrostatic forces towards the grounded collecting plate, which further decreases the diameter of the fabricated fiber.

seeding of cells along with allowing vascularization of the scaffold by invasion of blood vessels in the case of nutrient supply and waste removal to and from the transplanted cells (Lee *et al.*, 2005).

These fabricated electrospun fibrous mats also have high surface area to volume ratios which further resemble the fibrillary structure of the ECM. The structure of the scaffold changes over time, due to the degradation of fibers which could allow cell proliferation. Cell-populated, three-dimensional scaffolds must permit cell migration to occur throughout the depth of the fibrous-porous material in order to achieve the optimal response upon implementation (Zhu *et al.*, 2008).

In order for tissue to be successfully generated, there must be a sufficient amount of cell propagation and appropriate differentiation within the three-dimensional extracellular scaffold. These fibrous mats have a high porosity distribution throughout the membrane in all axes' directions. Current research has investigated a copolymer, (PPDO/PLLA-b-PEG), that has optimal qualities for being a biodegradable release matrix (Bhattarai *et al.*, 2004). The biological function during cell interaction of the resultant fibrous network allowed for the cells to maintain their original phenotype while adhering to the structure. The study was able to determine that this copolymer provided multifunctionality of drug release and a surgery-induced adhesion barrier matrix.

The complexity of fibrous-porous materials is dependent on the size and geometry of the fibers and fabrication technique utilized (Bhattarai *et al.*, 2004). Specifically, the pore size range of 10–1000 nm is optimal for nutrients and metabolic waste to pass through. Therefore, the electrospinning technique is preferred, due to providing continuous nonwoven fibers from 10–300-micron porosity. These scaffolds allow the attachment of human cells that then organize around the fibers of the network given that the diameter of the fibers is smaller than the cell themselves (Zhu *et al.*, 2008). The fibrous-porous membrane would be relatively robust, and thus optimal for maintaining structural integrity during cell growth and biodegradation (Lee *et al.*, 2005).

8.3.1 Core-Shell Fibers

Electrospinning has been labeled as one of the more common fabrication methods explored, as it relates to polymeric nanofibers in the pharmaceutical industry. Specifically, there has been recent complexity introduced to the process with the utilization of a concentric spinneret. This method is called coaxial electrospinning, which consists of one needle that is nested within a second needle yielding nanofibers of core/shell morphological structures (llangakoon *et al.*, 2015; Huang *et al.*, 2017). The experimental setup for coaxial electrospinning involves two syringe pumps that can be set a specific rate with a high voltage power supply. The outermost need of the concentric spinneret has a diameter large enough to have another needle with embedded within it. A positive charge is applied to the spinneret while a negative charge is applied to a collector plate yield a potential difference, similar to that of the traditional electrospinning setup. It is common for all experiments to be performed under ambient conditions.

When there is no voltage applied at given pump rate, there is no mixing of the outer solution (sheath) and inner solution (core). This is an indication that sol-gel solution parameters were optimized for nanofiber core/shell structure morphology. When a specific voltage is applied, there is a thinning jet that has a straight path from a compounded Taylor cone produced at the spinneret tip (Illangakoon *et al.*, 2015). The compounded solution undergoes bending and whipping motions, developing loops of enlarging sizes before it is deposited on a metallic collection surface. There is cored inner layer is surrounded by a shelled outer layer, shown schematically in Figure 8.2, yielding various characteristics that are applicable to delivery systems and tissue engineering. In particular, therapeutic molecules or cells can be protected within the core phase of the nanofiber during transport to targeted defects *in vivo* (Perez and Kim, 2015). This fabrication process also has tunability as it relates to material composition and experimental parameters to develop specific cross-sectional geometries and designs that give the ability to control therapeutic release kinetics for potential use in drug delivery and bio-doped tissue engineering scaffold systems.

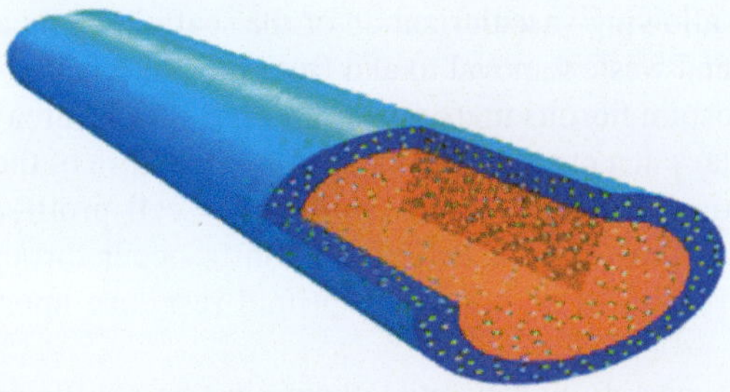

FIGURE 8.2 Diagram of a typical core-shell fiber (inner-core as orange and outer-core as blue) with different drugs loaded into each section (yellow and white spheres representing potential therapeutic agents).

The core and shell phases of the resulting nanofiber have two independent functional roles related to tissue engineering. The core is usually considered the load carrier for therapeutic molecules and/or tissue cells providing a three-dimensional, non-threatening environment (Perez and Kim, 2015). The shell is typically used to protect the inner biological ingredients (core) while also controlling the overall release kinetics of the inner core phase. Such nanofibrous structures are optimal for drug molecule incorporation and scaffolding for cell support as a therapeutic tissue matrix.

Recently, the investigations of coaxial electrospinning to develop pH-sensitive drug delivery systems for anti-cancer drugs, such as 5-fluorouracil (5-FU) has shown great promise (Illangakoon *et al.*, 2015). The nanofibers consisted of smooth and cylindrical-shape morphology where one could distinctively see the two phases with the utilization of a transmission electron microscopy. The active ingredient was determined to have an amorphous orientation within the core of the fibers. In this particular study, the shell phase of the fiber had a burst release immediate reaction within a stomach-mimicking environment; while the core phase released upon transfer into a specific pH which was an imitation of a small intestine. Due to the relatively low molecular weight of 5-FU, the solubility is increased in acidic or basic environments yielding and effective method for localization of drug release and delivery capabilities.

8.3.2 Janus Fibers

While electrospinning has consistently been used by numerous research groups due to its benefits previously mentioned, the fabrication of more complex nanofibrous structures has seen a surge in interest over the last decade. Among the many different structures, biphasic or triphasic nanofibers, also known as Janus fibers, are appealing due to their potential for dual functionality (Yu *et al.*, 2017). Like core-shell fibers, Janus fibers are commonly utilized within the field of tissue engineering and drug delivery (Geng *et al.*, 2016). Dissimilar to core or shell fibers, the Janus structure allows for direct contact of both polymeric hemispheres with the surrounding environment, which can be beneficial for developing multifunctional, nanofibrous scaffolds with modifiable release and degradation kinetics.

Scientific pioneers Gupta and Wilkes established the methodology for fabricating Janus fibers by implementing a side-by-side electrospinning setup. The composition of the fibers consisted of polyvinyl chloride/polyurethane and polyvinylchloride/polyvinylidiene fluoride (Yu *et al.*, 2016). Since then, much research has confirmed that this structure is optimal for drug delivery systems by providing adjustable, biphasic release profiles. These release profiles have the capability of effectively delivering a loading dose while producing an initial burst of the drug molecule concentration resulting to rapid patient relief. This structure also offers ideal surface chemistry for scaffolds for the selective coculturing of various cell types due to the biphasic nanofibrous network (Srivastava *et al.*, 2009).

A recent study reported the implementation of a Teflon-coated spinneret to optimize conditions to fabricate a distinct interface between the hemispheres of a Janus-structured nanofiber, as depicted in Figure 8.3, where the two biphasic hemispheres were composed of poly(vinylpyrrolidone) (PVP) and ethyl cellulose loaded with ketoprofen. By introducing the Teflon-coated spinneret, separating the two working solutions was effectively prevented, compared to traditional biaxial electrospinning. According to traditional electrospinning, the separation of fluids during cospinning rises repulsive forces such as Ft (between the two individual Taylor cones), Fs (between the two individual flow jets), and Fc (between the coils of the spinneret) (Yu *et al.*, 2016). This separation of spinning jets was possible because the Teflon component eliminated the repulsive charge distribution in the vicinity of the spinneret (Yu *et al.*, 2016). The charge neutralization was achieved due to the nonconductivity of the Teflon when the voltage was applied, coupled with Teflon's nonstick nature, yielding a lack of adhesion between the Teflon and the spinning fiber. All of the listed factors were pivotal in the formation of the Janus Taylor cone of electrospinning. This experimental setup results in fibers exhibiting biphasic drug release with an initial drug burst followed by a sustained rate of release, which was determined by the *in vitro* dissolution test. The fabrication method can be tunable to control the release rate and distance of diffusion into the second phase by doping small amounts of one of the sol-gel solutions into the other before spinning. With manipulation of experimental parameters, there are various structures that can be altered, such as the surface or volume of each side in the fiber's diameter, fiber width, and also the interfacial distinction. The angle port in which solution is ejected from the spinneret has a strong impact on the following morphology of the Janus fiber structure (Chen *et al.*, 2015). The port angle induces a convergence of both solutions once ejected from the spinneret that facilitates the Janus structure formation within the spun nanofiber. Larger port angles yield a relatively large diffusion of miscible solutions developing a gradient of distributions across the center of the fiber and larger fiber volumes. However, small diameter ports result to a more distinctive interface between both parts of the Janus fiber and smaller volumes. The structure dimensions and morphology can be tailored to have a specific volume on both sides, allowing for optimal functional performance as it relates to drug release rates.

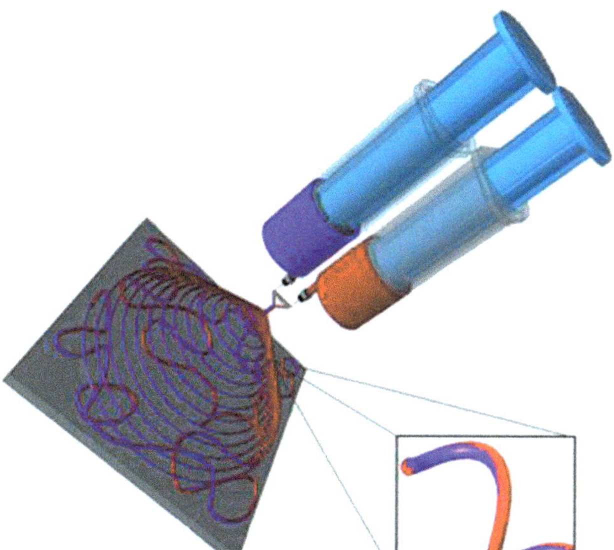

FIGURE 8.3 General setup and fabrication of Janus fibers, where separate polymer solutions are loaded into individual syringes and fed through a charged, single-axis mixing needle and drawn towards a collecting surface just as in the electrospinning process. The inset photos show the Janus fiber morphology with one polymer hemisphere shown in red, and the second polymeric hemisphere is shown in purple.

8.4 CELLULAR RESPONSE AND THE IMPORTANCE OF NANOARCHITECTURE

As we have shown, the biological, geometrical, and mechanical properties of scaffolds used in tissue engineering are all factors that must be considered in order to achieve ideal results *in vivo*. In Section 8.4, we will outline how these variables can be used in order to control and optimize cellular response for ultimately controlling the differentiation of stem cells into specific lineages, as well as maintaining then phenotype and morphological characteristics of various cell types. For instance, cells seeded onto stiff substrates and given appropriate growth factors can be guided into osteogenic differentiation, while those seeded on more pliable, aligned substrates can be directed toward neurogenic differentiation. To untrained researchers in the field of biology, the area of cellular response, proliferation, differentiation, and ultimate morphogenesis can seem daunting, therefore we can simplify parts of this process. The attachment and proliferation of a cell on a biomedical scaffold can be seen as analogous to a rock climber scaling a wall. The climber will follow a path towards the top of the wall that affords sufficient handholds that do not cause overreaching. This process will continue until the climber has used all of the handholds, leaving him or her at the top or side of the wall, depending on the positioning of such handholds. An in-depth overview of the effects of nanoarchitecture on the response of stem cells, specifically, can be found the recent work of Budhwani *et al.* (2016).

8.4.1 Fiber Diameter

As stated at the outset of this chapter, the ECM is composed of a water-rich matrix phase with a dispersed fibrillary network throughout. This fibrous web grants both structural rigidity and cellular guidance to the ECM, and is the Holy Grail for those wishing to produce optimal scaffolds for tissue engineering. When viewed on the nanoscale, electrospun fibrous substrates have only two variables of interest when it comes to their morphology: (1) diameter and (2) surface characteristics. Length is rarely a factor in regards to these substrates since, theoretically, nanofiber meshes are composed of one continuous fiber. While this may not be the case in reality, as far as what the cells actually "see" when seeded onto the structure, the fibers can be viewed as one continuous fiber. Other important factors of the scaffolds, porosity, and pore size, are an attribute of the fiber diameter and will be covered in detail in later sections. It is pivotal, therefore, that the understanding of how fiber diameter affects the response and directing of cells be known thoroughly. In regards to the second major variable, surface characteristics, we can briefly say that with increased surface roughness/porosity, comes an increase in surface area leading to an increase in cellular adhesion rates and growth directionality. The actual effect of surface functionality will be elaborated on when we look into the effect of fiber composition.

It has been widely accepted that nanofibers provide an increase in biocompatibility that improves with decreasing diameter, largely due to their increased surface area and ECM saliency (Boudriot *et al.*, 2006). To this end, a recent study on nanofibers ranging from 70 nm to 1120 nm found that the size, morphology, and actin (the major cytoskeletal protein which eventually forms actin filaments) organization in both seeded keratinocytes and skin fibroblasts were affected by changing the fiber diameter in the scaffold (Pelipenko *et al.*, 2014). The morphology of the cells had an optimum diameter with the average size of the keratinocytes decreasing with an increasing fiber diameter, eventually increasing with a continued increase in fiber diameter. The smallest cell size was found on fibers with a diameter of 305 nm. As the fiber diameter continued to increase, the cells appeared to see the increased inter-fiber pores and spacing, which led to increased cellular infiltration into the scaffold and a resulting increase in cell diameter. This result also highlights an important fact of the use of nanofibrous scaffolds in tissue engineering—the small pore size inhibiting cellular infiltration into the scaffold bulk. An in-depth study of this was carried out by seeding human mesenchymal stem cells onto poly(ε-caprolactone) (PCL) nanofibrous membranes while varying the fiber diameter (Bean and Tuan, 2015). Although cellular adhesion is typically enhanced when nanofibers are implemented, the results of this study revealed that chondrogenic gene expression, cellular proliferation, and the production of sulfated GAGs and collagen were enhanced upon the use of microfibers. This effect is largely attributable to the increase in pore size, which allows for cellular migration into the scaffold as well as decreasing the initial cell–cell interactions which can also inhibit the movement of cells towards the bulk interior.

With seemingly contradictory studies, where some researchers show that nanofibers are optimal for tissue engineering due to their fine diameter, while other studies point out the inherent inability for such scaffolds to stimulate an ideal cellular response, namely ingrowth and bulk proliferation, it is worth pointing out the studies in the middle ground. Such research efforts have not dismissed the contradictory reports, but have instead taken efforts to unite the two worlds by creating fibrous substrates containing both nanofibers and microfibers. In an effort to achieve promising results from fibrous scaffolds, the density of nanofibers within the scaffold needs to be controlled, since with an increasing density of fibers, pore size is drastically diminished. The nanofibers provide benefits by allowing for the better spreading of cells in two-dimensions, since interfiber spacing is reduced. Referring back to our analogy of a rock-climber, this reduced spacing is equivalent to having handholds that are closer together and therefore easier to grab and migrate along, which is equivalent to cellular attachment and spreading, respectively. By introducing fibrin nanofibers into a network of PCL microfibers, a significant increase in GAG content, ECM production, and cellular infiltration into the scaffold bulk was reported (Levorson *et al.*, 2013). Further work into the effect of thick fibers and increased pore size on arterial regeneration revealed that an increase in diameter and subsequent pore size resulted in enhanced cellular infiltration, patency, and an appreciable amount of ECM production (Wang *et al.*, 2014). These results are confirmed with by other studies of cellular response on microfibers and nanofibers separately. One other comparison of note was carried out by Yang *et al.* (2005), who investigated the performance of microfiber and nanofiber scaffolds with respect to the resultant cell morphology, differentiation, and neuritic outgrowth of neural stem cells (NSCs). Their results showed that while the orientation of the cells was not directly dependent on the fiber diameter, the rate of differentiation was increased for smaller diameter fibers. It is well known, and has been stated and explained here, that cellular attachment, spreading, and ECM production are enhanced with decreasing fiber diameter, while the scaffold infiltration of cells is enhanced with increasing fiber diameter and the resultant increasing pore size.

8.4.2 Fiber Composition

While fibrous materials that have been sourced from natural origins (such as collagen, cellulose, fibrin, silk, casein, etc.), spidroin fibers offer benefits due to their innate bioactivity and typical biodegradability, and they are also plagued with suboptimal mechanical properties, variance across specimens, and, in some cases, the risk of viral transfer. On the other hand, synthetically derived materials that are fabricated into a fibrous form offer enhanced mechanical response, long-term degradation profiles, and batch-to-batch comparability of properties, but typically lack inherent bioactivity. This is not to say that all naturally derived materials offer an ideal bioresponse, or that all synthetic materials offer the paradigm of mechanical characteristics, but that across the board, the majority of the materials do fall within the spectrum of this classification. It would thus appear obvious that an ideal tissue engineering scaffold would be a mixture of both natural and synthetic materials, but it would be presumptuous to not outline a baseline of responses for each type of system. While most tissue engineering applications involve the use of fibrous scaffolds and have been centered on the regeneration of soft tissue systems, the use of ceramic fibers and polymeric fibers doped with ceramic particulates have seen increased interest in the field of bone tissue engineering.

Fibrous scaffolds that have been fabricated from naturally sourced materials have gained recent and growing interest, not only due to the biological benefits mentioned above, but also in response to the increased interest in material processing methods that are both economically beneficial and environmentally benign. Typically, the source material is denatured thermally or through solvent dissolution to form a spinnable dope. Occasionally, as is the case with the soy protein isolate (SPI), these pre-fiber solutions must be mixed with a minimal amount of a high molecular weight polymer as the denatured/dissolved proteinaceous chains can become unspinnable. A review of naturally sourced materials commonly used in applications for tissue engineering can be found in numerous reviews previously published (Silva *et al.*, 2010; O'Brien, 2011; Muzzarelli *et al.*, 2012; He and Lu, 2016). Of note, by seeding endothelial cells onto a composite scaffold containing a mixture of collagen-GAG, improved microvascularization prior to *in vivo* implantation was achieved (O'Brien, 2011). In an interesting research paper, Zhou *et al.* used tilapia-derived collagen nanofibers to induce the proliferation and differentiation of keratinocytes (Zhou *et al.*, 2016). This biomimetic network of fibers not only allowed for rapid and efficient cellular adhesion and proliferation, but also showed the ability to expedite re-epithelialization generating a continuous layer of epidermis after only 14 days. This response to the nanofibrous environment composed of a derivative of fish shows not only the expected response when using natural material, but it also shows the benefits of using a novel source of material. This cellular response is one found across any type of cell and shows the major benefit of the use of natural materials.

By implementing carbonized cellulose-derived nanofibers, naturally derived materials were made conductive and thus enhanced the formation of a neural network within 10 days, at a time frame that is beneficial for the SH-SY5Y neuroblastoma cells which are used to evaluate neural function and differentiation (Kuzmenko *et al.*, 2016). These results were compared to that of a control group, as well as cellulose fibers with embedded multiwall carbon nanotubes (MWCNTs). It was found that the carbonized cellulose samples had comparable results to those with MWCNTs which is of interest due to the health concerns with handling CNTs. Although this improvement in bioactivity and subsequent cellular response is a leap forward in regenerative medicine, these particular scaffolds for skin regeneration had an ultimate tensile strength of only 6.72 ± 0.44 MPa—far below the typical value of 20 MPa typically attributed to human skin. In a comparable study on modified carbon nanofibers (mCNF) modified by oxidation, which was then cospun with PCL for bone tissue engineering, our group sought to elucidate the effects the fibrous phases had on mechanical and cellular properties of a scaffold (Deshpande *et al.*, 2012). Incorporation of mCNF not only increased the mechanical properties of the scaffold, but also improved the cell-spreading in a preliminary study to assess cytocompatibility hMSCs on both aligned PCL and PCL–mCNF scaffolds. Both the scaffolds showed indistinguishable spread morphology with a directional growth of the cells along the nanofiber orientation in PCL–mCNF scaffolds. Later, the conductive property of the PCL–CNF scaffolds was used to deposit hydroxyapatite by electrophoretic deposition (Deshpande *et al.*, 2011). While the use of carbon nanotubes has seen recent exponential interest in biomedical applications, precautions must be taken during handling to avoid toxic exposure.

In efforts to remedy the inherently low mechanical properties among naturally sourced fibrous materials, a number of research groups have turned to synthetic materials. These synthetically fabricated fibrous scaffolds offer advantages such as long-term biostability, improved mechanical characteristics, uniformity across samples, and ease of modification, but tend to lack the inherent biodegradability and high bioactivity of natural fibers. Common synthetic materials used towards this end include PCL, PLGA, PLA, and PU, among others. An in-depth review of synthetic materials and their use in tissue engineering scaffolds can be found in Thomas *et al.* (2006). Understanding the above-mentioned responses for natural and synthetic materials in the use of tissue engineering, we now are capable to fully understand the benefits found in a natural–synthetic composite scaffold.

8.4.3 Organic–Bioinorganic Hybrid Fibers

In an effort to compositionally mimic, to an extent, extracellular matrix (ECM) proteins of bone matrix and biodegradable polymers have been extensively coelectrospun with calcium phosphate minerals, such as particulate nanohydroxyapatite (nHA). Biphasic organic–inorganic composite fibers of nanoHA and bioglass or other calcium phosphates (TCP) have exhibited higher mechanical properties and cellular responses with mesenchymal stem cells. We have studies bioactive composite fibers scaffolds such PCl-nHA, PLGA-nHA, and collagen-nHA with nHA content 1 to 20%

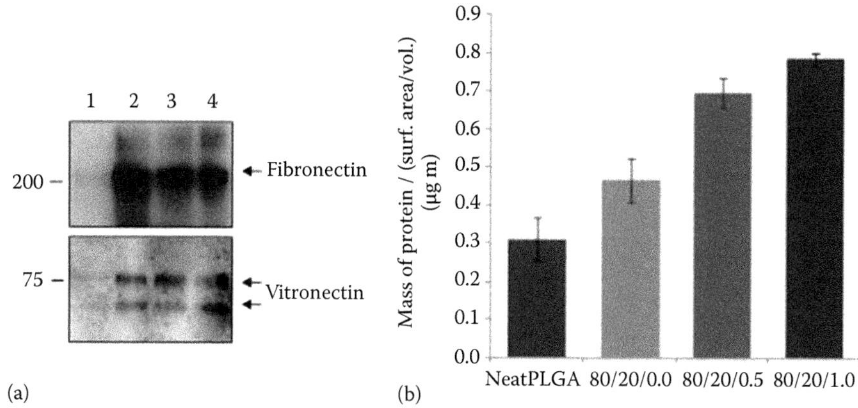

FIGURE 8.4 (a) Western blot analysis for fibronectin and vitronectin adsorbed on the multi-component system: 1- Neat PLGA, 2- 80.20, 3- 80/20/0.5, 4- 80/20/1.0. (b) Total protein adsorption on scaffold calculated from microBCA assay. (Jose MV, Thomas V, Xu Y, Bellis S, Nyairo E, Dean D: Aligned bioactive multi-component nanofibrous nanocomposite scaffolds for bone tissue engineering. *Macromol Biosci*. 2010. 10. 433–44. Copyright Wiley-VCH Verlag GmbH & Co. KGaA. Reproduced with permission.)

for potential use in bone tissue engineering (Jose *et al.*, 2009; Thomas *et al.*, 2007; Thomas *et al.*, 2006). A biomimetic tri-component fibrous scaffold of PCL-nHA-collagen, PLGA-nHA-collagen, and PLCL-nHA-gelatin, as well as nHA gradient-scaffold for bone–cartilage interface tissue engineering were developed and reported (Jose *et al.*, 2010; Bottino *et al.*, 2011; Phipps *et al.*, 2011; Ramalingham *et al.*, 2013). Detailed mechanical evaluations of our composite nanofibrous mat with various polymers (PCL, PLGA, PLCL) or collagen proteins and nanoHA has shown increased mechanical properties (tensile strength and modulus) with the increased incorporation of nanoHA particles via coelectrospinning or electrophoretic deposition.

Moreover, nHA incorporation with synthetic polymers has shown increased adhesion of cell-adhesive proteins such as fibronectin and vitronectin *in vitro* and *in vivo* (Deshpande *et al.*, 2011; Phipps *et al.*, 2011). Figure 8.4 shows a higher protein adsorption (fibronectin and vitronectin) on the blend and multicomponent system compared to the neat PLGA nanofibers. This was attributed to the collagen and nHA in the system. The presence of more durable PCL or PLCL (85/15) provided the scaffolds excellent dimensional stability with no significant change in mechanical properties *in vitro* culture condition and *in vivo*. ECM mimicking nanofibrous morphology and chemistry favored positive cell–matrix interactions, as evident from our earlier works. Moreover, our recent studies have shown that growth factors such as platelet-derived growth factor (PDGF) or bone morphogenic proteins (BMP-2) adsorbed passively on the nanoHA composite scaffolds. PCL/col/HA scaffolds electrospun scaffolds incorporating collagen I and nanoHA, compared with scaffolds composed of PCL alone, adsorbed and released greater quantities of PDGF (with sustained release extending over an 8-week interval), leading to enhanced MSC chemotaxis. Therefore, our bone-mimetic triphasic polymer/collagen/nanoHA electrospun scaffolds supported greater MSC adhesion, proliferation, and activation of integrin-related signaling cascades than scaffolds composed of PCL or collagen I alone, and has motivated researchers to use a similar composition at the bone-facing surface in a specially-layered periodontal membrane design (Phipps *et al.*, 2011; Bottino *et al.*, 2012). Figure 8.5 shows the effect of nHA and collagen on human mesenchymal stem cell growth.

8.4.4 Mechanical Properties

As covered in Section 8.4.1, the diameter and subsequent porosity of fibrous scaffolds for use in tissue engineering is of paramount importance in fabricating ideal substrates. These substrate

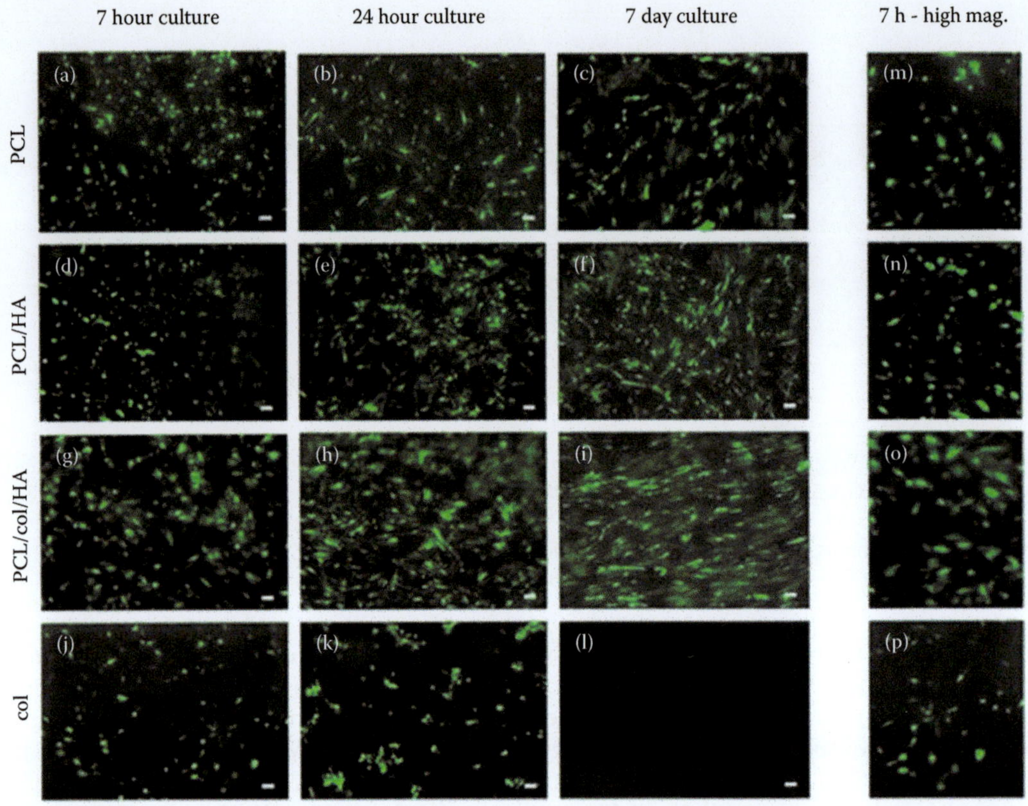

FIGURE 8.5 Live cell imaging of GFP-expressing MSCs seeded on electrospun scaffolds at different time points: (a–c) depicts PCL scaffolds; (d–f) shows the PCL/HA composite scaffolds; (g–i) shows PCL/col/HA scaffolds; while (j–l) shows neat collagen scaffolds. (Scale bar = 100 μm.) The higher magnification images (m–p) show the GFP-expressing MSCs at 7 h on the various electrospun scaffolds. (Reproduced with permission from Phipps MC *et al.*, *PLoS One*. 2011; 6: e16813. © 2016 PLoS ONE under Creative Commons Attribution License.)

properties are also influential of the mechanical properties of the substrate as with increasing fiber diameter, strength increases while an increasing porosity density results in a decrease in mechanical properties. The compositional makeup of the fibrous scaffold also is a direct relationship to the subsequent mechanical properties, as naturally-sourced materials have an inherently lower range of mechanical values than those that have been synthetically derived.

One such evaluation of these mechanical properties on the differentiation of mesenchymal stem cells was carried out using multiple substrates seeded with mesenchymal stem cells (Tse and Engler, 2011). By creating and seeding cells onto polyacrylamide substrates with various stiffness values, it was shown that the fate of the cells was directly related to the stiffness of the material surrounding them, as seen in Figure 8.6. This is to say that stem cells can be guided into specific lineages by introducing them to environments with mechanical properties similar to those of the desired regenerated tissue system. Further along this point, we can see that if we desired regeneration of a soft biological system, such as skeletal muscle or the dermis, stem cells could be guided by seeding them onto substrates with higher values of elasticity. This point seems like common knowledge when we examine the extremes previously stated above in the section regarding biomimicry. It would be more likely to choose a stiff ceramic or metal oxide material doped material for the scaffold system when attempting to induce osteogenic differentiation, while more flexible polymer or polymer-fiber-reinforced hydrogels (Wood and Thomas, 2016) would be chosen for soft tissue systems.

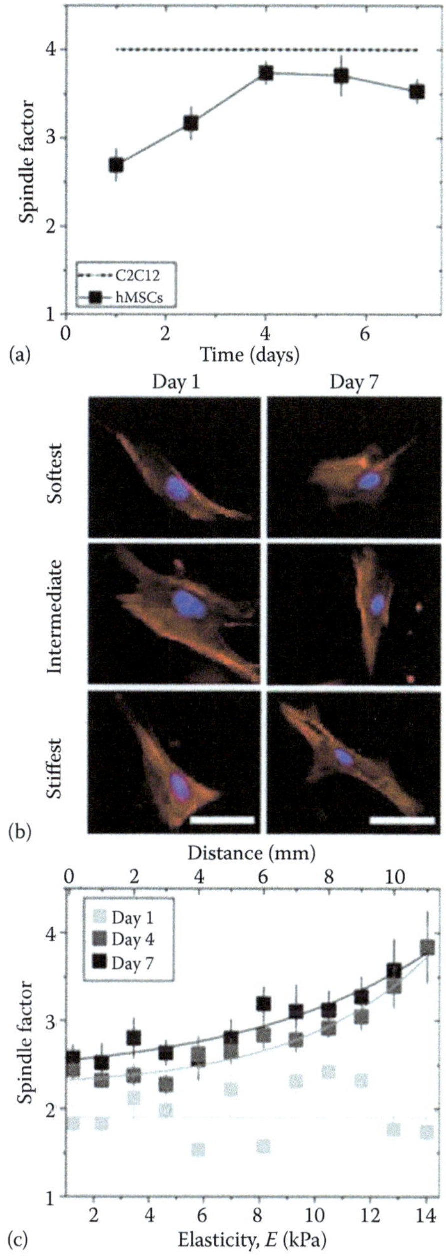

FIGURE 8.6 The impact of matrix elasticity on MSC morphological features: (a) MSC spindle factor on 11 kPa matrices increase their spindle factor while C2C12 myoblasts have a constant spindle factor; (b) stained with Hoescht 33342 (blue) and phallodin (red), the morphology of MSCs seeded on hydrogel matrices of a stiffness gradient showing increasing spindle-shape with increasing stiffness; (c) spindle factor as a factor of time. (Engler AJ *et al.*, *Cell.* 2006; 126: 677–89. © 2016 PLoS ONE under Creative Commons Attribution License.)

8.4.5 Fiber Alignment

In certain cases of tissue engineering and regenerative medicine, the directionality of cell proliferation ultimately controls the resulting morphology of the cells. By controlling the alignment of deposited PCL-Gelatin nanofibers, Fee *et al.* (2016) showed that by incorporating unidirectional-aligned nanofibers into a scaffold seeded with NIH3T3 cells, the fibroblasts were able to maintain their elongated, spindle-shaped morphology as opposed to multidirectional growth. These results can be seen in Figure 8.7, showing that not only did the cells become elongated, but also the nucleus was elongated on aligned fibers while being spherical on the randomly oriented fibrous substrate. Keeping with our previous analogy of the cells being comparable to a climber scaling a rock face, we can imagine the individual fiber surfaces as individual handholds. If these hand holds along the rock face are positioned so that the next one is directly above the one our climber is currently gripping, the path of least resistance is to reach straight up and grab the aligned handhold. If each successive handhold is positioned in the same manner and we track our climber, his or her mapped path will be more or less perfectly straight. Conversely, if the handholds are separated and our climber has to overextend to reach the next one, his or her mapped path will be scattered randomly about the rock face. This is perfectly pictured by the cellular reaction in Figure 8.7, where cells along aligned fibers grow along the path of least resistance which is the path that is parallel to the fibers. The cells seeded on scaffolds of randomly oriented fibers tend to be spread out across various fibers and do not achieve the same type of alignment. Since within the nervous system, electrical impulses are sent along the neural network, it would be beneficial for a scaffold for such a purpose to be aligned so as to direct the growth of seeded neural cells. This was investigated by Yang *et al.* (2005), who compared the differentiation of NSCs on aligned and randomly oriented PLLA fibers. It was shown, as we would now expect, that the alignment of PLLA fibers within the scaffold was able to direct

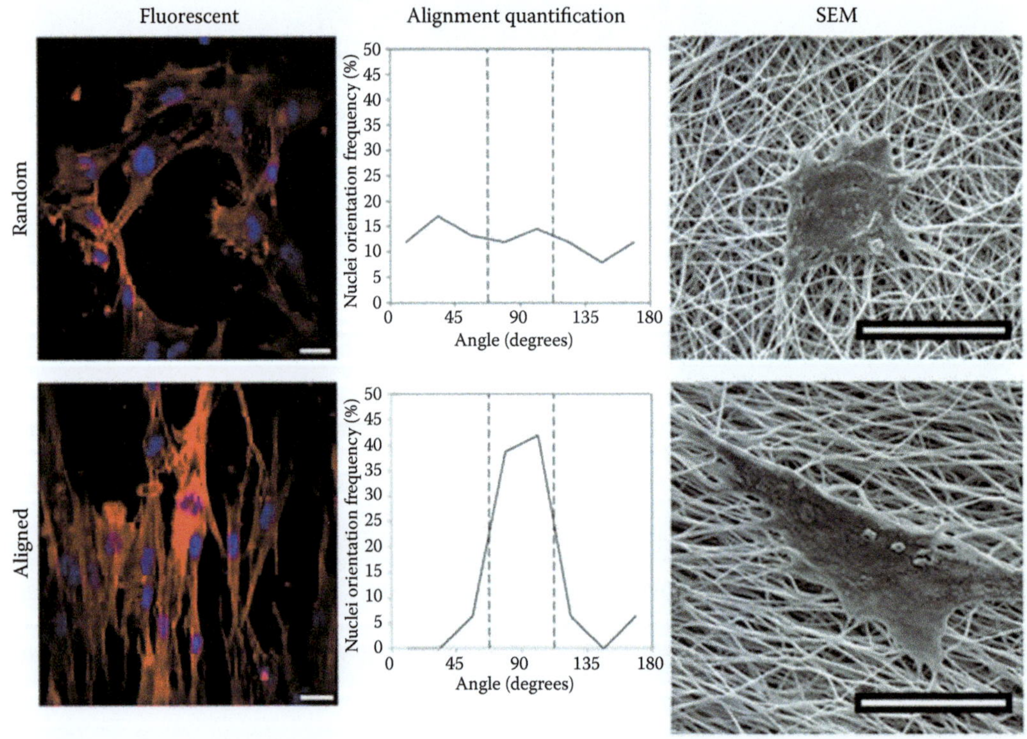

FIGURE 8.7 Fluorescent images, distribution of cell orientation, and SEM images of cell interactions with the fibrous substrate. Scale bars are 20 μm. (Reproduced with permission from Fee T *et al.*, *PLoS One*. 2016; 11: e0154806. © 2016 PLoS ONE under Creative Commons Attribution License.)

not only the elongation of NSCs, but also cause their neuritic outgrowth to extend parallel to the fiber direction.

8.5 FUTURE DIRECTIONS

The recent advances in tissue engineering technology include both the synthesis and fabrication of materials, and has shown its ability to rapidly and progressively move the field forward. With the enhanced understanding of material properties on the biological response, the fine-tuning of fabrication methods for production of scaffolds with high ECM saliency, and the resultant knowledge of the biological and mechanical properties of the scaffolds on the efficacy of the application, the road ahead is wide open. Further advancements in the tunability and control of fiber substrate characteristics, such as porosity density, mean fiber diameter, fiber alignment, and functionalization of synthetic fibrous materials to give them the desired biological response without losing their advantageous mechanical properties, will surely continue to push both the application side of this work, as well as future research endeavors, to new heights. The main drawbacks of electrospun scaffolds are their randomness, small pore-size for cell in-growth, and relatively poor mechanical properties. Developments such as cell microintegration or cell encapsulated spinning, melt-electrospinning, three-dimensional writing (Hochleitner *et al.*, 2017), and template-assisted spinning could be future directions for tailoring the properties of electrospun matrices.

ACKNOWLEDGMENTS

The authors acknowledge the financial support from UAB Faculty Development Grant Program from Provost Office to VT, Alabama Space Grant Consortium Pre-Doctoral Fellowship to ATW and Deep South Center Pilot Grant (pre-doctoral) to DE.

REFERENCES

Bean AC, Tuan RS. Fiber diameter and seeding density influence chondrogenic differentiation of mesenchymal stem cells seeded on electrospun poly(epsilon-caprolactone) scaffolds. *Biomed Mater.* 2015; 10: 015018.

Bhattarai SR, Bhattarai N, Yi HK, Hwang PH, Cha DI, Kim HY. Novel biodegradable electrospun membrane: scaffold for tissue engineering. *Biomaterials.* 2004; 25: 2595–602.

Bottino MC, Thomas V, Janowski GM. A novel spatially designed and functionally graded electrospun membrane for periodontal regeneration. *Acta Biomater.* 2011; 7: 216–24.

Bottino MC, Thomas V, Schmidt G, Vohra YK, Chu TM, Kowolik MJ *et al.* Recent advances in the development of GTR/GBR membranes for periodontal regeneration—a materials perspective. *Dent Mater.* 2012; 28: 703–21.

Boudriot U, Dersch R, Greiner A, Wendorff JH. Electrospinning approaches toward scaffold engineering—a brief overview. *Artif Organs.* 2006; 30: 785–92.

Budhwani KI, Wood AT, Gangrade A, Sethu P, Thomas V. Nanofiber and Stem Cell Enabled Biomimetic Systems and Regenerative Medicine. *J Nanosci Nanotechnol.* 2016; 16: 1–12.

Chen G, Xu Y, Yu DG, Zhang DF, Chatterton NP, White KN. Structure-tunable Janus fibers fabricated using spinnerets with varying port angles. *Chem Commun (Camb).* 2015; 51: 4623–6.

Deshpande HD, Dean DR, Thomas V, Clem WC, Jose MV, Nyairo E *et al.* Carbon nanofiber reinforced polycaprolactone fibrous meshes by electrostatic co-spinning. *Curr Nanosci.* 2012; 8: 753–61.

Deshpande HD, Schindler C, Dean DR, Clem WC, Bellis SL, Nyairo E *et al.* Nanocomposite scaffolds based on electrospun polycaprolactone/modified CNF/nanohydroxyapatite by electrophoretic deposition. *J Biomater Tiss Eng.* 2011; 1: 177–84.

Engler AJ, Sen S, Sweeney HL, Discher DE. Matrix elasticity directs stem cell lineage specification. *Cell.* 2006; 126: 677–89.

Fee T, Surianarayanan S, Downs C, Zhou Y, Berry J. Nanofiber alignment Regulates NIH3T3 cell orientation and cytoskeletal gene expression on electrospun PCL+gelatin nanofibers. *PLoS One.* 2016; 11: e0154806.

Frantz C, Stewart KM, Weaver VM. The Extracellular Matrix at a Glance. *J Cell Sci.* 2010; 123: 4195–200.

Geng Y, Zhang P, Wang Q, Liu Y, Pan K. Novel PAN/PVP Janus ultrafine fiber membrane and its application for biphasic drug release. *J Mater Chem. B.* 2017; 1–18.

He Y, Lu F. Development of synthetic and natural materials for tissue engineering applications using adipose stem cells. *Stem Cells Int.* 2016; 2016: 5786257.

Hochleitner G, Kessler M, Schmitz M, Boccaccini AR, Teßmar J, Groll J. Melt electrospinning writing of defined scaffolds using polylactide-poly (ethylene glycol) blends with 45S5 bioactive glass particles. *Mater Lett.* 2017; 15; 205: 257–60.

Holloway JL, Lowman AM, VanLandingham MR, Palmese GR. Interfacial optimization of fiber-reinforced hydrogel composites for soft fibrous tissue applications. *Acta Biomater.* 2014; 10: 3581–9.

Huang ZX, Wu JW, Wong SC, Qu JP, Srivatsan TS. The technique of electrospinning for manufacturing core-shell nanofibers. *Mater. Manuf. Processes*. 2018; 33(2): 202–19

Huot S, Rohman G, Riffault M, Pinzano A, Grossin L, Migonney V. Increasing the bioactivity of elastomeric poly(caprolactone) scaffolds for use in tissue engineering. *Biomed Mater Eng.* 2013; 23: 281–8.

Illangakoon UE, Yu DG, Ahmad BS, Chatterton NP, Williams GR. 5-Fluorouracil loaded Eudragit fibers prepared by electrospinning. *Int J Pharm.* 2015; 495: 895–902.

Jose MV, Thomas V, Johnson KT, Dean DR, Nyairo E. Aligned PLGA/HA nanofibrous nanocomposite scaffolds for bone tissue engineering. *Acta Biomater.* 2009; 5: 305–15.

Jose MV, Thomas V, Xu Y, Bellis S, Nyairo E, Dean D. Aligned bioactive multi-component nanofibrous nanocomposite scaffolds for bone tissue engineering. *Macromol Biosci.* 2010; 10: 433–44.

Krynauw H, Bruchmuller L, Bezuidenhout D, Zilla P, Franz T. Degradation-induced changes of mechanical properties of an electro-spun polyester-urethane scaffold for soft tissue regeneration. *J Biomed Mater Res B Appl Biomater.* 2011; 99: 359–68.

Kuzmenko V, Kalogeropoulos T, Thunberg J, Johannesson S, Hagg D, Enoksson P *et al.* Enhanced growth of neural networks on conductive cellulose-derived nanofibrous scaffolds. *Mater Sci Eng C Mater Biol Appl.* 2016; 58: 14–23.

Liao GY, Zhou XP, Xie XL, Mai YW. (2017). Electrospun Polymer Scaffolds: Their Biomedical and Mechanical Properties. In: Li Q and Mai YW, eds., *Biomaterials for Implants and Scaffolds.* Springer Series in Biomaterials Science and Engineering, Vol. 8, Springer, Berlin, Heidelberg, pp. 237–270.

Lee YH, Lee JH, An IG, Kim C, Lee DS, Lee YK *et al.* Electrospun dual-porosity structure and biodegradation morphology of Montmorillonite reinforced PLLA nanocomposite scaffolds. *Biomaterials.* 2005; 26: 3165–72.

Leipzig ND, Shoichet MS. The effect of substrate stiffness on adult neural stem cell behavior. *Biomaterials.* 2009; 30: 6867–78.

Leong MF, Chian KS, Mhaisalkar PS, Ong WF, Ratner BD. Effect of electrospun poly(D, L-lactide) fibrous scaffold with nanoporous surface on attachment of porcine esophageal epithelial cells and protein adsorption. *J Biomed Mater Res A.* 2009; 89: 1040–8.

Levorson EJ, Raman Sreerekha P, Chennazhi KP, Kasper FK, Nair SV, Mikos AG. Fabrication and characterization of multiscale electrospun scaffolds for cartilage regeneration. *Biomed Mater.* 2013; 8: 014103.

Macri-Pellizzeri L, Pelacho B, Sancho A, Iglesias-Garcia O, Simon-Yarza AM, Soriano-Navarro M *et al.* Substrate stiffness and composition specifically direct differentiation of induced pluripotent stem cells. *Tissue Eng Part A.* 2015; 21: 1633–41.

Muzzarelli RA, Greco F, Busilacchi A, Sollazzo V, Gigante A. Chitosan, hyaluronan and chondroitin sulfate in tissue engineering for cartilage regeneration: a review. *Carbohydr Polym.* 2012; 89: 723–39.

Norouzi M, Shabani I, Ahvaz HH, Soleimani M. PLGA/gelatin hybrid nanofibrous scaffolds encapsulating EGF for skin regeneration. *J Biomed Mater Res A.* 2015; 103: 2225–35.

O'Brien FJ. Biomaterials & scaffolds for tissue engineering. *Mater Today.* 2011; 14: 88–95.

Peh P, Lim NS, Blocki A, Chee SM, Park HC, Liao S *et al.* Simultaneous delivery of highly diverse bioactive compounds from blend electrospun fibers for skin wound healing. *Bioconjug Chem.* 2015; 26: 1348–58.

Pelipenko J, Kocbek P, Kristl J. Nanofiber diameter as a critical parameter affecting skin cell response. *Eur J Pharm Sci.* 2014; 66C: 29–35.

Perez RA, Kim HW. Core-shell designed scaffolds for drug delivery and tissue engineering. *Acta Biomater.* 2015; 21: 2–19.

Phipps MC, Clem WC, Catledge SA, Xu Y, Hennessy KM, Thomas V *et al.* Mesenchymal stem cell responses to bone-mimetic electrospun matrices composed of polycaprolactone, collagen I and nanoparticulate hydroxyapatite. *PLoS One.* 2011; 6: e16813.

Ramalingham M, Young MF, Thomas V, Sun L, Chow LC, Tison CK *et al.* Nanofiber scaffold gradients for interfacial tissue engineering. *J Biomater Appl.* 2013; 27: 695–705.

Silva SS, Mano JF, Reis RL. Potential applications of natural origin polymer-based systems in soft tissue regeneration. *Crit Rev Biotechnol.* 2010; 30: 200–21.

Srivastava Y, Marquez M, Thorsen T. Microfluidic electrospinning of biphasic nanofibers with Janus morphology. *Biomicrofluidics.* 2009; 3: 12801.

Sung HJ, Meredith C, Johnson C, Galis ZS. The effect of scaffold degradation rate on three-dimensional cell growth and angiogenesis. *Biomaterials.* 2004; 25: 5735–42.

Tetteh G, Khan AS, Delaine-Smith RM, Reilly GC, Rehman IU. Electrospun polyurethane/hydroxyapatite bioactive scaffolds for bone tissue engineering: the role of solvent and hydroxyapatite particles. *J Mech Behav Biomed Mater.* 2014; 39: 95–110.

Thomas V, Dean DR, Jose MV, Matthew B, Chowdhury S, Vohra YK. Nanostructured biocomposite scaffolds based on collagen coelectrospun with nanohydroxyapatite. *Biomacromolecules.* 2007; 8: 631–7.

Thomas V, Jagani S, Johnson K, Jose MV, Dean DR, Vohra YK *et al.* Electrospun bioactive nanocomposite scaffolds of polycaprolactone and nanohydroxyapatite for bone tissue engineering. *J Nanosci Nanotechnol.* 2006; 6: 487–93.

Tse JR, Engler AJ. Stiffness gradients mimicking in vivo tissue variation regulate mesenchymal stem cell fate. *PLoS One.* 2011; 6: e15978.

Vacanti C. The history of tissue engineering. *J Cell Mol Med.* 2006; 1: 569–76.

Wang Z, Cui Y, Wang J, Yang X, Wu Y, Wang K *et al.* The effect of thick fibers and large pores of electrospun poly(epsilon-caprolactone) vascular grafts on macrophage polarization and arterial regeneration. *Biomaterials.* 2014; 35: 5700–10.

Wood AT, Thomas V. Dimensionally stable fiber-reinforced hydrogels for tissue engineering scaffolds. *Curr Tissue Eng.* 2016; 5.

Xie J, MacEwan MR, Liu W, Jesuraj N, Li X, Hunter D *et al.* Nerve guidance conduits based on double-layered scaffolds of electrospun nanofibers for repairing the peripheral nervous system. *ACS Appl Mater Interfaces.* 2014; 6: 9472–80.

Yang F, Murugan R, Wang S, Ramakrishna S. Electrospinning of nano/micro scale poly(L-lactic acid) aligned fibers and their potential in neural tissue engineering. *Biomaterials.* 2005; 26: 2603–10.

Ye L, Cao J, Chen L, Geng X, Zhang AY, Guo LR *et al.* The fabrication of double layer tubular vascular tissue engineering scaffold via coaxial electrospinning and its 3D cell coculture. *J Biomed Mater Res A.* 2015; 103: 3863–71.

Yu DG, Yang C, Jin M, Williams GR, Zou H, Wang X *et al.* Medicated Janus fibers fabricated using a Teflon-coated side-by-side spinneret. *Colloids Surf, B.* 2016; 138: 110–6.

Zhou T, Wang N, Xue Y, Ding T, Liu X, Mo X *et al.* Electrospun tilapia collagen nanofibers accelerating wound healing via inducing keratinocytes proliferation and differentiation. *Colloids Surf, B.* 2016; 143: 415–22.

Zhu X, Cui W, Li X, Jin Y. Electrospun fibrous mats with high porosity as potential scaffolds for skin tissue engineering. *Biomacromolecules.* 2008; 9: 1795–801.

9 Nanocarrier-Mediated Therapeutic Protein Delivery

*Akhlesh Kumar Jain, Teenu Sharma, and Sunil Kumar Jain**

CONTENTS

9.1 Introduction 185
9.2 Limitations and Challenges for Protein Delivery 186
9.3 Vesicular Nanocarriers 187
9.3.1 Liposomes 187
9.3.2 Niosomes 188
9.3.2.1 Protein and Peptide Delivery Using Niosomes 189
9.4 Particulate Carriers 189
9.4.1 Polymeric Nanoparticles 189
9.4.2 Solid Lipid Nanoparticles 193
9.4.2.1 Protein/Peptide-Loading in SLNs 193
9.5 Inorganic Nanocarriers 195
9.5.1 Silica Nanoparticles 195
9.5.2 Gold Nanoparticles 196
9.5.3 Calcium Phosphate Nanoparticles 196
9.6 Conclusions 197
References 197

9.1 INTRODUCTION

Proteins are vital elements involved in all the essential activities of the biological system, such as signal transduction, enzyme catalysis, gene regulation and maintaining the classic balance of cell survival and programmed death (Walsh, 2010; Leader *et al.*, 2008). Advances in genetic engineering have led to proliferation in the enormous diversity of proteins with high biological purity. Despite their strength and specificity in physiological activities, most of the therapeutic proteins are tough to deliver clinically. Hence, an efficient means of transporting proteins to cells and organs of interest in an active state is a key aspect in tumor targeting, vaccination, regenerative medicine, treatment of functional loss and genetic disorders. Despite the administration of these peptides and proteins, these have very short biological half-life. The inherent problems to parenteral protein delivery are patient compliance, discomfort and high variability in bioavailability.

Noteworthy efforts in last decade have been carried out to synthesize nanoparticles (NPs) as delivery vehicles, since they provides appropriate ways of delivering low molecular weight therapeutics, as well as large bioactives, i.e. proteins, peptides, vaccines or nucleotides either by restricted or by tissue-specific delivery. Researchers, due to multiple reasons, are fascinated by nanocarriers as formulation vehicles, because the ratio of number of surface atoms or molecules to the total number of atoms or molecules is enhanced phenomenally, leading eventually to exponential augmentation of the effective surface area (Hadjipanayis *et al.*, 2010). Furthermore, nanocarriers can become much larger in number and, due to their tiny size, could reach regions of low accessibility such as tumor cells, damaged tissues and inflamed

* Corresponding author.

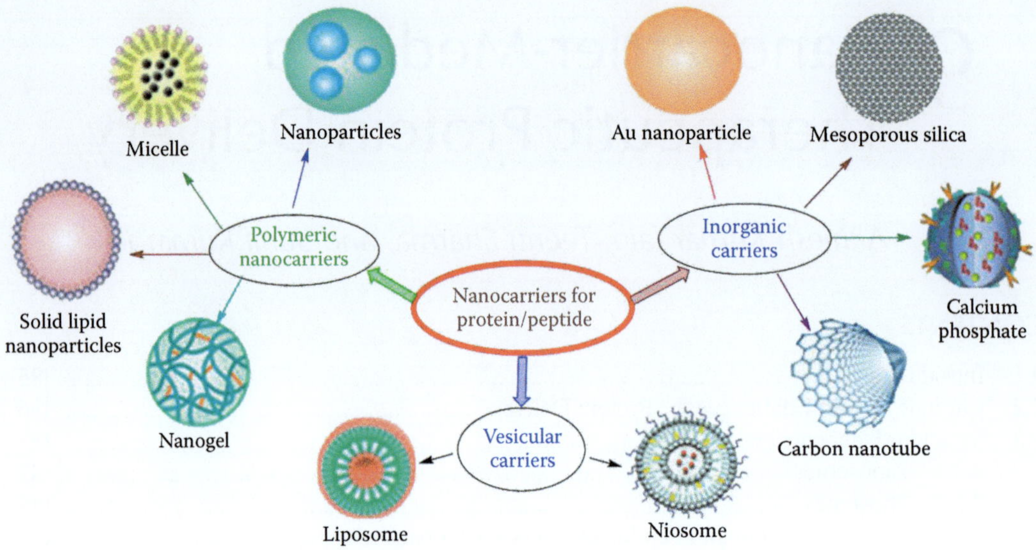

FIGURE 9.1 Schematic of various types of nanocarriers used for protein delivery.

organs. Owing to their enhanced permeability and retention effect (EPR) they also can weaken lymphatic drainage, and thus could be used as a peroral route for delivery of genes, peptides and proteins (Jong and Borm, 2008). Nanotechnology focuses on encapsulating drugs in bio-friendly nanocomposites such as polymeric NPs, nanoliposomes, solid lipid nanoparticles, micellar systems and bioconjugates. A schematic diagram of wide variety of nanocarriers used for protein delivery is depicted as Figure 9.1.

These carriers usually used to enhance oral bioavailability, to prolong drug/gene delivery in the desired organ, to solubilize drugs for intravascular delivery, to increase the stability of therapeutic agents against enzymatic degradation and/or to achieve targeted (cellular/organ) delivery of drugs. In addition, the release of an encapsulated cargo from NPs can be controlled in the organ of interest to produce the desired therapeutic level for desired period of time to generate maximum therapeutic benefits. Intracellular uptake of NPs is usually much higher *vis-à-vis* microparticles (Desai *et al.*, 1996). Protein/peptide-loaded nanocarriers have made a remarkable contribution in the treatment of several important human disorder–related diseases (Yu *et al.*, 2016). NPs have been employed as peroral drug carriers with diverse objectives, viz. (i) improvement of the bioavailability of agents with low absorption potential, (ii) delivery of vaccines to the gut associated lymphoid tissue (GALT), (iii) controlled or sustained release of the drug, (iv) reduction of the gastrointestinal (GI) mucosal irritation and (v) stability assurance of drugs in the GIT (gastrointestinal tract). A wide variety of nanocarriers for delivery of proteins and peptides delivery, i.e. nanoemulsions, NPs and liposomes, have been utilized to defend the bioactives against the cruel gastric environment (i.e. acidic pH with higher enzymatic activity) (Muheem *et al.*, 2016).

9.2 LIMITATIONS AND CHALLENGES FOR PROTEIN DELIVERY

Physicochemical properties of the drug are mainly responsible for governing its absorption in the body. Pharmaceutical proteins possess salient characteristics of large size, high hydrophilicity, susceptibility to physical and chemical degradation, structural fragility, and complexity, all significantly affecting their pharmacokinetic and pharmacodynamic behaviour *in vivo*. Moreover, these formulation conditions, such as exposure to organic solvents and change in temperature, further affect the stability of these bioactives. In a nutshell, these therapeutic proteins easily undergo physical, chemical, enzymatic degradation and denaturation leading eventually to their biological inactivation during storage and delivery. Proteolytic enzymes present in the GIT, nasal and other muocsae and restricted permeation through the mucosal barrier and rapid clearance from the body further limit protein bioavailability.

To addresses all these issues, noteworthy attempts have been made to encapsulate proteins/peptides in nanocapsules, hydrogels, micellar systems, microemulsions, nanoliposomes, niosomes, bioadhesive carriers and stealth NPs. Further, the conjugation of these carriers by specific ligands has been utilized in order to deliver protein therapeutics into the target organs in their active form. The critical factors considered during synthesizing nanocarriers in order to fulfil the specific objectives include management of the particle size, surface characteristics and effective delivery without degradation. Hence, characterization of these nanocarriers is a very critical issue to control their desired behavior *in vitro* as well as *in vivo*. Verily, stability aspects of these bioactives into the carrier system(s) are one of the major concerns during their successful delivery.

9.3 VESICULAR NANOCARRIERS

9.3.1 Liposomes

Liposomes are bilayered phospholipid vesicles with an enclosed aqueous compartment, and are used as carrier(s) for low molecular weight drugs, peptides, proteins, cytotxic drugs and nucleotides (Yousefi *et al.*, 2009; Patel *et al.*, 2011). Liposomes possess adaptable characteristics in terms of particle size, charge, bilayer composition and their encapsulation ability, making these carriers useful for protein delivery. The size of liposomes ranges from 25 nm to several micrometers, depending upon the number of bilayers present. Commonly used lipids for the preparation of liposomes are blends of distearoyl phosphatidylcholine, dimyristoyl phosphatidylglycerol and cholesterol (Anderson *et al.*, 1999), blend of phosphatidylcholine, castor oil and polyethylene glycol, coupled to distearoyl phosphatidylethanolamine (Song *et al.*, 1996) and blends of distearoyl phosphatidylglycerol, distearoyl phosphatidylcholine and cholesterol (Rezler *et al.*, 2007). Quite frequently, the hydrophilic agents are encapsulated in aqueous compartment, and liphophilic drugs are incorporated into the lipid bilayer (Patel *et al.*, 2009).

The stability of protein bioactives could be retained using the reversed-phase evaporation process, as this technique tends to avoid undesired exposure of protein to organic solvents. Smaller nanosized liposomes can also sustain the release of entrapped agents, ensuring a prolonged effect at the target site together with improved efficacy. Nevertheless, nanoliposomes are more prone to faster removal from circulation and are taken up by liver macrophages, and hence could be used for active targeting (Kumar *et al.*, 2010). Liposomes have been exploited for peptide/protein delivery via the pulmonary route, demonstrating the distinct advantage of being a substitute to frequent injection of drugs like insulin. Liposomal insulin dry powder inhalation via pulmonary delivery demonstrated a better blood glucose lowering effect for a prolonged period of time with relatively enhanced bioavailability (Bi *et al.*, 2008). Bak proteins and a voltage-dependent anionic channel (VDAC) have also been encapsulated into the liposomes exhibiting apoptosis through cytochrome C and caspases activation into mammalian cells (Ligouri *et al.*, 2008).

Liposomes as therapeutic carriers undergo clearance from the blood stream by size-dependent filtration, membrane fusion action, and interactions with specific serum proteins (opsonins). However, cell surface receptors identify liposomes as a xenobiotic via cellular and serum proteins, which are attached to a liposomal surface (Ishida *et al.*, 2002). The circulation time of liposomes can be prolonged by minimizing their macrophage uptake by means of PEGylation. Polyethylene glycol forms a hydrophilic layer of polymer on the liposome exterior, thereby sheltering the charge associated with the liposome surface. Thus, pegylated liposomes escape reticuloendothelial system (RES) due to anti-opsonizing properties of the hydrophilic PEG layer. Such designed liposomes are also popularly called as long-circulating liposomes or stealth liposomes. The optimized liposomal formulation of antidiuretic hormone, arg-8-vasopressin (VP) peptide, showed enhanced and prolonged biological activity because of long circulation characteristics (Woodle *et al.*, 1992). Another research group

(Kedar *et al.*, 1994) has demonstrated significantly enhanced immunomodulatory and antitumor effects of recombinant human interleukin-2 (IL-2), pegylated IL-2 (PEG-IL-2) and long circulatory liposomal encapsulated IL-2 (SSL-IL-2). The survival period of mice with metastatic carcinoma (earlier subjected to cyclophosphamide chemotherapy) was enhanced by two to six-fold followed by administration of SSL-IL-2, *vis-á-vis* IL-2, even at lower doses and fewer administrations. Efficient cellular uptake of TAT peptide (transactivator of transcription of human immunodeficiency virus) by incorporating it into the liposomes was successfully demonstrated by Carsten *et al.* (2004). Liposomes containing dioleoyl phosphatidylethanolamine (DOPE) and guanidinium-cholesterol, the cationic lipid, successfully coencapsulated the β-galactosidase enzyme and the anti-cytokeratin (K8) antibody and demonstrated higher cellular delivery of the enzyme without any loss in its activity (Chatin *et al.*, 2015). Recently, Ahmad *et al.* (2017) reported the applicability of bilosomes as a viable approach for oral administration of large molecular weight proteins and peptides with associated biopharmaceutical challenges.

9.3.2 Niosomes

Niosomes are the self-assembled non-ionic surfactants vesicles which were first reported in 1970 by the scientists. Alkyl or dialkyl polyglycerol ether surfactants are generally used for the preparation of niosomes (Malhotra and Jain, 1994). The layer of cholesterol and non-ionic surfactant(s) forms a closed bilayer vesicle after hydration, followed by the provision of some energy source, such as physical agitation. Configuration of niosomes involves an amphiphilic monomer surfactant consisting of a hydrophilic or polar head group and a lipophilic tail or nonpolar group. Niosomes, the bilayered structures, involve the spontaneous arrangement of a surfactant monomer in such a way that hydrophilic heads remain exposed to the aqueous solvent, whereas the lipophilic tails are oriented opposite to the aqueous solvent. Surfactant composition could affect the size, charge, lamellarity and tapped volume of the niosomes. Forces like van der Waals forces amongst surfactant molecules and repulsive forces between charged moieties of surfactant molecules are accountable for sustaining the vesicular structure of niosomes. Niosomes can either be multilamellar or unilamellar vesicles with an aqueous compartment suitable for encapsulation of hydrophilic drugs, though hydrophobic drugs are encapsulated into the hydrophobic bilayer (Udupa, 2002). Surfactants, being the key constituents of niosomes, offer notable advantage of higher stability over the liposomes. The composition of niosomes makes these quite stable towards both temperature and oxidation compared to the key components of liposomes, i.e. phospholipids. Hence, niosomes are much easier to handle without requiring any serious precautions during storage (Fang *et al.*, 2001).

Niosomes, on administration, serve as drug depots and provide sustained release of the encapsulated material in a controlled fashion through its bilayered structure. Selective delivery of the drug at the target site can also be attained via niosomes, further resulting in minimization of the toxic effects and reduction of the drug dose (Mujoriya and Bodla, 2011). Sustained release of the drug using niosomes is attributed to the enhanced retention time of the drug into the body owing to delayed clearance from the circulation (Baillie *et al.*, 1985). Since the physical state of the dosage form of niosomal formulation is aqueous suspension, patient compliance is much better than with oil-based carriers. Furthermore, niosomes are also reported to enhance the stability of sensitive drugs and the oral bioavailability of poorly soluble drugs. However, the only concern is the selection of surface active agents used for the preparation of niosomes. The surfactants employed must be biodegradable, biocompatible and nonimmunogenic. Alkyl esters, especially the sorbitan monoesters (Spans) composed of 12 to 18 carbon atoms (C12–C18), polyglycerol ethers, alkyl amides, fatty acids and amino acids are the frequently used surfactants for noisome preparations. *In vivo* niosomes act similarly to liposomes, they prolong the circulation of entrapped drug and alter its tissue distribution as well (Azmin *et al.*, 1985). The choice of the method of preparation of niosomes is based upon the size and charge requirement, number of bilayers required, encapsulation efficiency and the release of drugs. Most commonly used methods of preparation for protein encapsulation include thin film hydration, ethanol injection, microfluidization, reverse phase evaporation technique (REV) and multiple membrane extrusion method.

9.3.2.1 Protein and Peptide Delivery Using Niosomes

Vasoactive intestinal peptide (VIP) is therapeutically used for the management of Alzheimer's disease. However, its uses, like any other endogenous peptides, are restricted due to its inability to cross the blood–brain barrier (BBB) and faster clearance, followed by parenteral delivery. Encapsulation of VIP in glucose-decorated niosomes has shown significant brain transport of encapsulated peptide *vis-á-vis* control niosomes (up to 86%, in 5 min) (Biswal *et al.*, 2008). Elastic niosomes containing ethanol, when loaded with Tat-GFP fusion protein, have shown increased intracellular translocation and enhanced stability of the peptide against chemical degradation (Manosoroi *et al.*, 2011). Insulin niosomes are reported to significantly reduce blood glucose levels for prolonged period of time (Khaksa *et al.*, 2000). Niosomes have also been employed for oral vaccine delivery. Additionally, the ability of non-ionic surfactant vesicles has been investigated for encapsulation of bovine serum albumin (BSA) and on comparison with Freund's complete adjuvant (FCA) in the BALB/c mice, the niosomes were found to improve the immune response better than observed with the FCA (Rentel *et al.*, 1999).

9.4 PARTICULATE CARRIERS

9.4.1 Polymeric Nanoparticles

NPs are dispersions of solid particles, existing in sub colloidal size range of 10–1000 nm. The proteins may be solubilized, encapsulated and covalently bound to the NP matrix or polymer (Figure 9.2). Encapsulation of protein bioactives into NPs serve as a reliable method to increase the stability of the former by protecting these from the harsh conditions inside the biological system and improve their delivery at site of action (Niven *et al.*, 1994). These bioactives get released from the NPs by various mechanisms including diffusion, erosion, swelling and/or polymer degradation. Polymeric NPs exhibit impressive promise for effective treatment of various disorders, as these have striking physicochemical characteristics of size, surface charge, hydrophilicity and lipophilicity. Hence, these have been considered as efficient carriers for pharmaceutical proteins like, vaccines, peptides and anticancer drugs. Extensive use of polymeric NPs for oral administration is also intended to enhance the bioavailability of encapsulated materials associated with low absorption, GI irritation and instability issues in the biological milieu (Bajpai *et al.*, 2006). NPs offer stellar advantages to the encapsulated moieties, including the protection from unwanted reactions with biological moieties and breakdown, improving the absorption into a desired/or effected organ (tumors), and escalating the pharmacokinetics of the therapeutics. Besides, the release rate of the drug from NPs can be modulated easily to the required therapeutic concentration into the target organ for the desired period of time. NPs also have demonstrated higher intracellular uptake than microparticles. If developed suitably, NPs can serve as an ideal delivery vehicle by selectively targeting cancer cells or tumor mass and also can avoid early degradation of drugs during their transport. Furthermore, NPs may also act at the cellular level by their engulfment through endocytosis/phagocytsis for the intracellular delivery of the encapsulated drug.

Since the properties of a polymer could be easily modified, NPs comprise of a flexible carrier system that can be customized to develop the particles which are capable to cross through the biological barriers and transport the load into the cells and/or intracellular space. However, only a few polymers have been reported for the synthesis of NPs for *in vivo* protein delivery (Qiu and Bae, 2006; Vauthier and Bouchemal, 2009). Further, it is important to note that an appropriate polymer should also be rapidly eliminated from the biological system in order to tolerate its frequent administrations without any deposition of the carrier molecule. Nanocarriers themselves and their metabolites/by-products should be nontoxic and nonimmunogenic (Vauthier and Labarre, 2008). Depending upon

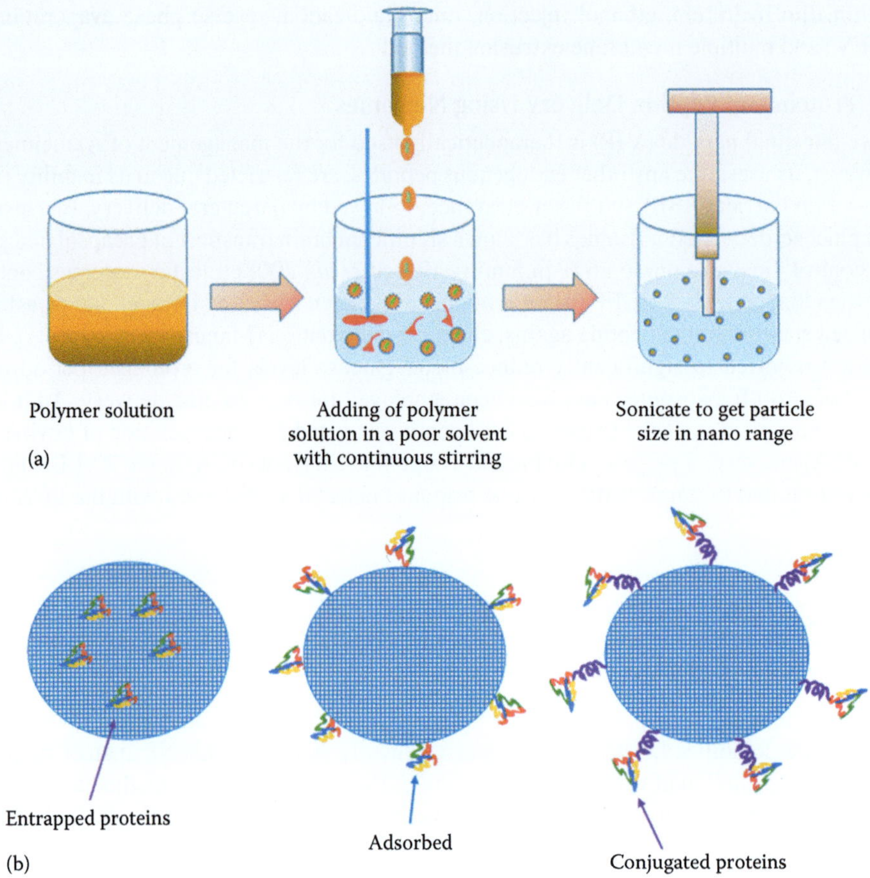

FIGURE 9.2 (a) Schematic representation of commonly used method for preparation of polymeric nanoparticles. (b) Schematic of entrapment of protein in nanocarrier composites.

the choice of release type, i.e. whether prolonged or controlled required for the incorporated protein, numerous natural and/or artificial polymers have been utilized for the development of NPs, such as starch, gelatin, dextran, albumin, chitosan, copolymer of maleic anhydride and 2-methoxyethyl vinyl ether, cyanoactrylates, poly(lactic acid) (PLA), poly(lactic acid-*co*-glycolic acid) (PLGA), PEG-PLA block copolymers and poly(*n*-hexadecyl cyanoacrylate) (Soppimath *et al.*, 2001; Solaro and Chiellini, 2006; Rytting *et al.*, 2008; Solaro, 2008; Duan *et al.*, 2009, Ulbrich *et al.*, 2016). A list of polymers which are frequently used for development of NPs as protein carriers is presented in Table 9.1.

The primary method of preparation of polymeric NPs reports two key steps involving the preparation of an emulsion followed by solvent evaporation or gelation/precipitation of polymer or polymerization. A schematic demonstration of commonly used techniques for development of the polymeric NPs is illustrated in Figure 9.2. The other method utilized for synthesis of NPs is precipitation of spontaneous dispersion of macromolecule by self-assembling. Great attention has been imparted to biodegradable polymers from the large number of polymers available, as these are self-eliminating. Hence, these bypass the problem of surgical removal of carrier system after their administration. Improved transmucosal protein transport of insulin and tetanus toxoid after oral and nasal delivery of their PEG and chitosan coated PLA/PLGA NPs, respectively (Vila *et al.*, 2002). PLA and PLGA NPs are usually prepared in aseptic conditions, so as to prevent any environmental contamination (Bala *et al.*, 2004; Freitas *et al.*, 2004; Damge *et al.*, 2007). These polymers undergo hydrolysis inside the biological system and get converted into the bio-friendly and biolabile by-products, i.e. lactic acid and glycolic acid, which are eliminated via the citric acid cycle. These polymers protect the

TABLE 9.1
Most Commonly Reported Polymers Used for Development of Nanoparticles as Drug Carriers

Artificial Homopolymers	Copolymers	Polysaccharides/ Proteins	Colloidal Agents
Poly(alkylcyanoacrylate)	Polyethlene glycol-PLA block copolymers	Pullulan	Dextran
Polyaspartic acid	Polyethylamine	Agarose	Gelatin
Poly(n-isppropylAcrylalmide)	Poly(hexadecylcyanoacrylate-co-poly(ethylene glycol) cyanoacrylate)	Alginate	Pluronic F68
Polyethlamine	Polylactide-poly(ethylene glycol)	Albumin	Poly(vinyl alcohol)
Poly(ε-caprolactone)	Poly(lactide-co-glycolide)-poly(ethylene glycol)	Chitosan	Poly oxyethylene lauryl ether
Poly(isobutylcyanoacrylate)	Poly(ε-caprolactone)-poly(ethylene glycol)	Guar Gum	Tween® 40
Poly(isohexylcyanoacrylate)	Poly(lactide-co-glycolide)	Gelatin	Tween® 60
Poly(n-butylcyanoacrylate)		Starch	
Polyacrylates and polymethacrylates	–	–	–
Polystyrene	–	–	–
Poly lactic acid (PLA)	–	–	–
Crosslinked poly(*N*-vinyl-2-pyrrolidinone	–	–	–

Sources: Vauthier C and Bouchemal K, *Pharm Res*. 2009; *26*: 1025–1058; Solaro R, *J Polym Sci*. 2008; 46: 1–11; Vyas SP and Khar RK. *Targeted and Controlled Drug Delivery*. 1st edition. New Delhi: CBS Publishers and Distributors; 2002; 332–334; Soppimath KS *et al*., *J Control Rel*. 2001; 70: 1–20.

encapsulated bioactives against breakdown and provide a controlled release rate of cargo, which is dependent of the polymer degradation rate.

Oral administration of insulin and its derivatives using NPs to diabetic rats demonstrated that biodegradable polymers such as poly(ε-caprolactone) (PCL) and apt blends of PLA and PLGA not only protect insulin from proteolytic degradation into the GIT but also fruitfully decrease blood glucose level (Damge *et al*., 2007). Tetanus toxoid encapsulated into PLA NPs as well as pegylated PLA naoparticles after nasal and oral administration to rats was found to result in higher concentrations in the bloodstream and lymph nodes (Vila *et al*., 2005). Polysaccharides mainly dextran, starch, chitosan and alginate, either alone or in mixtures, are also used for the preparation of NPs. Our research group has already established the usefulness of starch NPs for nasal delivery of insulin (Jain *et al*., 2008). The hypoglycaemic response of SC insulin (2 IU/kg), when compared to 10 IU/kg of nasal EE–NPs (epichlorhydrin crosslinked starch NPs made via emulsion method) is greater ($P < 0.05$) at 2 h compared to nasal NPs (Figure 9.3). However, the effect of insulin NPs administered through nasal route, sustained the release up to 6 h compared to plain nasal insulin solution used as a control. The relative pharmacological efficacy of EE–L3–NPs insulin was found to be 48.6%. Insulin-loaded alginate/chitosan NPs showed effectiveness in decreasing the blood glucose level to 40% of initial levels after oral administration to diabetic rats (Sarmento *et al*., 2007). Moreover, *N*-trimethylchitosan NPs were assessed *in vitro* and *in vivo* as potential protein/vaccine carriers for oral delivery (Chen *et al*., 2008). Receptor-mediated endocytosis by conjugation with various nanocarriers is also used as an effective approach for protein delivery (*vide infra*). Already, effective oral delivery of insulin using vitamin B12 (VB12) coupled dextran NPs has been reported by us (Chalalsani *et al*., 2007). Dextran NPs of medium cross-linking density, conjugated with O'hexyl derivative of VB12 showed a dose-dependent hypoglycaemic response (Figure 9.4). They have also demonstrated the prevalence of VB12 mediated

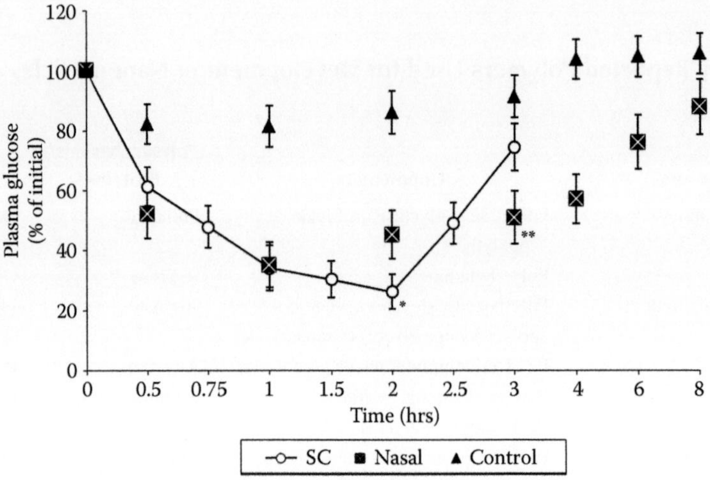

FIGURE 9.3 Hypoglycemic effect of EE–NPs in presence of Na glycocholate after nasal administration and plain SC insulin to STZ induced diabetic rats (mean ± SE, n = 5). *Hypoglycemic action of sc insulin is higher (P < 0.05) at 2 h compared to nasal NPs. **Hypoglycemic action of nasal insulin is significantly higher compared to SC and control. EE-NPs = epichlorhydrin crosslinked Starch NPs made via emulsion method. With kind permission from Elsevier *Eur J Pharm Bio.* Effective insulin delivery using starch nanoparticles as a potential transnasal mucoadhesive.

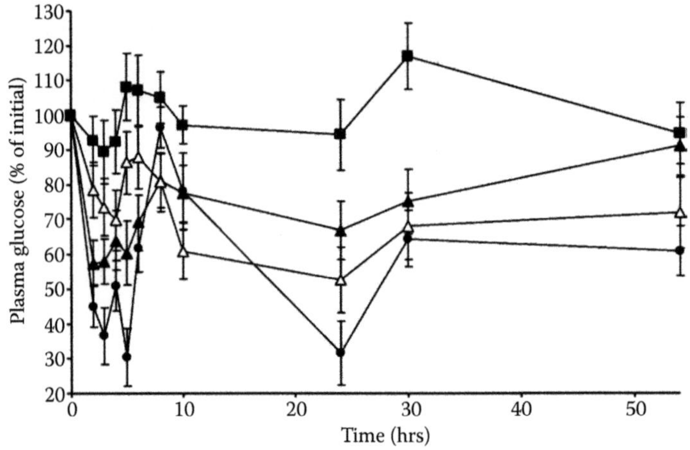

FIGURE 9.4 VB12-mediated uptake study of Dextran NPs of medium crosslinking density conjugated with O'Hexyl derivative of Vitamin B12 (NPs-M35'OH) in terms of plasma glucose reduction after oral administration to STZ diabetic rats (n = 3, mean ± S.E.) (△) VB 12 + 20 IU/kg. (▲) 10 IU/kg. (●) 20 IU/kg. (■) Plain insulin control. n = 6. (Reprinted from *J Control Rel.*, 122, Chalasani KB, Russel-Jones GJ, Jain AK, Jain SK, Diwan PV, Effective oral delivery of insulin in animal models using vitamin B12-coated dextran nanoparticles, 141–150, Copyright 2007, with permission from Elsevier.)

RME uptake of NPs by coadministering the large excess of VB12 to saturate the receptors an hour prior to dosing of VB12 conjugated nanocarriers (Chalalsani *et al.*, 2007). Bovine serum albumin (BSA), encapsulated into the carboxymethyl-β-cyclodextrin decorated chitosan (CMCD-g-CS) NPs as a model protein, was found to release faster and to a greater extent in simulated intestinal fluid as compared to gastric fluid (Song *et al.*, 2017). The conjugation of developed cell-penetrating peptide (CPP) to the surface of the chitosan NPs (15 nm; prepared by nanoemulsion method) has also been explored recently (Barbari *et al.*, 2017). These CPP–conjugated NPs could easily be loaded with nearly all proteins, peptides and nucleotides for oral delivery. In this regard, these formulations have been tested in CaCo-2 cell lines for translocation of insulin across the cell monolayer, and were found to be effective with 15%–19% increase in the insulin levels.

9.4.2 Solid Lipid Nanoparticles

Solid lipid nanoparticles (SLNs), prepared from solid lipids and stabilized using surfactants in an aqueous suspension, bear resemblance to nanoemulsions, except that liquid lipid in nanoemulsion is swapped with a solid lipid. However, controlled drug release can be attained in an outstanding manner by replacing a portion of solid lipids with oils, as solid lipids tend to lower the mobility of drug significantly (Martins *et al.*, 2007). SLNs, of late, have emerged as an attractive carrier system for the delivery of protein/peptide (Marcato and Duran, 2008). SLNs offer various advantages over many other delivery vehicles, i.e. NPs, liposomes, ethosomes and lipid emulsions (Garcıa-Fuentes *et al.*, 2003; Hou *et al.*, 2003; Liu *et al.*, 2008; Joshi and Muller, 2009) which include:

i. Drug targeting capability and controlled drug release ability
ii. Protection of sensitive and labile drugs from photochemical, chemical and oxidative degradation
iii. Safety to use over some polymeric NPs, since SLNs utilized physiological and biocompatible lipids
iv. Capability to load and deliver both hydrophilic and hydrophobic agents
v. Avoidance of organic solvents
vi. Amenability to administer using various routes, i.e. intravenous, ophthalmic, oral, pulmonary and dermal

Delivery of bioactives like proteins and peptides with established therapeutic efficacy are usually associated with problem of low entrapment, as most of these have hydrophobic part which has tendency to get adsorbed onto the surfaces (e.g. plastic and glass), leading to significant loss of the bioactives' content (Duncan *et al.*, 1995). Thus, successful encapsulation of bioactives into SLNs and their application has been widely studied during parenteral and oral delivery. The literature reports describe three models for incorporating drugs or proteins into the SLNs (Muller *et al.*, 2000; Mehnert and Mader, 2001), viz. (i) drug-enriched shell model, (ii) homogeneous matrix model and (iii) drug-enriched core model. These models can be differentiated based on the formulation composition, such as chemical composition of the lipid, bioactive and surfactants. Muller *et al.* (1995) were first to report the SLNs preparative techniques devoid of the potentially toxic organic solvents, which could have detrimental effects on pharmaceutical proteins, in contrast to most polymeric microspheres and NPs (Wissing *et al.*, 2004).

9.4.2.1 Protein/Peptide-Loading in SLNs

The production of SLNs is dependent upon solidification of the dispersed phase. The hydrophilic bioactives are supposed to be inadequately entrapped into the lipophilic matrix of SLNs which tends to further get dispersed into the aqueous phase during the course of preparation. Thus, the preparation process is promoted by the utilization of suitable emulsifier(s) (surfactants) and stabilizer(s). Furthermore, the SLNs can exhibit incomplete entrapment efficiency owing to drug seepage during storage. Lipids are flexible molecules that can develop diverse structured solid matrices, e.g. lipid–drug conjugate NPs (LDCs) and nanostructured lipid carriers (NLCs), which have been developed to increase drug encapsulation efficiency (Wissing *et al.*, 2004). In last three decades, scientists have unraveled encouraging results regarding encapsulation of numerous proteins and peptides in SLNs (Hu *et al.*, 2004; Fan *et al.*, 2014; Yi *et al.*, 2014). Various peptides such as lysozyme, calcitonin, somatostatin, LHRH, insulin, malarial antigens, HBsAg and BSA have been investigated for their stability, *in vitro* drug release kinetics and *in vivo* performance (Table 9.2).

TABLE 9.2
Different Method Used for Preparation of Solid Lipid Nanoparticles of Different Proteins and Peptides and Their Entrapment Efficiency

S. No.	Method of Development of Solid Lipid Nanoparticles	Protein/Peptide Entrapped into Solid Lipid Nanoparticles	Entrapment Efficiency	References
1.	Adsorption onto SLN	BSA	n.a.	(Gualbert *et al.*, 2003)
2.	Solvent evaporation (w/o/w)	Calcitonin	> 90%	(Garcia-Fuentes *et al.*, 2005b; Garcia-Fuentes *et al.*, 2005a)
3.	HPH hot dispersion Warm microemulsion (o/w)	CyA	96.6–97.8% 88.4% n.a. 96.1% 13%	(Olbrich *et al.*, 2000) (Zhang *et al.*, 2000) (Radtke and Muller 2001a; Radtke and Muller 2001b) (Muller *et al.*, 2006) (Ugazio *et al.*, 2002)
4.	Solvent displacement	Gonadorelin	50.4–69.4%	(Hu *et al.*, 2004)
5.	Adsorption onto SLN	HSA	12.4–32.4% (stealth)7–13.5%	(Cavalli *et al.*, 1999)
6.	Solvent displacement Solvent evaporation (w/o/w) Supercritical CO_2 (PGSS) Coacervation method	Insulin	26.8% n.a. 67.9% 75% 89%	(Zhang *et al.*, 2006) (Garcıa-Fuentes *et al.*, 2003) (Zhang *et al.*, 2006) (Caliceti *et al.*, 2006) (Gallarate *et al.*, 2011)
7.	Warm microemulsion (w/o/w)	[D-Trp-6] LHRH	90%	(Morel *et al.*, 1994)
8.	HPH cold dispersion Solid-in-oil-in-water (s/o/w) emulsification-solvent-evaporation	Lysozyme	43.2–59.2% 42.8%	(Almeida *et al.*, 1997) (Devrim and Bozkır, 2015)
9.	Adsorption onto SLN Melt-dispersion (o/w)	Ovalbumin	70–97% > 80%	(Videira *et al.*, 2002) (Videira *et al.*, 1998)
10.	Warm microemulsion (o/w with ionic pair or w/o/w)	Thymopentin	5.2% or 1.7%	(Morel *et al.*, 1996)
11.	Double emulsion solvent evaporation (w/o/w)	Yak interferon-α	83.7%	(Li *et al.*, 2010)
12.	Microemulsion technique	Streptavidin	~100	(Pedersen *et al.*, 2006)
13.	Hot HPH technique	TATc peptide	–	(Carsten *et al.*, 2004)
14.	Microemulsion technique	Human recombinant EPGb	–	(Pedersen *et al.*, 2006)
15.	Coacervation method	Leuprolide	50–80%	(Gallarate *et al.*, 2011)

Note: BSA = bovine serum albumin; CyA = cyclosporine A; HPH = high-pressure homogenization; HSA = human serum albumin; LHRH = luteinizing hormone-releasing hormone; n.a. = not available; PGSS = particles from gas saturated solution technique.

Different approaches have been utilized to load proteins in SLNs. Fan *et al.* (2014) have developed and evaluated ligand-decorated SLNs to increase oral bioavailability of proteins. The peptides-linked SLNs showed their activity via both caveolae- and clathrin-mediated endocytosis. However, mucus was found to be an obstacle in the translocation of SLNs. The absolute bioavailability of peptides was found to be increased by 2.45 to 1.98 times compared to naive SLNs, suggesting their significant potential for the improved oral bioavailability of proteins. In another report, the peptide-loaded SLNs were developed by solvent diffusion method. Gonadorelin (a model peptide) was incorporated into the SLNs and evaluated for various parameters. The average diameter and zeta potential of SLN were observed to be 421.7 nm and –21.1 mV, respectively. However, *in vitro* release followed biphasic pattern with slow release of gonadorelin (Hu *et al.*, 2004). SLNs were prepared by solubilizing lysozyme (a model peptide) into the molten lipid phase. Results revealed a solubility-dependent entrapment efficiency of the peptide in the hydrophobic phase of the final formulation. The lysozyme molecule remained intact during the process course without diminishing its activity (Almeida *et al.*, 1997). In another study, a protein such as whey protein isolate (WPI), or soy protein isolate (SPI), was employed as one of the components of beta-carotene (BC)-loaded SLNs and were assessed for their physiochemical stability, intracellular uptake by CaCco-2 cells and *in vitro* cytotoxicity. The formulations were found to be devoid of any cytotoxicity particularly after dilution (≥ 10 times). However, the intracellular delivery of BC was greatly enhanced by 3.4 and 1.7 times for WPI and SPI, respectively, compared to the unentrapped BC (Yi *et al.*, 2014). Goppert and Muller (2005) studied human adsorption pattern of plasma layer of SLNs following i/v administration and they anticipated SLNs as a potential targeted drug delivery carrier, based on their findings.

9.5 INORGANIC NANOCARRIERS

Current headway in nanotechnology has led to the application of various inorganic NPs in drug delivery, e.g. calcium phosphate, carbon nanotubes, mesoporous silica NPs and gold NPs (Malmsten and Zauscher, 2013). Recently, carbon nanotubes, gold NPs and nanospheres have been widely studied as drug carriers owing to their smaller size and high surface area, enabling these to move easily inside the body. The therapeutic agents are either embedded inside the nanotube or bound to the surface. The major advantages of inorganic nanocarriers include their hydrophilic nature, low toxicity profile, biocompatibility, resistance to microbial growth and higher stability.

9.5.1 Silica Nanoparticles

Coassembly of silicates and surfactant produces mesoporous silica, a category of surfactant-templated inorganic compound. Mesoporous silica nanoparticles (MSNs) are comparatively biocompatible, making them ideal for biological use (Radin *et al.*, 2002; Lai *et al.*, 2003; Fang and Trewyn, 2012; Deodhar *et al.*, 2017). However, these are not bioresorbable. Pore size and pore structure could be easily controlled by the selection of apt surfactant and coassembly conditions. MSNs have a porous surface, offering huge effective surface areas for the encapsulation of bioactive molecules. Nevertheless, protein release could be controlled by modifying their size, shape and surface functionalization of MSNs with different ligands. MSNs encapsulating FITC–cytochrome C were prepared using a diffusion-mediated process. Their effective delivery into the HeLa cells was demonstrated by confocal laser scanning microscopic images. FITC–cytochrome C was observed to effectively run away from the endosomal degradation pathways (Tan *et al.*, 2004; Lu *et al.*, 2007). This cellular uptake and endosomal break out processes of MSNs are energy-dependent, whereas surface modification of MSNs by amine and guanidinium make them capable to penetrate the cells by a clathrin- and covalent-free mechanism (Xing *et al.*, 2005; Slowing *et al.*, 2006). Prasetyanto *et al.* (2016) demonstrated successful cellular delivery of organo-silica matrices encapsulating a highly cytotoxic protein into the C6 glioma cells by disintegration of the matrix upon exposure to a chemical response. Alternatively, ligands and antibodies govern the cellular uptake of MSNs by receptor-mediated endocytosis. Moreover, MSNs coated with release-retarding agents such as chitosan–PEG copolymers have been documented to modulate the release rate of entrapped material (Tan *et al.*, 2004; Lu *et al.*, 2007). Encapsulation efficiency of ovalbumin (OVA) was increased by 2.5-fold after surface modification of MSN with positively charged amine because of electrostatic interactions (Mahony *et al.*, 2013). OVA was predominantly released through passive diffusion mechanism and found to be 7.9% in initial 30 min. Amino-modified silica nanospheres, along with photoluminescent CaF2:Tm,Yb nanocrystals, have shown higher protein encapsulation efficiency and prolonged release of the encapsulated cargo, establishing their usefulness in cellular uptake (Li *et al.*, 2017). MSNs are being also investigated as immune boosters, eliciting immune responses even at smaller OVA levels. These NPs are synthesized under controlled conditions, allowing the modification of pore shape and surface groups (Vallet-Regi *et al.*, 2007), which could be later on conjugated with various monoclonal antibodies (mAb) structure. Larger pores offer chances for higher loading of mAbs. Surface functionalization of MSNs has been encapsulated with a mAb against tumor cells surface protein and administered to mice with melanoma (Lei *et al.*, 2010). Administration of particles encapsulating mAb demonstrated enhanced suppression of tumor when compared to pure mAb injection. The findings suggest that encapsulation of mAbs into MSNs did not change the therapeutic effect or immunological potential, though by assisting prolonged release of antibody, results in enhancement of the half-life of the mAbs in the tumor tissue. Mice with malignant mesotheliomas on treatment with intraperitoneal injection of doxorubicin (DOX) encapsulated into the MSNs, tethered with mesothelin-specific antibodies that corroborated the high efficacy of the MSNs as carrier systems (Marcura *et al.*, 2013). The surface-decorated MSNs showed higher effectiveness than plain DOX at lower doses, thus, lowering the undesired effects of DOX. MSNs were also surface-modified using antibodies directed against epidermal growth factor in order to treat lung cancer cells, with cytotoxic drug pyrrolidone-2. These demonstrated 38% reduction in tumor growth with low systemic toxicity (Sundarraj *et al.*, 2014). Silica nanospheres, composed of a core pooled with pH sensitive chitosan were used to transport encapsulated protein specifically into breast cancer cells by means of targeting antibody ErbB2, and found higher levels in tumors because of lower local pH at tumors (pH 4.0 versus physiological pH) upon administration to mouse model (Deng *et al.*, 2011).

9.5.2 Gold Nanoparticles

Gold nanoparticles (GNPs), owing to their bioinertness, biocompatiblity, flexibility for surface modifications and capability of cellular imaging, have shown their wide applications in the biomedical domain. GNPs of 2.5 nm size have been explored as useful transporters for intracellular delivery of β-galactosidase into various cell lines, especially HeLa cells (Ghosh *et al.*, 2010). Internalization mechanism of GNPs depends on the surface properties, i.e. surface charge and/or size. Modified GNPs with antibodies are taken up by receptor-mediated endocytic pathways, whereas GNPs *per se* are transported by endocytosis (Tkachenko *et al.*, 2003; Connor *et al.*, 2005). Protein antigens, covalently conjugated with GNPs, have been explored as a vaccine for cancer immunotherapy. Balb/c mouse treated with OVA-loaded 5 nm GNPs were found to be highly effective in producing anti-OVA IgG antibodies (Tang *et al.*, 2013). Gold nanocarriers, including nanorods and multifunctional nanocarriers conjugated with remedial mAbs on investigation, have been found to be highly specific to the target cells without losing any antibody functionality (Shao *et al.*, 2011; Bisker *et al.*, 2012; Cho *et al.*, 2014; Lee *et al.*, 2014; Shen *et al.*, 2014).

9.5.3 Calcium Phosphate Nanoparticles

Calcium phosphate nanoparticles (CaPNPs) with diameters of 40–50 nm have also been explored for the protein delivery. CaPNPs modified using PEG, have zeta potential close to zero. Insulin encapsulated in CaPNPs is protected against the gastric degradation by covering the nanocarriers with a pH-responsive material, which solubilize in the intestinal pH. *In vitro* release studies demonstrated that insulin tends to be released for a period of 8 h in intestinal pH with hardly any release observed in simulated gastric fluid (Ramachandran *et al.*, 2008). Zinc is reported to retard the insulin release (long-acting insulins). Thus, calcium phosphate and zinc phosphate are attractive contenders for developing ceramic-based insulin carriers. The oral route of administration is the most preferred one, owing to its ease in administration, leading to improved patient compliance. Preferably, the absorption of CaPNPs takes place via the Peyer's patches region which joins the lymphatic system, thereby avoiding the first pass metabolism during which insulin gets degraded significantly. BioSante Pharmaceuticals, a company based in the United States, has successfully synthesized CaPNPs and has entered into the first stage of toxicity studies (www.wallstreetresearch.org).

9.6 CONCLUSIONS

Protein delivery has been explored as an artistic approach with immense potential of treatment of the complex diseases requiring safer administration. With recent advancements in the recombinant technology, leading to availability of a huge variety of therapeutically effective protein bioactives, protein delivery has gained significant momentum. The endurance of these agents in the biological environment, particularly in the absence of an ideal carrier, is critical due to their possible denaturation or enzymatic degradation. The selected carrier(s) must assure the protection of the enclosed protein(s) in the biological milieu and ability to translocate it as per their desired needs. Hence, it is vital to design a suitable carrier with apt size, composition, surface behavior and biocompatibility.

Chapter 9 has attempted to provide a holistic overview on various strategies used to stabilize proteins during their delivery into the biological fluids. Among the wide range of the carriers investigated, nanocarriers have emerged as an excellent choice. Hence, these have been investigated extensively for effective proteins delivery via invasive and noninvasive routes. Nanoliposomes and SLNs have shown higher biocompatibility because of their lipidic nature and have the capability to prolong the release of their entrapped protein. Moreover, lipid-based carriers act as efficient immunological adjuvants, capable of eliciting cellular and humoral immunity against a gamut of antigens related to human disease. However, owing to their lower stability, these suffer from limitations, such as inadequate shelf life and shorter half time. Nevertheless, polymeric NPs have emerged as promising carriers of proteins which offer high stability. Further, the surface modification(s) of these NPs employing conjugation with specific ligands helps in avoiding RES, which is a major constraint with these carriers. On the other hand, organic solvents are generally employed in the development of polymeric nanocarriers, which could denature and/or degrade the protein. Despite these challenges, the limitations could be successfully surmounted by rational selection of protein-friendly techniques like particle synthesis, employing self-assembling water-soluble polymers.

REFERENCES

Ahmad J, Singhal M, Amin S, Rizwanullah M, Akhter S, Kamal MA *et al.* Bile salt stabilized vesicles (bilosomes): a novel nano-pharmaceutical design for oral delivery of proteins and peptides. *Curr Pharm Des.* 2017; 23(11):1575–1588.

Almeida AJ, Runge S, Muller RH. Peptide-loaded solid lipid nanoparticles (SLN): Influence of production parameters. *Int J Pharm Sci.* 1997; 149(2): 255–265.

Anderson KE, Stevenson BR, Rogers JA. Folic acid-PEO-labeled liposomes to improve gastrointestinal absorption of encapsulated agents. *J Control Rel.* 1999; 60: 189–198.

Azmin MN, Florence AT, Handjani-Vila RM, Stuart JF, Vanlerberghe G, Whittaker JS. The effect of non-ionic surfactant vesicle (niosome) entrapment on the absorption and distribution of methoterxate in mice. *J Pharm Pharmacol.* 1985; 37: 237–242.

Baillie AJ, Florence AT, Hume LR, Muirhead GT, Rogerson A. The preparation and properties of niosomes-non-ionic surfactant vesicles. *J Pharm Pharmacol.* 1985; 37: 863–868.

Bajpai AK, Choubey J. Design of gelatin nanoparticles as swelling controlled delivery system for chloroquine phosphate. *J Mater Sci.* 2006; 17: 345–358.

Bala I, Hariharan S, Ravi KMNV. PLGA nanoparticles in drug delivery: The state of the art. *Crit Rev Ther Drug Carrier Syst.* 2004; 21: 387–422.

Barbari GR, Dorkoosh FA, Amini M, Sharifzadeh M, Atyabi F, Balalaie S *et al.* A novel nanoemulsion-based method to produce ultrasmall, water-dispersible nanoparticles from chitosan, surface modified with cell-penetrating peptide for oral delivery of proteins and peptides. *Int J Nanomedicine.* 2017; 12: 3471–3483.

Bi R, Shao W, Wang Q, Zhang N. Spray-freeze-dried dry powder inhalation of insulin loaded liposomes for enhanced pulmonary delivery. *J Drug Target.* 2008; 16: 639–648.

BioSante Pharmaceuticals, Inc. News. Available online at: www.wallstreetresearch.org/conf070814/profile_bpa.pdf. (Accessed on 12 November, 2017).

Bisker G, Yeheskely-Hayon D, Minai L, Yelin D. Controlled release of rituximab from gold nanoparticles for phototherapy of malignant cells. *J Control Rel.* 2012; 162: 303–309.

Biswal S, Murthy PN, Sahu J, Sahoo P, Amir F. Vesicles of non-ionic surfactants niosomes and drug delivery potential. *Int J Pharm Sci Nanotechnology.* 2008; 1: 1–10.

Caliceti P, Brossa A, Salmaso S, Bersani S, Elvassore N, Bertucco A. Preparation of protein loaded solid lipid nano-particles by compressed fluid process. *Proc Intern Symp Control Rel Bioact Mater.* 2006; 33: 383.

Carsten R, Ulrike S, Aurora O *et al.* Application of novel solid lipid nanoparticle (SLN)-gene vector formulations based on a dimeric HIV-1 TAT-peptide in vitro and *in vivo. Pharm Res.* 2004; 21: 1662–1669.

Cavalli R, Bocca C, Miglietta A, Caputo O, Gasco M. Albumin adsorption on stealth and non-stealth solid lipid nanoparticles. *STP Pharma Sci.* 1999; 9(2): 183–189.

Chalasani KB, Russel-Jones GJ, Jain AK, Jain SK, Diwan PV. Effective oral delivery of insulin in animal models using vitamin B12-coated dextran nanoparticles. *J Control Rel.* 2007; 122: 141–150.

Chatin B, Mevel M, Devalliere J, Dallet L, Haudebourg T, Peuziat P *et al.* Liposome-based formulation for intracellular delivery of functional proteins. *Mol Ther Nucleic Acid.* 2015; 4: 244.

Chen F, Zhang ZR, Yuan F, Qin X, Wang M, Huang Y. *In vitro* and *in vivo* study of N-trimethyl chitosan nanoparticles for oral protein delivery. *Int J Pharm.* 2008; 349: 226–233.

Cho SK, Emoto K, Su LJ, Yang X, Flaig TW, Park W. Functionalized gold nanorods for thermal ablation treatment of bladder cancer. *J Biomed Nano.* 2014; 10: 1267–1276.

Connor EE, Mwamuka J, Gole A, Murphy CJ, Wyatt MD. Gold nanoparticles are taken up by human cells but do not cause acute cytotoxicity. *Small.* 2005; 1: 325–327.

Damge C, Maincent P, Ubrich N. Oral delivery of insulin associated to polymeric nanoparticles in diabetic rats. *J Control Rel.* 2007; 117: 163–170.

Deng Z, Zhen Z, Hu X, Wu S, Xu Z, Chu PK. Hollow chitosan-silica nanospheres as pH-sensitive targeted delivery carriers in breast cancer therapy. *Biomaterials.* 2011; 32: 4976–4986.

Deodhar GV, Adams ML, Trewyn BG. Controlled release and intracellular protein delivery from mesoporous silica nanoparticles. *Biotechnol J.* 2017 Jan; 12(1). doi: 10.1002/biot.201600408.

Desai MP, Labhasetwar V, Amidon GL, Levy RJ. Gastrointestinal uptake of biodegradable microparticles effect of particle size. *Pharm Res.* 1996; 13: 1838–1845.

Devrim B, Bozkır A. Design and evaluation of hydrophobic ion-pairing complexation of lysozyme with sodium dodecyl sulfate for improved encapsulation of hydrophilic peptides/proteins by lipid-polymer hybrid nanoparticles. *J Nanomed Nanotech.* 2015; 6: 259.

Duan J, Zhang Y, Chen W, Shen C, Liao M, Pan Y *et al.* Cationic polybutyl cyanoacrylate nanoparticles for DNA delivery. *J Biomed Biotechnol.* 2009; 149–254.

Duncan MR, Lee JM, Warchol MP. Influence of surfactants upon protein/peptide adsorption to glass and polypropylene. *Int J Pharm Sci.* 1995; 120(2): 179–188.

Fan T, Chen C, Guo H, Xu J, Zhang J, Zhu X *et al.* Design and evaluation of solid lipid nanoparticles modified with peptide ligand for oral delivery of protein drugs. *Eur J Pharm Biopharm.* 2014; 88(2): 518–528.

Fang IJ and Trewyn BG. Application of mesoporous silica nanoparticles in intracellular delivery of molecules and proteins. *Methods Enzymol.* 2012; 508: 41–59.

Fang JY, Hong CT, Chiu WT, Wang YY. Effect of liposomes and niosomes on skin permeation of enoxacin. *Int J Pharm.* 2001; 219: 61–72.

Freitas S, Hielscher G, Merkle HP, Gander B. A fast and simple method for producing biodegradable nanospheres. *Eur Cell Mater.* 2004; 7: 28.

Gallarate M, Battaglia L, Peira E, Trotta M. Peptide-loaded solid lipid nanoparticles prepared through coacervation technique. *J Chem Eng.* 2011; 2011, Article ID 132435.

Garcia-Fuentes M, Prego C, Torres D, Alonso MJ. A comparative study of the potential of solid triglyceride nanostructures coated with chitosan or poly(ethylene glycol) as carriers for oral calcitonin delivery. *Eur J Pharm Sci.* 2005a; 25(1): 133–143.

Garcıa-Fuentes M, Torres D, Alonso M. Design of lipid nanoparticles for the oral delivery of hydrophilic macromolecules. *Colloids Sur B Interfaces.* 2003; 27(2): 159–168.

Garcia-Fuentes M, Torres D, Alonso MJ. New surface-modified lipid nanoparticles as delivery vehicles for salmon calcitonin. *Int J Pharm Sci.* 2005b; 296(1–2): 122–132.

Ghosh XP, Yang R, Arvizo ZJ, Zhu SS, Agasti Z, Rotello VJ. Intracellular delivery of a membrane-impermeable enzyme in active form using functionalized gold nanoparticles. *Am Chem Soc.* 2010; 132: 2642–2645.

Goppert TM, Muller RH. Protein adsorption patterns on poloxamer- and poloxamine-stabilized solid lipid nanoparticles (SLN). *Eur J Pharm Biopharm.* 2005; 3: 361–372.

Gualbert J, Shahgaldian P, Coleman AW. Interactions of amphiphilic calix[4]arene-based solid lipid nanoparticles with bovine serum albumin. *Int J Pharm.* 2003; 257 (1-2): 69–73.

Hadjipanayis CG, Machaidze R, Kaluzova M, Wang L, Schuette AJ, Chen H *et al.* EGFRvIII antibody-conjugated iron oxide nanoparticles for magnetic resonance imaging-guided convection-enhanced delivery and targeted therapy of glioblastoma. *Cancer Res.* 2010; 70(15): 6303–6312.

Hou D, Xie C, Huang K, Zhu C. The production and characteristics of solid lipid nanoparticles (SLNs). *Biomaterials.* 2003; 24(10): 1781–1785.

Hu FQ, Hong Y, Yuan H. Preparation and characterization of solid lipid nanoparticles containing peptide. *Int J Pharm Sci.* 2004; 273(1–2): 29–35.

Ishida T, Harashim H, Kiwada H. Liposome clearance. *Biosci Rep.* 2002; 22: 197–224.

Jain, AK, Khar RK, Ahmed FJ, Diwan PV. Effective insulin delivery using starch nanoparticles as a potential trans-nasal mucoadhesive carrier. *Eur J Pharm Bio.* 2008; 69(2): 226–435.

Jong WHD, Borm PJA. Drug delivery and nanoparticle applications and hazards. *Int J Nanomedicine.* 2008; 3(2): 133–149.

Joshi MD, Muller RH. Lipid nanoparticles for parenteral delivery of actives. *Eur J Pharm Biopharm.* 2009; (2): 161–172.

Kedar E, Braun E, Rutkowski Y, Emanuel N, Barenholz Y. Delivery of cytokines by liposomes II. Interleukin-2 encapsulated in long-circulating sterically stabilized liposomes: Immunomodulatory and anti-tumor activity in mice. *J Immunother Emphasis Tumor Immunol.* 1994; 16: 115–124.

Khaksa G, D'Souza R, Lewis S, Udupa N. Pharmacokinetic study of niosome encapsulated insulin. *Indian J Exp Biol.* 2000; 38: 901–905.

Kumar A, Badde S, Kamble R, Pokharkar VB. Development and characterization of liposomal drug delivery system for nimesulide. *Int J Pharm Pharm Sci.* 2010; 2(4): 87–89.

Lai CY, Trewyn BG, Jeftinija DM. A mesoporous silica nanosphere-based carrier system with chemically removable CdS nanoparticle caps for stimuli-responsive controlled release of neurotransmitters and drug molecules. *J Am Chem Soc.* 2003; 125: 4451–4459.

Leader B, Baca QJ, Golan DE. Protein therapeutics: A summary and pharmacological classification. *Nat Rev Drug Discovery.* 2008; 7: 21–39.

Lee H, Lee MY, Bhang SH. Hyaluronate-gold nanoparticle/tocilizumab complex for the treatment of rheumatoid arthritis. *ACS Nano.* 2014; 8: 4790–4798.

Lei C, Liu P, Chen B. Local release of highly loaded antibodies from functionalized nanoporous support for cancer immunotherapy. *J Am Chem Soc.* 2010; 132: 6906–6907.

Li S, Zhao B, Wang F, Wang M, Xie S, Wang S *et al.* Yak interferon-alpha loaded solid lipid nanoparticles for controlled release. *Res Vet Sci.* 2010; 88: 48–153.

Li Y, Chen X, Liu H, Mou X, Ren Z, Ahmad Z *et al.* Silica nanospheres entrapped with ultra-small luminescent crystals for protein delivery. *Chem Eng J.* 2017; Article In press.

Liguori L, Marques B, Villegas-Mendez A, Rothe R, Lenormand JL. Liposomes mediated delivery of pro-apoptotic therapeutic membrane proteins. *J Control Rel.* 2008; 126: 217–227.

Liu W, He Z, Liang J, Zhu Y, Xu H, Yang X. Preparation and characterization of novel fluorescent nanocomposite particles: CdSe/ZnS core-shell quantum dots loaded solid lipid nanoparticles. *J Biomed Mater Res.* 2008; 84(4): 1018–1025.

Lu J, Liong M, Sherman S, Xia T, Kovochich M, Nel AE *et al.* Mesoporous silica nanoparticles for cancer therapy: energy-dependent cellular uptake and delivery of paclitaxel to cancer cells. *Nanobiotech.* 2007; 3: 89–95.

Macura SL, Steinbacher JL, MacPherson MB. Microspheres targeted with a mesothelin antibody and loaded with doxorubicin reduce tumor volume of human mesotheliomas in xenografts. *BMC Cancer.* 2013; 13: 400.

Mahony D, Cavallaro AS, Stahr F, Mahony TJ, Qiao SZ, Mitter N. Mesoporous silica nanoparticles act as a self-adjuvant for ovalbumin model antigen in mice. *Small.* 2013; 9(18): 3138–3146.

Malhotra M, Jain NK. Niosomes as drug carriers. *Indian Drugs.* 1994; 31: 81–86.

Malmsten M, Zauscher S. Colloids and surfaces in biology. *Curr Opin Colloid in Sci.* 2013; 18: 468.

Manosroi A, Lohcharoenkal W, Gotz F, Werner RG, Manosroi W, Manosroi J. Cellular uptake enhancement of Tat-GFP fusion protein loaded in elastic niosomes. *J Biomed Nanotechnol.* 2011; 7: 366–376.

Marcato PD, Duran N. New aspects of nanopharmaceutical delivery systems. *J Nanosci Nanotechnol.* 2008; 8(5): 2216–2229.

Martins S, Sarmento B, Ferreira DC, Souto EB. Lipid-based colloidal carriers for peptide and protein delivery – Liposomes versus lipid nanoparticles. *Int J Nanomedicine.* 2007; 2(4): 595–607.

Mehnert W, Mader K. Solid lipid nanoparticles: Production, characterization and applications. *Adv Drug Deliv Rev.* 2001; 47(2–3): 165–196.

Morel S, Rosa Gasco M, Cavalli R. Incorporation in lipospheres of [d-Trp-6]LHRH. *Int J Pharm Sci.* 1994; 105(2): R1–R3.

Morel S, Ugazio E, Cavalli R, Gasco MR. Thymopentin in solid lipid nanoparticles. *Int J Pharm Sci.* 1996; 132(1): 259–261.

Muheem A, Shakeel F, Jahangir MA, Anwar M, Mallick N, Jain GK *et al.* A review on the strategies for oral delivery of proteins and peptides and their clinical perspectives. *Saudi Pharm J.* 2016; 24(4): 413–428.

Mujoriya RZ, Bodla RB. Niosomes – Challenge in preparation for pharmaceutical scientist. *Int J App Pharm.* 2011; 3: 11–15.

Muller R, Mehnert W, Lucks J-S, Schwarz C, Zur Mühlen A, Meyhers H *et al.* Solid lipid nanoparticles (SLN): an alternative colloidal carrier system for controlled drug delivery. *Eur J Pharm Biopharm.* 1995; 41(1): 62–69.

Muller RH, Mader K, Gohla S. Solid lipid nanoparticles (SLN) for controlled drug delivery – A review of the state of the art. *Eur J Pharm Biopharm.* 2000; 50(1):161–177.

Muller RH, Runge S, Ravelli V, Mehnert W, Thunemann AF, Souto EB. Oral bioavailability of cyclosporine: Solid lipid nanoparticles (SLN) versus drug nanocrystals. *Int J Pharm Sci.* 2006. 317(1): 82–89.

Niven RW, Lott FD, Ip AY, Cribbs JM. Pulmonary delivery of powders and solutions containing recombinant human granulocyte colony-stimulating factor (rhG-CSF) to the rabbit. *Pharm Res.* 1994; 11: 1101–1109.

Olbrich C, Runge SA, Mehnert W, Thunemann AF, Muller RH. Entrapment efficiency and biodegradation of cyclosporine loaded solid lipid nanoparticles (SLN) for peroral administration. 3rd World Meeting APV/APGI, Berlin, *In Proc.* 2000; 425–426.

Patel RP, Patel H, Baria AH. Formulation and evaluation of liposomes of ketokonazole. *Int J Drug Deliv Tech.* 2009; 1(1): 16–23.

Patel S, Bhirde AA, Rusling JF, Chen X, Gutkind JS, Patel V. Nano delivers big: Designing molecular missiles for cancer therapeutics. *Pharmaceutics.* 2011; 3(1): 34–52.

Pedersen N, Hansen S, Heydenreich AV. Solid lipid nanoparticles can effectively bind DNA, streptavidin and biotinylated ligands. *Eur J Pharm Biopharm.* 2006; 62: 155–162.

Prasetyanto EA, Bertucci A, Septiadi D, Corradini R, Castro-Hartmann P, De Cola L. Breakable hybrid organosilica nanocapsules for protein delivery. *Angew Chem Int Ed Engl.* 2016; 55(10): 3323–3327.

Qiu LY, Bae YH. Polymer architecture and drug delivery. *Pharm Res.* 2006; 23: 1–30.

Radin S, Falaize S, Lee MH, Ducheyne P. In vitro bioactivity and degradation behavior of silica xerogels intended as controlled release materials. *Biomaterials.* 2002; 23: 3113–3122.

Radtke M, Muller RH. Novel concept of topical cyclosporine delivery with supersaturated SLN creams. *Proc Intern Symp Control Rel Bioact Mater.* 2001a; 28: 470–471.

Radtke M, Muller RH. Stability study of creams containing cyclosporine SLN. *Proc Intern Symp Control Rel Bioact Mater.* 2001b; 28: 472–473.

Ramachandran R, Paul W, Sharma CP. Synthesis and characterization of *J Biomed Mater Res B.* 2008; 18: 699–703.

Rentel CO, Bouwstra JA, Naisbett B, Junginger HE. Niosomes as a novel peroral vaccine delivery system. *Int J Pharm.* 1999; 186: 161–167.

Rezler EM, Khan DR, Lauer-Fields J, Cudic M, Baronas-Lowell D, Fields GB. Targeted drug delivery utilizing protein-like molecular architecture. *J Am Chem Soc.* 2007; 129: 4961–4972.

Rytting E, Nguyen J, Wang X, Kissel T. Biodegradable polymeric nanocarriers for pulmonary drug delivery. *Expert Opin Drug Deliv.* 2008; 5: 629–639.

Sarmento B, Ribeiro A, Veiga F, Sampaio P, Neufeld R, Ferreira D. Alginate/chitosan nanoparticles are effective for oral insulin delivery. *Pharm Res.* 2007; 24: 2198–2206.

Shao X, Zhang J, Rajian R. ^{125}I-labeled gold nanorods for targeted imaging of inflammation. *ACS Nano.* 2011; 5: 8967–8973.

Shen J, Li K, Cheng L, Liu Z, Lee ST, Liu J. Specific detection and simultaneously localized photothermal treatment of cancer cells using layer-by-layer assembled multifunctional nanoparticles. *ACS Appl Mater Interfaces.* 2014; 6: 6443–6452.

Slowing I, Trewyn B G, Lin VSY. Effect of surface functionalization of MCM-41-type mesoporous silica nanoparticles on the endocytosis by human cancer cells. *J Am Chem Soc.* 2006; 128: 14792–14793.

Solaro R. Targeted delivery of proteins by nanosized carriers. *J Polym Sci.* 2008; 46: 1–11.

Solaro R, Chiellini F. Nanoparticles for the targeted delivery of peptides and proteins. In *Handbook of Particulate Drug Delivery*; Ravi Kumar MNV, eds., American Scientific Publisher New York, USA, 2006; Vol 2: pp. 193–222.

Song M, Zhang Y, Chen K, Wang H, Gong R. Carboxymethyl-β-cyclodextrin grafted chitosan nanoparticles as oral delivery carrier of protein drugs. *Reac Funct Poly.* 2017; 117: 10–15.

Song YK, Liu D, Maruyama K, Takizawa T. Antibody mediated lung targeting of long circulating emulsions. *J Pharm Sci Tech.* 1996; 50: 372–377.

Soppimath KS, Aminabhavi TM, Kulkarni AR, Rudzinski WE. Biodegradable polymeric nanoparticles as drug delivery devices. *J Control Rel.* 2001; 70: 1–20.

Sundarraj S, Thangam R, Sujitha MV, Vimala K, Kannan S. Ligand-conjugated mesoporous silica nanorattles based on enzyme targeted prodrug delivery system for effective lung cancer therapy. *Tox App Pharm.* 2014; 275: 232–243.

Tan W, Wang K, He X, Zhao XJ, Drake T, Wang L *et al.* Bionanotechnology based on silica nanoparticles. *Med Res Rew.* 2004; 24: 621–638.

Tang H, Kobayashi H, Niidome Y, Mori T, Katayama Y, Niidome T. CW/pulsed NIR irradiation of gold nanorods: Effect on transdermal protein delivery mediated by photothermal ablation. *J Control Rel.* 2013; 171(2): 178–183.

Tkachenko AG, Xie H, Coleman D, Glomm W, Ryan J, Anderson MF *et al.* Multifunctional gold nanoparticle-peptide complexes for nuclear targeting. *J Am Chem Soc.* 2003; 125: 4700–4701.

Udupa N. Niosomes as drug carriers. In: Jain NK, editor. *Controlled and Novel Drug Delivery.* 1st edition. New Delhi: CBS Publishers and Distributors; 2002.

Ugazio E, Cavalli R, Gasco MR. Incorporation of cyclosporin A in solid lipid nanoparticles (SLN). *Int J Pharm Sci.* 2002; 241: 341–344.

Ulbrich K, Hola K, Subr V, Bakandritsos A, Tucek J, Zboril R. Targeted drug delivery with polymers and magnetic nanoparticles: covalent and noncovalent approaches, release control, and clinical studies. *Chem Rev.* 2016; 116: 5338–5431.

Vallet-Regi M, Balas F, Arcos D. Mesoporous materials for drug delivery. *Angew Chem.* 2007; 46: 7548–7558.

Vauthier C, Bouchemal K. Methods for the preparation and manufacture of polymeric nanoparticles. *Pharm Res.* 2009; 26: 1025–1058.

Vauthier C, Labarre D. Modular biomimetic drug delivery systems. *J Drug Del Sci Tech.* 2008; 18: 59–68.

Videira M, Azevedo AF, Almeida AJ. Entrapment of a high molecular weight protein into solid lipid nanoparticles. 1998; APV/APGI: 629–630.

Videira M, Florin do H, Almeida AJ. Preparation of solid lipid nanoparticles (SLN): A potential protein delivery system. *V Spanish–Portuguese Con Control Drug Del.* 2002; 69–70.

Vila A, Sanchez A, Evora CI, Soriano I, McCallion O, Alonso MJ. PLA-PEG particles as nasal protein carriers: The influence of the particle size. *Int J Pharm.* 2005; 292: 43–52.

Vila A, Sanchez A, Tobio M, Calvo P, Alonso MJ. Design of biodegradable particles for protein delivery Bouchemal. *J Control Rel.* 2002; 78: 15–24.

Vyas SP, Khar RK. *Targeted and Controlled Drug Delivery.* 1st edition. New Delhi: CBS Publishers and Distributors; 2002; pp. 332–334.

Walsh G. Biopharmaceutical benchmarks 2010. *Nat Biotechnol.* 2010; 28(9): 917–924.

Wissing SA, Kayser O, Muller RH. Solid lipid nanoparticles for parenteral drug delivery. *Adv Drug Deliv Rev.* 2004; 56(9): 1257–1272.

Woodle MC, Storm G, Newman MS, Jekot JJ, Collins LR, Martin FJ *et al.* Prolonged systemic delivery of peptide drugs by long-circulating liposomes: Illustration with vasopressin in the Brattleboro rat. *Pharm Res.* 1992; 9: 260–265.

Xing X, He X, Peng J, Wang K, Tan W. Uptake of silica coated nanoparticles by HeLa cells. *J Nanosci Nanotechnol.* 2005; 5: 1688–1693.

Yi J, Lam TI, Yokoyama W, Cheng LW, Zhong F. Cellular uptake of beta-carotene from protein stabilized solid lipid nanoparticles prepared by homogenization-evaporation method. *J Agri Food Chem.* 2014; 62(5): 1096–1104.

Yousefi A, Esmaeili F, Rahimian S, Atyabi F, Dinarvand R. Preparation and *in vitro* evaluation of a pegylated nano-liposomal formulation containing docetaxel. *Sci Pharm.* 2009; 77: 453–464.

Yu M, Wu J, Shi J, Farokhzad OC. Nanotechnology for protein delivery: Overview and perspectives, *J Contr Rel.* 2016; 240: 24–37.

Zhang N, Ping Q, Huang G, Xu W, Cheng Y, Han X. Lectin-modified solid lipid nanoparticles as carriers for oral administration of insulin. *Int J Pharm Sci.* 2006; 327: 153–159.

Zhang Q, Yie G, Li Y, Yang Q, Na gai T. Studies on the cyclosporin A loaded stearic acid nanoparticles. *Int J Pharm Sci.* 2000; 200: 153–159.

10 Nanotechnology-Based Prodrug Conjugates for Site-Specific Antineoplastic Therapy

*Nishi Mody, Rajeev Sharma, Surbhi Dubey, Atul Jain, and Suresh P. Vyas**

CONTENTS

10.1 Introduction 203
10.2 Prodrugs: Structure and Classification 205
  10.2.1 Carrier-Linked Prodrugs 205
  10.2.2 Bioprecursor Prodrugs 205
10.3 The Role of Linker in Prodrugs 205
10.4 Targeting-Ligand Conjugated Prodrugs 206
  10.4.1 Antibody–Drug Conjugate 206
  10.4.2 Peptide–Drug Conjugates 207
  10.4.3 Aptamer–Drug Nanoconjugates 209
  10.4.4 Folic Acid–Drug Conjugates 209
10.5 Polymeric Prodrugs 210
  10.5.1 PEG–Drug Conjugates 210
  10.5.2 Other Polymer–Drug Conjugates 210
10.6 Enzyme-Activated Prodrugs 211
  10.6.1 Antibody-Directed Enzyme Prodrug Therapy 212
  10.6.2 Gene-Directed Enzyme Prodrug Therapy 212
10.7 Stimuli-Sensitive Prodrugs 212
10.8 Lipid-Based Prodrugs 214
10.9 Nanostructured Prodrugs 214
10.10 Conclusion 216
References 217

10.1 INTRODUCTION

Last two decades have witnessed sporadic implementation of the prodrug approach for improving the different physiological, pharmacokinetic and/or biopharmaceutical properties of bioactives. A prodrug can simply be defined as a shrouded form of an active moiety which would be converted to active one within the biological milieu. Figure 10.1 illustrates the basic concept of prodrug activation upon its administration (Rautio *et al.*, 2008; Rautio, 2010; Vincent, 2007; Stella, 2010). Initially, the prodrug strategy was developed to alter the pharmacokinetics of the drug in a favourable manner. However, these days, the strategy has also been applied for site-specific delivery that

* Corresponding author.

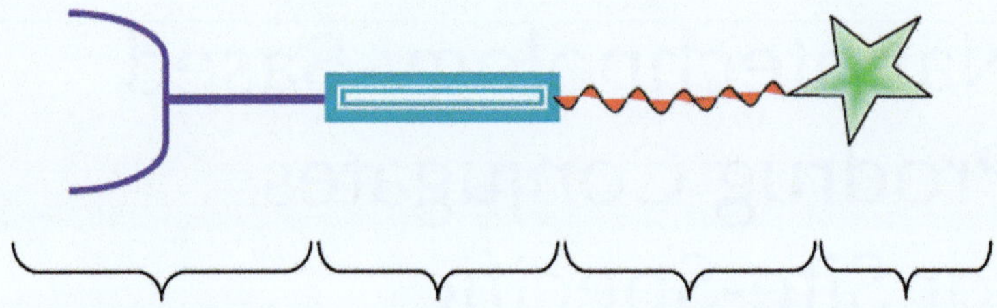

FIGURE 10.1 Schematic representation of prodrug.

improves therapy with lesser systemic and/or organ-specific side effects. As a prodrug can alter the pharmacokinetics and toxicological (ADMET) characteristics of a bioactive/drug, a comprehensive study on a drug candidate can help in designing an apt prodrug in a better way.

The feasibility of altering an existing bioactive to overcome its drawbacks could be faster and more economical than looking for a better but completely new drug entity (Rautio, 2010; Stella, 2010). Acetanilide (Antifebrin®) was the first compound considered as a prodrug, falling under the category of antipyretic agents. It is an inactive form of acetaminophen and it gets converted into acetaminophen in the body by hydroxylation. Similarly, phenacetin is another instance of this category, where the activation results by O-dealkylation process. Aspirin (acetylsalicylic acid) is another classical prodrug, first synthesized by Felix Hoffman (Bayer, Germany) in 1897, and was later used as a therapeutic agent by Dreser in 1899 (Bertolini *et al.,* 2006; Zawilska *et al.,* 2013), getting converted to salicylic acid within the body. However, these were the moieties that were not intentionally designed as prodrugs, but their mechanisms of drug action claimed these to be prodrugs in later stages of development.

It is assessed that at present about 10% of globally marketed drugs can be placed under the category of prodrugs (Srinivas, 2011). However, it was in the 20th century when the prodrug theory was deliberately used for modifying drug properties. Chloramphenicol sodium succinate and chloramphenicol palmitate are the examples of two intentionally designed prodrugs of chloramphenicol (Zawilska *et al.,* 2013). Cancer is a disease, wherein a number of physiological changes tend to occur in the tumour environment, and by making use of these alteration in behaviour of tumour tissue, prodrugs are now seeking increased interest in the management of cancer biology. Despite the merits of prodrugs, their full potential has not yet been exploited in the present drug development era. It is expected that effectiveness of therapy and bioavailability of a drug should be increased using prodrugs. Testa (2009) enlisted three vital objectives in prodrug research:

1. Pharmaceutical Objectives:
   a. To enhance organoleptic properties, stability and solubility of the bioactive
   b. To improve the formulation-related problems in order to develop suitable dosage form for drug administration
2. Pharmacokinetic Objectives:
   a. To reduce the presystemic metabolism, to improve site-specific release absorption of the therapeutic agents
3. Pharmacodynamic Objectives:
   a. Better therapeutic index and low toxicity can be accomplished by the prodrug concept and brain- and cancer-specific delivery are the most prominent achievements in prodrug synthesis during the last a couple of decades

10.2 PRODRUGS: STRUCTURE AND CLASSIFICATION

Based on the promoiety, prodrugs can be broadly categorized as:

1. Carrier-linked prodrugs
2. Bioprecursor prodrugs

10.2.1 Carrier-Linked Prodrugs

As the name suggests, a carrier-linked prodrug is formed when the bioactive is linked temporarily to a carrier *via* reversible covalent linkage. Upon administration into the body, these are transformed into the active compounds. Ideally, the carrier should be stable, non-immunogenic and biodegradable in nature (Huttunen *et al.*, 2011; Srinivas, 2011). Another category of prodrugs is where, instead of bioactive(s) and carrier(s), two active moieties are conjugated and converted into a single molecule, with each moiety serving as a promoiety for the other one. There are a vast number of examples in this category like sulfamethoxazole-nalidixic acid, indomethacin-paracetamol, L-DOPA-enthacapone, gabapentin-pregabalin, sulfapyridine-5-aminosalicylic acid, ampicilin sulbactram, 5-fluorouracil-dexamethasone triamcinolone and 5-fluorouracil-cytarabine (Das *et al.*, 2010). Owing to their ease of cleavability, ester and amide are two groups that are commonly found in the carrier-linked prodrug category. Other than these two, carbonates, carbamate, oximes, phosphate, *N*-mannich and imines base are also reported.

10.2.2 Bioprecursor Prodrugs

Bioprecursors lack promoiety, but these are formed by introducing a molecular alteration of the parent active agent itself. The bioprecursor prodrug is metabolically converted to the active compound by oxidation, reduction and hydration, depending upon the availability of the functional group (Graf and Lippard, 2012; Kokil and Rewatkar, 2010; Shi *et al.*, 2012; Teicher, 2009).

Apart from this, on the basis of their site of biotransformation into pharmacologically active species, prodrugs can also be classified into two additional categories, i.e. metabolized intracellularly and metabolized extracellularly. Some of the prodrugs belonging to more than one class of prodrugs are called as mixed-type prodrugs (Wu, 2009).

10.3 ROLE OF LINKER IN PRODRUGS

To formulate a prodrug *via* chemical linkage, primarily there should be a free or open active site(s) in the parent moiety, which can be utilized to modify with the promoiety. In general, the linker should be easily cleavable or should undergo cleavage in response to some stimulus like change in the pH or enzyme action (Albert, 1998). Among several functional groups, ester is one of the most commonly employed groups for linkage in designing of a prodrug. Most of the promoieties, including parent drugs, contain carboxyl and/or hydroxyl groups and are, therefore, easy to form ester bonds. Additionally, esterase enzymes are distributed all over in the body. Upon administration, the ester bond undergoes hydrolysis in the presence of esterase, ultimately resulting into the delivery of the parent drug at the desired site. Apart from esters, the amide bond is also a good alternative to design prodrugs as amides, and is comparatively more stable than ester in the presence of enzymes. Nonetheless, most of the amide linkages in prodrug synthesis are engineered to be cleaved by a particular enzyme, either to improve the site-specific targetability and/or safety (Bhandari *et al.*, 2007). Apart from ester and amide bonds, oximes/imines (Bobeck *et al.*, 2010), disulfide bonds (Charkoudian *et al.*, 2006) or uncleavable thioether bonds are also reported for prodrug formation.

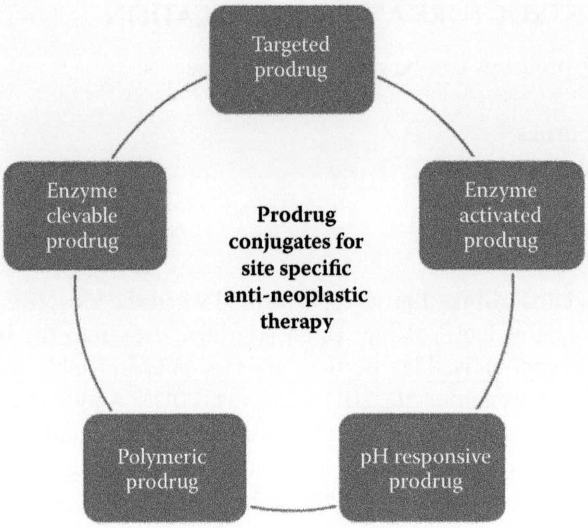

FIGURE 10.2 Different approaches for targeted antineoplastic therapy.

The bond formed between the promoiety and parent drug should be stable enough in the blood circulation, and should cleave at the target site. Thus, the linkage optimization is strongly recommended for the process of prodrug development so as to attain maximum efficacy and stability. Figure 10.2 compiles different approaches for targeted antineoplastic therapy using prodrug approach.

10.4 TARGETING-LIGAND CONJUGATED PRODRUGS

A tumour cell differs from the normal cell in terms of biology. These cells overexpress specific receptors, transporters and antigens. These overexpressed moieties can be prudently used for selective delivery of the chemotherapeutic agent across the tumour cell by making use of targeting molecules like antibodies, aptamers, ligands and even peptide and antibody fragments. Targeting molecules could be attached to the chemotherapeutic drug to form a prodrug, which will be selectively taken up by the tumour cell. These types of prodrugs are discussed in the subsequent section:

10.4.1 Antibody–Drug Conjugate

Monoclonal antibodies (mAbs) are the foremost targeting moiety employed for prodrug approach for the management of tumours. They are highly selective and site-specific to the antigen overexpressed on cancer cells, and deliver therapeutic agents to tumour cells without altering the cellular environment of nearby normal tissue cells. The antibody–drug conjugates (ADCs) are internalized by receptor-mediated endocytosis and remain inactive prior to any transformation into the body. Once internalized, the ADC gets converted into the active form releasing the anticancer drugs in the tumour cell environment. In case of ADCs, it is imperative to calculate the ratio of antibody to drug in an optimal manner so as to achieve maximal safety and efficacy (Schrama *et al.*, 2006). Nevertheless, immunological consequences, tumour permeability and specific delivery are the raising concerns that need to be addressed prior to the fruitful clinical application. Deficiency of cell permeability is the foremost obstacle for ADCs (Stohrer *et al.*, 2000), which can be achieved by employing fragments of antibodies like scFv and Fab and internalized more efficiently into cancerous cells as compared to the intact antibodies. Instances of recently used tumour-specific antigens include prostate membrane specific antigen (PSMA), TAG-72, αv integrin, CD19, CD22, MUC16 and epidermal growth factor receptor (EGFR). Expression of MUC16 is much more profound in ovarian cancer cells rather than tumours of different origins (Gires and Seliger, 2009). Likewise, the expression of PSMA is preferentially associated with prostate cancerous tissues (Wang *et al.*, 2009). On the other hand, regardless of origin, some antigens show high degree of affinity to tumour cells, e.g. TAG-72 is preferentially associated in human adenocarcinomas, encompassing gastric, breast, colorectal, pancreatic, lung (non-small cell), endometrial and ovarian cancers. Several anticancer bioactives have also been modified and conjugated to tumour-specific antibodies. These includes, antifolates (Schrama *et al.*, 2006), vinca alkaloids (Spearman *et al.*, 1987), anthracyclines (Senter, 2009), taxanes (Quiles *et al.*, 2010), monomethyl auristatin E (Senter, 2009), calichemicin (Boghaert *et al.*, 2008) and mytansine derivative DM1 (Krop *et al.*, 2010). Although the clinical trials are being conducted on about 20 ADCs, only one out of them is Food and Drug Administration (FDA)-approved ADC, i.e. Gemtuzumab ozogamicin (marketed by Wyeth as Mylotarg), employed for the management of acute myeloid leukaemia (AML) (Teicher, 2009). More than 90% of the AML patients have overexpressed CD33 antigen, which is conversely absent in normal human tissues and pluripotent stem cells. In Mylotarg, the antineoplastic agent, calicheamicin, is conjugated to the humanized anti-CD33 mAb (clone P67.6), through a bifunctional hydrolabile linker (Sievers, 2003). Specifically, it adheres to the CD33 antigen present on cancer cells, followed by prodrug internalization (van der Velden *et al.*, 2001). The FDA approved the prodrug in 2000, followed by post-marketing surveillance in 2004. In the year 2010, unfortunately, Mylotarg was withdrawn from the market, because of the severe adverse and toxic effects observed after the post-approval clinical trial (www.fda.gov). Nonetheless, there are several other ADC prodrugs, which had been undergoing clinical trials in advanced phases (DiJoseph *et al.*, 2006; Krop *et al.*, 2009). Trastuzumab-amide-MCC-DM1 (Trastuzumab-DM1) and inotuzumab ozogamicin (CMC-544) have been approved by the FDA for human epidermal growth factor receptor 2 (HER2)-positive breast cancer cell and β-lymphoid malignancies in 2013 and recently in 2017, respectively (www.breastcancer.org; www.fda.gov). Later, on HER2 was found to be associated to a greater incidence of breast cancer (20–25%), reflecting poor diagnosis and violent cancer phenotypes, thus making it a favourable theronostic target. Trastuzumab (humanized monoclonal HER2 antibody), was approved by the FDA in 1998 for the management of HER2-positive metastatic breast cancer (Tai *et al.*, 2010). Even though trastuzumab appeared as most remarkable mAb in cancer management, a noteworthy population of cancer patient does not respond to trastuzumab, and this calls for adjuvant chemotherapeutic agents for trastuzumab therapy (Phillips *et al.*, 2008). The direct chemical conjugation of anticancer bioactives to trastuzumab has therefore been considered as a better alternative to antibody monotherapy. An instance of such conjugate is trastuzumab-DM1 in which trastuzumab is used to deliver DM1 (i.e. a derivative of maytansine 1), an effective anticancer agent, specifically at HER2-positive cancer cells. Phillips *et al.* (2008) have studied different linkages between trastuzumab and DM1 conjugates, and found that the highest activity was shown by the irreducible thioether linkage as compared to the disulphide linkage. Another ADC which has been under the preliminary studies is CMC-544, in which N-acetyl gamma calicheamicin dimethylhydrazide (CalichDMH) is linked to the humanized anti-CD22 mAb, linked through acid-hydrolysable linkage. CalichDMH is a DNA damaging antibiotic agent and CMC 544 uses CD22 as a targeting moiety. In general, CD22 is upregulated on the external surface of B lymphocytes only. After adhering to CD22 on malignant B-cells, CMC-544 undergoes rapid internalization to deliver Calich DMH into the cells (DiJoseph *et al.*, 2006). Table 10.1 summarizes a few FDA-approved therapeutic mAbs for cancer therapy along with their mechanism of action (Firer and Gellerman, 2012; Scott *et al.*, 2012).

10.4.2 Peptide–Drug Conjugates

Similar to antibodies, peptides are also reported as a suitable site-specific moiety owing to its excellent permeability across the cell membrane, ease of synthesis, small molecular weight and flexibility in chemical conjugation (Tai *et al.*, 2010). Peptidic-targeting molecules can be successfully developed against overexpressed tumour-specific antigens. Such peptidic ligands can be conjugated directly to antineoplastic bioactives, so as to achieve the targeted drug therapy for cancer cells.

TABLE 10.1
FDA Approved Therapeutic Monoclonal Antibodies for Cancer Therapy with Their Mechanism of Action

Generic Name	Proprietary Name	Target	Mechanism	Technology	Isotype	Additional Manipulations	Year FDA Approved	Approved Clinical Indication
Rituximab	Rituxin®/ Mabther®	CD20	ADCC, direct induction of apoptosis and CDC	Mouse Hybridoma	IgG1-kappa	Chimeric	1997	NHL; later CD20+CLL, FL, RA Transtuzumab Herceptin®
Transtuzumab	Herceptin®	HER-2	Inhibition of ERBB2 signalling and ADCC	Mouse Hybridoma	IgG1-kappa	Humanized	1998	HER-2+ MBC
Alemtuzumab	Campath®/ Mabcampath®	CD52	Direct induction of apoptosis and CDC	Rat Hybridoma	IgG1-kappa	Humanized	2001	CL L, T-cell Lymphoma Ibritomomab tiuxitan Zevalin®
Ibritomomab tiuxitan	Zevalin®	CD20	Delivery of the radioisotope90Y	Mouse monoclonal	IgG1-kappa	Conjugated to Yittrium-90	2002	NHL
Tositumomab	Bexxar®	CD20	Delivery of the radioisotope131I, ADCC and direct induction of apoptosis	Mouse monoclonal	IgG2a-lambda	Conjugated to I-131	2003	NHL
Cetuximab	Erbitux®	GRF, HER-1	Inhibition of EGFR signalling and ADCC	Mouse monoclonal	IgG1-kappa	Chimeric	2004	EGRF+ MCC
Bevacizumab	Avastin®	VEGF	Inhibition of VEGF signalling	Mouse monoclonal	IgG1-kappa	Humanized	2004	MCC
Panitumumab	Vectibix™	GRF, HER-1	Inhibition of EGFR signalling	Human monoclonal	IgG1-kappa	Human	2006	MCC
Ofatumumab	Arzerra™	CD20	ADCC and CDC	Human monoclonal	IgG1-kappa	Human	2009	Refractory CLL
Ipilimumab	Yervoy™	CTLA-4	Inhibition of CTLA4 signalling	Human monoclonal	IgG1-kappa	Human	2011	MMel
Pertuzumab	Perjeta™	GRF, HER-2	Inhibits HER2:HER3 dimer formation and downstream signaling, while also mediating ADCC	Mouse monoclonal	IgG1-kappa	Humanized	2012	BC

Lutenizing hormone releasing hormone (LHRH) protein conjugates were reported as a prodrug approach for cancer management (Khandare *et al.*, 2006). Several cancer cells of prostate, ovarian, breast and endometrium tumour sites over-express LHRH receptors, despite the fact that this receptor is present on the healthy cells as well. Camptothecin was conjugated to the LHRH peptide using a spacer molecule like PEG. The results showed a momentous enhancement in the effectiveness of the peptide-conjugated prodrugs along with minimum adverse effects on normal tissue organs. Overexpression of αv β3 integrin receptors is observed in activated endothelia cells and metastatic cancer cells, where αv β3 integrin acts as the binding site for RGD peptide. Paclixatel RGD cyclic peptide conjugate $E[c(RGDyK)]_2$ (cRGD) have been reported to be synthesized. Yoneda *et al.* (2008), explored the antigenic properties of glucose-regulated protein 78 (GRP78) in tumour targeting for site-specific delivery of doxorubicin and paclitaxel. GRP78 specifically overexpressed on several cancer cells, such as skin, breast, colon and prostate, but not in the normal cells (Shin *et al.*, 2003). A cyclic short peptide (Pep 42) binds *via* Val-Cit motif to GPR78, leading to endocytosis. The drug conjugate is then carried to the lysosome, where the Val-Cit motif is cleaved by cathepsin, resulting eventually in the release of free drug. Besides, cytotoxicity of the Pep42-prodrug has been studied in osteosarcoma cells. SJSA-1, Pep42-paclitaxel and Pep42-doxorubicin showed increased toxicity *vis-á-vis* their free counterparts.

10.4.3 Aptamer–Drug Nanoconjugates

Aptamer is a three-dimensional single-stranded DNA or RNA that specifically binds to the target site with great affinity. The three-dimensional structure of the aptamer further adds to binding specificity (Nimjee *et al.*, 2005). Aptamers also possess low molecular weight, ease of synthetic accessibility, low immunogenicity, better stability and suppleness in conjugation, and therefore, they appear as an important targeting moiety (Zhou *et al.*, 2009). Aptamers can serve as nonprotein-based alternative to antibody for site-specific delivery of various bioactives to tumour cells (Sefah *et al.*, 2010). Additionally, aptamers also serve as suitable therapeutic agents for several high-quality therapeutic applications because of desirable synthesis in the laboratory. PSMA-specific aptamers (A10 RNA aptamer, Apt) were conjugated with cisplatin NPs, and was used for targeting neovasculature of the non-prostate solid tumour, together with lungs, breast and prostate cancer cells (Milowsky *et al.*, 2007; Morris *et al.*, 2007; Dhar *et al.*, 2008; Kolishetti *et al.*, 2010). The aptamer-bioconjugated cisplatin polymeric nanoparticulate system (Pt-NP-Apt) showed IC_{50} value of 0.03 μM, while the unmodified nanoparticles showed the IC_{50} value of 0.13 μM, indicating higher toxicity of the drug conjugate in PSMA-positive prostate cancer cells. Moreover, efficiency of Pt-NP-Apt NP was found to be approximately 10-fold higher than that of free drugs (Dhar *et al.*, 2008). Another aptamer drug conjugate was reported by Huang *et al.* (2009), in which they have conjugated a DNA aptamer, sgc8c, with doxorubicin for treating human T-cell acute lymphoblastic leukaemia (T-cell ALL). The sgc8c aptamer tends to identify the protein tyrosine kinase 7 (PTK7), a transmembrane receptor over-expressed on T-cell ALL cells, like CCRFCEM (Human T cell lymphoblast-like cell line) (Shangguan *et al.*, 2006).

10.4.4 Folic Acid–Drug Conjugates

Folate receptors are highly-expressed in various tumour tissues, including kidney, breast, ovarian, lung and endometrial cancers, as compared to the normal tissues. Folic acid, a member of the vitamin B family, devoid of immunogenity, is most widely used ligand to deliver drugs, nanoscaled systems and imaging agents to folate receptors, overexpressed on the tumour cells (Jaracz *et al.*, 2005). A number of anticancer bioactives have been coupled with folic acid in order to bind specifically to the folate receptors on cells. Folic acid-conjugated prodrug binds to folate receptors undergoing receptor-mediated endocytosis. Folic acid was conjugated to campthothecin by Philips *et al.* (2008) through a peptide (hydrophilic) containing a disulfide bond. Specificity of the prodrug was augmented by folic acid, while the solubility of poorly soluble camptothecin was increased by the use of cleavable spacer. Disulphide linker was cleaved by the disulphide reduction and camptothecin was released within the vicinity of tumour cells. Furthermore, treating KB cells prior with excess of folic acid resulted in decreased activity of the prodrug, indicating that the prodrug was internalized *via* folate receptors. Doxorubicin, another widely used chemotherapeutic agent, is associated with drawbacks like poor solubility, short half-life and low therapeutic index, which limit its application in anticancer therapy. To overcome these drawbacks, it was conjugated with D-α-tocopheryl polyethylene glycol succinate (TPGS), along with folic acid. This resulted in increased solubility which enhanced the cell permeability *via* folate receptors. The TPGS-Dox-FOL prodrug exhibited significant increase in half-life, leading to improved anticancer potential (i.e., 45-fold more effectiveness than the naïve doxorubicin), and low cardiac accumulation, the major side effect associated with doxorubicin (Anbharasi *et al.*, 2010).

10.5 POLYMERIC PRODRUGS

10.5.1 PEG–Drug Conjugates

Polyethylene glycol (PEG) is an FDA-approved, amphiphilic, water-soluble polymer used for parenteral, oral and topical administrations. It is excreted out from the body by kidneys and liver, and is thus extremely safe for administration (Yamaoka *et al.*, 1994). PEGylation was initially employed to alter the physicochemical characteristics of macromolecules, including growth factors, enzymes and antibodies. The latest developments suggest that PEG can also be employed to conjugate the small molecule drugs in order to improve their site-specific targetability. Based on the molecular weight, low and high molecular weight PEG (LM-PEG and HM-PEG) are used for synthesis of a prodrug. The HM-PEG-conjugated prodrug encompasses PEG ranging from 30 to 60 kDa or above. Verily, PEG is used to increase the molecule size in order to extend the circulation half-life and alter the pharmacokinetic profile, as needed. The increased contact time of a prodrug with the tumour cells can be attributed to the enhanced permeation and retention (EPR) effect, which takes the advantage of abnormal tumour vasculature. A number of PEGylated prodrugs are being tested under clinical trials, e.g. NKTR-102 is PEGylated derivative of irinotecan, (the drug) mainly employed in the management of colorectal cancer. Irinotecan, modified with HM-PEG through a cleavable linkage, will release the drug slowly *in vivo* (Eldon *et al.*, 2007a). It is also evident from preclinical studies that NKTR-102 has demonstrated the antitumor effect in phase II trials for colorectal cancer (clinicaltrials.gov). Besides NKTR-102, large PEG has also been conjugated to other drugs such as docetaxel (i.e., NKTR-105; currently under phase III trials), camptothecin (Pegamotecan) and SN-38 (EZN-2208) (Pasut and Veronese, 2009). The structural design (linear, branched or even dendritic) of PEG is a vital aspect considered in drug loading capacity. However, the branched structure of the PEG not only raises the steric hindrance and improves the half-life of the prodrug, but also provides more active conjugation sites for the parent drug. When the PEG used is of molecular weight less than 5000 Da, such prodrugs fall under the category of low-molecular-weight ones.

10.5.2 Other Polymer–Drug Conjugates

Even though PEG is effectively employed as a polymer of choice for prodrugs, its drug loading capacity is limited due to limited number of functional groups available for drug conjugation. Hence, some other different polymers like, N-(2-hydroxypropyl) methacrylamide (HPMA) and polyglutamic acid (PGA), with a number of active site/groups, can be used to achieve high drug loading up to 37% of weight (Sabbatini *et al.*, 2004). A peptide linker (Gly-Phe-Leu-Gly, GFLG) (8 wt %) is present in FCE28068 (PK1), which is a HPMA-doxorubicin prodrug. Anthracycline-associated drawbacks were overcome by FCE28068, including cardiotoxicity. Polyglutamic acid (PGA) is another polymer which is reported for preparing polymeric drug conjugates. It is biodegradable in nature, and thus its longer chain with high molecular weight can be used safely for formulation of prodrug. Moreover, PGA-conjugated prodrugs have a high drug loading capacity, as revealed in cases of PGA-paclitaxel conjugate (CT-2103, Xyotax, currently in phase III trials) and PGA-camptothecin (CT-2106; currently in initial phase trials) with 37% and 33–35% drug loading capacity, respectively (www.drugs.com; Homsi *et al.,* 2007; Lin *et al.,* 2007).

10.6 ENZYME-ACTIVATED PRODRUGS

Another approach to impart selectivity to the drug release is site-specific activation of prodrugs with the help of the enzymes. A high dose of drug, (i.e., at least 50-fold of the maximum recommended dose), can be administered by these prodrugs, making use of those enzymes which are either specifically expressed in tumours but not in normal tissues, or whose concentration in tumour cells is considerably high. However, only a few human tumours express notably high levels of activating enzymes. To overcome this problem, two strategies *viz.* antibody-directed enzyme prodrug therapy (ADEPT) and gene-directed enzyme prodrug therapy (GDEPT) were developed. Figure 10.3 represents the pictorial image of ADEPT. In this therapy, the enzyme that is responsible to cleave the bond in the prodrug, is adhering/binding to the antibody associated with specific tumour, so that it is directed towards the targeted cancerous mass. After administration of conjugated enzyme, the prodrug is administered, which upon reaching to the target site, gets dissociated by the enzyme in the vicinity of the tumour tissue. Another way is to transport the gene encrypting of the activating enzyme into the tumour cells, earlier than the administration of prodrugs. Since gene is involved in the therapy, this is termed as gene-directed therapy, GDEPT. Figure 10.4 represents the pictorial image of GDEPT. In tumour cells, enzyme- based therapies like GDEPT and ADEPT specifically activate prodrugs, thus minimizing the systemic toxicity allied with the majority of the chemotherapeutic drugs (Mahato *et al.,* 2011).

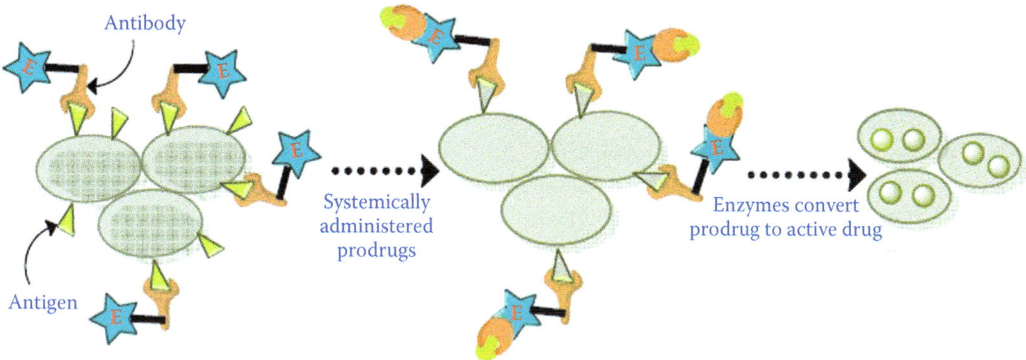

FIGURE 10.3 Schematic diagram of ADEPT.

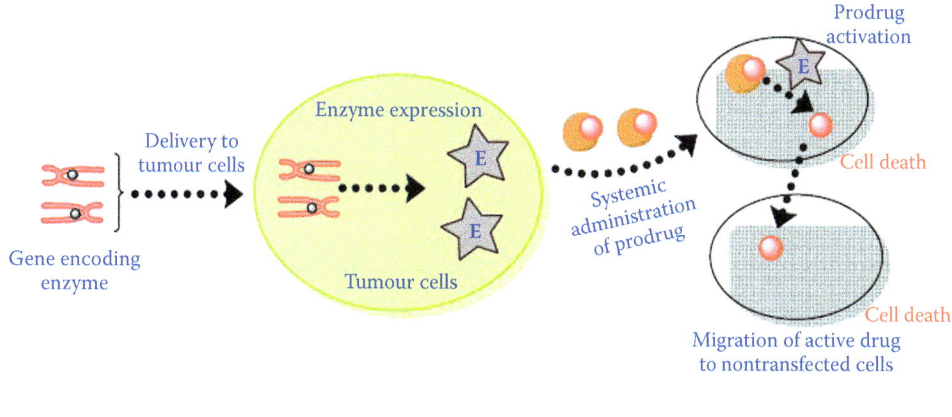

FIGURE 10.4 Schematic diagram of GDEPT.

10.6.1 Antibody-Directed Enzyme Prodrug Therapy

As the name suggests, activating enzyme is specially transported to tumours, followed by administration of prodrug, by making use of a tumour-specific antibody. A fusion protein, comprising of antibody and enzyme, is produced either by chemical means or by making use of recombinant DNA technology. After enzymatic action, the prodrug was converted into an active agent inside the tumour cells enzyme–antibody conjugates. Primarily, the benefit of this strategy is that only single enzyme can generate hundreds of active molecules from a prodrug. Secondly, since the released enzyme can diffuse to the adjacent tumour environment, uptake *via* a single tumour cell can also serve the purpose. Several groups of antibodies, enzymes and prodrugs have been fruitfully implemented for management of different types of human cancers, which are otherwise fairly resistant to the conventional chemotherapy. Besides, there are some drawbacks associated with this therapy like immunogenicity, heterogeneity and the non-site-specific release of active molecule by activation of prodrugs due to unbound enzyme antibodies which are available in the blood. The first shortfall of immunogenicity can be overcome by replacing the murine antibodies and bacterial enzymes with humanized antibodies. This was evident from the studies carried out with the murine anti-TAG-72 mAb (CC49), which was approved by FDA in 1992 (Satummab®) for detection of colorectal and ovarian cancer (www.actip.org). It was found that human anti-mouse-antibody response was induced by murine anti-TAG- 72 mAb (CC49), but no immunogenic response was observed with humanized antiTAG-72 (Fang *et al.*, 2006; Xiao *et al.*, 2005). Additionally, the same group showed that geldanamycin prodrug (17-AG-C2_Gal) is preferentially activated *via* anti-TAG-72 humanized antibody (HuCC49 ΔCH2) conjugated β-galactosidase. The released bioactive was bound to Hsp90 and induced AKT degradation by up to 70% after penetrating into the cancer cells, that improves the anticancer activity nearly by 25-fold *vis-á-vis* the prodrug (Francis *et al.*, 2002).

10.6.2 Gene-Directed Enzyme Prodrug Therapy

Gene-directed enzyme prodrug therapy (GDEPT), also known as gene prodrug activation therapy (GPAT) or virus-directed enzyme/prodrug therapy (VDEPT), is an efficient prodrug approach employed to target the tumour site by transporting an activating enzyme-encoding gene to tumour tissue, prior to the systemic administration of a prodrug. The prodrug gets transformed to the active moiety using the enzyme expressed *via* gene expression. GDEPT and ADEPT are similar in the manner that both make use of a nonendogenous enzyme for the activation of a prodrug. The variation lies in the activation of the prodrug. The prodrug is activated extracellularly in ADEPT, but intracellularly in GDEPT (Han and Amidon, 2000). Therefore, the enzyme–prodrug combination, which is appropriate for ADEPT may not be applicable for GEDPT. GDEPT approach makes use of bacterial or viral enzymes, because a negligible intracellular immune response is induced with these enzymes. These do not possess any corresponding activity in humans and have a few requirements for post-translational modifications. Results from the Clinical trial (phase I) studies for managing breast cancer proved that GDEPT could be safe therapeutic strategies for human cancer patients. The proximal erbB-2 promoter is present in the gene expression vector, which particularly expresses the cytosine deaminase enzyme in erbB-2 positive cancer cells, followed by systemic infusion of the prodrug 5-fluorcytosine. 5-Fluorcytosine gets transformed to 5-fluorouracil, a toxic product within the cancer cells in the presence of cytosine deaminase, which is specifically and selectively found in the cancer cells, followed by significant tumour regression in the patients receiving GEDPT. Moreover, no side effects like, anti-DNA antibody formation, tumour nodule ulceration or local infection insertional mutagenesis were observed following this treatment.

10.7 STIMULI-SENSITIVE PRODRUGS

Stimuli-responsive polymeric systems have emerged as an attractive alternate for prodrug formulations. Self-assembling amphiphilic block copolymers can be used for delivering drug conjugates to the targeted site. The conjugate remains inert throughout the circulation, but is cleaved in the tumour environment in response to specific stimulus. Tumour tissues have lower pH levels (pH 4-6.5) *vis-á-vis* the normal tissues (pH 7.4) (Zhao *et al.*, 2015). Also, lysosomes and intracellular endosomes exhibit slightly acidic pH (5.0–6.0) compared to the extracellular milieu. Inside the cancer cells, the cleavage of bond is initiated due to acidic environment created by endosomes and lysosomes. One such study was carried out by Tu *et al.*, 2015 to improve the delivery of doxorubicin, a hydrophobic drug. The researchers have developed a self-assembled polymeric drug delivery system in which PEG constituted the core, TAT (cell-penetrating peptide), matrix metalloproteinase 2 (MMP2) (peptide linker), doxorubicin as model drug. The formulation, so obtained, possessed several key features, including:

1. Self-assembling of the polymer conjugate in form of nanoparticles
2. Prevention of RES uptake and loss of site specificity *via* PEGylation
3. MMP2-mediated tumour targeting *via* PEG deshielding and exposing TAT
4. Cell penetration *via* TAT
5. TAT-induced endosomal escape
6. TAT-mediated nuclear translocation
7. Prevention of P-glycoprotein mediated drug efflux

These features significantly improved cancer cell internalization, intracellular distribution, tumour targetability and anticancer activity in drug sensitive as well as resistant cancer cells. Moreover, because of high stability of the self-assembled nanoconstructs in the plasma, the PEG-ppTAT-DOX may act as a model for the fabrication and improvement of other anticancer nanomedicines as well. The mechanism of drug release from PEG-ppTAT DOX is shown in Figure 10.5. This nanoconstruct appears to be a potential cargo for site-specific drug delivery. Another example of stimuli-sensitive

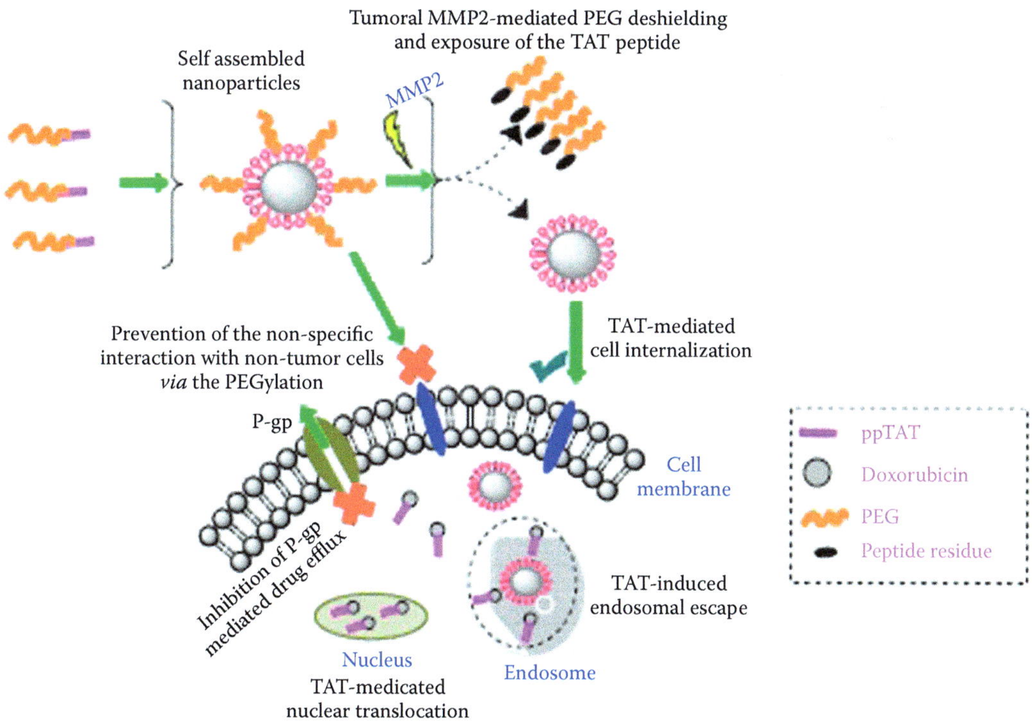

FIGURE 10.5 Drug delivery mechanisms of PEG-ppTAT-DOX.

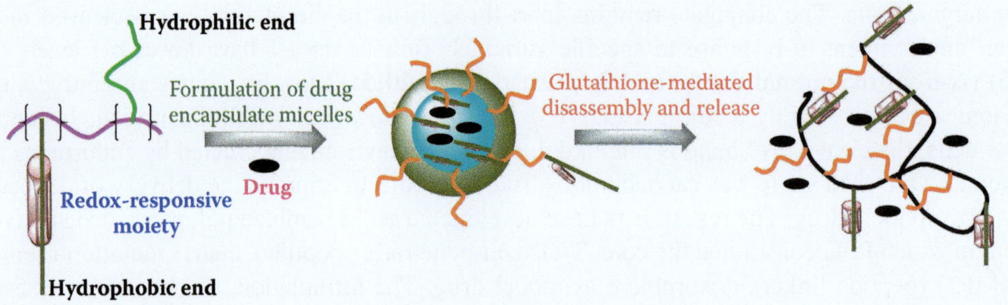

FIGURE 10.6 Amphipilic block polymer with redox-responsive linkage.

doxorubicin prodrug conjugate is reported from Ma *et al.* (2012), who synthesized acid-sensitive labile bond between DOX and Pluronic F127-chitosan (F127-CS). Later, a pH-sensitive polymeric micellar system was formulated, based upon on the conjugated prodrugs (F127-CS-DOX) to deliver paclitaxel (PTX) and DOX simultaneously. The DOX was conjugated with the polymeric backbone and PTX was simultaneously entrapped physically within the lipophilic inner core of the micelles. Spectroscopic studies like 1H-NMR and UV–visible spectrum were used to analyse the conjugated structures, followed by their size and surface morphology analysis. The micelles were nanometres with good polydispersity, and the drug release was observed to follow acid-triggered manner. The mechanism is shown in Figure 10.6.

10.8 LIPID-BASED PRODRUGS

Apart from polymers, fatty acids and cholesterol derivatives, phospholipids or triglycerides are the common lipid materials that are used for covalent conjugation of the drugs in order to form promoieties. Owing to their biocompatibility, these are present as the materials of choice for fabricating nanocarrier materials. Besides, the structural similarity between different lipid-based nanocarriers, lipidic prodrugs and may assist prodrug loading/insertion, e.g. by a superficial nanocarrier membranes–anchoring, as discussed here very briefly:

Methotrexate (MTX) is an antimetabolite anticancer bioactive, which possesses systemic toxicity as well as insurgence of cellular resistance, and thus the efficacy of the treatment involving MTX is compromised. In order to overcome its systemic toxicity, MTX was conjugated torac-1,2-dioleoylglycerol through ester linkages. The consequent prodrug, which is hydrophobic in nature, was formulated using phosphatidylcholine and phosphatidylinositol, ultimately resulting in the formulation of mixed liposomes with a size range of 100 and 150 nm. These mixed liposomes were highly effective in terms of cellular resistance as compared to free MTX, when studied for human leukaemia cells. However, reduction in plasma clotting ability was observed in a dose-dependent manner with these liposomes (Kuznetsova *et al.*, 2012).

Gemcitabine, another drug of choice for treating solid tumours, is associated with certain limitations like (i) short biological half-life, (ii) poor diffusion and (iii) induction of resistance at many levels. Gemcitabine is capped with lipophilic 4-(N)-acyl-derivatives in order to protect it against deamination. This derivatization leads to the formation of the inactive difluorodeoxyuridine (dFdU) metabolite, which allowed incorporation of the drug with high loading into the lipidic liposomal bilayers, whereas free gemcitabine easily escaped from the lophilic compartment (Immordino, 2004; Mura, 2015).

10.9 NANOSTRUCTURED PRODRUGS

Apart from the different types of prodrug systems discussed in the previous sections, nanostructured prodrugs are lately drawing increased attention in the present era, as these combine the advantages

TABLE 10.2
Different Nanostructured Prodrugs for Cancer Therapeutics

Nanostructure(s)	Drug/Complex(s)	Type of Cancer/Model	Remark(s)	Reference
Nanostructured lipid carrier(s)	Gemcitabine (GEM) and Baicalein (BCL)	Prostate Cancer	Nanostructured lipid-based targeted delivery system of HA-GEM-BCL can be considered for the treatment of cancers.	Lu *et al.*, 2017
Prodrug micelles	Deoxycholic acid-doxorubicin-conjugated dextran	Solid tumour	Biocompatable efficient, effective nanocarrier system for the cancer treatment.	Jin *et al.*, 2017
Liposomes	Cisiplatin (IV)-prodrug conjugated phospholipid forming CAT@pt(IV)		Findings suggested that cisplatin-based chemotherapy and catalase induced tumour hypoxia relief together for combined chemo-radiotherapy with great synergistic efficacy; may be a promising system for clinical translation in cancer therapy.	Zhang *et al.*, 2017
Micellar nanoparticles	Cisplatin-poly(ethylene glycol)-block-polycaprolactone	Lung cancer	A novel approach to improve currently available anticancer therapy.	Zhu *et al.*, 2017
Lipid-based hybrid nanoparticle	(DSPE)-doxorubicin	Breast cancer	Result showed that this approach is suitable to systemically combat resistant cancers.	Zhao *et al.*, 2017
Amphiphilic prodrug delivery system	Phosphatidyl-doxorubicin (PX)	Multi-drug resistance tumour	Findings suggested that conjugated system may be used as a promising strategy to formulate nanoscale delivery system for cancer management.	Tao *et al.*, 2017
Theranostic Prodrug Vesicles	Camptothecin and siRNA		The assembled system was utilized as a drug/gene vector to achieve combinational gene therapy and chemotherapy with a synergistic effect, providing an alternative strategy to deliver both prodrugs and therapeutic genes.	Chen *et al.*, 2017
Spherical nanoassemblies (NAs)	Cytarabine (Ara-C)	Leukaemia	Self-assembled HA-Ara spherical nanoassemblies showed immense potential for site specific delivery for leukaemia therapy.	Liu *et al.*, 2017
PEGylated PLGA Nanoparticles	Cisplatin and docetaxel	Lung cancer	Coencapsulation of both drugs in nanoparticles shows significant improvement in cancer therapy as compare to free drug or in combination *in vivo*.	Tian *et al.*, 2017

(Continued)

TABLE 10.2 (CONTINUED)
Different Nanostructured Prodrugs for Cancer Therapeutics

Nanostructure(s)	Drug/Complex(s)	Type of Cancer/Model	Remark(s)	Reference
Nanoparticles	Doxorubicin		Targeted chemo-photodynamic combination therapeutic platform; may be a promising candidate for enhanced cancer therapy.	Zhang *et al.*, 2017
Dendrimer	Gemcitabine	Breast cancer	PEG-peptide dendritic system may be used as safe and efficient nanodrug delivery system in cancer nanomedicine.	Zhang *et al.*, 2017
Self-assembled nanoaggregates	Camptothecin (CPT)		CPT prodrug can be used for the diagnosis and recognition of cancerous cells. Besides, it also provides a capable modality towards safe and effective cancer management.	Xu *et al.*, 2017
Mesoporous silica nanoparticles (MSN)	Ursolic acid	Hepatocellular carcinoma cancer	Ursolic acid engineered MSN prodrug-based approach can be a safe and efficient candidate for anticancer therapy.	Li *et al.*, 2016
Polymersomal mertansine prodrug system	Mertansine (DM1)		Smart nano-prodrug platform with higher drug loading, excellent stability and triggered drug release is reported for site directed melanoma therapy.	Meng *et al.*, 2017

of both nanotechnology and prodrugs. Even though the field of nanotechnology-based chemotherapeutics is being continuously explored by the scientists and researchers, the effective delivery of anticancer bioactives is still a demanding task. Prodrug-based nanostructures possess a number of advantages and therefore, serve as a compelling platform for delivery of chemotherapeutic agents. Table 10.2 summarises various prodrug-based nanocarriers for cancer therapeutics, taking into its ambit the nanosystems based upon self-assembling small molecular weight prodrugs, polymer-drug conjugates and prodrug-encapsulated nanoparticles.

10.10 CONCLUSION

The prodrug approach has emerged as the one of the most effective and promising strategies to augment the therapeutic efficacy, and/or reduces the unfavourable effects of anticancer bioactives *via* different mechanisms. With the successful advent and usage of nanocarriers, the nanostructured prodrug strategy has become an attractive and indispensable part of the drug discovery paradigms, of late. Their importance could be documented by the escalating proportion of the approved new drug entities which, by themselves, are prodrugs. Prodrugs are providing versatile platform to improve the pharmacokinetic and biopharmaceutical properties of diverse nature of drugs. Also, the selective chemical conjugations/modifications are improving the site-directed drug delivery. Despite the phenomenal headway made in the domain of prodrug design, the targeted prodrug issue is still in its infancy. Further extensive studies are, accordingly, called for especially during early stages of the drug discovery, to ascertain their role in cancer management. Immunogenicity, cell specificity and cell permeability are the three major drawbacks confronting the development of efficacious targeted prodrugs. The use of humanized antibodies is confronting in order to address the efficacious and to increase cell permeability; small ligands or antibody fragments can be used to serve the purpose. Furthermore, recognizing target molecules is still a major challenge. Down to the numerous restrictions of intact antibodies, small artificial biomolecules, like aptamers or antibody fragments and peptides, are favoured better for the targeted prodrug approach. Nevertheless, screening for high affinity aptamers and peptides is considered an arduous and time-consuming task. Future identification of molecules that are specific for tumours, with high stability and affinity for the target site, better permeability and lower immunogenicity is pivotal for the successful development of prodrugs for market applications.

REFERENCES

Albert A. Chemical aspects of selective toxicity. *Nature.* 1998; 182: 421–22.

Anbharasi V, Cao N, Feng SS. Doxorubicin conjugated to D-alpha-tocopheryl polyethylene glycolsuccinate and folic acid as a prodrug for targeted chemotherapy. *J Biomed Mater Res A.* 2010; 94(3): 730– 43.

Bertolini A, Ferreri A, Guerzoni S, Tacchi R, Leone S. Paracetamol: New vista of an old drug. *CNS Drug Rev.* 2006; 12: 250–75.

Bhandari KH, Newa M, Yoon SI, Kim JS, Jang KY, Kim JA *et al.* Evaluation of physicochemical properties, skin permeation and accumulation profiles of ketorolac fatty ester prodrugs. *Biol Pharm Bull.* 2007; 30: 2211–16.

Bobeck DR, Schinazi RF, Coats SJ. Advances in nucleoside monophosphate prodrugs as anti-HCV agents. *Antivir Ther.* 2010; 15: 935–50.

Boghaert ER, Sridharan L, Khandke KM, Armellino D, Ryan MG, Myers K. The oncofetal protein, 5T4, is a suitable target for antibodyguided anti-cancer chemotherapy with calicheamicin. *Int J Oncol.* 2008; 32: 221–34.

Charkoudian LK, Pham DM, Franz KJ. A prochelator triggered by hydrogen peroxide inhibits iron-promoted hydroxyl radical formation. *J Am Chem Soc.* 2006; 128: 12424–25.

Chen H, Jia H, Tham HP, Qu Q, Xing P, Zhao J *et al.* Theranostic prodrug vesicles for imaging guided co-delivery of camptothecin and siRNA in synergetic cancer therapy. *ACS Appl Mater Interfaces.* 2017; 9(28): 23536–43.

Das N, Dhanawat M, Dash B, Nagarwal RC, Shrivastava SK. Codrug: An efficient approach for drug optimization. *Eur J Pharm Sci.* 2010; 41: 571–88.

de Groot Vincent FMH. Targeting–Cancer–Small Molecules. (ISBN: 978-0-387-49785-3). *In:* VJ Stella, RT Borchardt, MJ Hageman, R Oliyai, H Maag, JW Tilley, eds., *Prodrugs: Challenges and Rewards Part 1,* Springer, 2007; pp. 447–506.

Dhar S, Gu FX, Langer R, Farokhzad OC, Lippard SJS. Targeted delivery of cisplatin to prostate cancer cells by aptamer functionalized Pt (IV) prodrug-PLGA-PEG nanoparticles. *Proc Natl Acad Sci.* 2008; 105: 17356– 61.

DiJoseph JF, Dougher MM, Kalyandrug LB, Armellino DC, Boghaert ER, Hamann PR. Antitumor efficacy of a combination of CMC-544 (inotuzumabozogamicin), a CD22-targeted cytotoxic immunoconjugate of calicheamicin, and rituximab against non-Hodg kin's B-cell lymphoma. *Clin Cancer Res.* 2006; 12: 242–49.

Drug.com know more. Be sure; Development Status and FDA Approval Process for Opaxio. Available from: https://www.drugs.com/history/opaxio.html

Eldon MA, CMS, Viegas T, Bentley M. NKTR-102 a novel PEGylated-irinotecan conjugate, results in sustained tumor growth inhibition in mouse models of human colorectal and lung tumours that is associated with increased and sustained tumor SN38 exposure. Proceedings of *AACR-NCI-EORTC International Conference on Molecular Targets and Cancer Therapeutics,* October 22–26, 2007a, San Francisco, California, USA, pp. 45.

Eldon MA, Theresa A, Neumann RWo, Cheng L, Viegas TX, Bentley MD. NKTR-118 (oral PEGNaloxol), a PEGylated derivative of Naloxone: Demonstration of selective peripheral opioid antagonism after oral administration in preclinical models. Proceedings of *The American Academy of Pain Management 18th Annual Clinical Meeting.* 2007b; pp. 28.

Fang L, Battisti RF, Cheng H, Reigan P, Xin Y, Shen J *et al.* Enzyme specific activation of benzoquinone ansamycin prodrugs using HuCC4 9DeltaCH2-beta-galactosidase conjugates. *J Med Chem.* 2006; 49 (21): 6290– 97.

FDA approves inotuzumab ozogamicin for relapsed or refractory B-cell precursor ALL, U.S. Food and Drug Administration; U.S. Department of Health and Human Services. Available from: https://www.fda.gov/Drugs/InformationOnDrugs/ApprovedDrugs/ucm572133.htm

Firer MA and Gellerman G. Targeted drug delivery for cancer therapy: The other side of antibodies. *J Hematol Oncol.* 2012; 5: 1–16.

Francis RJ, Battisti RF, Cheng H, Reigan P, Xin Y, Shen J *et al.* A phase I trial of antibody directed enzyme prodrug therapy (ADEPT) in patients with advanced colorectal carcinoma or other CEA producing tumours. *Br J Cancer.* 2002; 87(6): 600–607.

Gires O, Seliger B. Taas and Their Usefulness In: *Tumor-Associated Antigens: Identification, Characterization, and Clinical Applications.* Wiley-Blackwell. 2009; pp. 383.

Graf N, Lippard SJL. Redox activation of metal-based prodrugs as a strategy for drug delivery. *Adv Drug Deliv Rev.* 2012; 64: 993–1004.

Han HK, Amidon GL. Targeted prodrug design to optimize drug delivery. *AAPS Pharm Sci.* 2000; 2: E6.

Homsi J, Simon GR, Garrett CR, Springett G, De Conti R, Chiappori AA *et al.* Phase I trial of poly-L-glutamate camptothecin (CT-2106) administered weekly in patients with advanced solid malignancies. *Clin Cancer Res.* 2007; 13(19): 5855–61.

Huang YF, Shangguan D, Liu H, Phillips JA, Zhang X, Chen Y *et al.* Molecular assembly of an aptamer-drug conjugate for targeted drug deli very to tumor cells. *Chem Biochem.* 2009; 10(5): 862–68.

Huttunen KM, Raunio H, Rautio J. Prodrugs – From serendipity to rational design. *Pharmacol Rev.* 2011; 63: 750–71.

Immordino ML, Brusa P, Rocco F, Arpicco S, Ceruti M, Cattel L. Preparation, characterization, cytotoxicity and pharmacokinetics of liposomes containing lipophilic gemcitabine prodrugs. *J Control Rel.* 2004; 100: 331–46.

Jaracz S, Chen J, Kuznetsova LV, Ojima I. Recent advances in tumor-targeting anticancer drug conjugates, *Bioorg Med Chem.* 2005; 13(17): 5043–54.

Jin R, Guo X, Dong L, Xie E, Cao A. Amphipathic dextran-doxorubicin prodrug micelles for solid tumor therapy. *Colloids Surf B Biointerfaces.* 2017; 158: 47–56.

Khandare JJ, Chandna P, Wang Y, Pozharov VP, Minko T. Novel polymeric prodrug with multivalent components for cancer therapy. *J Pharmacol Exp Ther.* 2006; 317: 929–37.

Kokil GR, Rewatkar PV. Bioprecursor prodrugs: Molecular modification of the active principle. *Mini Rev Med Chem.* 2010; 10: 1316–30.

Kolishetti N, Dhar S, Valencia PM, Lin LQ, Karnik R, Lippard SJ. Engineering of self-assembled nanoparticle platform for precisely controlled combination drug therapy. *Proc Natl Acad Sci.* 2010; 107: 17939–44.

Krop IE, Beeram M, Modi S, Jones SF, Holden SN, Yu W *et al.* A phase 2 study of trastuzumab-dm 1 (t-dm 1), a novel her 2 antibody–drug conjugate, in her 2+ metastatic breast cancer (mbc) patients previously treated with conventional chemotherapy, lapatinib and trastuzumab. *J Clinic Oncol.* 2009; 28: 1–16.

Krop IE, Beeram M, Modi S, Jones SF, Holden SN, Yu W *et al.* Phase I study of trastuzumab-DM1, an HER2 antibody–drug conjugate, given every 3 weeks to patients with HER2-positive metastatic breast cancer. *J Clin Oncol.* 2010; 28: 2698–704.

Kuznetsova NR, Sevrin C, Lespineux D, Bovin NV, Vodovozova EL, Meszaros T *et al.* Hemocompatibility of liposomes loaded with lipophilic prodrugs of methotrexate and melphalan in the lipid bilayer. *J Control Release.* 2012; 160: 394–400.

Li T, Chen X, Liu Y, Fan L, Lin L, Xu Y *et al.* pH-Sensitive mesoporous silica nanoparticles anticancer prodrugs for sustained release of ursolic acid and the enhanced anti-cancer efficacy for hepatocellular carcinoma cancer. *Eur J Pharm Sci.* 2016; 96: 456–463.

Lin NU, Parker LM, Come SE, Burstein HJ, Haldoupis M, Ryabin N *et al.* Phase II study of CT-2103 as first- or second-line chemotherapy in patients with metastatic breast cancer: Unexpected incidence of hypersensitivity reactions *Invest New Drugs.* 2007; 25(4): 369–75.

Liu J, Zhao D, He W, Zhang H, Li Z, Luan Y. Nanoassemblies from amphiphilic cytarabine prodrug for leukemia targeted therapy. *J Colloid Interface Sci.* 2017; 487: 239–49.

Lu Z, Su J, Li Z, Zhan Y, Ye D. Hyaluronic acid-coated, prodrug-based nanostructured lipid carriers for enhanced pancreatic cancer therapy. *Drug Dev Ind Pharm.* 2017; 43(1):160–70.

Mahato R, Tai W, Cheng K. Prodrugs for improving tumor targetability and efficiency. *Adv Drug Deliv Review.* 2011; 63: 659–70.

Meng H, Zou Y, Zhong P, Meng F, Zhang J, Cheng R *et al.* A smart nano-prodrug platform with reactive drug loading, superb stability, and fast responsive drug release for targeted cancer therapy. *Macromol Biosci.* 2017; 17(10): 1600518.

Milowsky MI, Nanus DM, Sheehan CE, Vallabhajosula S, Goldsmith SY, Ross JS *et al.* Vascular targeted therapy with anti-prostate-specific membrane antigen monoclonal antibody J591 in advanced solid tumors. *J Clin Oncol.* 2007; 25: 540–47.

Monoclonal Antibodies Approved by the EMA and FDA for Therapeutic Use (status 2017). Animal cell technology industrial platform. Available from: http://www.actip.org/products/monoclonal-antibodies-approved-by-the-ema-and-fda-for-therapeutic-use/

Morris MJ, Taskat NP, Divgi CR, Bender S, O'Donoghue JA, Nacca A *et al.* Phase I evaluation of J591 as a vascular targeting agent in progressive solid tumors. *Clin Cancer Res.* 2007; 13: 2707–13.

Mura S, Bui DT, Couvreur P, Nicolas J. Lipid prodrug nanocarriers in cancer therapy. *J Control Rel.* 2015; 208: 25–41.

Mylotarg (gemtuzumab ozogamicin): Market Withdrawal cited, Available from: http://www.fda.gov/Safety/MedWatch/SafetyInformation/SafetyAlertsforHumanMedicalProducts/ucm216458.htm (Accessed on 14th June 2017).

Nimjee SM, Rusconi CP, Sullenger BA. Aptamers: An emerging class of therapeutics. *Annu Rev Med.* 2005; 56: 555–83.

NKTR-102 Versus Irinotecan in Patients with Second-Line, Irinotecan-Naïve, KRAS Mutant, Colorectal Cancer; ClinicalTrials.gov Identifier: NCT00856375; US National Library of medicine. Available from: https://clinicaltrials.gov/ct2/show/NCT00856375

Pasut G, Veronese FM. PEG conjugates in clinical development or use as anticancer agents: An overview. *Adv Drug Deliv Rev.* 2009; 61: 1177–88.

Phillips GD, Li G, Dugger DL, Crocker LM, Parsons KL, Mai E *et al.* Targeting HER2-positive breast cancer with trastuzumab-DM1, an antibody-cytotoxic drug conjugate. *Cancer Res.* 2008; 68: 9280–90.

Quiles S, Raisch KP, Sanford LL, Bonner JA, Safavy A. Synthesis and preliminary biological evaluation of high-drug load paclitaxel-antibody conjugates for tumor-targeted chemotherapy. *J Med Chem.* 2010; 53: 586–94.

Rautio J, Kumpulainen H, Heimbach T, Oliyai R, Oh D, Jarvinen T *et al.* Prodrugs: Design and clinical applications. *Nat Rev Drug Discov.* 2008; 7: 255–70.

Rautio J. Prodrug strategies in drug design, in Prodrugs and Targeted Delivery: Towards Better ADME Properties. *Rautio J ed.* Wiley-VCH Verlag GmbH & Co. KGaA, Weinheim. 2010; 1–30.

Research news, Breastcancer.org. Available from: http://www.breastcancer.org/research-news/20130222-5

Sabbatini P, Aghajanian C, Dizon D, Anderson A, Dupont J, Brown JV *et al.* Phase II study of CT-2103 in patients with recurrent epithelial ovarian, fallopian tube, or primary peritoneal carcinoma. *J Clin Oncol.* 2004; 22(22): 4523–31.

Schrama D, Reisfeld RA, Becker JC. Antibody targeted drugs as cancer therapeutics. *Nat Rev Drug Discov.* 2006; 5: 147–59.

Scott MS, Wolchok JD, Lloyd JO. Antibody therapy of cancer. *Nat Rev Cancer.* 2012; 12: 278–87.

Sefah K, Shangguan D, Xiong X, O'Donoghue MB, Weihong Tan. Development of DNA aptamers using Cell-SELEX. *Nat Protoc.* 2010; 5: 1169–85.

Senter PD. Potent antibody drug conjugates for cancer therapy. *Curr Opin Chem Biol.* 2009; 13: 235–44.

Shangguan D, Li Y, Tang Z, Cao ZC, Chen HW, Mallikaratchy P *et al.* Aptamers evolved from live cells as effective molecular probes for cancer study. *Proc Natl Acad Sci.* 2006; 103: 11838–43.

Shi Y, Liu SA, Kerwood DJ, Goodisman J, Dabrowiak JC. Pt (IV) complexes as prodrugs for cisplatin. *J Inorg Biochem.* 2012; 107: 6–14.

Shin BK, Wang H, Yim AM, Naour F, Brichory F, Jang JH *et al.* Global profiling of the cell surface proteome of cancer cells uncovers an abundance of proteins with chaperone function. *J Biol Chem.* 2003; 278: 7607–16.

Sievers EL. Antibody-targeted chemotherapy of acute myeloid leukemia using gemtuzumab ozogamicin (Mylotarg). *Blood Cells Mol Dis.* 2003; 31: 7–10.

Spearman ME, Goodwin RM, Kau D. Disposition of the monoclonal antibody vinca alkaloid conjugate, KS1/4-DAVLB (LY256787), in Fischer 344 rats and rhesus monkeys. *Drug Metab Dispos.* 1987; 15: 640–47.

Srinivas NR. The rationality for using prodrug approach in drug discovery programs for new xenobiotics: Opportunities and challenges. *Eur J Drug Metab Pharmacokinet.* 2011; 36: 49–59.

Stella VJ. Prodrugs: Some thoughts and current issues. *J Pharm Sci.* 2010; 99: 4755–65.

Stohrer M, Boucher Y, Stangassinger M, Jain RK. Oncotic pressure in solid tumors is elevated. *Cancer Res.* 2000; 60: 4251–55.

Tai W, Mahato R, Cheng K. The role of HER2 in cancer therapy and targeted drug delivery. *J Control Rel.* 2010; 146: 264–75.

Tao X, Jia N, Cheng N, Ren Y, Cao X, Liu M, Wei D, Wang FQ. Design and evaluation of a phospholipase D based drug delivery strategy of novel phosphatidyl-prodrug. *Biomaterials.* 2017; 131: 1–14.

Teicher BA. Antibody–drug conjugates targets. *Curr Cancer Drug Targets.* 2009; 9: 82–1004.

Testa B. Prodrugs, bridging pharmacodynamic/pharmacokinetic gaps. *Curr Opin Chem Biol.* 2009; 13: 338–44.

Tian J, Min Y, Rodgers Z, Wan X, Qiu H, Mi Y. Nanoparticle delivery of chemotherapy combination regimen improves the therapeutic efficacy in mouse models of lung cancer. *Nanomedicine.* 2017; 13(3): 1301–7.

Tu Y, Zhu L. Enhancing cancer targeting and anticancer activity by a stimulus-sensitive multifunctional polymer-drug conjugate. *J Control Rel.* 2015; 212: 94–102.

Van Der Velden VH, te Marvelde JG, Hoogeveen PG, Bernstein ID, Houtsmuller AB, Berger MS *et al.* Targeting of the CD33-calicheamicin immunoconjugate Mylotarg (CMA-676) in acute myeloid leukemia: In vivo and in vitro saturation and internalization by leukemic and normal myeloid cells. *Blood.* 2001; 97(10): 3197–204.

Wang W, Tavora F, Sharma R, Eisenberger M, Netto GJ. PSMA expression in Schwannoma: A potential clinical mimicker of metastatic prostate carcinoma. *Urol Oncol.* 2009; 27: 525–28.

Wu KM. A new classification of prodrugs: Regulatory perspective. *Pharmaceutical.* 2009; 2: 77–81.

Xiao J, Horst S, Hinkle G, Cao X, Kocak E, Fang J *et al.* Pharmacokinetics and clinical evaluation of 125I-radio labelled humanized CC49 monoclonal antibody (HuCC49deltaC(H)2) in recurrent and metastatic colorectal cancer patients. *Cancer Biother Radiopharm.* 2005; 20(1): 16–26.

Xu Z, Hou M, Shi X, Gao YE, Xue P, Liu S *et al.* Rapidly cell-penetrating and reductive milieu-responsive nanoaggregates assembled from an amphiphilic folate-camptothecin prodrug for enhanced drug delivery and controlled release. *Biomater Sci.* 2017; 5(3): 444–54.

Yamaoka T, Tabata Y, Ikada Y. Distribution and tissue uptake of poly(ethylene glycol) with different molecular weights after intravenous administration to mice. *J Pharm Sci.* 1994; 83: 601–6.

Yoneda Y, Sebastian CJ, Čapková SK, Mee JM, Ying Liu, Kaufmann GF, Janda KD. A cell-penetrating peptidic GRP78 ligand for tumor cell-specific prodrug therapy. *Bioorg Med Chem Lett.* 2008; 18: 1632–36.

Zawilska JB, Wojcieszak J, Olejniczak AB. Prodrugs: A challenge for the drug development. *Pharmacolo Reports.* 2013; 65: 1–14.

Zhang C, Pan D, Li J, Hu J, Bains A *et al.* Guys N Enzyme-responsive peptide dendrimer-gemcitabine conjugate as a controlled-release drug delivery vehicle with enhanced antitumor efficacy. *Acta Biomater.* 2017; 55: 153–62.

Zhang R, Song X, Liang C, Yi X, Song G, Chao Y *et al.* Catalase-loaded cisplatin-prodrug-constructed liposomes to overcome tumor hypoxia for enhanced chemo-radiotherapy of cancer. Biomaterials. 2017; 138: 13–21.

Zhang Y, Huang F, Ren C, Yang L, Liu J, Cheng Z *et al.* Targeted chemo-photodynamic combination platform based on the dox prodrug nanoparticles for enhanced cancer therapy. *ACS Appl Mater Interfaces.* 2017; 9(15): 13016–28.

Zhao C, Shao L, Lu J, Deng X, Tong Y, Wu Y. Hybrid prodrug nanoparticles with tumor penetration and programmed drug activation for enhanced chemoresistant cancer therapy. *ACS Appl Mater Interfaces.* 2017; 9(22): 18450–61.

Zhou J, Swiderski P, Li H, Zhang J, Neff CP, Akkina R, Rossi JJ. Selection, characterization and application of new RNA HIV gp 120 aptamers for facile delivery of Dicer substrate siRNAs into HIV infected cells. *Nucleic Acids Res.* 2009; 37: 3094–109.

Zhu YH, Sun CY, Shen S, Khan MIU, Zhao YY, Liu Y *et al.* A micellar cisplatin prodrug simultaneously eliminates both cancer cells and cancer stem cells in lung cancer. *Biomater Sci.* 2017; 5(8):1612–21.

11 Osteomyelitis: Therapeutic Management using Nanomedicines

Ashish Ranjan, Teenu Sharma, Atul Jain, Sumant Saini, Shantanu Bandyopadhyay, Shreeniwas Singh, and Bhupinder Singh**

CONTENTS

11.1 Introduction 221
11.1.1 Types of Wounds 221
11.1.1.1 Acute Wounds 221
11.1.1.2 Chronic Wound 222
11.2 Features of Chronic Wound Microenvironment 222
11.3 Osteomyelitis (OSM) 223
11.3.1 Progression of OSM 225
11.3.2 Diagnosis of OSM 225
11.3.3 Antimicrobial Therapy to Treat OSM 226
11.4 Localized Disruption of Biofilms and Triggered Antibiotic Delivery by Hyperthermia 227
11.5 Animal Models of OSM 232
11.6 Nanomedicines for Treating OSM 232
11.7 Conclusions 235
References 235

11.1 INTRODUCTION

Persistent wounds are caused by colonization of injured skin and internal tissues with pathogenic bacteria, resulting in recurring infection (Bowler *et al.*, 2001).

11.1.1 Types of Wounds

11.1.1.1 Acute Wounds

Acute wounds are caused by shallow scratches to deep injuries that harm vein, nerves, muscles or other body parts. Examples include:

- Scraping and rubbing of the skin against rough surfaces
- Poking or jabbing of sharp pointed objects into body tissue
- Cuts with sharp edges or blades
- Tissue tearing with a shear force

* Corresponding authors.

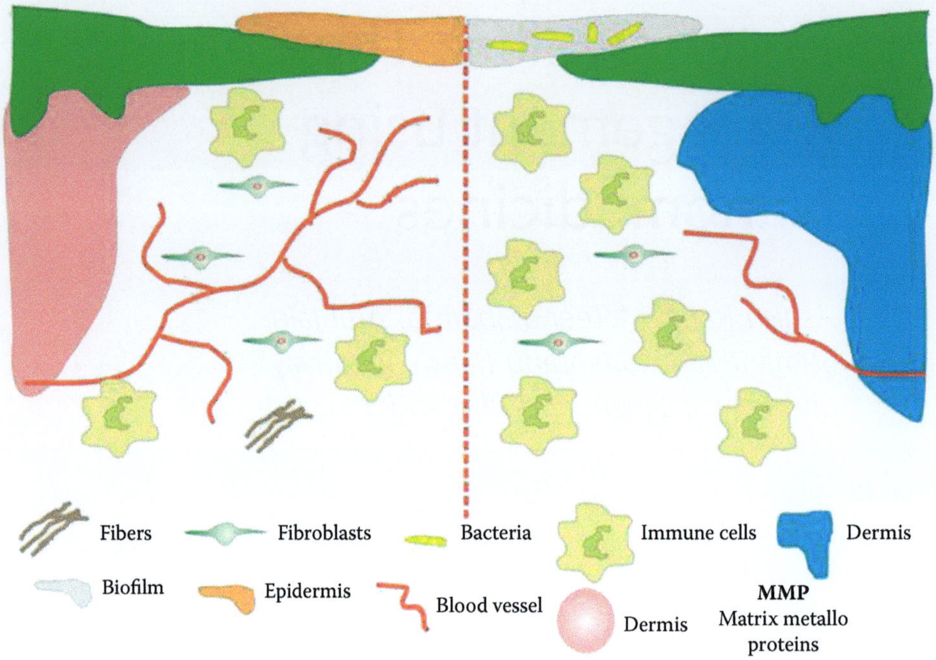

FIGURE 11.1 Pathophysiological comparison of acute and chronic wounds. Left panel shows a schematic of acute wound that typically is well vascularised with efficient tissue repair and immunoclearance capabilities. Right panel shows a chronic wound that typically is minimal to avascular with a high rate of fibrosis, immunomodulation, and biofilm infestations.

11.1.1.2 Chronic Wound

Chronic wounds are defined as infections that fail to clear with conventional antimicrobial treatment (Mustoe *et al.,* 2006). Typically, acute injuries progress to the chronic wound during venous insufficiency, low perfusion, diabetes, and immunosuppression (Menke *et al.,* 2007; Werdin *et al.,* 2009). Figure 11.1 depicts the pathophysiological differences between acute and chronic wounds.

11.2 FEATURES OF CHRONIC WOUND MICROENVIRONMENT

Chronic wound is characterized by raised, hyperproliferative, and a fibrosed microenvironments (Werdin *et al.,* 2009). The microenvironment contains inflammatory cells, cytokines, and mostly lattice metalloproteases (Zgheib *et al.,* 2014). Thus, the mechanisms governing the inflammatory and fibrotic environment is key to instituting new treatment strategies against chronic wounds (Rao *et al.,* 2011).

Lower extremity ulcers are the most frequently occurring chronic wounds, which can either be of vascular or diabetic origin, as depicted in Figure 11.2. The ulcerative infection can migrate to deeper regions such as bone (Bowler *et al.,* 2001), leading to an elevated risk (20–66%) of developing osteomyelitis (OSM) (Malhotra *et al.,* 2014). Chronic wounds require recurring treatment and frequent hospitalizing, significantly raising treatment cost (Sen *et al.,* 2009; Julian *et al.,* 2015).

Chronic infection of bones are multistaged and subdivided into three discrete stages: inflammation, proliferation, and maturation (Sinno and Prakash, 2013). Each phase manifests distinct inflammatory mediators, produced by the specific cells in the wound bed as described below.

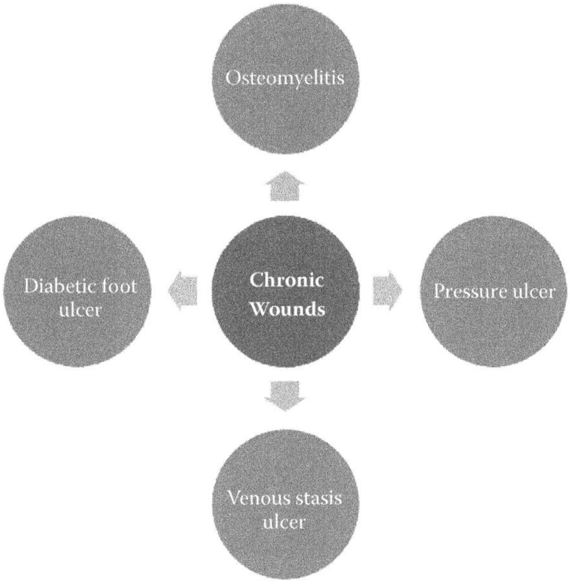

FIGURE 11.2 Chronic wound-related complications.

11.3 OSTEOMYELITIS (OSM)

OSM is the infectious inflammation of the bone. It is one of the oldest documented diseases, with the earliest descriptions dating back to Hippocrates (5th century BC). OSM is an illness that is particularly prevalent among the elderly, diabetics, children, and in 3rd world countries (Schmitt, 2017). Before the advent of antibiotics, the mortality rate following OSM was 25–45%. This rate has reduced significantly in the post-penicillin era, but bacterial clearance in a minority of patients still remain a challenge (Uskokovi, 2015). In 1773, Bromfield portrayed OSM as a bone that "may become carious, first in its internal parts and then from external injury, as well as from a vitiated state of the animal fluids" (Klenerman, 2007). Later, Smith (1831) observed that the most common patients with OSM demonstrate "protracted suffering and great thinness, and in a few cases, fatality". He assigned the term necrosis to indicate bone death. Nelaton (1844) coined the term "osteomyelitis" and defined it as a progressive inflammatory damage of the bone, followed by the obstruction of new bone as part of the reparative process. The word "osteomyelitis" is acquired from the old Greek words *osteon* (which means "bone") and *muelinos* (which means "marrow"), implying it as a contamination of the medullar segment of the bone (Mallikarjun *et al.*, 2011). OSM commonly happens when bacterial inoculation happens with injury, putrefaction, or ischemia of tissue, as well as the nearness of remote material, as shown in Figure 11.3.

OSM usually affects the long bones but may also be found in other locations such as joints, and feet (Mader and Calhoun, 1996). Etiologically, it can:

- Spread from a contiguous source of infection
- Spread from the bloodstream
- Cause secondary OSM resulting from vascular insufficiency

OSM typically accompanies patients suffering from diabetes mellitus (Calhoun *et al.*, 2009) and is characterized by bacterial biofilm formation by *Staphylococcus aureus* and *Pseudomonas aeruginosa* (Bjarnsholt, 2013).

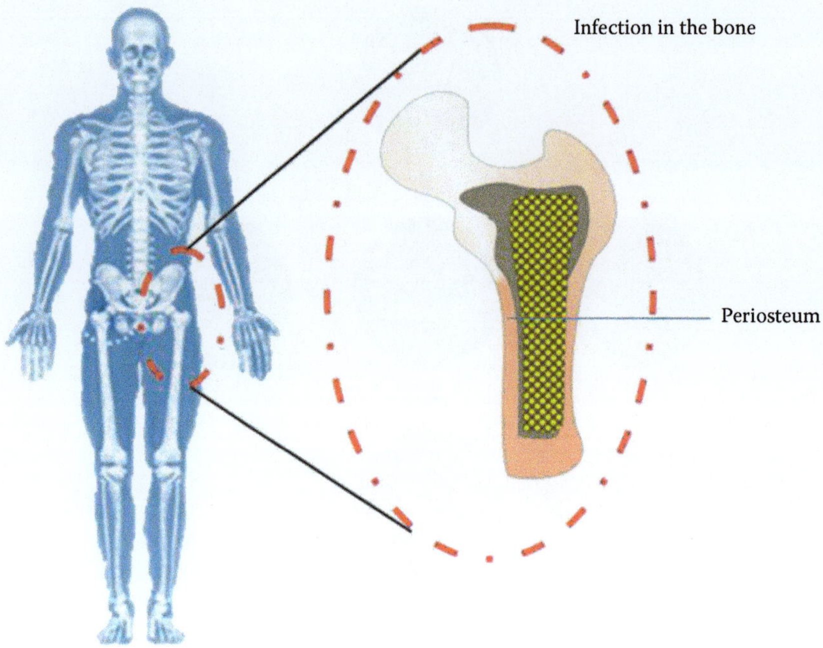

FIGURE 11.3 Infection of the periostium of the shaft of long bones in OSM.

The pathogens remain attached to the necrotic bone and surgical implants, and proliferate rapidly in the glycocalyx matrix, creating degeneration and ambulatory issues, and eventual transition to non-healing OSM (Gogia *et al.*, 2009, Marais *et al.*, 2013).

OSM biofilm is a highly complex environment. Gristina *et al.* (1988) suggested that "host cells set up and incorporate defensive cell layer with practical obstruction mechanism (counting opsonification, phagocytosis, and complement-mediated lysis), while the attacking microbes enter their default development design and set up a biofilm." The layer-like aggregate of microbial cells and extracellular polymeric substances give space to bacterial communication and multiplication, eventually resulting in biofilm formation (Gristina *et al.*, 1988). The biofilm formation occurs in the five different stages: (1) adhesion to the substrate, (2) production of the extracellular matrix, (3) colonization, (4) ripening, and (5) dispersion of bacteria. Bacterial attachment occurs by means of molecules such as adhesins or MSCRAMMs (microbial surface segments perceiving cement network particles), which are particular to host proteins (e.g., fibronectin, laminin, sialoglycoproteins, fibrinogen, and collagen) (Krishnan and Narayana, 2011). Besides, surface tension gradients, hydrophobicity, and electrostatic forces are nonspecific methods that may be involved in biofilm formation. Further, when the microbial density reaches a saturation point, genes are activated to produce an exocellular polysaccharide or glycocalyx. This is defined as quorum sensing, where the microbial settlement is detected by modifying gene expression. This stage of biofilm formation is typically observed in the orthopedic metal implants (Yarwood and Schlievert, 2003). At the end of biofilm formation, dispersion of planktonic bacteria usually takes place (Kaplan, 2010).

Staphylococcus aureus (*S. aureus*), *Enterobacteriaceae*, and *Pseudomonas* species are commonly observed in bone wounds (Gogia *et al.*, 2009). The prevalence of *S. aureus* in such infections is attributed to its several virulence factors, including secreted products, avoidance of the immune system, and adherence factors causing biofilm formation (Zhao *et al.*, 2013). In OSM, bacteria do not differentiate during the initial phase and act as interactive communities rather than as a multicellular organism (Bjarnsholt, 2013). A bacterial sub-populace may develop a phenotypic resistant state to express a biofilm-specific antimicrobial resistance gene. Local tissue invasion is then followed by the production of hydrolases and exotoxins, and growth is supported by antimicrobial resistance and immunosuppressed immune defenses, as shown in Figure 11.4 (Mah, 2012). The mechanisms of biofilm formation

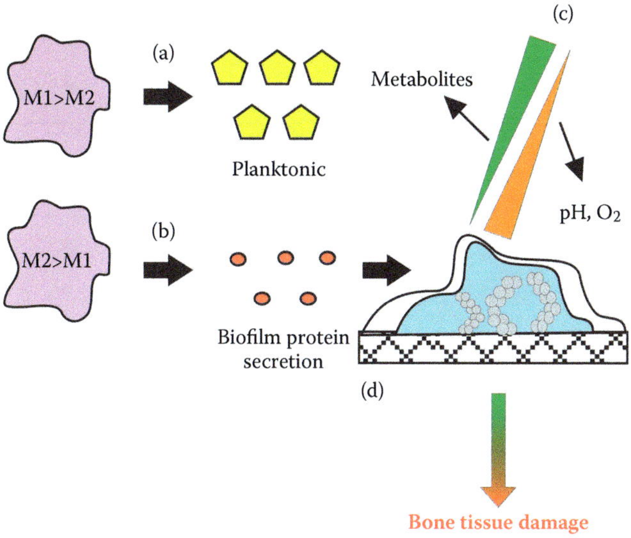

FIGURE 11.4 Process of biofilm formation. Depending on the local cytokine levels, macrophages are directed towards either classical M1 (planktonic pathway), or alternatively activated M2 phenotypes with highly potent antimicrobial and pro-fibrotic properties (M1-type macrophages release cytokines that inhibit the proliferation of surrounding cells and damage contiguous tissue, and M2-type macrophages release cytokines that promote the proliferation of contiguous cells and tissue repair).

are the game of hide and seek. Depending upon the levels of M1 (planktonic pathway; a), and M2 (biofilm forming protein secretion; b). There is an increase in the metabolite concentration accompanied by the fall in oxygen and pH (c). The biofilm as such induces a persistent, anti-inflammatory immune response that favors bacterial persistence (d), which ultimately leads to bone tissue damage.

11.3.1 Progression of OSM

Antibiotics face mechanical and osmotic challenges during penetration in a biofilm, and with the progression of the diseased state, the biofilm-based bacteria become even more resilient (Donlan and Costerton, 2002). Biofilm formation allows for immune evasion *via* the persistence of a small colony variant bacteria within host cells (Grant and Hung, 2013). The persistor bacterial subpopulation is shielded from host defenses and antibiotics, thereby causing recurrence in chronic OSM after the initial infection (Ciampolini and Harding, 2000). OSM is characterized by subtle morphological changes in the bone structure as shown in Figure 11.5. Typically, the interruption of the blood supply to the bone and the death of bone cells is observed (Fondi and Franchi, 2007). Clinical symptoms of OSM is generally nonspecific, and can include, chronic pain, a persistent sinus tract or wound drainage, poor wound healing, malaise, and sometimes fever (Ciampolini and Harding, 2000).

11.3.2 Diagnosis of OSM

Proper diagnosis of bone infections is often cumbersome. The clinical symptoms and inflammatory markers are nonspecific. Examinations *via* imaging are unable to differentiate acute and chronic lesions. Cultures are rarely accurate, and thus bone biopsy is generally considered as a gold standard (Hatzenbuehler and Pulling, 2011). Diagnosis of OSM (Hatzenbuehler and Pulling, 2011) typically begins with the medical examination, and patients are recorded for presentation (intense or low pain), contamination site, and type of bone infected (Calhoun *et al.*, 2009). In acute OSM and in those of hematogenous origin, edema, hyperemia, and fever show up to two weeks after the initial

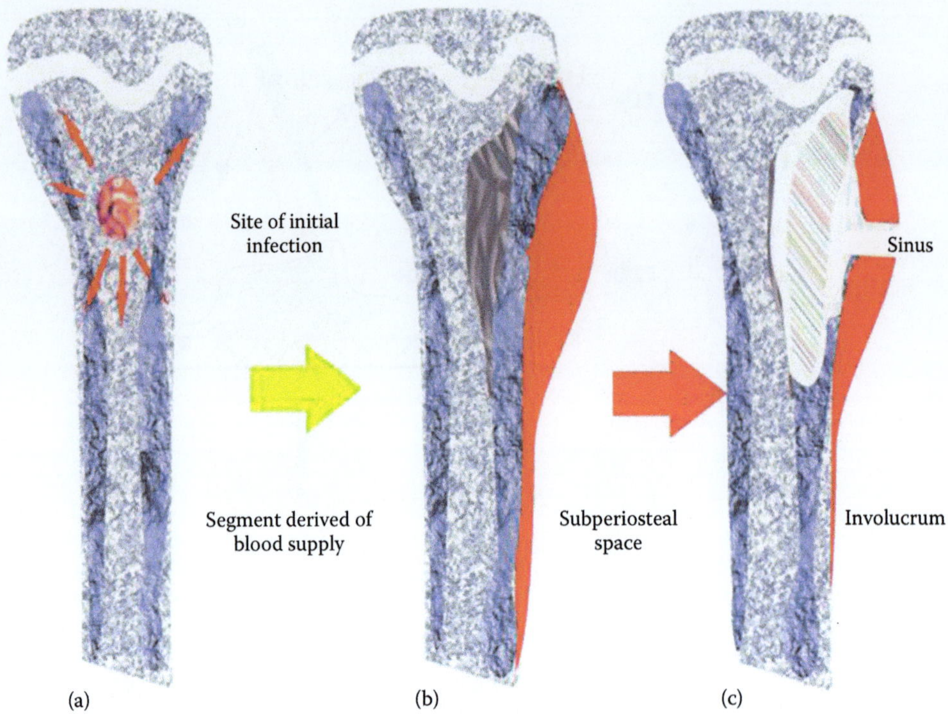

FIGURE 11.5 Progression of osteomyelitis. (a) Lesion initiation in the metaphysis; (b) abscess formation and hypervascularization; (c) the hypervascularized regions eventually sequester, while new bone starts to emerge in the periosteum that is typically perforated by sinuses. Such sinuses can also serve as tract through which the pus escapes in chronic osteomyelitis.

infection (Fan *et al.,* 2014). However, in chronic OSM, overt symptoms are typically absent, and hyperemia, edema, and fistulation show up irregularly or years after the initial bone infection.

11.3.3 Antimicrobial Therapy to Treat OSM

Aptly designed antibiotic regimes are critical for the management of OSM, although targets of OSM therapy may differ in different types and stages of infection. The most significant consideration for antibiotic selection is the mode of action (Fraimow, 2009; Gomes *et al.,* 2013). Treatment targets include remission of the illness for up to one year after the antimicrobial treatment, as mentioned in Table 11.1 (Rao *et al.,* 2011). Infiltration of an antimicrobial into infected tissues relies upon its pharmacological properties, the level of vascularization, physiological states of delicate tissues, and presence of healthy tissues (Gomes *et al.,* 2013; Berbari *et al.,* 2015; Li *et al.,* 2015). With treatment, the prognosis for chronic OSM is generally good, though the outlook is worse for those with long-term chronic OSM, even with surgery. Bone amputation may eventually be required, particularly in individuals with diabetes or poor blood course. Specifically, patients with implants and its subsequent removal may depend on:

- Health status of the person
- Infection type
- Condition of infected prosthesis

Regardless, antibiotics must penetrate and eradicate bacteria residing within the biofilm up to 100-fold greater than the standard bactericidal concentration for complete clearance (Olson *et al.,* 2002), as tabulated in Table 11.2.

TABLE 11.1
Antimicrobials for which Oral Dosing is Bioequivalent to Parenteral Dosing

Broad-spectrum cephalosporin	Ceftriaxone	1–2 g q 24 h
	Cefepime	1–2 g q 8–12 h
	Ceftazidime	2 g q 8 h
Monobactam	Aztreonam	2 g q 8 h
Carbapenem	Imipenem	500 mg q 6 h
	Meropenem	1–2 g q 8 h
	Ertapenem	1 g q 24 h
β-Lactamase inhibitors	Ampicillin-sulbactam	1.5–3 g q 6 h
	Piperacillin-tazobactam	4.5 g q 6 h
Glycopeptide	Vancomycin	1 g q 12 h
	Teicoplanin	400 mg q 24 h
Lipopeptide	Daptomycin	6 mg/kg q 24 h
Glycylcycline	Tigecycline	50 mg q 12 h
Aminoglycoside	Gentamicin	5 mg/kg q 24 h
	Tobramycin	5 mg/kg q 24 h
	Amikacin	15 mg/kg q 24 h
Streptogramin	Quinupristin-dalfopristin	7.5 mg/kg q 8 h
Colistin	Polymyxin	2.5–5 mg kg^{-1} day^{-1}

To improve drug penetration, antibiotic-impregnated beads are particularly utilized for inoculation at the site of debridement in chronic OSM (Walenkamp *et al.*, 1998). Local antibiotic treatment is also less expensive, compared to systemic therapy (Gogia *et al.*, 2009). However, beads are non-biodegradable and require surgical removal, thus increasing the cost of treatment and impacting the quality of life of the patients.

11.4 LOCALIZED DISRUPTION OF BIOFILMS AND TRIGGERED ANTIBIOTIC DELIVERY BY HYPERTHERMIA

Many studies have established that the combination of hyperthermia and chemotherapy promotes the additive or synergistic killing of infecting bacteria (Mackowiak *et al.*, 1982). For instance, Mackowiak and Mading-Cersen (1983) found a progressive increase in antimicrobial activities against both gram-positive and gram-negative bacteria with increasing temperature either in broth medium or serum obtained from patients within the physiological range (37–43 °C). Hajdu *et al.* (2010) reported that incubation with antibiotics (daptomycin, vancomycin, tigecycline, fosfomycin, and cefamandole) at > 40 °C significantly reduced bacterial growth in *S. aureus* biofilms vs. those seen at normal body temperature. Against wounds, a variety of approaches are utilized to achieve targeted heating of *S. aureus*. Table 11.3 enlists various novel approaches used for the same. For instance, heating with superparamagnetic iron oxide nanoparticles (SPIONs) achieves localized heating when combined with an alternating magnetic field. SPIONs (γ-Fe_2O_3, or magnetite, Fe_3O_4) are single biocompatible particles of about 5–20 nm in diameter and can be physically targeted to biofilms with a magnetic field. SPIONs can absorb electromagnetic radiation and efficiently transmit energy in the form of highly localized heat (Kim *et al.*, 2013). Park *et al.* (2011) added a SPION solution to a *Pseudomonas aeruginosa* PA01 biofilm, and temperature was increased by placing the nanoparticles (NP) containing biofilm in the magnetic field. This procedure resulted in a 4-log inactivation of the biofilm, caused mainly by the disintegration of bacterial cell membranes during therapy. Such a drastic reduction was ascribed by the decrease in the intracellular secondary

TABLE 11.2
Antimicrobials Used in OSM along with Oral Dosing and Equivalent Parenteral Dosing

Drug	Class	Usual i.v. Dose	Usual Oral Dose
Ciprofloxacin	Fluoroquinolone	400 mg q 12 h	750 mg q 12 h
Levofloxacin	Fluoroquinolone	500–750 mg q 24 h	500–750 mg q 24 h
Moxifloxacin	Fluoroquinolone	400 mg q 24 h	400 mg q 24 h
Clindamycin	Lincomycin	600–900 mg q 8 h	300–450 mg q 6 h
Linezolid	Oxazolidinone	600 mg q 12 h	600 mg q 12 h
Trimethoprim-sulfamethoxazole	Sulfa-drugs	8–20 mg/kg/day IV q6-12hr	160 mg q 8–12 h
Minocycline	Tetracycline	100 mg q 12 h	100 mg q 12 h
Metronidazole	Nitroimidazole	500 mg q 8 h	500 mg q 8 h
Rifampin	Rifamycin	300 mg q 12 h	600 mg q 24 h or 300 mg q 12 h
Fluconazole	Azole antifungal	400 mg q 24 h	400 mg q 24 h

Antibiotics for which Oral Therapy Is Not Usually Equivalent to Parenteral Dosing

Parenteral Agent	Class	Usual Intravenous Dose	Oral Alternatives
Penicillin G	Penicillin	3–4 million units q 4–6 h	Penicillin, Amoxicillin
Nafcillin	Penicillin	1–2 g q 4–6 h	Dicloxacillin, Cloxacillin
Ampicillin	Penicillin	2 g q 4–6 h	Ampicillin, Amoxicillin
Ampicillin-sulbactam	β-Lactamase inhibitor	1.5–3 g q 6 h	Amoxicillin–Clavulanic acid
Cefazolin	First-generation cephalosporin	2 g q 8 h	Cephalexin, Cefadroxil

messengers such as cyclic di-GMP (c-di-GMP) following temperature increase, which decreased the biomass and increased the bacterial dispersal (Nguyen *et al.*, 2015). In addition to hyperthermia generation, the SPIONs also enhanced treatment sensitivity. In one study, SPIONs were conjugated to metals such as iron, zinc, and silver, and demonstrated superioractivity to that of the antibiotics or metal salts alone against *S. aureus* antibiotic-resistant biofilms (Taylor *et al.*, 2012). These activities were further potentiated in the presence of fructose metabolites that allowed greater uptake of nanoparticles (NPs) by bacteria (Durmus *et al.*, 2013). Thus, SPIONs may be appropriate for remote hyperthermia generation and improving the sensitivity of bacterial killing in wounds.

Other approaches for localized delivery of thermal energy in an organ range from warm water bath immersion (which is anatomically limited and cannot provide fine control of warming), to invasive applicators (which are limited by costly surgical procedures) (Locke *et al.*, 2005). High-intensity focused ultrasound (HIFU) transducers deliver ultrasound (US) waves noninvasively and provides a flexible and powerful tool for clinical control of anatomically- specified heating and drug delivery (Maples *et al.*, 2015).

Our laboratory has combined HIFU with both MRI– and US-guided drug delivery methods (Negussie *et al.*, 2010; Partanen *et al.*, 2012; Ranjan *et al.*, 2012; Bing *et al.*, 2015; Maples *et al.*, 2015). We and others have shown that HIFU effectively and accurately achieve well-tolerated mild local tissue hyperthermia (40–42 °C) within a region of interest (Staruch *et al.*, 2010; Partanen *et al.*, 2012). Rigorous studies have revealed that HIFU–induced local warming decreased resistance within vascular beds, relieved interstitial pressure, and increased vascular permeability, which together, elevated the local intravascular concentrations of systemically administered drugs. Specifically, this approach improved drug coverage and penetration from low temperature–sensitive liposomes (LTSLs). LTSLs are lysolecithin-based nanosystems, and achieve targeted release in the wound, following warming at 40–42 °C (Manzoor *et al.*, 2012; Ranjan *et al.*,2012; de Smet *et al.*, 2013).

TABLE 11.3
Varieties of Approaches Utilized to Achieve Targeted Heating of Osteomyelitis

Drug	Delivery System	Preparation Technique	Attributes	Reference(s)
Aragonite	Cockle shell Nanoparticles	Chemical Synthesis	Release profile exhibited high antibacterial effect against methicillin-resistant *S. aureus* ATCC 29213. The cell proliferation assay showed 80% cell viability of human fetal osteoblast cell line along with optimal antibacterial efficacy, good bone resorbability, and biocompatibility.	(Saidykhan *et al.*, 2016)
Ciprofloxacin	Hydroxyapatite-polycaprolactone nanocomposites	Solvent evaporation	The *in vitro* cell-line study showed a good biocompatibility and bioactivity of the developed nanocomposite film.	(Nithya and Sundaram, 2015)
Gentamycin	PLGA nanoparticles	Precipitation method	Good drug delivery system with sustained antibiotic effect *in vivo*. It was an effective method for the treatment of osteomyelitis.	(Posadowska *et al.*, 2015)
Ciprofloxacin	Hydroxyapatite nanoparticles	*In situ* precipitation	High antibacterial activity against *S. aureus* and *E. coli* that causes osteomyelitis.	(Kumar *et al.*, 2014)
Hydroxyapatite	Chitosan nanoparticles	Ultrasound-assisted sequential precipitation	Positive effect of chitosan coating on the drug elution profile of HAp nanoparticles as carriers for the controlled delivery of antibiotics in the treatment of osteomyelitis.	(Uskoković and Desai, 2014)
Tetracycline	*O*-carboxymethyl chitosan nanoparticles	Ionic crosslinking	*S. aureus* binds and aggregates with Nps, increasing drug concentrations at the infection site. Nps were 6-fold more effective in killing intracellular *S. aureus* compared to Tet alone in HEK-293 and differentiated THP1 macrophage cells.	(Maya *et al.*, 2012)
Teicoplanin	Poly(ethylene glycol) monomethyl ether (mPEG) and PLGA copolymeric thermosensitive hydrogel nanoparticles	Double-emulsification	Implantation of the mPEG–PLGA hydrogel containing teicoplanin was effective for treating osteomyelitis.	(Peng *et al.*, 2010)
Nafcillin	Poly(DL-lactide-co-glycolide) nanoparticles	Single emulsion–solvent evaporation	Effective killing of intracellular *S. aureus* is possible by treating the infected osteoblasts with nanoparticles loaded with nafcillin.	(Pillai *et al.*, 2008)

(*Continued*)

TABLE 11.3 (CONTINUED)
Varieties of Approaches Utilized to Achieve Targeted Heating of Osteomyelitis

Drug	Delivery System	Preparation Technique	Attributes	Reference(s)
Anti-protein A antibodies.	Gold Nanoparticles + Laser killing of bacteria	______	Selective killing of the gram-positive bacterium *S. aureus* by targeting the bacterial surface	(Zharov *et al.*, 2006)

The combination of HIFU and biodegradable LTSLs represents a translational technology for triggered induction of liposomal drug release *via* mild local elevations in tissue temperature (> 40 °C). In cancer models, LTSL/HIFU achieved 5- to 10-fold greater drug delivery in solid tumors with hyperthermia compared to treatment with non-thermosensitive NPs (Negussie *et al.*, 2010; Staruch *et al.*, 2010; Manzoor *et al.*, 2012; Partanen *et al.*, 2012; Ranjan *et al.*, 2012; de Smet *et al.*, 2013; Bing *et al.*, 2015; Maples *et al.*, 2015). Thus, a translation of LTSLs for triggered antibiotic therapy of bone wounds has significant potential (Figure 11.6). To test this idea, we designed ciprofloxacin LTSLs (Wardlow *et al.*, 2016) using a remote loading procedure described by Oh *et al.* (2011). The ability of LTSLs to achieve marked reduction in bacterial viability at 42 °C compared to body temperature (37 °C) was evaluated using an MBEC 96-well plate biofilm system. Between 2.7 and 22 µM ciprofloxacin, LTSLs plus hyperthermia resulted in 60–70% killing of *S. aureus* biofilms compared to LTSLs alone (Figure 11.7). LTSLs alone had relatively lower killing rates (~ 10%) compared to free ciprofloxacin (37 and 42 °C) or LTSLs plus heating (~ 42 °C), thereby suggesting stable encapsulation. The addition of LTSL treatment at 42 °C also induced changes in the biofilm matrix, as shown by scanning electron microscopy (SEM) of *S. aureus* biofilms. The matrix of LTSL–treated biofilms at 37 °C was rough and the bacterial membrane appeared to be uniformly spherical (Figure 11.8a). In contrast, *S. aureus* biofilms that were exposed to 42 °C had areas of the matrix that appeared very smooth, whereas some areas around the edges of the biofilm appeared to be granular and rough (Figure 11.8b). Additionally, the cocci bacteria looked deformed or possibly damaged, as these exhibited a less-than-uniform spherical shape. Although the mechanisms are not known, it is plausible that higher physiological temperature-induced changes in the biofilm matrix and altered bacterial structure and ribosomal subunit function, which, in turn, produced a bactericidal response (Mackowiak *et al.*, 1982). Finally, this approach can be very effective for target-specific antimicrobial delivery, as is evidenced by the 4-fold higher ciprofloxacin concentration in

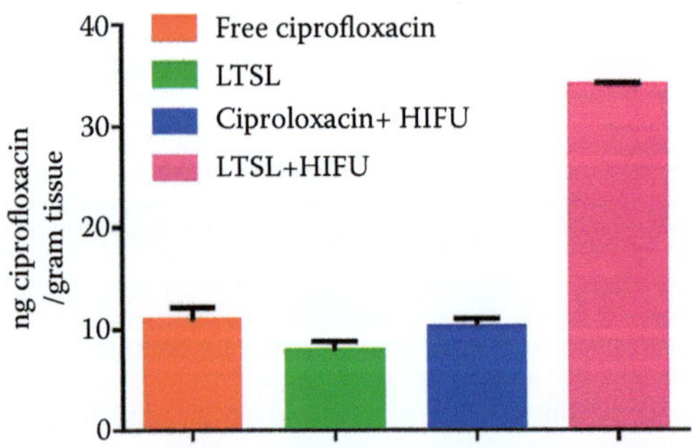

FIGURE 11.6 Ciprofloxacin concentrations from heated and unheated regions of thigh muscles in rats following treatment, either with free ciprofloxacin ± heat (~ 40–42 °C) or LTSL ± heat at a dose of 10 mg ciprofloxacin/kg body weight. Data are shown as mean ciprofloxacin concentration in the muscle with standard error of mean (n = 3, $^{**}p < 0.05$) (Reproduced with permission from Wardlow *et al.*, 2016).

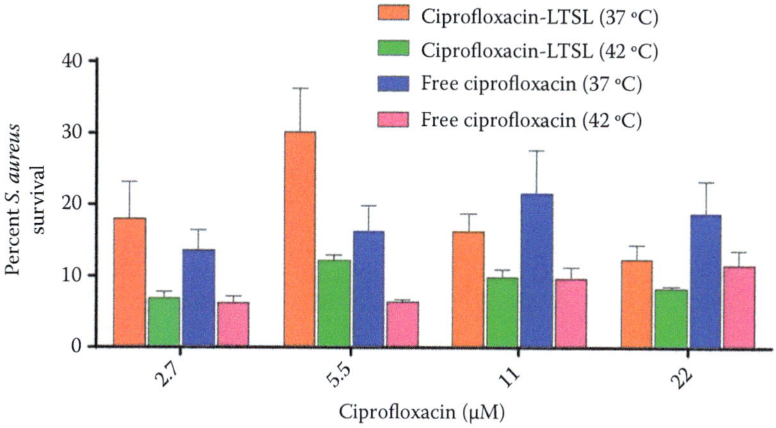

FIGURE 11.7 *S. aureus* survival following treatment of 5-day-old Biofilms with free ciprofloxacin and LTSL at 37 and 42 °C (2.7 –22 μM). Significant killing of *S. aureus* was noted for Cipro-LTSL and free ciprofloxacin at 42 °C compared to body temperature ($P < 0.05$, Tukeys). (Reproduced from Wardlow R *et al.*, *Int J Hyperthermia.* 2016; 32: 254–64. With permission.)

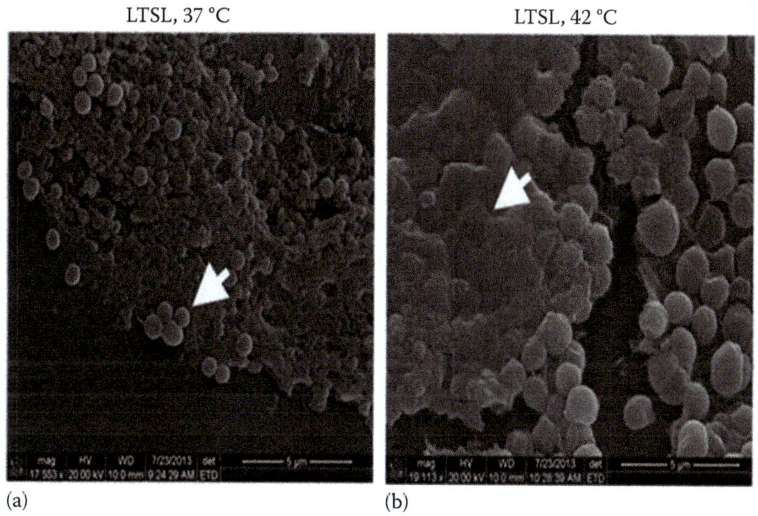

FIGURE 11.8 (a) Biofilms treated at 37 °C with LTSL shows the presence of extracellular matrix surrounding and encasing spherical *S. aureus* (arrow); (b) biofilms treated at 42 °C with LTSL suggest a loss in the granularity of the matrix (arrow), and deformation of bacterial membrane. (Reproduced from Wardlow R *et al.*, *Int J Hyperthermia.* 2016; 32: 254–64. With permission.)

heated muscle *vis-à-vis* the contralateral unheated muscle following HIFU. These findings illustrate the value of LTSLs for localized antimicrobial delivery in chronic wounds, and this area should be actively investigated in future.

Biofilms in wounds can also be disrupted by irradiation with laser shock waves. These waves are created by cavitation bubbles in liquids, as the boiling point in wounds is dramatically decreased in the tensile forces of the shock wave (causing the liquid to vaporize). This process induces asymmetric collapse of bubbles with supersonic speeds, resulting in highly energetic forces for bacterial lysis (Horn *et al.*, 2009). For instance, the treatment of a clinical ear infection isolated from *P. aeruginosa* with fiber-based Q-switched Nd:YAG shockwave laser probes achieved instant oscillation and powerful stress that resulted in removal from the surface where they were attached (Krespi *et al.*, 2008; Taylor *et al.*, 2010; Kasimanickam *et al.*, 2013). Similarly, periapical lesions of post-endodontic therapy were effectively eliminated with a Er:YAG laser at 2.94-μm wavelength (Araki *et al.*, 2006). High-energy extracorporeal shock waves (ESWs) are also effective against resistant *S. aureus* in the infected bones, especially when combined with antimicrobials such as gentamicin to achieve synergistic killing (Wagner *et al.*, 2010). It is likely that ESWs change transmembrane permeability that then triggers greater uptake of antimicrobials by the bacteria. Lasers in combination with an indocyanin green–based photosensitizer (ICG–based PS) can be effective for photodynamic therapy (PDT). Saffarpour *et al.* (2016) estimated the impact of PDT on an *Aggregatibacteractinomycetemcomitans* biofilm on an implant surface. Following exposure to diode laser radiation (810-nm wavelength and 300 mW power) and ICG–based PS, a significant decrease in bacterial counts on the implant surface was noted. This combination was also found to be effective in endodontic treatment in which exposure to ICG with a laser at a specific wavelength produced reactive oxygen species (ROS) that directly damaged subcellular components of *Enterococcus faecalis* (*E. coli*) biofilms (Pourhajibagher *et al.*, 2016). Like ICG, lasers combined with gold (Au) NPs can mediate photothermal therapy (PT). Unlike PDT, PT relies on the targeted destruction of bacteria by heat because the short lifetimes of ROS sometimes are not adequate to damage bacteria in the host (e.g., bone infection). Meeker *et al.* (2016) combined Au nanocages (AuNCs) coated with polydopamine (PDA) containing daptomycin (Dap) to achieve selective killing of methicillin-resistant *S. aureus*. These NPs were also focused on *S. aureus* by conjugating antibodies against staphylococcal protein A (aSpa), generating a photoactivatable, exceptionally specific nanodrug with enhanced efficacy. In summary, spatially controlled delivery of hyperthermia and distribution of drugs within wounds *in vivo* has significant potential for enhancing the efficacy of chronic wound treatment.

11.5 ANIMAL MODELS OF OSM

In order to study OSM clinically, different animal models have been developed for in vivo investigation. Small animal models are easier in handling and are cost-effective, though these are unable to survive therapeutic antibiotic doses and multiple procedures. Larger animals are clinically relevant to therapeutic antibiotic doses and multiple procedures but are limited by their accessibility, costs, and potentially aggressive behavior (Patel *et al.*, 2009). Table 11.4 illustrates the different models adopted for OSM using various causative agents.

11.6 NANOMEDICINES FOR TREATING OSM

Nanotechnology has an important role in diagnostic and therapeutic medicine. Nanocarriers can be engineered to target the diseased cells, which allow direct treatment of those cells (Bowler *et al.*, 2001), thus assisting in reduced toxicity to healthy cells (Wei *et al.*, 2007). Targeted drug delivery using nanoparticles can be translated for the bone disorder. Gelperina *et al.* (2005) enlist the key advantages offered by NPs as follows:

- Deposition of the drug at the site of infection
- Increase in the circulation time of encapsulated antimicrobials
- Enhancement in the solubility profile of hydrophobic drugs
- Surface functionalization of targeting molecules to achieve bacteria-specific delivery

NPs, used for delivery of drugs to treat bone diseases, can be classified as organic and inorganic NPs. Organic NPs mostly include polymer-based (e.g., poly-L-lysine and poly[lactic-co-glycolic acid] PLGA) (Tautzenberger *et al.*, 2012), whereas the inorganic NPs consist of silica-based mesoporous NPs and layered double hydroxides (Oh *et al.*, 2011; Hartono *et al.*, 2012). The drugs used

TABLE 11.4
Different Animal Model Developed of OSM

S. No	Animal	Causative agent	Result	Reference
1.	Rabbit Model	Intravenous injections of *S. aureus*	High mortality and low yields of osteomyelitis.	(Rodet, 1884)
		Sodium morrhuate (5%) and a *S. aureus* suspension	Highly reproducible model with chronic progressive osteomyelitis.	(Scheman *et al.*, 1941)
		Sodium morrhuate (5%) and *S. aureus* suspension	Confirm the necessity of both agents in the formation of osteomyelitis.	(Norden and Kennedy, 1970)
		Giorgio staphylococci and insertion of a stainless-steel pin	Approximately 90% percent of rabbits with both bacterial inoculation and pinning produced clinical, evidence of chronic osteomyelitis.	(Andriole *et al.*, 1973)
		Anaerobic organisms, sponges soaked with 1 mL lactated Ringer solution in their left femurs and 1 mL of 107 colony forming units (cfu) of Bacteroides fragilis.	Bilaterally osteomyelitis confirmed the development of contiguous and hematogenous osteomyelitis.	(Johansson *et al.*, 1991)
		A biopsy needle was inserted into the intramedullary cavity of the proximal femur and 1 mL of bone marrow was replaced with 0.1 mL 5% sodium morrhuate and 0.1 mL of a *Staphylococcus* suspension (Kanin strain, 3_106 cfu/mL)	Osteomyelitis were developed as a result of bone marrow lesion and use of asclerosing agent in combination with a foreign body (leaving the pin *in situ*).	(Schulz *et al.*, 2001)
2.	Rat Model	Sodium morrhuate (5%) followed by a *S. aureus* suspension	Established a chronic infection, yielding a large and stable number of organisms for an extended period of time.	(Zak *et al.*, 1982)
		Fibrin glue and *S. aureus* or fibrin glue alone	Animals treated with bacteria with fibrin glue developed similar to human osteomyelitis in 90% of cases.	(Spagnolo *et al.*, 1993)
		Stainless steel rods were coated with biofilm bacteria (*S. aureus* strain ATCC 29213 slime-producing variety)	Osteomyelitis were developed in all the animal showing that the bacterial suspension used by Garcia is not necessary. In addition, study confirmed the effectiveness of cefuroxime in eradicating biofilm bacteria.	(Garcia *et al.*, 1998)
		Sterile solid K-wires	Acute destructive osteomyelitis was created with as little as 102 cfu/10 mL. Also, the use of sterile solid K-wires simulated a clinical situation in which biofilm formation occurs *in vivo* after foreign body implantation.	(Lucke *et al.*, 2003)

(Continued)

TABLE 11.4 (CONTINUED)
Different Animal Model Developed for OSM

S. No	Animal	Causative Agent	Result	Reference
3.	Mouse Model	Intravenous injection of *S. aureus* (strain LS-1) along the fracture	Explored the knowledge and function of the immune system in osteomyelitis were correlated.	(Chadha *et al.*, 1999)
4.	Avian/ Chick Models	Injection of *S. aureus* (coagulase positive, phage 6/42E/53/77/83A/84)	This study provides a basis to study the natural histologic history of the development of acute hematogenous osteomyelitis. It also provides a method of studying osteomyelitis without the presence of associated trauma. Inexpensive and easily reproducible. These models, though beneficial to study posttraumatic osteomyelitis as seen in adults.	(Emslie and Nade, 1983)
5.	Canine Models	Injection of staphylococci	This model created a disease that was assume to be different from human osteomyelitis and did not allow for the evaluation of infection secondary to foreign body placement.	(Deysine *et al.*, 1976)
		S. aureus and filled with polymethylmethacrylate (PMMA) cement	Evaluate the role of surgical procedures, foreign body reactions, and the effectiveness of local antimicrobial therapy.	(Fitzgerald Jr, 1983)
		S. aureus or *E. coli* was followed by the insertion of a foreign body cylinder (stainless-steel alloy, chromium-cobalt alloy, high-density polyethylene, or PMMA)	The medullary canal of the proximal femur was inoculated with *S. aureus*, *S. epidermidis*, or *Escherichia coli* and was followed by the insertion of a foreign body cylinder (stainless-steel alloy, chromium-cobalt alloy, high-density polyethylene, or PMMA). It was found that insertion of any foreign material significantly increased the risk of infection with *S. aureus*, although only PMMA increased the risk of infection when less than 1000 cfu of *S. aureus* were used. Again, only PMMA increased the risk of infection in femurs infected with only *S. epidermidis* or *E. coli*.	(Petty *et al.*, 1985)
6.	Sheep Models	Tetra decylsodiumsulfate (1 mL 3%) solution and *S. aureus* (PS 8368, phage type I) soaked in gelatin sponge strips	This model uses a sclerosing agent that allows researchers to investigate factors influencing surgical therapy or drug delivery systems.	(Kaarsemaker *et al.*, 1997)
7.	Goat Model	In 28 goats, a 3-mm drill hole was made and injected with 1 mL 5% sodium morrhuate, followed by an injection 10 minutes later with *S. aureus*.	Reproducible and effective in studying chronic osteomyelitis.	(Salgado *et al.*, 2005)

to treat bone diseases mostly include traditional drugs such as antibacterial and chemotherapeutic agents as well as gene therapy reagents. Currently, targeted delivery to bones is mostly achieved using bisphosphonates (BPs) (Clezardin *et al.*, 2010; Sun *et al.*, 2010). These are bone-specific molecules that are anchored onto the NPs for targeted delivery of therapeutic agents to bone tissues (Gu *et al.*, 2013). Fine-tuned bioresorbable nanomaterials for supporting bone recovery has been also attempted to improve bone regeneration (Dvir *et al.*, 2010; Gogia *et al.*, 2009).

As our comprehension of the pathogenesis of bone illnesses at the molecular and cellular level is advancing, utilization of molecular mechanisms and gene therapies with plasmid DNA are viable alternatives (Lu *et al.*, 2011; Rose and Uludag, 2013). NPs are an efficient nonviral vector for gene delivery to chondrocytes, however, in vivo testing are required to establish the feasibility of such delivery system.

The major concern for orthopedic surgeons in OSM is the bone defects and malformations caused by trauma, infection, tumor resection, congenital deformity, and physical and pathological degeneration. Nanotechnology can address these limitations to improve bone regeneration required for healing (Wang and Li, 2016). Typically, nanostructured scaffolds are created by polymer composites or bioactive glass with well-ordered nanosized pores. Bioactive NPs can fundamentally enhance the mechanical quality, mineralization capacity, debasement, and cyto-compatibility of polymer platforms (Polini *et al.*, 2013). Thus, translations of nanocomposites as an alternative to microsized composites can be a handy approach to enhance bone recovery applications.

Alternatively, Yan *et al.* (2006) arranged mesoporous bioactive glasses (MBGs) by combining sol-gel strategy and supramolecular surfactants. Compared with regular non-mesoporous bioactive glasses, the MBGs has an ideal surface region, pore volume, capacity to actuate in vitro apatite mineralization in mimicked body liquids, and superb cytocompatibility (Xia and Chang, 2006). MBGs prepared as 3D porous scaffolds offered improved bone tissue engineering and drug delivery applications than conventional MBGs.

11.7 CONCLUSIONS

OSM is an infective and inflammatory condition of long bones, joints, and the feet. OSM is typically colonized by biofilms, thereby requiring a long duration of antimicrobial treatment and even amputation in some cases. To improve the conventional antimicrobial therapy and achieve biofilm-specific targeting of bone wounds, development of the nano-drug delivery system can be an effective alternative approach. Nanoparticulate drug delivery has several advantages, including high antimicrobial loading efficiencies, stimuli-sensitive release, targeted accumulation in a diseased lesion, and reduced toxicity. Increasing evidence suggests that the combination of nanomedicine with hyperthermia applicators (laser, magnetic, and focused ultrasound devises) can augment the bacterial killing effects, and induce faster clearance of drug resistant and sensitive pathogens. Prior to clinical use, it is critical that nanomedicines are characterized in detail for healthy organ toxicity, as nanomaterials based on silver, zinc oxide, iron oxide, cerium oxide, silicon dioxide, and silica are known to produce free radicals at high dosage to alter DNA. Thus, appropriate surface functionalization and modifications are needed in future works to overcome such barriers. Because of the close proximity of the osteomyelitic tissue to the body surface, nanomedicines applied locally with combinatorial regimens can offer newer possibilities in augmenting bone regeneration and removal of dead tissues through resorption. In general, nanotechnology has significant potential to ease therapy of OSM, yet safe use in humans must be appropriately addressed prior to clinical translation.

REFERENCES

Araki AT, Ibraki Y, Kawakami T, Lage-Marques JL. Er:Yag laser irradiation of the microbiological apical biofilm. *Braz Dent J.* 2006; 17: 296–9.

Berbari EF, Steckelberg JM, Osmon DR, editors. *Osteomyelitis. 8th ed.* Philadelphia, PA: Elsevier Saunders; 2015.

Bing C, Nofiele J, Staruch R, Ladouceur-Wodzak M, Chatzinoff Y, Ranjan A *et al.* Localised hyperthermia in rodent models using an MRI-compatible high-intensity focused ultrasound system. *Int J Hyperthermia.* 2015; 31: 813–22.

Bjarnsholt T. The role of bacterial biofilms in chronic infections. *APMIS Suppl.* 2013; 136: 1–51.

Bowler PG, Duerden BI, Armstrong DG. Wound microbiology and associated approaches to wound management. *Clin Microbiol Rev.* 2001; 14: 244–69.

Calhoun JH, Manring M, Shirtliff M. Osteomyelitis of the long bones. *Semin Plast Surg.* 2009; 23: 59–72.

Ciampolini J, Harding K. Pathophysiology of chronic bacterial osteomyelitis. Why do antibiotics fail so often? *Postgrad Med J.* 2000; 76: 479–83.

Clezardin P, Benzaid I, Croucher PI. Bisphosphonates in preclinical bone oncology. *Bone.* 2010; 49: 66–70.

De Smet M, Hijnen NM, Langereis S, Elevelt A, Heijman E, Dubois L *et al.* Magnetic resonance guided high-intensity focused ultrasound mediated hyperthermia improves the intratumoral distribution of temperature-sensitive liposomal doxorubicin. *Invest Radiol.* 2013; 48: 395–405.

Donlan RM, Costerton JW. Biofilms: Survival mechanisms of clinically relevant microorganisms. *Clin Microbiol Rev.* 2002; 15: 167–93.

Durmus NG, Taylor EN, Kummer KM, Webster TJ. Enhanced efficacy of superparamagnetic iron oxide nanoparticles against antibiotic-resistant biofilms in the presence of metabolites. *Adv Mater.* 2013; 25: 5706–13.

Dvir T, Timko BP, Kohane DS, Langer R. Nanotechnological strategies for engineering complex tissues. *Nat Nanotechnol.* 2010; 6: 13–22.

Fan H, Kim SM, Cho YJ, Eo MY, Lee SK, Woo KM. New approach for the treatment of osteoradionecrosis with pentoxifylline and tocopherol. *Biomater Res.* 2014; 18: 1–10.

Fondi C, Franchi A. Definition of bone necrosis by the pathologist. *Clin Cases Miner Bone Metab.* 2007; 4: 21–6.

Fraimow HS. Systemic antimicrobial therapy in osteomyelitis. *Semin Plast Surg.* 2009; 23: 90–9.

Gelperina S, Kisich K, Iseman MD, Heifets L. The potential advantages of nanoparticle drug delivery systems in chemotherapy of tuberculosis. *Am J Respir Crit Care Med.* 2005; 172: 1487–90.

Gogia JS, Meehan JP, Di Cesare PE, Jamali AA. Local antibiotic therapy in osteomyelitis. *Semin Plast Surg.* 2009; 23: 100–7.

Gomes D, Pereira M, Bettencourt AF. Osteomyelitis: An overview of antimicrobial therapy. *Braz J Pharm Sci.* 2013; 49: 13–27.

Grant SS, Hung DT. Persistent bacterial infections, antibiotic tolerance, and the oxidative stress response. *Virulence.* 2013; 4: 273–83.

Gristina AG, Naylor P, Myrvik Q. Infections from biomaterials and implants: A race for the surface. *Med Prog Technol.* 1988; 14: 205–24.

Gu W, Wu C, Chen J, Xiao Y. Nanotechnology in the targeted drug delivery for bone diseases and bone regeneration. *Int J Nanomedicine.* 2013; 8: 2305–17.

Hajdu S, Holinka J, Reichmann S, Hirschl AM, Graninger W, Presterl E. Increased temperature enhances the antimicrobial effects of daptomycin, vancomycin, tigecycline, fosfomycin, and cefamandole on staphylococcal biofilms. *Antimicrob Agents Chemother.* 2010; 54: 4078–84.

Hartono SB, Gu W, Kleitz F, Liu J, He L, Middelberg AP *et al.* Poly-L-lysine functionalized large pore cubic mesostructured silica nanoparticles as biocompatible carriers for gene delivery. *ACS Nano.* 2012; 6: 2104–17.

Hatzenbuehler J, Pulling TJ. Diagnosis and management of osteomyelitis. *Am Fam Physician.* 2011; 84(9): 1027–33.

Horn C, Gerdesmeyer L, von Eiff C, Gradinger R, Gollwitzer H. Energy-dependent stimulatory and inhibitory effects of extracorporeal shock waves on bacteria and on gentamicin activity. *Med Sci Monit.* 2009; 15: MT77–83.

Jabr-Milane L, van Vlerken L, Devalapally H, Shenoy D, Komareddy S, Bhavsar M *et al.* Multi-functional nanocarriers for targeted delivery of drugs and genes. *J Control Release.* 2008; 130: 121–8.

Julian F, Nadia A, Tracey M, Ijeoma U, Alyson G, Diana W *et al.* Health economic burden that wounds impose on the National Health Service in the UK. *BMJ Open.* 2015; 5: e009283.

Kaplan J. Biofilm Dispersal: Mechanisms, clinical implications, and potential therapeutic uses. *J Dent Res.* 2010; 89: 205–18.

Kasimanickam RK, Ranjan A, Asokan GV, Kasimanickam VR, Kastelic JP. Prevention and treatment of biofilms by hybrid- and nanotechnologies. *Int J Nanomedicine.* 2013; 8: 2809–19.

Kim MH, Yamayoshi I, Mathew S, Lin H, Nayfach J, Simon SI. Magnetic nanoparticle targeted hyperthermia of cutaneous *Staphylococcus aureus* infection. *Ann Biomed Eng.* 2013; 41: 598–609.

Klenerman L. A history of osteomyelitis. *J Bone Joint Surg Br.* 2007; 89: 667–70.

Krespi YP, Stoodley P, Hall-Stoodley L. Laser disruption of biofilm. *Laryngoscope.* 2008; 118: 1168–73.

Krishnan V, Narayana SV. Crystallography of gram-positive bacterial adhesins. *Adv Exp Med Biol.* 2011; 715: 175–95.

Li HK, Scarborough M, Zambellas R, Cooper C, Rombach I, Walker AS *et al.* Oral versus intravenous antibiotic treatment for bone and joint infections (OVIVA): Study protocol for a randomized controlled trial. *Trials.* 2015; 16: 583.

Locke J, Zeug A, Thompson DJ, Allan J, Mazzarella K, Novak P *et al.* Localized versus regional hyperthermia: Comparison of xenotransplants treated with a small animal ultrasound system and water bath limb immersion. *Int J Hyperthermia.* 2005; 21: 271–81.

Lu HD, Zhao HQ, Wang K, Lv LL. Novel hyaluronic acid-chitosan nanoparticles as non-viral gene delivery vectors targeting osteoarthritis. *Int J Pharm.* 2011; 420: 358–65.

Mackowiak PA, Marling-Cason M. Hyperthermic enhancement of serum antimicrobial activity: Mechanism by which fever might exert a beneficial effect on the outcome of gram-negative sepsis. *Infect Immun.* 1983; 39: 38–42.

Mackowiak PA, Marling-Cason M, Cohen RL. Effects of temperature on antimicrobial susceptibility of bacteria. *J Infect Dis.* 1982; 145: 550–3.

Mader JT, Calhoun J (1996). Bone, joint, and necrotizing soft tissue infections. In: Baron S, ed., *Medical Microbiology* (4th edition). Galveston (TX): University of Texas Medical Branch at Galveston; Chapter 100.

Mah TF. Biofilm-specific antibiotic resistance. *Future Microbiol.* 2012; b7: 1061–72.

Malhotra R, Chan CSY, Nather A. Osteomyelitis in the diabetic foot. *Diabet Foot Anklea.* 2014; 5: 10.

Mallikarjun K, Kohli A, Kumar A, Tanwar A. Chronic suppurative osteomyelitis of the mandible. *J Indian Soc Pedod Prev Dent.* 2011; 29: 176–9.

Manzoor AA, Lindner LH, Landon CD, Park JY, Simnick AJ, Dreher MR *et al.* Overcoming limitations in nanoparticle drug delivery: Triggered, intravascular release to improve drug penetration into tumors. *Cancer Res.* 2012; 72: 5566–75.

Maples D, McLean K, Sahoo K, Newhardt R, Venkatesan P, Wood B *et al.* Synthesis and characterisation of ultrasound imageable heat-sensitive liposomes for HIFU therapy. *Int J Hyperthermia.* 2015; 31: 674–85.

Marais LC, Ferreira N, Aldous C, Le Roux TLB. The pathophysiology of chronic osteomyelitis. *SA Orthop J.* 2013; 12: 14–8.

Meeker DG, Jenkins SV, Miller EK, Beenken KE, Loughran AJ, Powless A *et al.* Synergistic photothermal and antibiotic killing of biofilm-associated staphylococcus aureus using targeted antibiotic-loaded gold nanoconstructs. *ACS Infect Dis.* 2016; 2: 241–50.

Menke NB, Ward KR, Witten TM, Bonchev DG, Diegelmann RF. Impaired wound healing. *Clin Dermatol.* 2007; 25: 19–25.

Mustoe TA, O'Shaughnessy K, Kloeters O. Chronic wound pathogenesis and current treatment strategies: A unifying hypothesis. *Plast Reconstr Surg.* 2006; 117: 35S–41S.

Negussie AH, Yarmolenko P, Partanen A, Ranjan A, Jacobs G, Woods D *et al.* Formulation and characterization of magnetic resonance imaginable thermally sensitive liposomes for use with magnetic resonance-guided high intensity focused ultrasound. *Int J Hyperthermia.* 2010; 27: 140–55.

Nelaton A. Elements de pathologiechirurgicale. 1844; 1844–59.

Nguyen TK, Duong HT, Selvanayagam R, Boyer C, Barraud N. Iron oxide nanoparticle-mediated hyperthermia stimulates dispersal in bacterial biofilms and enhances antibiotic efficacy. *Sci Rep.* 2015; 5: 18385.

Oh JM, Park CB, Choy JH. Intracellular drug delivery of layered double hydroxide nanoparticles. *J Nanosci Nanotechnol.* 2011; 11: 1632–5.

Olson ME, Ceri H, Morck DW, Buret AG, Read RR. Biofilm bacteria: Formation and comparative susceptibility to antibiotics. *Can J Vet Res.* 2002; 66: 86–92.

Park H, Park HJ, Kim JA, Lee SH, Kim JH, Yoon J *et al.* Inactivation of Pseudomonas aeruginosa PA01 biofilms by hyperthermia using superparamagnetic nanoparticles. *J Microbiol Methods.* 2011; 84: 41–5.

Partanen A, Yarmolenko PS, Viitala A, Appanaboyina S, Haemmerich D, Ranjan A *et al.* Mild hyperthermia with magnetic resonance-guided high-intensity focused ultrasound for applications in drug delivery. *Int J Hyperthermia.* 2012; 28: 320–36.

Patel M, Rojavin Y, Jamali AA, Wasielewski SJ, Salgado CJ. Animal models for the study of osteomyelitis. In: Seminars in plastic surgery (Vol. 23, No. 2). Thieme Medical Publishers; 2009; pp.148.

Polini A, Bai H, Tomsia AP. Dental applications of nanostructured bioactive glass and its composites. *Wiley Interdiscip Rev Nanomed Nanobiotechnol.* 2013; 5: 399–410.

Pourhajibagher M, Chiniforush N, Shahabi S, Ghorbanzadeh R, Bahador A. Sub-lethal doses of photodynamic therapy affect biofilm formation ability and metabolic activity of Enterococcus faecalis. *Photodiagnosis Photodyn Ther.* 2016; 15: 159–66.

Ranjan A, Jacobs GC, Woods DL, Negussie AH, Partanen A, Yarmolenko PS *et al.* Image-guided drug delivery with magnetic resonance guided high intensity focused ultrasound and temperature sensitive liposomes in a rabbit Vx2 tumor model. *J Control Release.* 2012; 158: 487–94.

Rao N, Ziran BH, Lipsky BA. Treating osteomyelitis: Antibiotics and surgery. *Plast Reconstr Surg.* 2011; 127(1): 177S–87S.

Rose L, Uludag H. Realizing the potential of gene-based molecular therapies in bone repair. *J Bone Miner Res.* 2013; 28: 2245–62.

Saffarpour A, Fekrazad R, Heibati MN, Bahador A, Saffarpour A, Rokn AR *et al.* Bactericidal effect of erbium-doped yttrium aluminum garnet laser and photodynamic therapy on aggregatibacteractinomycetemcomitans biofilm on implant surface. *Int J Oral Maxillofac Implants.* 2016; 31: e71–8.

Schmitt SK. Osteomyelitis. *Infect Dis Clin North Am.* 2017; 31: 2: 325–38.

Sen CK, Gordillo GM, Roy S, Kirsner R, Lambert L, Hunt TK *et al.* Human skin wounds: A major and snowballing threat to public health and the economy. *Wound Repair Regen.* 2009; 17: 763–71.

Sinno H, Prakash S. Complements and the wound healing cascade: An updated review. *Plast Surg Int.* 2013; 7: 1–7.

Smith N. Medical and surgical memoirs. Published by William A. Francis. Baltimore, 1831:97.

Staruch R, Chopra R, Hynynen K. Localised drug release using MRI-controlled focused ultrasound hyperthermia. *Int J Hyperthermia.* 2010; 27: 156–71.

Sun M, Iqbal J, Singh S, Sun L, Zaidi M. The crossover of bisphosphonates to cancer therapy. *Ann NY Acad Sci.* 2010; 1211: 107–12.

Tautzenberger A, Kovtun A, Ignatius A. Nanoparticles and their potential for application in bone. *Int J Nanomedicine.* 2012; 7: 4545–57.

Taylor EN, Kummer KM, Durmus NG, Leuba K, Tarquinio KM, Webster TJ. Superparamagnetic iron oxide nanoparticles (SPION) for the treatment of antibiotic-resistant biofilms. *Small.* 2012; 8: 3016–27.

Taylor ZD, Navarro A, Kealey CP, Beenhouwer D, Haake DA, Grundfest WS *et al.* Bacterial biofilm disruption using laser generated shockwaves. *ConfProc IEEE Eng Med Biol Soc.* 2010; 2010: 1028–32.

Uskokovi V. Nanostructured platforms for the sustained and local delivery of antibiotics in the treatment of osteomyelitis. *Crit Ther Drug Carrier Syst.* 2015; 32: 1–59.

Uskoković V, Desai TA. In vitro analysis of nanoparticulate hydroxyapatite/chitosan composites as potential drug delivery platforms for the sustained release of antibiotics in the treatment of Osteomyelitis. *J Pharm Sci.* 2014; 103(2): 567–79.

Wagner MS, Ernst J, Maier M, Chiquet M, Joos H, Müller PE, Jansson V, Sievers B, Hausdorf J. The effect of high-energy extracorporeal shock waves on hyaline cartilage of adult rats in vivo. *J Orthop Res.* 2010; 28(8):1050-6.

Walenkamp GH, Kleijn LL, deLeeuw M. Osteomyelitis treated with gentamicin-PMMA beads: 100 patients followed for 1-12 years. *Acta Orthop Scand.* 1998; 69: 518–22.

Wang X, Li W. Biodegradable mesoporous bioactive glass nanospheres for drug delivery and bone tissue regeneration. *Nanotechnology.* 2016; 27: 225102.

Wardlow R, Bing C, VanOsdol J, Maples D, Ladouceur-Wodzak M, Harbeson M *et al.* Targeted antibiotic delivery using low temperature-sensitive liposomes and magnetic resonance-guided high-intensity focused ultrasound hyperthermia. *Int J Hyperthermia.* 2016; 32: 254–64.

Wei C, Yamato M, Wei W, Zhao X, Tsumoto K, Yoshimura T *et al.* Genetic nanomedicine and tissue engineering. *Med Clin North Am.* 2007; 91: 889–98.

Werdin F, Tennenhaus M, Schaller HE, Rennekampff HO. Evidence-based management strategies for treatment of chronic wounds. *Eplasty.* 2009; 9: e19.

Xia W, Chang J. Well-ordered mesoporous bioactive glasses (MBG): A promising bioactive drug delivery system. *J Control Release.* 2006; 110: 522–30.

Yan X, Huang X, Yu C, Deng H, Wang Y, Zhang Z *et al.* The in-vitro bioactivity of mesoporous bioactive glasses. *Biomaterials.* 2006; 27: 3396–403.

Yang L, Webster TJ. Nanotechnology controlled drug delivery for treating bone diseases. *Expert Opin Drug Deliv.* 2009; 6: 851–64.

Yarwood JM, Schlievert PM. Quorum sensing in Staphylococcus infections. *J Clin Invest.* 2003; 112: 1620–5.

Zgheib C, Xu J, Liechty KW. Targeting inflammatory cytokines and extracellular matrix composition to promote wound regeneration. *Adv Wound Care.* 2014; 3: 344–55.

Zhao G, Usui ML, Lippman SI, James GA, Stewart PS, Fleckman P *et al.* Biofilms and inflammation in chronic wounds. *Adv Wound Care (New Rochelle).* 2013; 2: 389–99.

12 Nanostructured Lipid Carrier-Mediated Methotrexate Delivery Evokes Transcription Factors to Induce Selective Apoptosis in Rheumatoid Arthritis

Naisargee Patel, Rajeev K. Tyagi, and Neeraj K. Garg*

CONTENTS

12.1 Introduction 239
12.1.1 Rheumatoid Arthritis: An Inflammatory Disorder 239
12.1.2 Causes of Rheumatoid Arthritis 240
12.1.3 Pathophysiology of Rheumatoid Arthritis 240
12.1.4 Molecular Understanding of Rheumatoid Arthritis 240
12.1.5 Available Drugs and Vaccines 241
12.2 Definition of the Problem 241
12.3 Lipid Nanoparticles As Drug Delivery Carriers in Inflammatory Diseases 242
12.4 Nanostructured Lipid Carriers Developed for Transdermal Delivery of MTX 242
12.4.1 Preparation of MTX-Loaded NLCs 242
12.4.2 Characterization of MTX-Loaded Nanostructured Lipid Carriers 243
12.4.3 Incorporation of MTX-Loaded NLCs in Hydrogel for Transdermal Delivery 244
12.4.4 MTX-Loaded NLCs: An Effective Delivery Vehicle 244
12.5 Conclusion and Future Perspectives 244
Acknowledgments 244
References 244

12.1 INTRODUCTION

12.1.1 Rheumatoid Arthritis: An Inflammatory Disorder

Rheumatoid arthritis (RA; i.e., watery, inflamed joints) is an autoimmune disorder, leading to chronic inflammation that causes swelling of the tissue around the joints, also termed as immune-mediated inflammatory disease. The disease-induced inflammation in deep-seated tissues, leads to deformity through loosening joint ligaments and the destruction of cartilage and bone.

As per the current statistics, approximately 1.3 million people in United States are afflicted by the disease, and 0.5–1% of the total world population (about 50 in 100,000 people) are reportedly

* Corresponding author.

affected annually (Jacobsson *et al.*, 1994; Cimmino *et al.*, 1998; Simonsson *et al.*, 1999; Riise *et al.*, 2000; Symmons *et al.*, 2002; Carbonell *et al.*, 2008). The pathology can cause permanent damage to the joints. Occurrence of rheumatoid arthritis is three times more prevalent in women than men, and commonly starts between the ages of 40–60 (Srivastava *et al.*, 2017). It can also affect children in a condition known as juvenile idiopathic arthritis.

The framework of disease progression includes immunological as well as inflammatory pathways that include feed-forward pathways that are autopropagating in nature (Jacobsson *et al.*, 1994).

12.1.2 Causes of Rheumatoid Arthritis

The exact causes and mechanisms of the development of RA is still unknown. The presence of "rheumatoid factor," which is a group of autoantibodies in patients' blood is a characteristic feature of RA. Scientific studies have revealed various factors contributing in the development of the condition to which genetic factors contribute almost 50% (Scott *et al.*, 2010). Some studies show the causative factor as some herpes virus infections: Epstein-Barr virus (EBV) and human herpes virus 6 (HHV-6) (Alvarez-Lafuente *et al.*, 2005). The other nongenetic risk factors include smoking and vitamin D deficiency (Scott *et al.*, 2010; Gatenby *et al.*, 2013).The presence of *Porphyromonas gingivalis* in the oral mucosa in periodontal disease is also associated with RA (Koziel *et al.*, 2014). The enzyme class peptidyl arginine deaminase expressed in these bacteria causes "citrullination" of the peptides in the mouth, significantly contributing to the pathogenesis of the condition. Increase in the number of *Prevotella* species in the gut, altering the overall gut microbiome, is one of the consequences of early RA (Scher *et al.*, 2013).

The bacterial environment of the gut is reported to affect the susceptibility and severity of arthritis in preclinical models (Block *et al.*, 2016). The increase in proinflammatory cytokines due to epigenetic changes, such as hypomethylation of DNA, dysregulated histone marks, or expression of microRNAs, may lead to increased risk of RA (Bottini and Firestein, 2013).

12.1.3 Pathophysiology of Rheumatoid Arthritis

TNF-α-, IL1- and IL17-induced bone destruction is a characteristic feature of RA. When immune cells are activated, autoantibodies, as well as inflammatory cytokines, are produced, and a cascade of inflammation occurs that results in the formation of pannus. This pannus invades and destroys cartilage and bone. Inflammatory mediators released in the affected joints cause joint damage and other systemic complications (Rindfleisch and Müller 2005; Dewing *et al.*, 2012). Hypertrophy of the synovium, along with the heavy infiltration of immune cells causes tumor-like conditions and bone erosion. The synovial lining cells and chondrocytes cause degradation of tissue and cartilage by production of proteolytic enzymes (collagenase, stromelysin), leading to impaired integrity and water content (Dewing *et al.*, 2012).

12.1.4 Molecular Understanding of Rheumatoid Arthritis

The decreased apoptosis of synoviocytes (macrophages, fibroblasts and lymphocytes) may be considered as one of the most critical factor contributing to the pathogenesis of RA. It leads to synovial hyperplasia, further resulting in inflammation. The activation of signalling pathways, including nuclear factor kappa-B(NF-kB), phosphatidylinositol 3 kinase/Akt-1 and signal transducer, and activator of transcription-3 pathways, lead to the expression of genes that cause inflammation and variety of antiapoptotic molecules, including FLICE inhibitory protein, Bcl-2 and Mcl-1 (Djerbi *et al.*, 2001). Therefore, the induction of apoptosis of synoviocytes by suppressing these signalling pathways or inhibiting expression of antiapoptotic molecules may be introduced to develop a potential therapeutic interventional approach for RA.

In RA joints, TNF-α contributes to inflammation and acts as an antiapoptotic signal in conjunction with FasL. Subsequently, production of proinflammatory cytokines (IL1, IL17, TNF-α) by synoviocytes, further triggers the cascade of inflammatory events (Pelletier *et al.*, 2016).

Fas and FasL, which are highly expressed on synovial lining macrophages, contribute in the recurrence of the disease *via* insufficient apoptosis. One of the other factors that may contribute in insufficient apoptosis in the joints is the presence of Flip in synovial fibroblasts. Moreover, NF KB, PI3K/Akt1 and STAT3 are activated and lead to expression of inflammatory mediators resulting in the expression of antiapoptotic molecules (Hongtao *et al.*, 2003).

One of the other common pathophysiological dysfunctions observed in RA is anaemia of chronic disease (ACD). The patients with RA and ACD show reduced numbers of $CD34^+$ progenitor cells, increased apoptotic $CD34^+$ cells, increased $FasL^+$ cells, and increased TNF-α in the supernatants from cultured cells of bone-marrow; which can further be reversed by incorporation of anti-TNF-α antibodies (Papadaki *et al.*, 2002a, 2002b). Thus, it can be concluded that the ACD observed in RA is the resultant of increase in TNF- α and FasL induced apoptosis of progenitor cells. The development of more effective therapies against RA requires better understanding of the mechanisms that regulates apoptosis (Pelletier *et al.*, 2016).

12.1.5 Available Drugs and Vaccines

So far, therapeutic interventional strategies have not proven effective against RA, as various treatments to control inflammation including medications, exercise and surgery are verily the available prophylactic strategies. Recently, there was report which claimed for a vaccine against RA, after successful clinical trials (www.healthline.com). This vaccine provides a ray of hope for safe and effective treatment of RA.

One of the major classes of drugs used to subside inflammation in RA is nonsteroidal anti-inflammatory drugs (NSAIDs). These drugs are widely used to reduce inflammation mounted by RA, which reportedly inhibit cyclooxygenase (Cox) enzymes (Pelletier *et al.*, 2016). Some of the commonly used NSAIDs include drugs like aspirin, celecoxib, diclofenac, diflunisal, etodolac, fenoprofen, flurbiprofen and ibuprofen. However, the rise in the risk of heart attack due to the prolonged use of NSAIDs has limited their use for treating RA (Pelletier *et al.*, 2016).

Another class of drugs used for slowing down the progression of joint damage in RA are called disease-modifying drugs (DMARDs). These drugs slow down the progression of RA by acting on the immune system. Some of the milder DMARDs with minimum side effects include hydroxychloroquine, methotrexate (MTX) and sulfasalazine. MTX is the most commonly used DMARD, but causes side effects such as rash and stomach upset similar to other DMARDs. Also, it can prove toxic to the liver or bone marrow, may cause birth defects and causes shortness of breath in a few cases. Regular blood examination is necessary while taking MTX. However, use of folic acid helps to reduce some of its side effects. It can, however, be used for extended periods safely and can be given to children as well, which proves to be its greatest advantage. The biologics are administered to the patients who do not show the effects of these drugs. These are genetically-engineered proteins obtained from human genes. They inhibit the components of the immune system which induce inflammation. Further, steroids are also given to the patients to inhibit inflammation. However, cost effectivity and availability of methotrexate proves it as the preferred drug for treating RA (Jain *et al.*, 2015).

12.2 DEFINITION OF THE PROBLEM

Methotrexate (MTX) is well known for its anti-inflammatory, anti-neoplastic, as well as folate antimetabolite activity, and is used to treat psoriasis, cancer and rheumatoid arthritis (RA) (Paulus, 1990; Amarji *et al.*, 2015; Jain *et al.*, 2015). The oral administration of MTX has been commonly used for treating RA (Paulus, 1990; Weinblatt *et al.*, 1991).

It inhibits the production of proinflammatory cytokines such as interferon (IFN-γ), tumour necrosis factor (TNF-α), interleukin (IL)-1β, IL-6, IL-15, and IL-18, as well as joint-destructive enzymes *viz.* inducible nitric oxide synthase (iNOS), NOS2 and cyclooxygenase-2 (COX-2) during chronic inflammation in joints (Feldmann *et al.*, 1996; Choy and Panayi, 2001; Zwerina *et al.*, 2005; Lee *et al.*, 2012;) and therefore, slows down the rate of joint destruction (Feldmann *et al.*, 1996; Choy and Panayi, 2001; Cutolo *et al.*, 2001; Zwerina *et al.*, 2005). However, continued usage of MTX has been reported to have adverse effects, including mucosal ulceration, stomatitis, bone marrow suppression, loss of appetite, drug-induced hepatic fibrosis and cirrhosis, as already discussed (Frank and Alan, 2004). These adverse effects are reported to a decrease of number of patients receiving the treatment approximately by 30% (van Rossum, 1998).

These limitations of MTX treatment have led to its transdermal mode of delivery, bypassing the gastrointestinal tract, leading to improved MTX therapeutic potential by its accumulation at the inflammation sites (Prausnitz *et al.*, 2004; Prausnitz and Langer, 2008). However, stratum corneum (SC) which is the hardest outermost layer of the skin, served as the main physical barrier for delivery of MTX (hydrophilic molecule, log P = –1.85) for this mode of delivery (Steele *et al.*, 1979; Bos and Meinardi, 2000).

12.3 LIPID NANOPARTICLES AS DRUG DELIVERY CARRIERS IN INFLAMMATORY DISEASES

Various novel drug delivery systems like liposomes (transepidermal) (Ali *et al.*, 2008), deformable liposomes (dermal) (Trotta *et al.*, 2004), niosomes (Abdelbary and AbouGhaly, 2015), ethanolic liposomes (Dubey *et al.*, 2007), solid-lipid-oil nanocarriers (Yang *et al.*, 2012) and temperature-responsive lipidic nanogels (Samah *et al.*, 2010; Singka *et al.*, 2010) have been developed for the optimal transdermal delivery of MTX (Antunes *et al.*, 2017). Methods to enhance penetration of MTX by transdermal mode, like iontophoresis (Alvarez-Figueroa *et al.*, 2001; Prasad *et al.*, 2007) and electroporation (Wong *et al.*, 2006), have also been attempted. However, the methods employing lipidic nanocarriers are gaining more importance as drug delivery options for the delivery of the active agents for the management of anti-inflammatory disorders, including RA (Modi *et al.*, 2017).

12.4 NANOSTRUCTURED LIPID CARRIERS DEVELOPED FOR TRANSDERMAL DELIVERY OF MTX

Solid lipid nanoparticles (SLNs) are reported to show a few limitations like poor drug loading, drug leakage during storage and lipid polymorphisms. The driving force to overcome these limitations, resulted in the advent of nanostructured lipid carriers (NLCs), which have now become prevalent for improved skin penetration and permeation of drugs (Müller *et al.*, 2002; Souto *et al.*, 2004; Souto and Müller, 2005; Müller *et al.*, 2007). Besides, NLCs are also reported for lower systemic cytotoxicity in comparison to SLNs (Müller *et al.*, 2007; Andrade *et al.*, 2014). Recent studies have shown further improved transdermal drug delivery through engineered NLCs (Fan *et al.*, 2013; Garg *et al.*, 2016a; Garg *et al.*, 2017). The summary of the same has been shown in the graphical abstract (Figure 12.1).

12.4.1 Preparation of MTX-Loaded NLCs

MTX-loaded NLCs were prepared by the hot microemulsion method *viz.* the heat-mediated melting of MTX and a lipid mixture (SA+Gelucire®50/13+Transcutol® P), followed by addition of ethanolic solution of phospholipid and Tween-80. Water was added to form primary microemulsion, and heated with stirring at 60–70°C. The suspension was sonicated for 15 seconds at 3 W by a probe sonicator (Sonicator 3000, Misonix) to form hot primary microemulsion. This was then added to surfactant solution and homogenized using high shear homogenizer (Heidolph, Silent Crusher M, Germany), followed by magnetic stirring (Remi, Mumbai, India). Any residual-free drugs were removed by dialysis of NLCs dispersion against double distilled

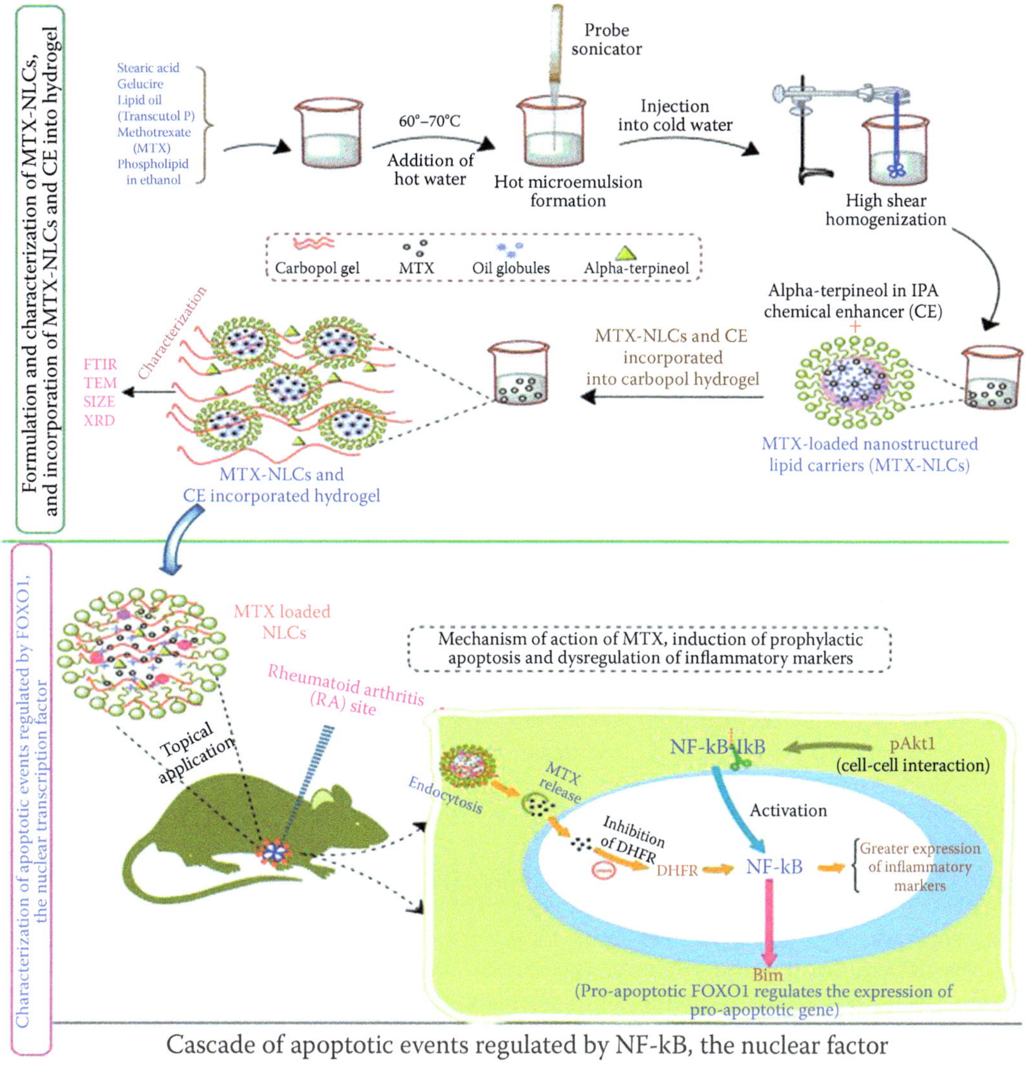

FIGURE 12.1 Diagrammatic representation showing the formulation and use of nanostructured lipid carriers for transdermal delivery of methotrexate to treat rheumatoid arthritis. (Adapted from Garg *et al.*, 2016.)

water mixed with dimethyl formamide (DMF) in the volume ratio 2:1 in a cellulose dialysis bag (MWCO 10 kDa).

12.4.2 Characterization of MTX-Loaded Nanostructured Lipid Carriers

The prepared NLCs were characterized for their size and polydispersity index by a dynamic light scattering method, as well as for their zeta potential, based on their mobility under an electric field using Malvern Zetasizer. The transmission electron microscopy (TEM) was conducted for morphological characterization the formulated nanoparticles. Drug encapsulation efficiency (EE) was calculated using direct lysis method. Further, chemical structure of MTX-loaded NLCs was validated by spectroscopy Fourier transform infrared (FTIR) spectroscopy and X-ray diffraction (XRD) analysis.

12.4.3 Incorporation of MTX-Loaded NLCs in Hydrogel for Transdermal Delivery

The MTX-loaded NLCs were incorporated in a hydrogel form, which can be applied topically for effective transdermal delivery. A 80:20 mixture of MTX-NLCs dispersion and 2% (v/v) solution of α-terpineol in 70% isopropanol was added into concentrated, previously hydrated and neutralized (with tri-ethanolamine) Carbopol®934 gel 1% (w/v) gel (Jain *et al.*, 2015). The rheological properties of NLC-gel were studied using a rheometer (Anton Paar, MCR-102; Austria) and the texture profile analysis was done using a TA.XT Plus texture analyser (Stable Microsystems, UK).

12.4.4 MTX-Loaded NLCs: An Effective Delivery Vehicle

Drug release study of MTX-loaded NLCs and MTX-NLC gel formulation was done by *in vitro* dialysis method. The analysis of cellular uptake of NLCs, MTX-loaded NLCs and MTX-NLC gel formulation was done using cell uptake studies in HaCaT cells. The expression profile of different genes involved in apoptosis was estimated by PCR micro array in U937 cells mimicking inflammation (human monocytic cell line) to assess the therapeutic potential of NLCs, MTX-NLCs and MTX-NLC gel formulation. Induction of apoptosis by NF-kB and FOXO1 pathways that regulated the expression of gene Bim (proapoptotic) was studied by immunocytochemistry (ICC) and immunofluorescence assay (IFA). Further, inhibition of apoptosis of RA cells by cell–cell interaction was shown. Association of NF-kB with its inhibitor IkB remains intact in cytoplasm without cell–cell interaction. On the other side, FOXO1 in the nucleus regulates Bim. We had observed the greater rate of apoptosis of RA cells which was regulated by nuclear factors, NF-kB and FOXO1 (Garg *et al.*, 2016b).

12.5 CONCLUSION AND FUTURE PERSPECTIVES

The present study was aimed for development of simple method based on microemulsion to develop MTX-loaded lipid-based nanoformulation for effective transdermal delivery of MTX and selective apoptosis of RA cells. A novel apoptotic pathway regulated by NF-kB and FOXO1 to subside inflammation mounted by RA pathology, when MTX was delivered through NLCs following transdermal route was discovered. Here, we report for the first time NLCs prepared by novel lipid mixture (SA+Gelucire and transcutol P) for transdermal delivery of MTX. Delivery of MTX via NLCs may be explored to see whether production of regulatory T cells is associated with the suppression of host immunity during RA pathology.

ACKNOWLEDGMENTS

The corresponding author would like to acknowledge Department of Biotechnology (DBT), India, for providing research fellowship as well as Georgia Regents University, USA for financial support to carry out the research.

REFERENCES

Abdelbary A, AbouGhaly M. Design and optimization of topical methotrexate loaded niosomes for enhanced management of psoriasis: Application of Box-Behnken design, *in-vitro* evaluation and *in-vivo* skin deposition study. *Int J Pharm.* 2015; 485: 235–243.

Ali MF, Salah M, Rafea M, Saleh N. Liposomal methotrexate hydrogel for treatment of localized psoriasis: preparation, characterization and laser targeting. *Med Sci Monit.* 2008; 14: 66–74.

Alvarez-Figueroa MJ, Delgado-Charro MB, Blanco-Mendez J. Passive and iontophoretic transdermal penetration of methotrexate. *Int J Pharm.* 2001; 212: 101–107.

Alvarez-Lafuente R, Fernández-Gutiérrez B, de Miguel S, Jover JA, Rollin R, Loza E, Clemente D, Lamas JR. Potential relationship between herpes viruses and rheumatoid arthritis: Analysis with quantitative real time polymerase chain reaction. *Ann Rheum Dis.* 2001; 64(9): 1357–1359.

Amarji B, Garg NK, Singh S, Katare OP. Microemulsions mediated effective delivery of methotrexate hydrogel: More than a tour de force in psoriasis therapeutics. *J Drug Target.* 2015; 1–14.

Andrade LM, de Fatima Reis C, Maione-Silva L, Anjos JL, Alonso A, Serpa RC, Marreto RN, Lima EM, Taveira SF. Impact of lipid dynamic behavior on physical stability, *in vitro* release and skin permeation of genistein-loaded lipid nanoparticles. *Eur J Pharm Biopharm.* 2014; 88: 40–47.

Antunes A, Pereira P, Reis C, Rijo P, Reis C. Nanosystems for skin delivery: from drugs to cosmetics. *Curr Drug Metab.* 2017; 18(5): 412–425.

Block KE, Zheng Z, Dent AL, Kee BL, Huang H. Gut microbiota regulates K/BxN autoimmune arthritis through follicular helper T but Not Th17 cells. *J Immunol.* 2016; 196: 1550.

Bos JD, Meinardi MM. The 500 Dalton rule for the skin penetration of chemical compounds and drugs. *Exp Dermatol.* 2000; 9: 165–169.

Bottini N, Firestein GS. Epigenetics in rheumatoid arthritis: A primer for rheumatologists. *Curr Rheumatol Rep.* 2013; 15: 372.

Carbonell J, Cobo T, Balsa A, Descalzo MA, Carmona L. The incidence of rheumatoid arthritis in Spain: Results from a nationwide primary care registry. *Rheumatology.* 2008; 47: 1088–1092.

Choy EH, Panayi GS. Cytokine pathways and joint inflammation in rheumatoid arthritis. *N Engl J Med.* 2001; 344: 907–916.

Cimmino MA, Parisi M, Moggiana G, Mela GS, Accardo S. Prevalence of rheumatoid arthritis in Italy: The Chiavari Study. *Ann Rheum Dis.* 1998; 57: 315–318.

Cutolo M, Sulli A, Pizzorni C, Seriolo B, Straub RH. Anti-inflammatory mechanisms of methotrexate in rheumatoid arthritis. *Ann Rheum Dis.* 2001; 60: 729–735.

Dewing K, Setter S, Slusher B. Osteoarthritis and rheumatoid arthritis 2012: Pathophysiology, diagnosis, and treatment. *Clinical advisor.* 2012; 15: 44.

Djerbi M, Darreh-Shori T, Zhivotovsky B, Grandien A. Characterization of the human FLICE-inhibitory protein locus and comparison of the anti-apoptotic activity of four different flip isoforms. *Scand J Immunol.* 2001; 54: 180–189.

Dubey V, Mishra D, Dutta T, Nahar M, Saraf DK, Jain NK. Dermal and transdermal delivery of an antipsoriatic agent via ethanolic liposomes. *J Control Release.* 2007; 123: 148–154.

Fan X, Chen J, Shen Q. Docetaxel-nicotinamide complex-loaded nanostructured lipid carriers for transdermal delivery. *Int J Pharm.* 2013; 458: 296–304.

Feldmann M, Brennan FM, Maini RN. Role of cytokines in rheumatoid arthritis. *Annu Rev Immunol.* 1996; 14: 397–440.

Frank CS, Alan MM. Methotrexate and psoriasis in the era of new biologic agents. *J Am Acad Dermatol.* 2004; 50: 301–309.

Garg NK, Singh B, Tyagi RK, Sharma G, Katare OP. Effective transdermal delivery of methotrexate through nanostructured lipid carriers in an experimentally induced arthritis model. *Colloids Surf B Biointerfaces.* 2016a; 147: 17–24.

Garg NK, Tyagi RK, Sharma G, Jain A, Singh B, Jain S *et al.* Functionalized lipid-polymer hybrid nanoparticles mediated codelivery of methotrexate and aceclofenac: a synergistic effect in breast cancer with improved pharmacokinetics attributes. *Mol. Pharmaceutics.* 2017; 14(6): 1883–1897.

Garg NK, Tyagi RK, Singh B, Sharma G, Nirbhavane P, Kushwah V *et al.* Nanostructured lipid carrier mediates effective delivery of methotrexate to induce apoptosis of rheumatoid arthritis via NF-rm kappa b and FOXO1. *Int. J. Pharm.* 2016b; 499(1–2): 301–320.

Gatenby P, Lucas R, Swaminathan A. Vitamin D deficiency and risk for rheumatic diseases: An update. *Curr Opin Rheumatol.* 2013; 25(2): 184–191.

Healthline News. Available online at: http://www.healthline.com/health-news/vaccines-could-be-answer-for-conquering-rheumatoid-arthritis-062315 (Accessed on 12 July, 2017).

Jacobsson LT, Hanson RL, Knowler WC, Pillemer S, Pettitt DJ, McCance DR *et al.* Decreasing incidence and prevalence of rheumatoid arthritis in Pima Indians over a twenty-five-year period. *Arthritis Rheum.* 1994; 37: 1158–1165.

Jain A, Jain A, Garg NK, Tyagi RK, Singh B, Katare OP, Webstere TJ, Soni V. Surface engineered polymeric nanocarriers mediate the delivery of transferrin methotrexate conjugates for an improved understanding of brain cancer. *Acta Biomater.* 2015; 24: 140–151.

Koziel J, Mydel P, Potempa J. The link between periodontal disease and rheumatoid arthritis: An updated review. *Curr Rheumatol Rep.* 2014; 16: 408.

Lee SW, Kim JH, Park MC, Park YB, Chae WJ, Morio T, LeeDH, Yang SH, Lee SK. Alleviation of rheumatoid arthritis by cell-transducible methotrexate upon transcutaneous delivery. *Biomaterials.* 2012; 33: 1563–1572.

Modi Y, Xiangru F, Jianxun D, Fei C, Xuesi C. Nanotherapeutics relieve rheumatoid arthritis. *Journal of Controlled Release.* 2017; 252: 108–124.

Müller RH, Petersen RD, Hommoss A, Pardeike J. Nanostructured lipid carriers (NLC) in cosmetic dermal products. *Adv Drug Deliv Rev.* 2007; 59: 522–530.

Müller RH, Radtke M, Wissing SA. Nanostructured lipid matrices for improved microencapsulation of drugs. *Int J Pharm.* 2002; 242: 121–128.

Papadaki HA, Kritikos HD, Gemetzi C, Koutala H, Marsh JC, Boumpas DT, Eliopoulos GD. Bone marrow progenitor cell reserve and function and stromal cell function are defective in rheumatoid arthritis: Evidence for a tumor necrosis factor alpha-mediated effect. *Blood.* 2002a; 99: 1610–1619.

Papadaki HA, Kritikos HD, Valatas V, Boumpas DT, Eliopoulos GD. Anemia of chronic disease in rheumatoid arthritis is associated with increased apoptosis of bone marrow erythroid cells: Improvement following anti-tumor necrosis factor-alpha antibody therapy. *Blood.* 2002b, 100: 474–482.

Paulus HE. The use of combinations of disease-modifying antirheumatic agents in rheumatoid arthritis. *Arthritis Rheum.* 1990; 33: 113–120.

Pelletier JP,Pelletier JM, Rannou F, Cooper C. Efficacy and safety of oral NSAIDs and analgesics in the management of osteoarthritis: Evidence from real-life setting trials and surveys. *Seminars in Arthritis and Rheumatism.* 2016; 45: 22–27.

Prasad R, Koul V, Anand S, Khar RK. Effect of DC/mDC iontophoresis and terpenes on transdermal permeation of methotrexate: *In vitro* study. *Int J Pharm.* 2007; 333: 70–78.

Prausnitz MR, Langer R. Transdermal drug delivery. *Nat Biotechnol.* 2008; 26: 1261–1268.

Prausnitz MR, Mitragotri S, Langer R. Current status and future potential of transdermal drug delivery. *Nat Rev DrugDiscov.* 2004; 3: 115–124.

Riise T, Jacobsen BK, Gran JT. Incidence and prevalence of rheumatoid arthritis in the county of Troms, northern Norway. *J Rheumatol.* 2000; 27: 1386–1389.

Rindfleisch JA, Muller D. *Am Fam Physician.* 2005; 72: 1036–1047.

Samah NA, Williams N, Heard CM. Nanogel particulates located within diffusion cell receptor phases following topical application demonstrates uptake into and migration across skin. *Int J Pharm.* 2010; 401: 72–78.

Scher JU, Sczesnak A, Longman RS, Segata N, Ubeda C, Bielski C *et al.* Expansion of intestinal *Prevotella copri* correlates with enhanced susceptibility to arthritis. *Elife.* 2013; 2:e01202.

Scott DL, Wolfe F, Huizinga TW. Rheumatoid arthritis. *Lancet.* 2010; 376(9746): 1094–108.

Simonsson M, Bergman S, Jacobsson LT, Petersson IF, Svensson B. The prevalence of rheumatoid arthritis in Sweden. *Scand J Rheumatol.* 1999; 28: 340–343.

Singka GS, Samah NA, Zulfakar MH, Yurdasiper A, Heard CM. Enhanced topical delivery and anti-inflammatory activity of methotrexate from an activated nanogel. *Eur J Pharm Biopharm.* 2010; 76: 275–281.

Souto EB, Müller RH. SLN and NLC for topical delivery of ketoconazole. *J Microencapsul.* 2005; 22: 501–510.

Souto EB, Wissing SA, Barbosa CM, Müller RH. Development of a controlled release formulation based on SLN and NLC for topical clotrimazole delivery. *Int J Pharm.* 2004; 278: 71–77.

Srivastava S, Patel S, Singh D, Singh M. Rationalized insights on causes of rheumatoid arthritis in the elderly and women: special emphasis on treatment strategies. *Crit Rev Ther Drug Carrier Syst.* 2017; 34: 97–147.

Steele WH, Stuart JF, Lawrence JR, McNeill CA, Sneader WE, Whiting B *et al.* Enhancement of methotrexate absorption by subdivision of dose. *Cancer Chemother Pharmacol.* 1979; 3: 235–237.

Symmons D, Turner G, Webb R, Asten P, Barrett E, Lunt M *et al.* The prevalence of rheumatoid arthritis in the United Kingdom: New estimates for a new century. *Rheumatology.* 2002; 41: 793–800.

Trotta M, Peira E, Carlotti ME, Gallarate M. Deformable liposomes for dermal administration of methotrexate. *Int J Pharm.* 2004; 270: 119–125.

Van Rossum MA, Fiselier TJ, Franssen MJ, Zwinderman AH, ten Cate R, van Suijlekom-Smit LW *et al.* Sulfasalazine in the treatment of juvenile chronic arthritis: A randomized, double-blind, placebo-controlled, multicenter study. *Arthritis Rheum.* 1998; 41: 808–816.

Weinblatt ME, Kaplan H, Germain BF, Merriman RC, Solomon SD, Wall B *et al.* Methotrexate in rheumatoid arthritis: Effects on disease activity in a multicenter prospective study. *J Rheumatol.* 1991; 18: 334–338.

Wong TW, Chen CH, Huang CC, Lin CD, Hui SW. Painless electroporation with a new needle-free microelectrode array to enhance transdermal drug delivery. *J Control Release.* 2006; 110: 557–565.

Yang F, Kamiya N, Goto M. Transdermal delivery of the anti-rheumatic agent methotrexate using a solid-in-oil nanocarrier. *Eur J Pharm Biopharm.* 2012; 82: 158–163.

Zwerina J, Redlich K, Schett G, Smolen JS. Pathogenesis of rheumatoid arthritis: Targeting cytokines. *Ann N Y Acad Sci.* 2005; 1051: 716–729.

13 Superparamagnetic Iron Oxide Nanoparticles for Magnetic Hyperthermia Applications

Yogita Patil-Sen and Vikesh Chhabria*

CONTENTS

13.1 Introduction 247
13.2 SPIONs 249
13.2.1 Synthesis of SPIONs 249
13.2.2 Physicochemical Properties of SPIONs 250
13.2.2.1 Size 250
13.2.2.2 Shape 251
13.2.2.3 Surface Chemistry and Charge 251
13.2.2.4 Magnetic Properties 252
13.3 Hyperthermia 253
13.3.1 Cellular Mechanism 254
13.3.2 Thermal Mechanism 255
13.3.3 Hyperthermia-Mediated Drug Delivery 257
13.4 Conclusions and Future Perspectives 257
Acknowledgments 258
References 258

13.1 INTRODUCTION

The last few decades have shown considerable progress in early detection, diagnosis and treatment of cancer, which has led to improved patient survival (Urruticoechea *et al.*, 2010). Although the mainstream anticancer treatments such as chemotherapy, radiation and surgery have provided ways to fight the fatal and usually terminal disease, these treatments are unpleasant, lack specificity, pose a risk of recurrence of disease and suffer from severe side effects. Cancer researchers are therefore looking for new strategies which will minimise the drawbacks associated with the current therapies and bring hope to patients. Some of the therapies that have recently been explored are immunotherapy, gene therapy, laser treatment and hyperthermia. Moreover, the design of novel anticancer drugs and a better understanding of their biological interactions on a cellular level would be beneficial to overcome the limitations of current therapies (Fahs *et al.*, 2014, 2015).

Recently, nanotechnology has arrested attention in the field of biomedicine as a promising device especially for smart drug delivery applications (Peer *et al.*, 2007; Zhang *et al.*, 2008). Nanotechnology deals with particles which are in the size range of 1–1000 nm; for specific applications in biomedicine, particles in the size range 10–200 nm are found to be the most effective, as these can evade the immune system and remain in blood circulation for extended periods. The properties that make nanoparticles an attractive choice are their extremely small size, high surface area to volume ratio and ability to encapsulate various types of molecules. Thus, the size, morphology

* Corresponding author.

and surface charge are considered the most important parameters, which need to be cleverly controlled while considering application of nanoparticles.

Different types of nanoparticles, which include those made from lipids, polymers, silica and metals have been investigated for their therapeutic applications in diverse ailments (Gupta and Wells, 2004; Gupta and Gupta, 2005; Hofheinz *et al.*, 2005; Immordino *et al.*, 2006; Cho *et al.*, 2008; Zhang *et al.*, 2008; Enlow *et al.*, 2011; Dykman and Khlebtsov, 2012; Li *et al.*, 2012; Mehta, 2017). Moreover, quantum dots, carbon and titanium nanotubes and peptide-based nanoparticles have also been studied in this regard (Resch-Genger *et al.*, 2008; Bawa *et al.*, 2012; Gaunt *et al.*, 2014; Kulkarni *et al.*, 2015a; Kulkarni *et al.*, 2015b; Patil-Sen *et al.*, 2016; Majid *et al.*, 2017). However, the role of most of these nanoparticles is still at the research stage, as only a handful of these have been approved by the U.S. Food and Drug Administration (FDA) for clinical use (Hofheinz *et al.*, 2005; Immordino *et al.*, 2006) (Table 13.1). Among the various types of nanoparticles, lipid-based nanoparticles, polymeric nanoparticles and metallic nanoparticles such as superparamagnetic iron oxide nanoparticles (SPIONs) are the dominant types which have received clinical approval for medical applications (Zhang *et al.*, 2008; Thuy *et al.*, 2012).

Lipids, being an integral part of membranes (Patil-Sen *et al.*, 2004), present an ideal biocompatible material to generate nanostructures. Lipid nanoparticles, such as liposomes, can encapsulate both hydrophilic as well as hydrophobic molecules which provide an additional advantage for these systems as a drug delivery vehicle. Nevertheless, the major challenges associated with liposomes, such as a lack of structural integrity, result in leakage of the content and instability during storage (Maurer *et al.*, 2001). This has led researchers to look for alternative nanoparticulate systems that would overcome the above-mentioned problems. Polymeric nanoparticles have gained increasing attention, as these present high structural integrity and stability during storage (Peer *et al.*, 2007). However, toxicity and biocompatibility are the main issues when considering polymeric nanoparticles as drug carriers. SPIONs, on the other hand, have demonstrated low toxicity, good biocompatibility and are 'hard' nanoparticles in contrast with aforementioned nanoparticles (Gupta and Gupta, 2005; Zhang *et al.*, 2008). In addition to these properties, the superparamagnetic behaviour of SPIONs have attributed to their ever-increasing potential.

In order to overcome the limitations of individual nanoparticulate system, new generation of hybrid or multifunctional nanoparticles have recently been developed by combining two or more types of nanoparticles mentioned above. Lipid–polymer hybrid nanoparticles, SPIONs coated with silica or polymer or lipid, lipid–carbon nanotube mixed system and multifunctional nanoparticles are proving to provide superior properties for theranostic applications (Gupta and Gupta, 2005; Gao *et al.*, 2009; Gaunt *et al.*, 2014; Hervault and Thanh, 2014; Krishnamurthy *et al.*, 2015; Kulkarni *et al.*, 2015b; Patil-Sen *et al.*, 2016).

TABLE 13.1
Examples of Nanoparticles and Their Stages in Clinical Use

Nanoparticle Type	Compound (Trade name)	Application/Target	Status	Reference(s)
Liposome	Doxorubicin (Doxil/Caelyx) Daunorubicin (DaunoXome)	Breast cancer, ovarian cancer, Kaposi's sarcoma Cancer therapy	FDA approved FDA approved	In market (Hofheinz *et al.*, 2005) In market (Immordino *et al.*, 2006)
Polymer	Pegaspargase (Oncaspar) Doxorubicin Paclitaxel	Acute lymphoblastic leukemia Breast/lung cancer Breast cancer	FDA approved Phase II Phase II	In market (Dinndorf *et al.*, 2007) (Gao and Fain, 2004) (Oerlemans *et al.*, 2010)
Metallic nanoparticles	Ferumoxides (Feridex) Iron oxide (NanoTherm)	MRI contrast agent Glioblastoma	FDA Approved EU Approved	In market (Wang, 2011) (Maier-Hauff *et al.*, 2011)

13.2 SPIONs

Magnetite (Fe_3O_4), maghemite (γ-Fe_2O_3) and hematite (α- Fe_2O_3) are the most widely researched SPIONs because of their good biocompatibility and low toxicity. Due to the unique magnetic behaviour exhibited by SPIONs, these have been employed in both diagnostic and therapeutic applications which include, magnetic separation of biomolecules, magnetic resonance imaging (MRI), targeted drug delivery and hyperthermia cancer therapy (Figure 13.1) (Gupta and Gupta, 2005; Gupta *et al.*, 2007; Hedayatnasab*et al.*, 2017; Owais Ansari *et al.*, 2017).

However, these metallic particles are chemically very active and are easily oxidized in air, resulting in a loss of magnetism. Moreover, due to their ultra-small size, these particles have a high surface area to volume ratio and thus have tendency to aggregate. The hydrophobic interactions and the interparticulate van der Waals forces lead to agglomeration of the particles, which reduces the total surface and interfacial energy. Agglomeration reduces the dispersibility, which in turn influences their intrinsic stability over longer periods of time. Stability is of the utmost importance for any *in vivo* application of SPIONs. In order to improve stability, SPIONs are usually coated with biocompatible materials such as lipids, silica and polymers to form core-shell type of nanoparticles. These coating shells help prevent aggregation of SPIONs and thereby improve their stability over time. More importantly, coatings provide multifunctionality, minimize side effects and enhance the efficacy of SPIONs for drug delivery and other bioapplications (Table 13.2).

13.2.1 Synthesis of SPIONs

The most common methods of preparation of SPIONs that are reported in the literature are co-precipitation, microemulsion and thermal decomposition (Tartaj *et al.*, 2005). The two approaches to synthesize SPIONs by the co-precipitation method are (i) partial oxidation of ferrous hydroxide suspension by oxidizing agent such as nitrate, as reported by Sugimoto and Matijević (1980) and (ii) from aqueous solution containing Fe^{2+}/Fe^{3+} ions in 1:2 molar ratio by addition of base in an oxygen free environment as prepared by Massart (1981). In the microemulsion method, the aqueous phase which is entrapped in surfactant-stabilized micelles, is dispersed in the continuous oil phase. Iron oxide is precipitated in the aqueous core from iron precursors, ferrous chloride and ferric chloride (Tartaj *et al.*, 2005). In the thermal decomposition method of the synthesis of iron oxide nanoparticles, thermal degradation of organometallic precursors of iron, such as iron acetyl acetate in high boiling organic solvent containing stabilizing surfactants, is carried out (Sun and Zeng, 2002). In addition, other synthesis methods such as solvothermal synthesis, sol–gel method, electrochemical deposition, sonochemical method, polyols synthesis, spray pyrolysis,

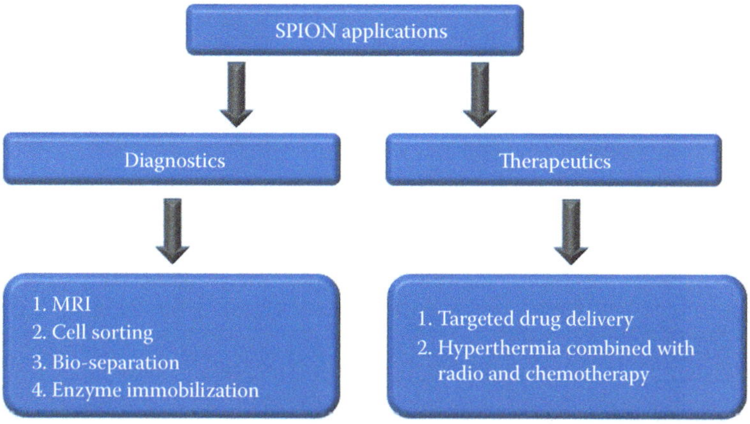

FIGURE 13.1 Biomedical applications of SPIONs.

TABLE 13.2
Materials Used for Coating the SPION Surface and Their Applications

Material	Application	Advantage	References
Amorphous silica	Drug delivery, isolation of biomolecules such as DNA, extraction of nucleic acid from soil	Does not require organic solvents, eliminates the need for the repeated centrifugation, vacuum filtration or separation	(Majid *et al.*, 2017)
Mesoporous silica	Controlled drug delivery, support for enzyme immobilisation for biocatalysis	Uniform pore size, large surface area	
Polymer	Drug delivery, *in vivo* imaging, contrasting, hyperthermia	Improves biocompatibility and stability, enhances blood circulation time and internalization efficiency	
Lipid	Imaging, contrasting, drug delivery, hyperthermia	Improves blood circulation time, biocompatibility and stability	

laser evaporation synthesis and biomineralization have also been reported in the literature, which are summarized in Table 13.3 (Pascal *et al.*, 1999; Abu Mukh-Qasem and Gedanken, 2005; Tartaj *et al.*, 2005; Tartaj *et al.*, 2006; Cai and Wan, 2007; Joseyphus *et al.*, 2007; Lu *et al.*, 2007; Xu *et al.*, 2007).

13.2.2 Physicochemical Properties of SPIONs

Physicochemical properties should be considered while designing SPIONs for various clinical applications. Size, shape, charge and surface chemistry of nanoparticles govern their biodistribution, stability and metabolism (Wang *et al.*, 2001). The method of synthesis of SPIONs and the chemistry involved in the preparation control the physicochemical properties of SPIONs, as described in Table 13.3.

13.2.2.1 Size

The size and size distribution are the significant parameters which affect the residual time of SPIONs in blood and thus their uptake by cells and elimination from the body (Chouly *et al.*, 1996). Nanoparticles greater than 200 nm are captured by the spleen and liver, whereas those which are less than 10 nm in diameter are selectively removed from the body by the renal system (Gupta and Gupta, 2005; Thorek *et al.*, 2006). Moreover, magnetic properties of SPIONs are strongly influenced by their size (Mahmoudi *et al.*, 2009a). SPIONs exhibit excellent superparamagnetic properties (see Section 2.2.4), though these particles tend to aggregate. Therefore, particles in the size range of 10–100 nm are preferred, as these are comparable to viruses (20–450 nm), proteins (5–50 nm) and genes (2–100 nm). Moreover, such small-sized particles can penetrate through small capillaries, can easily escape through immune system, and have longer circulation time which is essential for the bioapplications of SPIONs (Gupta and Wells, 2004). Although it is desirable to produce monodisperse particles, control over the size is a major challenge. All SPION synthesis methods tend to generate particles of varying size showing polydispersity which poses an important concern for their biodistribution and bioelimination (Andrade *et al.*, 2011).

The particle size of SPIONs can be measured either in a suspension or in a dry state. The techniques most commonly used to determine the particle size of SPIONs are dynamic light scattering (DLS) and transmission electron microscope (TEM). DLS measures the size of particles in supsension whereas TEM provides the size in dry state.

TABLE 13.3
Summary of Different Routes of Synthesis of SPIONs

Method Employed	Synthesis	Reaction Temperature (°C)	Solvent	Particle size (nm)	Size Distribution	Shape
Coprecipitation	Very simple and efficient method	20–90	Water	1–200	Relatively narrow	Spherical or rhombic
Microemulsion	Complicated, ambient condition	20–50	Organic compound	4–12	Relatively low	Spherical or cubic
Thermal decomposition	Complicated, inert atmosphere	100–320	Organic compound	3–20	Very narrow	Spherical
Solvothermal	Simple, high pressure	220	Water or organic solvents	200–500	Very narrow	Spherical
Sol–gel	Complicated and uses toxic reagents	200–400	Organic compound	20–200	Broad	Spherical
Electrochemical deposition	High cost of production	70–100	Organic compound	3–8	Very narrow	Spherical
Sonochemical method	Utilises organometallic precursors which can cause in vivo toxicity	25	Water	10–30	Very narrow	Spherical or rod-shaped
Polyols	Nonagglomerated particles with well-defined shape and size can be obtained	120–280	Organic compound	5–40	Broad	Spherical
Spray pyrolysis	Simple and rapid process but requires exhaustive control of the experimental conditions	400–700	Organic compound	5–60	Broad	Spherical
Laser evaporation	High cost of production Requires exhaustive control of the experimental conditions	1100	Organic compound	5–30	Narrow	Spherical
Biomineralization	Biocompatible nanoparticles produced	–	–	50–100	Very narrow	Spherical cluster, cubooctahedral

13.2.2.2 Shape

The shape of nanoparticles is thought to affect their biodistribution. Although no extensive studies have been reported on the effects of shape of SPIONs on their biodistribution, literature suggests a strong correaltion between the two for other types of nanoparticles. Spherical or short-rod shaped mesoporous silica nanoparticles were found to accumulate in liver whereas long rod-shaped particles were distributed in the spleen (Huang *et al.*, 2011). Moreover, circulation time in blood for non-spherical or rod-shaped particles has been reported to be longer compared to spherical particles which are cleared rapidly via the excretory system (Fadeel and Garcia-Bennett, 2010). Studies suggest that the shape of SPIONs also has impact on the cell toxicity, as the spherical SPIONs exhibit greater cellular toxocity than rod-shaped SPIONs (Mahmoudi *et al.*, 2009b). The shape of the particles has a significant effect on their magnetic properties, which could impact their bioapplication.

13.2.2.3 Surface Chemistry and Charge

Surface characteristics such as charge and hydrophilic/hydrophobic behaviour of SPIONs depend upon the method of synthesis, chemicals and reaction conditions employed as well as the pH.

Surface chemistry and charge of nanoparticles play a significant role in their colloidal stability and hence their biological fate (Hühn *et al.*, 2013). Surface charge can be quantitatively measured as an electric potential in the interfacial double layer on the surface of nanoparticles in suspension. Nanoparticles with positive and negative zeta potential show dispersion stability due to the electrostatic interaction and, as a result, lower agglomeration tendency on storage.

Charge also determines the distribution of SPIONs in the body and thus affects their internalization in target cells and hence their cellular uptake (Chouly *et al.*, 1996). Although the cellular uptake of these particles depends upon the cell type, it has been reported that positively charged SPIONs are better internalized than those with negative surface charge (Osaka *et al.*, 2009; Wotschadlo *et al.*, 2009; Schweiger *et al.*, 2012). However, positively charged SPIONs show higher toxicity compared to negatively charged particles, which could be due to their high affinity to cell membranes which are negatively charged. Furthermore, particles with hydrophobic surface are easily adsorbed at the protein surface, resulting their quick clearance from plasma and therefore low circulation time (Armstrong *et al.*, 1997; Moghimi *et al.*, 2001; Veiseh *et al.*, 2010). Engineering the size, shape and surface characteristics of SPIONs with suitable functional groups would allow for the tailoring and tuning of physicochemical properties and hence improve their bioapplicability.

13.2.2.4 Magnetic Properties

Magnetic materials, such as iron oxide nanoparticles, have a magnetic field created by the motion of their electrons. When the magnetic moments of the electrons align with each other, a unified magnetic field can be produced in materials. Depending upon the alignment of magnetic moments, materials are classified into following types: ferromagnetic, antiferromagnetic, ferrimagnetic and paramagnetic materials (Figure 13.2).

In ferromagnetic materials, all the individual magnetic moments are aligned with each other, and a permanent magnetic field is created in the material. On the other hand, in antiferromagnetic materials, the individual magnetic moments have same magnitude but opposite direction, and hence in such materials the net magnetic moment is zero. In ferrimagnetic materials, although most of the magnetic moments are aligned, some are anti-aligned and hence such materials possess a permanent magnetic field but of less magnitude compared to ferromagnetic materials. Ferri and ferromagnetic materials both can be considered as multi-domains of magnetic moments and when an external magnetic field is applied to these materials, the magnetic moments tend to align themselves in the same direction as that of the applied field. As soon as the magnetic moment in each domain aligns

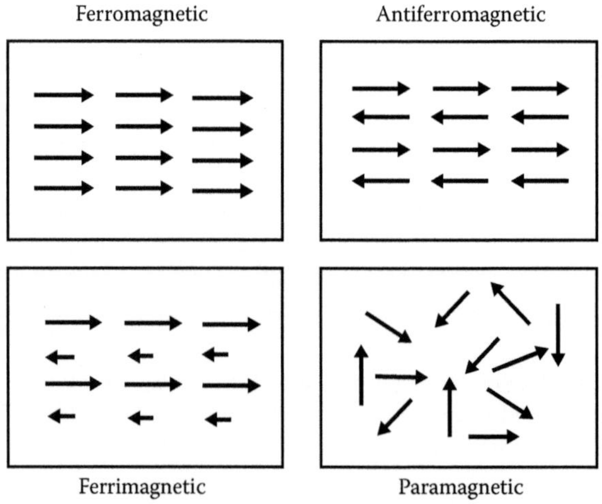

FIGURE 13.2 Alignment of magnetic moments in ferromagnetic, antiferromagnetic, ferrimagnetic and paramagnetic materials in the absence of external magnetic field.

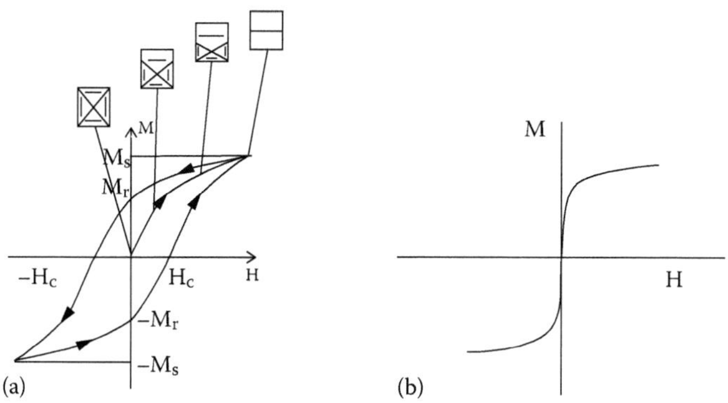

FIGURE 13.3 (a) Hysteresis loop of a multi-domain magnetic material such as ferro and ferrimagnet. H is the magnetic field amplitude; M is the magnetization of the material. Domain wall displacement in such materials is shown where squares represent multi-domain material with magnetization of each domain; arrows on the cycle indicate the way cycle is described when increasing or decreasing the field amplitude (adapted from Mornet *et al.*, 2004) and (b) Magnetization curve for paramagnetic or superparamagnetic material, no hysteresis is seen which indicates zero coercivity and no magnetic remanence.

in the direction of the magnetic field, saturation magnetization (M_s) is reached. When the magnetic field is removed, magnetization does not go back to zero and the material possesses magnetic field called remanent magnetization (M_r). A magnetic field of a precise intensity is required in order to bring the magnetization to zero and this field is called coercive field or coercivity (H_c). The magnetization curve of magnetic materials is represented by hysteresis loop as shown in Figure 13.3.

In paramagnetic materials, the magnetic moments are randomly oriented and hence no net magnetization is exhibited by such materials. However, when these materials are exposed to external magnetic field, the individual magnetic moments align, causing a net magnetization which disappears upon removal of the magnetic field. Superparamagnetic materials are basically ferro or ferrimagnetic in nature which are sufficiently small in size such as nanoparticles. Due to their small size, the grain boundaries disappear and the material becomes one single magnetic domain. When such materials are exposed to an external magnetic field, the magnetic moment of the single domain aligns with that of the applied external magnetic field but when the magnetic field is removed, no remanent magnetization is left in the material. Superparamagnetic materials are prefered over ferri and ferromagnetic materials for therapeutic applications because these materials do not retain any magnetization once the magnetic field is removed.

The magnetic properties of SPIONs depend on not only particle size and shape but also on crystallanity of particles, temperature, surface coating and surface properties. Vibrating sample magnetometer (VSM) is the most common technique used to measure the magnetic properties of magnetic materials (Foner, 1959).

13.3 HYPERTHERMIA

Hyperthermia is defined as elevation in temperature. The concept of using heat to treat disease dates back to over 5000 years ago and has been under practice in many ancient cultures such as Greek, Egyptian, Roman, Chinese, Japanese and Indian (Baronzio and Hager, 2006; Horsman and Overgaard, 2007; Moyer and Delman, 2008). Over the last a few years, the use of hyperthermia for cancer treatment has been taken more seriously and has proven to be an effective treatment complementary to conventional surgery, radiotherapy and chemotherapy (van der Zee, 2002). It has been shown that, compared to normal cells, tumour cells are more sensitive to hyperthermia, which could be due to the physiological differences between the two. Healthy cells can tolerate high temperatures (42–45°C), whereas cancer cells undergo apoptosis (cell death) at these temperatures. Blood flow remains greater in tumour cells than that in the healthy cells (Song, 1984), but this trend is reversed when hyperthermia is applied at temperatures over 42°C. Due to the decreased blood flow the heat dissipation rate lowers and the temperature increases faster in tumour cells, while in normal cells the physiological temperature is easily maintained (Song *et al.*, 1980; Song, 1984). The difficulty in dissipating heat may be the cause of apoptosis in cancer cells (Storm *et al.*, 1979). In principle, all types of tumour cells can be treated using hyperthermia (Cavaliere *et al.*, 1967; Gordon *et al.*, 1979; van der Zee, 2002). The further advantage of using hyperthermia as a complementary treatment to the standard therapies is that it can kill drug-resistant and radio-resistant cancer cells.

There are three types of hyperthermia that can be used to treat cancer, *viz.* local, regional and whole-body hyperthermia. Depending upon the location, depth and stage of cancer, the hyperthermia technique is chosen (Chicheł *et al.*, 2007). Local hyperthermia, which is a less invasive technique, uses microwaves, radio waves or ultrasound waves and is generally used to treat localised tumours. Regional hyperthermia is used in advanced stage cancer and thermal perfusion or microwave antennas are used in this technique. In whole-body hyperthermia, the entire body is under heat stress using hot water baths, infrared radiators or thermal chambers. This technique is employed in case of propagated metastases or when tumours are seated deep inside the body. Compared to local or regional hyperthermia, the risk of side effects in whole body hyperthermia is greater and therefore careful control of the skin temperature is necessary.

13.3.1 Cellular Mechanism

Despite the considerable number of studies carried out on hyperthermia over the years, the exact mechanism of its action on the cells of the body is poorly understood (Milleron and Bratton, 2007). Hyperthermia causes many changes in the cell membrane which leads to alteration in cellular homeostasis (Hildebrandt *et al.*, 2002; Roti Roti, 2008). The thermal energy required for apoptosis has been found to be similar to the energy needed for protein denaturation. Therefore, denaturation and aggregation of membrane and cytoplasmic proteins is thought to be a key event in hyperthermia-based cell cytotoxicity (Figure 13.4). Other main cellular effects of hyperthermia are inactivation of protein synthesis, inhibition of DNA synthesis and repair processes, membrane cytoskeletal damage

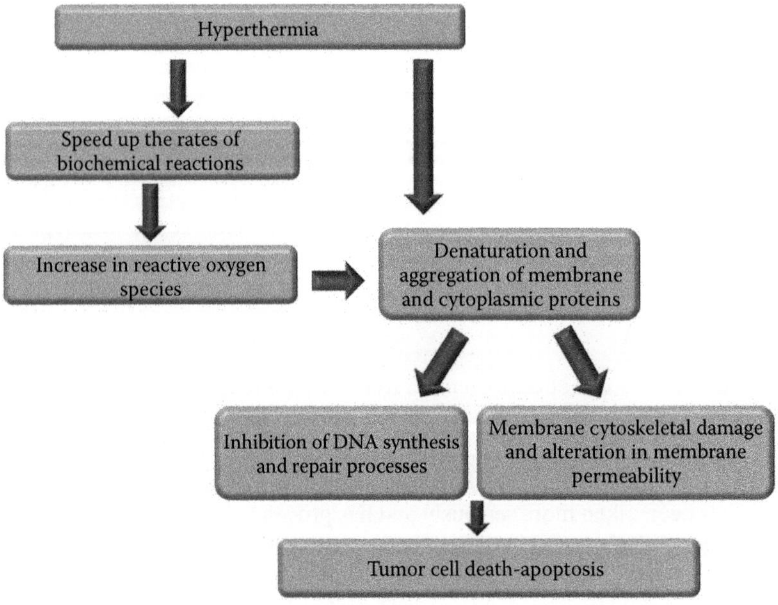

FIGURE 13.4 Possible cellular changes in hyperthermia-induced cancer cell death.

and alteration in membrane permeability (Lepock, 2005). Increase in temperature can speed up the rates of biochemical reactions which, in turn, can increase cell metabolism and can cause increase in oxidative stress. Exposure to a temperature higher than 40°C has shown to increase the levels of reactive oxygen species (ROS) possibly due to dysfunction of the mitochondrial respiratory chain (Bettaieb and Averill-Bates, 2008; Pallepati and Averill-Bates, 2011).

13.3.2 Thermal Mechanism

During *in vitro* hyperthermia measurements SPIONs are dispersed in a liquid to form a ferrofluid whereas in *in vivo* applications, the particles are introduced into the desired tissue where these disperse in blood. Thus, SPIONs are in a continuous state of motion and rotate randomly in the fluid. When SPIONs are subjected to an alternating magnetic field, the magnetic energy of particles is converetd into thermal energy. The heat generation can be explanined on the basis of two main mechanisms that occur during the movement of particles: Brownian and Néel relaxation (Mornet *et al.*, 2004; Suto *et al.*, 2009; Deatsch and Evans, 2014).

Brownian relaxation is caused by the friction arising from the rotation of the particle itself in the fluid while Néel relaxation is due to the reorientation of the magnetic moment of the particle in the same direction of the applied magnetic field (Figure 13.5). The Brownian (τ_B) and Néel (τ_N) relaxation times of a particle are given by the following equations:

$$\tau_B = \frac{3\eta V'}{kT} \tag{13.1}$$

$$\tau_N = \tau_0 \exp\left(\frac{KV}{kT}\right) \tag{13.2}$$

where, τ_B = Brownian relaxation time, η = viscosity of the liquid, V' = volume of the particles, k = the Boltzmann constant and T = temperature, τ_N = Néel relaxation time, $\tau_0 = 10^{-9}$ s, K = anisotropy constant and V = hydrodynamic volume of the particle.

In practice, both the mechanisms occur simultaneously and contribute to the magnetization of the particles. Hence, the overall effective relaxation time of the particle is

$$\tau_{eff} = \frac{\tau B\, \tau N}{\tau B + \tau N} \tag{13.3}$$

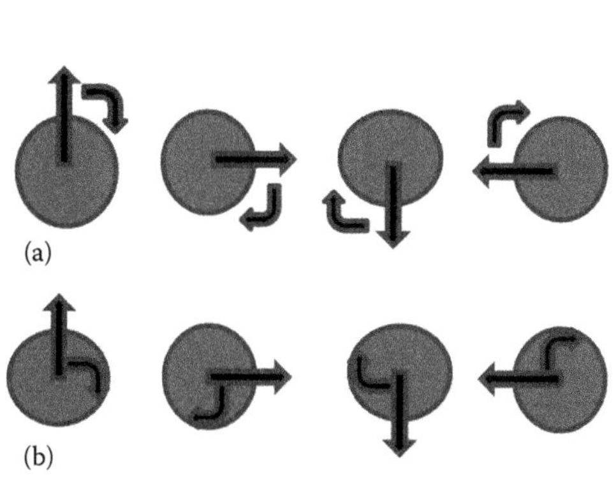

FIGURE 13.5 (a) Brownian rotation of a magnetic particle (the particle rotates as a whole); (b) Néel rotation of magnetic moment in a magnetic particle (the particle does not rotate).

The effective relaxation time strongly depends on the size and shape of the particle as well as the viscosity of the fluid (Kötitz *et al.*, 1995; Fortin *et al.*, 2008). The shape of the particle will affect the anisotropy constant, whereas the size of the particle will change its hydrodynamic volume. Moreover, the viscosity of the liquid will have a strong effect on the rotation of particles; the higher the viscosity of the fluid, the slower the rotation of particles. When magnetic particles are exposed to an alternating magnetic field which oscillates faster than the relaxation times of the particles, heat dissipation is caused by the delay in the relaxation of their magnetic moment either through Brownian or Néel rotation. The heat dissipation is given as Equation 13.4:

$$P = \mu_0 \chi'' f H^2 \tag{13.4}$$

where, P = heat dissipation value, μ_0 = the permeability of free space, χ'' = the alternating current (AC) magnetic susceptibity (imaginary part), f = frequency of the applied magnetic field and H = the strength of the applied AC magnetic field.

A typical experimental set up for SAR measurement, as shown in Figure 13.6, consists of an induction coil which generates magnetic field. The magnetic nanoparticles dispsered in a liquid are put in a glass vial and are subjected to the alternating magnetic field. The heat generated by the nanoparticles is measured using the fibre optic temperature sensor.

The heat generated is quantified as specific absorption rate (SAR) (Mornet *et al.*, 2004) which is given by the Equation 13.5.

$$SAR = \left(\frac{C}{m}\right)\left(\frac{dT}{dt}\right)\Bigg|_{t=0} \tag{13.5}$$

where, C = the heat capacity of the magnetic particle suspension, m = mass of the magnetic phase, $\left(\frac{dT}{dt}\right)\Big|_{t=0}$ = initial slope of the temperature increase and T as a function of time t.

SPIONs with higher SAR values are desired for biomedical applications because higher SAR value means higher heating ability which, subsequently, means less amount of nanoparticles required for application and treatment. SAR depends on the magnetic properties of the nanoparticles which, in turn, depend on their physicochemical characteristics as well as on other experimental parameters such as frequency and amplitute of the magnetic field applied.

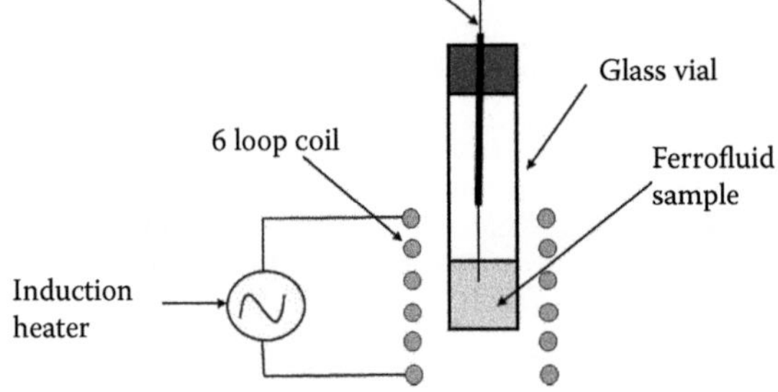

FIGURE 13.6 Diagrammatic representation of experimental set-up for SAR measurement. (Adapted from Huang S *et al.*, *Meas Sci Technol.* 2012; 23(3): 035701–035706).

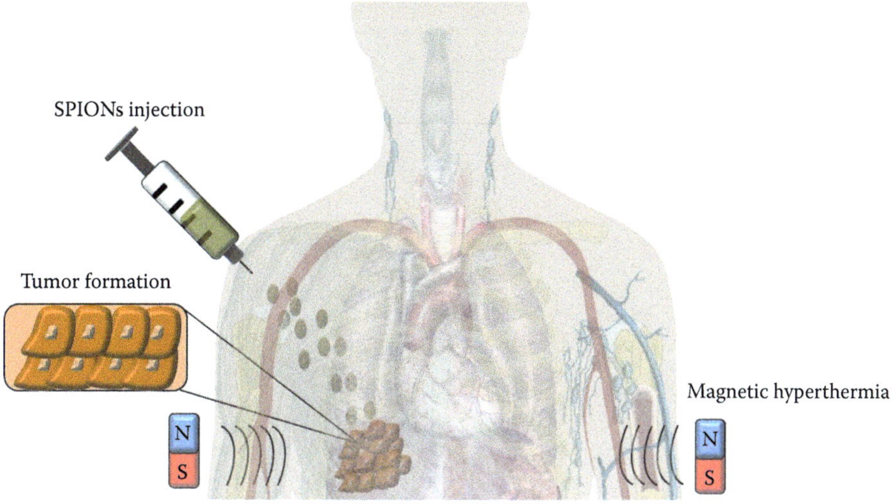

FIGURE 13.7 Sketch showing the intravenous injection of magnetic nanoparticles into tumour cell and induction of hyperthermia by applying external alternating magnetic field. (Adapted from Andrade A *et al.*, Coating Nanomagnetic Particles for *Biomedical Engineering – Frontiers and Challenges* (ISBN 978-953-307-309-5). In: R Fazel, ed., Intech. 2011).

13.3.3 Hyperthermia-Mediated Drug Delivery

Anticancer drugs are highly toxic to both the healthy and malignant cells. Therefore, it is highly desirable to develop strategies which will not only deliver the drug to a localised targeted area, but also release it in a controlled manner. The potential of magnetic nanoparticles for hyperthermia drug delivery was first proposed in the latter part of 1970s (Widder *et al.*, 1978; Gordon *et al.*, 1979). For hyperthermia-mediated drug delivery applications, core-shell type nanoparticles containing SPIONs and an anticancer drug can be directly injected intravenously and then guided by an external magnetic field. Figure 13.7 shows a diagrammatic representation of induction of hyperthermia by applied alternating magnetic field. Thus, via an external magnetic field, SPIONs can be driven magnetically inside the body and hence, the distribution of drug to healthy tissue can be minimized. Therefore, it is possible to achieve high and localised concentration of SPIONs, and hence carry high dose of drug to the target tissue.

13.4 CONCLUSIONS AND FUTURE PERSPECTIVES

In spite of the many advantages associated with SPIONs, there is still no drug delivery product in the market cosisting of these particles. Although considerable success has been achieved in *in vivo* applications of SPIONs, many fundamental limitations need to be overcome with repsect to their drug delivery and clinical applications. Much still needs to be done to improve the methodologies related to size-controlled SPION synthesis and their stability in suspension. Further studies are needed to investigate the biocompatibility, toxicology, pharmacokinetics and biodistribution of SPIONs in *in vivo*. However, the positive results so far indicate the tremendous potential of these metallic nanoparticles for a vaiety of applications in the field of medical science. Over the last few decades, the potential of hyperthermia as an efficient complementary cancer treatment to chemotherapy and radiation therapy has been strongly proved. Hyperthermia is currently under study in many clinical trials in the United States, Europe and Japan, in order to get clearer understanding of the physics and bioengineering involved in this promising technique. Despite good clinical results, further improved understanding of hyperthermia, more financial resources and awareness of successful high quality clinical trials are necessary in order to make the technique more popular and viable for cancer therapy and drug delivery applications.

ACKNOWLEDGMENTS

YPS is grateful to the Daphne Jackson Trust for the fellowship which is jointly funded by the Royal Society of Chemistry and the University of Central Lancashire, Preston, UK. The authors also thank Dr. Saurabh Prabhu, School of Pharmacy and Biomedical Sciences, University of Central Lancashire, for help with the reference manager.

REFERENCES

Abu Mukh-Qasem R, Gedanken A. Sonochemical synthesis of stable hydrosol of Fe_3O_4 nanoparticles. *J Colloid Interf Sci.* 2005; 284: 489–494.

Andrade A, Ferreira R, Fabris J, Domingues R. (2011). Coating nanomagnetic particles for biomedical applications. *Biomedical Engineering - Frontiers and Challenges* (ISBN 978-953-307-309-5). In: R Fazel, ed., InTech.

Armstrong TI, Davies, MC, Illum L. Human serum albumin as a probe for protein adsorption to nanoparticles: relevance to biodistribution. *J Drug Target.* 1997; 4: 389–398.

Baronzio GF, Hager ED. *Hyperthermia In Cancer Treatment: A Primer.* Springer US; 2006, pp. 366.

Bawa R, Fung S-Y, Shiozaki A, Yang H, Zheng G, Keshavjee S *et al.* Self-assembling peptide-based nanoparticles enhance cellular delivery of the hydrophobic anticancer drug ellipticine through caveolae-dependent endocytosis. *Nanomed Nanotech Biol Med.* 2012; 8: 647–654.

Bettaieb A, Averill-Bates DA. Thermotolerance induced at a fever temperature of 40 degrees C protects cells against hyperthermia-induced apoptosis mediated by death receptor signalling. *Biochem Cell Biol.* 2008; 86: 521–538.

Cai W, Wan J. Facile synthesis of superparamagnetic magnetite nanoparticles in liquid polyols. *J Colloid Interf Sci.* 2007; 305(2): 366–370.

Cavaliere R, Ciocatto EC, Giovanella BC, Heidelberger C, Johnson RO, Margottini M *et al.* Selective heat sensitivity of cancer cells. Biochemical and clinical studies. *Cancer.* 1967; 20: 1351–1381.

Chicheł A, Skowronek J, Kubaszewska M, Kanikowski M. Hyperthermia – Description of a method and a review of clinical applications. *Reports Pract Oncol Radiother.* 2007; 12: 267–275.

Cho K, Wang X, Nie S, Chen ZG, Shin DM. Therapeutic nanoparticles for drug delivery in cancer. *Clin Cancer Res.* 2008; 14: 1310–1316.

Chouly C, Pouliquen D, Lucet I, Jeune JJ, Jallet P. Development of superparamagnetic nanoparticles for MRI: Effect of particle size, charge and surface nature on biodistribution. *J Microencapsul.* 1996; 3: 245–255.

Deatsch AE, Evans BA. Heating efficiency in magnetic nanoparticle hyperthermia. *J Magn Magn Mater.* 2014; 354: 163–172.

Dinndorf PA, Gootenberg J, Cohen MH, Keegan P, Pazdur R. FDA drug approval summary: Pegaspargase (oncaspar) for the first-line treatment of children with acute lymphoblastic leukemia (ALL). *The Oncologist.* 2007; 12(8): 991–998.

Dykman L, Khlebtsov N. Gold nanoparticles in biomedical applications: Recent advances and perspectives. *Chem Soc Rev.* 2012; 41: 2256–2282.

Enlow EM, Luft J, Christopher N, Mary E, DeSimone JM. Potent engineered PLGA nanoparticles by virtue of exceptionally high chemotherapeutic loadings. *Nano Lett.* 2011; 11: 808–813.

Fadeel B, Garcia-Bennett AE. Better safe than sorry: Understanding the toxicological properties of inorganic nanoparticles manufactured for biomedical applications. *Adv Drug Deliv Rev.* 2010; 62: 362–374.

Fahs S, Patil-Sen Y, Snape TJ. Foldamers as anticancer therapeutics: Targeting protein-protein interactions and the cell membrane. *ChemBioChem.* 2015; 16: 1840–1853.

Fahs S, Rowther FB, Dennison SR, Patil-Sen Y, Warr T, Snape TJ. Development of a novel, multifunctional, membrane-interactive pyridinium salt with potent anticancer activity. *Bioorg Med Chem Lett.* 2014; 24: 3430–3433.

Foner S. Versatile and sensitive vibrating-sample magnetometer. *Rev Sci Instrum.* 1959; 30: 548–557.

Fortin J-P, Gazeau F, Wilhelm C. Intracellular heating of living cells through Néel relaxation of magnetic nanoparticles. *Eur Biophys J.* 2008; 37: 223–228.

Gao J, Gu H, Xu B. Multifunctional magnetic nanoparticles: Design, synthesis, and biomedical applications. *Acc Chem Res.* 2009; 42: 1097–1107.

Gao Z, Fain HD, Rapoport N. Ultrasound enhanced tumor targeting of polymeric micellar drug carriers. *Mol Pharm.* 2004; 1(4): 317–330.

Gaunt NP, Patil-Sen Y, Baker MJ, Kulkarni CV. Carbon nanotubes for stabilization of nanostructured lipid particles. *Nanoscale.* 2014; 7: 1090–1095.

Gordon RT, Hines JR, Gordon D. Intracellular hyperthermia a biophysical approach to cancer treatment via intracellular temperature and biophysical alterations. *Med Hypotheses.* 1979; 5: 83–102.

Gupta AK, Gupta M. Synthesis and surface engineering of iron oxide nanoparticles for biomedical applications. *Biomaterials.* 2005; 26: 3995–4021.

Gupta AK, Naregalkar RR, Vaidya VD, Gupta M. Recent advances on surface engineering of magnetic iron oxide nanoparticles and their biomedical applications. *Nanomedicine.* 2007; 2: 23–39.

Gupta AK, Wells S. Surface-modified superparamagnetic nanoparticles for drug delivery: Preparation, characterization, and cytotoxicity studies. *IEEE Trans Nanobioscience.* 2004; 3: 66–73.

Hedayatnasab Z, Abnisa F, Daud WMAW. Review on magnetic nanoparticles for magnetic nanofluid hyperthermia application. *Mater Des.* 2017; 123: 174–196.

Hervault A, Thanh NTK. Magnetic nanoparticle-based therapeutic agents for thermo-chemotherapy treatment of cancer. *Nanoscale.* 2014; 6: 11553–11573.

Hildebrandt B, Wust P, Ahlers O, Dieing A, Sreenivasa G, Kerner T *et al.* The cellular and molecular basis of hyperthermia. *Crit Rev Oncol Hematol.* 2002; 43: 33–56.

Hofheinz RD, Gnad-Vogt SU, Beyer U, Hochhaus A. Liposomal encapsulated anti-cancer drugs. *Anticancer Drugs.* 2005; 16: 691–707.

Horsman MR, Overgaard J. Hyperthermia: A potent enhancer of radiotherapy. *Clin Oncol (R Coll Radiol).* 2007; 19: 418–426.

Huang S, Wang SY, Gupta A, Borca-Tasciuc DA, Salon SJ. On the measurement technique for specific absorption rate of nanoparticles in an alternating electromagnetic field. *Meas Sci Technol.* 2012; 23(3): 035701–035706.

Huang X, Li L, Liu T, Hao N, Liu H, Chen D *et al.* The shape effect of mesoporous silica nanoparticles on biodistribution, clearance, and biocompatibility *in vivo. ACS Nano.* 2011; 5: 5390–5399.

Hühn D, Kantner K, Geidel C, Brandholt S, De Cock I, Soenen SJH *et al.* Polymer-coated nanoparticles interacting with proteins and cells: Focusing on the sign of the net charge. *ACS Nano.* 2013; 7: 3253–3263.

Immordino ML, Dosio F, Cattel L. Stealth liposomes: Review of the basic science, rationale, and clinical applications, existing and potential. *Int J Nanomedicine.* 2006; 1: 297–315.

Joseyphus RJ, Kodama D, Matsumoto T, Sato Y, Jeyadevan B, Tohji K. Role of polyol in the synthesis of Fe particles. *J Magn Magn Mater.* 2007; 310: 2393–2395.

Kötitz R, Fannin PC, Trahms L. Time domain study of Brownian and Néel relaxation in ferrofluids. *J Magn Magn Mater.* 1995; 149: 42–46.

Krishnamurthy S, Vaiyapuri R, Zhang L, Chan JM. Lipid-coated polymeric nanoparticles for cancer drug delivery. *Biomater Sci.* 2015; 3: 923–936.

Kulkarni M, Patil-Sen Y, Junkar I, Kulkarni CV, Lorenzetti M, Iglič A. Wettability studies of topologically distinct titanium surfaces. *Colloids Surf B.* 2015a; 129: 47–53.

Kulkarni CV, Moinuddin Z, Patil-Sen Y, Littlefield R, Hood M. Lipid-hydrogel films for sustained drug release. *Int J Pharm.* 2015b; 479: 416–421.

Lepock JR. How do cells respond to their thermal environment? *Int J Hyperthermia.* 2005; 21: 681–687.

Li Z, Barnes JC, Bosoy A, Stoddart JF, Zink JI. Mesoporous silica nanoparticles in biomedical applications. *Chem Soc Rev.* 2012; 41(7): 2590–2605.

Lu AH, Salabas EL, Schüth F. Magnetic nanoparticles: Synthesis, protection, functionalization, and application. *Angew Chemie Int Ed.* 2007; 46: 1222–1244.

Mahmoudi M, Shokrgozar MA, Simchi A, Imani M, Milani AS, Stroeve P *et al.* Multiphysics flow modeling and in vitro toxicity of iron oxide nanoparticles coated with Poly(vinyl alcohol). *J Phys Chem C.* 2009a; 113: 2322–2331.

Mahmoudi M, Simchi A, Milani AS, Stroeve P. Cell toxicity of superparamagnetic iron oxide nanoparticles. *J Colloid Interf Sci.* 2009b; 336: 510–518.

Maier-Hauff K, Ulrich F, Nestler D, Niehoff H, Wust P, Thiesen B *et al.* Efficacy and safety of intratumoral thermotherapy using magnetic iron-oxide nanoparticles combined with external beam radiotherapy on patients with recurrent glioblastoma multiforme. *J. Neuro-Oncology* 2011; 103(2): 317–324.

Majid A, Patil-Sen Y, Ahmed W, Sen T. Tunable self-assembled peptide structure: A novel approach to design dual-use biological agents. Mater. *Today Proc.* 2017; 4(1): 32–40.

Massart R. Preparation of aqueous magnetic liquids in alkaline and acidic media. *IEEE Trans Magn.* 1981; 17: 1247–1248.

Maurer N, Fenske DB, Cullis PR. Developments in liposomal drug delivery systems. *Expert Opin Biol Ther.* 2001; 1: 923–947.

Mehta RV. Synthesis of magnetic nanoparticles and their dispersions with special reference to applications in biomedicine and biotechnology. *Mater Sci Eng C.* 2017; 79: 901–916.

Milleron RS, Bratton SB. "Heated" debates in apoptosis. *Cell Mol Life Sci.* 2007; 64: 2329–2333.

Moghimi SM, Hunter AC, Murray JC. Long-circulating and target-specific nanoparticles: Theory to practice. *Pharmacol Rev.* 2001; 53: 283–318.

Mornet S, Vasseur S, Grasset F, Duguet E. Magnetic nanoparticle design for medical diagnosis and therapy. *J Mater Chem.* 2004; 14: 2161–2175.

Moyer HR, Delman KA. The role of hyperthermia in optimizing tumor response to regional therapy. *Int J Hyperthermia.* 2008; 24: 251–261.

Oerlemans C, Bult W, Bos M, Storm G, Nijsen JFW, Hennink WE. Polymeric micelles in anticancer therapy: Targeting, imaging and triggered release. *Pharm. Res.* 2010; 27(12): 2569–2589.

Osaka T, Nakanishi T, Shanmugam S, Takahama S, Zhang H. Effect of surface charge of magnetite nanoparticles on their internalization into breast cancer and umbilical vein endothelial cells. *Colloids Surf B.* 2009; 71(2): 325–330.

Owais Ansari M, Fahim Ahmad M, Parveen N, Ahmad S, Jameel S, Shadab GGHA. Iron oxide nanoparticles-synthesis, surface modification, applications and toxicity: A review. *Mater Focus.* 2017; 6(3): 269–279.

Pallepati P, Averill-Bates DA. Mild thermotolerance induced at 40°C protects HeLa cells against activation of death receptor-mediated apoptosis by hydrogen peroxide. *Free Radic Biol Med.* 2011; 50: 667–679.

Pascal C, Pascal JL, Favier F, Elidrissi Moubtassim ML, Payen C. Electrochemical synthesis for the control of γ-Fe_2O_3 nanoparticle size. Morphology, microstructure, and magnetic behavior. *Chem Mater.* 1999; 11: 141–147.

Patil-Sen Y, Tiddy GJT, Brezesinski G, DeWolf C. A monolayer phase behaviour study of phosphatidylinositol, phosphatidylinositol 4-monophosphate and their binary mixtures with distearoylphosphatidylethanolamine. *Phys Chem Chem Phys.* 2004; 6: 1562–1565.

Patil-Sen Y, Sadeghpour A, Rappolt M, Kulkarni CV. Facile preparation of internally self-assembled lipid particles stabilized by carbon nanotubes. *J Vis Exp.* 2016; 108: e53489–e53489.

Peer D, Karp JM, Hong S, Farokhzad OC, Margalit R, Langer R. Nanocarriers as an emerging platform for cancer therapy. *Nat Nanotechnol.* 2007; 2: 751–760.

Resch-Genger U, Grabolle M, Cavaliere-Jaricot S, Nitschke R, Nann T. Quantum dots versus organic dyes as fluorescent labels. *Nat Methods.* 2008; 5: 763–775.

Roti Roti JL. Cellular responses to hyperthermia (40–46 °C): Cell killing and molecular events. *Int J Hyperth.* 2008; 24: 3–15.

Schweiger C, Hartmann R, Zhang F, Parak WJ, Kissel TH, Rivera Gil P. Quantification of the internalization patterns of superparamagnetic iron oxide nanoparticles with opposite charge. *J Nanobiotechnology.* 2012; 10: 28–38.

Song CW. Effect of local hyperthermia on blood flow and microenvironment: A review. *Cancer Res.* 1984; 44: 4721s–4730s.

Song CW, Rhee JG, Levitt SH. Blood flow in normal tissues and tumors during hyperthermia. *J Natl Cancer Inst.* 1980; 64: 119–124.

Storm FK, Harrison WH, Elliott RS, Morton DL. Normal tissue and solid tumor effects of hyperthermia in animal models and clinical trials. *Cancer Res.* 1979; 39: 2245–2251.

Sugimoto T, Matijević E. Formation of uniform spherical magnetite particles by crystallization from ferrous hydroxide gels. *J Colloid Interf Sci.* 1980; 74: 227–243.

Sun S, Zeng H. Size-controlled synthesis of magnetite nanoparticles. *J Am Chem Soc.* 2002; 124: 8204–8205.

Suto M, Hirota Y, Mamiya H, Fujita A, Kasuya R, Tohji K *et al.* Heat dissipation mechanism of magnetite nanoparticles in magnetic fluid hyperthermia. *J Magn Magn Mater.* 2009; 321: 1493–1496.

Tartaj P, Morales MP, González-Carreño T, Veintemillas-Verdaguer S, Serna CJ. Advances in magnetic nanoparticles for biotechnology applications. *J Magn Magn Mater.* 2005; 290: 28–34.

Tartaj P, Morales MP, Veintemillas-Verdaguer S, Gonzalez-Carreño T, Serna CJ. Synthesis, properties and biomedical applications of magnetic nanoparticles. *Handb Magn Mater.* 2006; 16: 403–482.

Thorek DLJ, Chen AK, Czupryna J, Tsourkas A. Superparamagnetic iron oxide nanoparticle probes for molecular imaging. *Ann Biomed Eng.* 2006; 34: 23–38.

Thuy TT, Maenosono S, Thanh, NTK. (2012). Next Generation Magnetic Nanoparticles for Biomedical Applications, (ISBN 9781439869321). In: NTK Thanh, ed., *Magnetic Nanoparticles: From Fabrication and Clinical Applications*, CRC Press, Taylor and Francis, Boca Raton London New York, pp. 98–128.

Urruticoechea A, Alemany R, Balart J, Villanueva A, Vinals F, Capella G. Recent advances in cancer therapy: An overview. *Curr Pharm Des.* 2010; 16: 3–10.

Van der Zee J. Heating the patient: A promising approach? *Ann Oncol.* 2002; 13: 1173–1184.

Veiseh O, Gunn JW, Zhang M. Design and fabrication of magnetic nanoparticles for targeted drug delivery and imaging. *Adv Drug Deliv Rev.* 2010; 62: 284–304.

Wang YXJ, Hussain SM, Krestin GP. Superparamagnetic iron oxide contrast agents: Physicochemical characteristics and applications in MR imaging. *Eur Radiol.* 2001; 11: 2319–2331.

Wang YXJ. Superparamagnetic iron oxide based MRI contrast agents: Current status of clinical application. *Quant. Imaging Med Surg.* 2011; 1(1): 35–40.

Widder KJ, Senyei AE, Scarpelli DG. Magnetic Microspheres: A model system for site specific drug delivery in vivo. *Exp Biol Med.* 1978; 158: 141–146.

Wotschadlo J, Liebert T, Heinze T, Wagner K, Schnabelrauch M, Dutz S *et al.* Magnetic nanoparticles coated with carboxymethylated polysaccharide shells—Interaction with human cells. *J Magn Magn Mater.* 2009; 321: 1469–1473.

Xu J, Yang H, Fu W, Du K, Sui Y, Chen J *et al.* Preparation and magnetic properties of magnetite nanoparticles by sol–gel method. *J Magn Magn Mater.* 2007; 309: 307–311.

Zhang L, Gu FX, Chan JM, Wang AZ, Langer RS, Farokhzad OC. Nanoparticles in medicine: Therapeutic applications and developments. *Clin Pharmacol Ther.* 2008; 83: 761–769.

14 Development of In-House Nano-Hydroxyapatite Particles for Dental Applications

*Sunpreet Singh and Rupinder Singh**

CONTENTS

14.1 Introduction 263
14.2 Experimental Procedures 266
 14.2.1 Methodology 266
 14.2.2 Synthesis of Hydroxyapatite from Eggshell Waste 267
 14.2.3 Selection of Required Proportion of HAp/Nylon through Melt Flow Investigation 267
 14.2.4 Development of Alternative FDM Filament 268
 14.2.5 Obtaining Master Cast of Patient and Conversion in .STL Format 268
 14.2.6 Printing of PD 270
14.3 Results and Discussion 271
 14.3.1 Tensile Strength 271
 14.3.2 Wear test 272
14.4 Conclusions 274
Acknowledgments 275
References 275

14.1 INTRODUCTION

Loss of teeth has always been major concern of dental patients, since they can't eat, speak and even smile properly. Missing only few teeth that were scattered over either upper or lower arch, can be replaced inexpensively with a partial denture (PD). However, in case of an edentulous patient, the use of PD is not helpful, as the patient needs a complete set of artificial teeth known as a complete denture (CD). The historical backdrop of the development of dental implants is an exceptionally rich and captivating travelogue for the duration of the time. Since the start of humankind, people have utilized dental embeds in some shape to supplant missing teeth. Around 2500 BC, the ancient Egyptians used ligature wire made of gold (Abraham, 2014). About 500 BC, the Etruscans customized soldered gold bands from animals to restore oral function in humans. They also fashioned replacements for teeth from oxen bones, (Ring, 1985). In the early eighteenth century, Samuel Stockton, a goldsmith in terms of professional career, started fabricating top quality porcelain dentures mounted on 18-karat gold plates. Later, dentures from the 1850s onwards were made of vulcanite, a type of solidified elastic, in which porcelain teeth were set. In the twentieth century, acrylic pitch and different plastics were used (Eden *et al.*, 2002). Since then, dental implants and the technology surrounding this treatment has significantly evolved.

Nowadays, dental implant treatments are viewed as the debut treatment choice for individuals with missing teeth. Most of the clinical practices are dependent upon the use of standard materials such as thermoplastic resins, polymers and acrylic materials. Current dental utilization of these materials include preformed fractional denture catch, adaptable tooth conceived incomplete denture

* Corresponding author.

structure, single cast halfway dentures, impermanent crowns and scaffolds, temporary crowns and extensions, occlusal apparatuses, embed projections, orthodontic and rest apnea machines (Negrutiu *et al.*, 2005). Physiological, aesthetic and functional variables are associated with successful conventional complete denture therapy (Jendrensen *et al.*, 1997; Campbell *et al.*, 2017). In spite of the advancement of dentures bolstered by osseointegrated inserts, the recovery of resorbed remaining edges is still under test (Henry *et al.*, 1998). Fabricating patient-specific dentures commonly requires five visits, and reduction in these visits would be welcomed by both dentists and patients (Rodrigues and Morgano, 2007). However, this needs modifications in the existing dentistry practice, which may change its aspects of prosthodontic treatment in terms of results obtained from new treatment options and evidence from studies (Janus *et al.*, 2007). The main focus to computerize the creation of dental rebuilding efforts started over 20 years back (Strub *et al.*, 2006). Along these lines, computer-aided designing (CAD) and computer-aided manufacturing (CAM) advancements were acquainted with the dental group in mid-1980s (Beuer *et al.*, 2008). To create a physical model in industry or drug, two diverse methodologies have been used, i.e., subtractive and added substance. In a subtractive strategy, the material is subtracted from an underlying piece of material to leave the coveted formed part, (e.g., dental reclamation). Early CAD/CAM frameworks dependent solely on cutting reclamation from a preassembled obstruct with the utilization of pods, precious stones or jewel plates. This is typically proficient by regular numeric control (NC) machining, e.g., processing. Subtractive procedures utilize deliberately arranged device developments to cut material, (Strub *et al.*, 2006). NC machining is used typically in small model-making machines, for which they are used to fabricate metallic and/or ceramic crowns in dentistry (Chen *et al.*, 2006).

Additive manufacturing (AM) techniques, the today's generation of CAD/CAM (Yousef *et al.*, 2017), have been exploited to build complex three-dimensional (3D) models in medicine since the 1990s. AM has recently proposed as an effective applications in different dental fields, for example, creation of embed surgical aides, structures for settled and removable partial dentures, wax designs for the dental prosthesis, zirconia prosthesis and molds for metal castings, and maxillofacial prosthesis lastly for finished dentures (Lee and Lee, 2015). On the other hand, AM is a methodology wherein in which the last required part is fabricated by including various layers of material in one on another way. The key idea of this innovative method is that the 3D-CAD model is sliced into many thin layers and the manufacturing equipment uses this geometric data to build each layer sequentially until the part is completed. Hence, the additive fabrication is often referred as "layered manufacturing," "direct digital manufacturing," "three-dimensional printing," or "solid freeform fabrication." This additive technology scans yield arbitrarily complex shapes with cavities and undercuts; frequently the case in human anatomy structures (Torabi *et al.*, 2015).

AM simply consists of two phases: virtual phase (modelling and simulating) and physical phase (i.e., fabrication). Virtual prototyping is the development of model by dynamic and interactive simulation. The course of forming the physical model is formation of 3D physical model by CAD (Liu *et al.*, 2006). Stereolithography (SLA), inkjet-based system, selective laser sintering (SLS) and fused deposition modelling (FDM) are amongst the most frequently used technologies in dental practice. Plastics, ceramics and metals can be employed in these technologies, reported in the several studies (Azari and Nikzad, 2009; Andonović *et al.*, 2010). FDM is one of many cost-effective AM techniques in which a thermoplastic is first heated (inside the head) and then polymer melt is extruded in the form of thin layer through a nozzle, layer by layer, on the build platform. With FDM, it is possible to fabricate exact dimensions of dentures by utilizing patient specific scan data. The procedure begins with a cast delivered from an impression of a patient's mouth is carefully filtered and changed over to a CAD file, which is utilized to outline the structure and produce a standard triangulation language (.STL) file, appropriate for FDM processing (Wu *et al.*, 2012). The FDM process starts with importing a .STL file of a model into the preprocessing software. Process parameters like, tip size, material type and machining speed are automatically added and the control file for the machine is generated (Greul *et al.*, 1995). This model is oriented and mathematically sliced into horizontal layers varying from 0.127 to 0.331 mm in thickness. Figure 14.1 represents the schematic of FDM (Zein *et al.*, 2002).

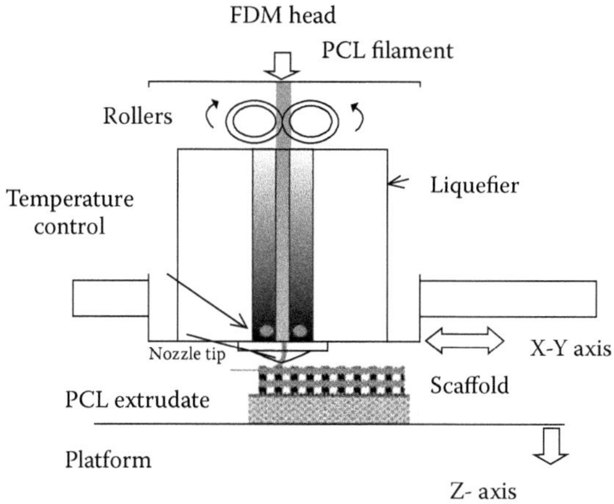

FIGURE 14.1 Schematic of FDM. (From Zein I *et al.*, *Biomaterials*. 2002; 23: 1169–85.)

FDM is the only AM technology that uses production grade thermoplastic materials to build the most of the durable parts (Attaran, 2017). Quality function deployment (QFD) analysis based comparison of FDM with seven other types of AM processes showed that FDM is the best AM method in terms of ability to create prototyped component for testing purpose (Peres and Martin, 1999). The research in the field of FDM-assisted dental science is very vast and reportedly various successful practices have been performed until now (Bibb *et al.*, 2006; Gibson *et al.*, 2006; Liu *et al.*, 2006; Espalin *et al.*, 2010; Guo and Leu, 2013). Since most of the commercial material used in existing FDM systems is not suitable for dental applications, research efforts have been continuously going on in the development of newer materials. Haq (2015) characterized the development of polymer nanocomposite filament wire for FDM. The polycaprolactone (PCL)-filled montmorillonite (MMT) and HAp composites were set up by dissolving and a solution utilizing a screw extruder. The mechanical characteristics were evaluated by ductile, flexural and charpy affect tests while the thermal characteristics were considered by means of differential scanning calorimetry (DSC) and thermogravimetric analyzer (TGA). A mimicked body liquid test was conducted to evaluate the effect of bioactivity. The properties of FDM system demonstrated that the specimens with ideal dimensional precision and relative density were observed at room temperature and 80 °C of build stage and nozzle temperature individually. Earlier, our group (Singh *et al.*, 2016) highlighted the development of in-house filament for FDM in order to enhance the application domain of the process.

In this study, wear-safe polymer composite filament (including, Nylon and Al_2O_3) was manufactured for grinding jobs of clinical dentistry. The outcomes of this study proposed that FDM system with composite feedstock filament fibre was within ± 4.5 sigma(s) constraint with respect to wear of prepared tooling. From the work presented by Li *et al.* (2008) it has been found that the use of ceramics like hydroxyapatite (HAp) could be considered as reinforcements, due to their chemical similarities with bones and jaw. Hontsu *et al.* (2011) arranged an unsupported HAp sheet of 5~10 μm thickness by a novel technique. Upon deposition of the thin HAp film on a dissolvable substrate by pulsed laser deposition (PLD) system, the HAp film was gathered as a detached sheet by dissolving the substrate by utilizing a solvent. After this, HAp sheets were straightforwardly joined to lacquer and dentin surfaces of extricated human teeth. Simulated salivation was showered each third day, and the course was watched for 90 days. Perception of the enamel-HAp and dentin-HAp interface by SEM uncovered that the HAp sheet had adhered to the polish and dentin surface. The literature is vast in the field of dentistry, but fewer studies have been reported for the development of cost-effective material for FDM in order to meet the required characteristics of PD/CD. In this work, an attempt has been made in the synthesis of in-house n-Hap particles by using eggshells. Next to synthesis, these n-HAp particles were used for the development of alternative FDM feedstock filaments (consisting of different %wt. of n-Hap particles) through a single screw extruder. The filaments were then used for prototyping (in the existing FDM setup) to test specimens in order to check the wear resistance properties of the resulting parts.

14.2 EXPERIMENTAL PROCEDURES

14.2.1 Methodology

The methodology of the present research work is illustrated in Figure 14.2. During the first step of experimentation, eggshell waste was collected and used to produce HAp powder through the conventional grinding process. In the second step, n-HAp powder has been synthesized by following standard procedure as illustrated by (Roy and Linnehan, 1974; Araiza *et al.,* 1999). The synthesized n-HAp powder was segregated by passing it through the sieve shaker in the third step, and verified through SEM and XRD analysis. Afterwards, in the fourth step, suitable blends of n-HAp-reinforced nylon have been established through melt flow test (MFT). Alternative feedstock filament of FDM was prepared with selected filament proportion in the fifth step. During the sixth step, the developed FDM filament was used for the preparation of sacrificial denture patterns/test specimens. Forthcoming research will be focused to investigate *in vitro* and metallurgical properties of the dentures in order to study the effect of the HAp reinforcement for improving denture characteristics. At the back end, activities like scanning of patient data, conversion into .STL format and filament testing were carried out, simultaneously.

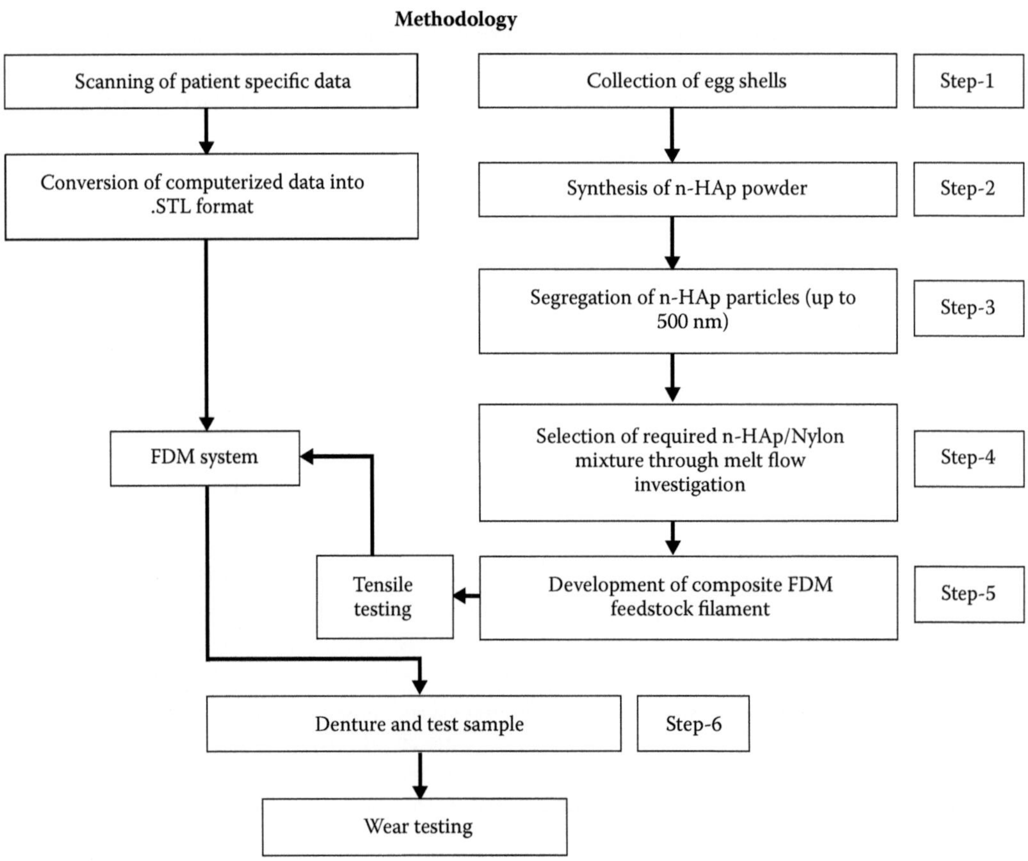

FIGURE 14.2 Methodology adopted.

14.2.2 Synthesis of Hydroxyapatite from Eggshell Waste

An egg shell residual was comprised of the fingernail skin and lamellar layer. The cuticle skin layer on the outermost surface contains various proteins, and lamellar layer shapes grid containing protein strands attached to calcite calcium carbonate crystals (Araiza *et al.*, 1999). The eggshell waste was collected, crushed into powder form, heated at various temperatures and chemically processed to obtain HAp ($Ca_{10}(PO_4)_6OH_2$) particles (Roy and Linnehan, 1974; Araiza *et al.*, 1999). After cleaning the eggshells in distilled water and subsequently grinding, the eggshell powder is dried in electric oven at 100 °C for about an hour. Then the powder is again heated at 400 °C for two hours in order to remove the organic product. Afterwards, eggshell powder is heated in a muffle furnace at 900 °C for 2.5 hrs. With this heating, calcium carbonate presented in the powder was converted into calcium oxide due to the elimination of carbon dioxide. In the next step, calcium phosphate tri-basic and calcium oxide is mixed in a specially designed container that produces a moist atmosphere at high temperature (1050 °C). The container was sealed to lock the steam inside the furnace. After continuous heating for about 3 hours, the reaction between calcium phosphate tri-basic and calcium oxide results into formation of HAp as per Equation 14.1.

$$3Ca_3(PO_4)_2 + CaO + H_2O \Rightarrow Ca_{10}(PO_4)(OH)_2 \quad (14.1)$$

Figure 14.3 shows the pictorial view of finally produced HAp particles whereas Figure 14.4 shows the SEM and XRD plots for resulted HAp particles.

14.2.3 Selection of Required Proportion of HAp/Nylon through Melt Flow Investigation

The present research work was focused on the development of HAp-reinforced Nylon filament of FDM, to be used for the fabrication of dentures. This alternative FDM filament was going to replace commercial ABS-P430 (used in Stratasys Inc. uPrintSE) filament with making any modification in the existing FDM system. So, for the successful run of alternative filament, its rheological properties were matched with original equipment manufacturer filament by performing melt flow analysis on melt flow indexer (MFIer) as shown in Figure 14.5.

Various combinations of HAp and Nylon were tested as per ASTM-D1238 standard, i.e., at 230 °C barrel temperature and 3.8 kg extrusion force. Table 14.1 shows the results of MFI.

It has been found that the mixture with 30% of HAp (%wt.) and 70% of Nylon-6 (%wt.) possessed melt flow index (MFI) similar to ABS P430 (i.e., 2.41gm/10min). So, this proportion is suitable for developing a filament for uPrintSE FDM system. However, presently we were also having

FIGURE 14.3 Synthetic HAp powder. (Courtesy of Unpublished Lab experimental work, Manufacturing Research Lab, Dept. of Production Engineering, GNDEC, Ludhiana, India.)

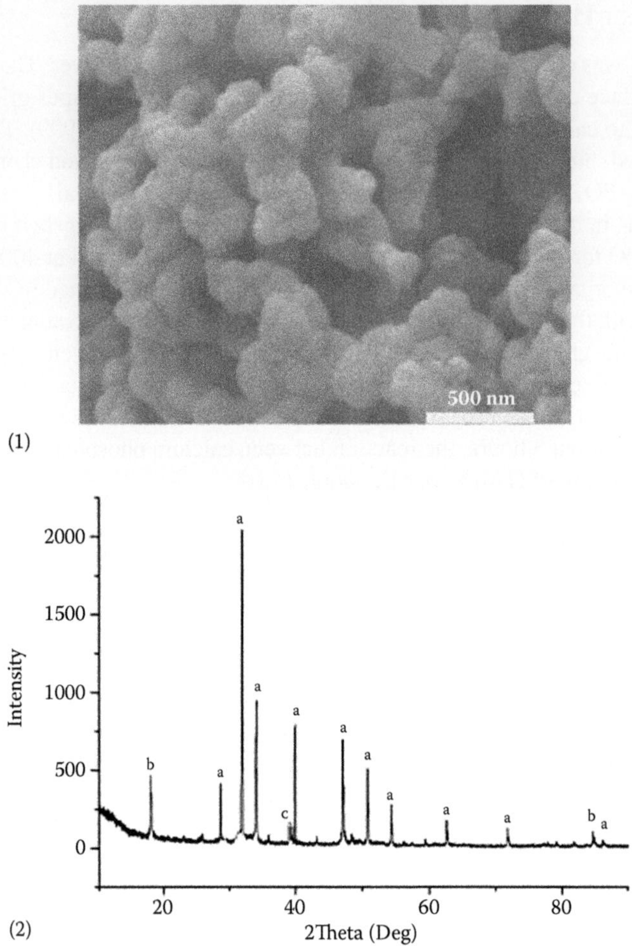

FIGURE 14.4 (1) SEM image (X1500) of n-HAp particles and (2) XRD plot showing peaks for n-HAp (a), $Ca(OH)_2$ (b) and CaO (c). (Courtesy of Unpublished Lab experimental work, Manufacturing Research Lab, Dept. of Production Engineering, GNDEC, Ludhiana, India.)

Accucraft 250i 3D printer (make: Divide by Zero) so it was easy to change the process parameters of FDM printers such as speed of printing, temperature of extrusion head, etc. Presently, first three filament proportions of Table 14.1 were used in the filament development.

14.2.4 Development of Alternative FDM Filament

The selected filament proportion was extruded in the form of wire (diameter ranging between 1.70–1.80 mm) on a single screw extruder of L/D 20. The extrusion parameters such as barrel temperature, die temperature and screw speed were kept at 160 °C, 190 °C and 15–18 rpm, respectively, whereas the speed of take up unit was changed to meet the requirements of wire diameter. Figure 14.6 shows the extrusion of alternative FDM filament wire.

14.2.5 Obtaining Master Cast of Patient and Conversion in .STL Format

One of the most important steps in denture construction is the generation of negative impression of the patient cavity for missing teeth. This consisted of two phases, i.e., primary impression and

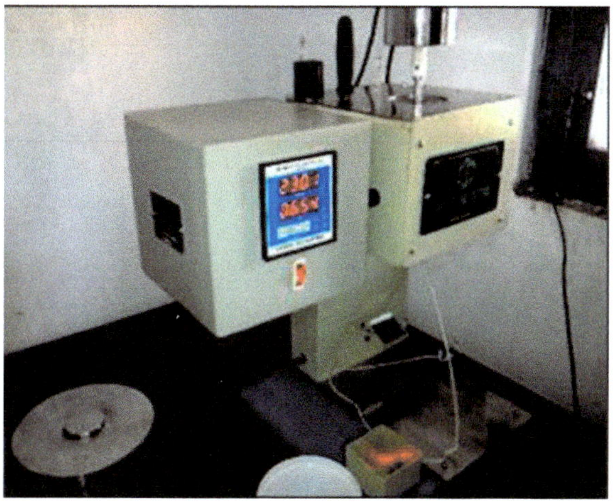

FIGURE 14.5 Pictorial view of MFIer. (Courtesy of Unpublished Lab experimental work, Manufacturing Research Lab, Dept. of Production Engineering, GNDEC, Ludhiana, India.)

TABLE 14.1
Results of MFI

S. No.	HAp (%wt.)	Nylon (%wt.)	MFI (gm/10min)
1	5	95	9.053
2	10	90	8.163
3	15	85	6.895
4	20	80	5.485
5	25	75	3.865
6	30	70	2.971
7	40	60	1.35

Source: Courtesy of Unpublished Lab experimental data, Manufacturing Research Lab, Dept. of Production Engineering, GNDEC, Ludhiana, India.

FIGURE 14.6 Extrusion of n-HAp-reinforced nylon filament wire. (Courtesy of Unpublished Lab experimental work, Manufacturing Research Lab, Dept. of Production Engineering, GNDEC, Ludhiana, India.)

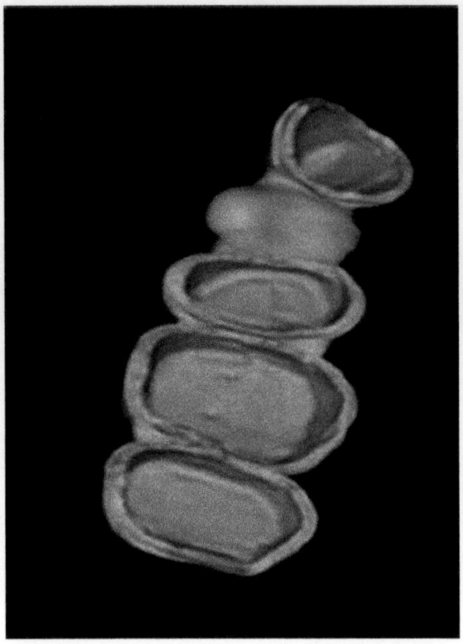

FIGURE 14.7 Firing of .STL format in CatalystEX® software of FDM. (Courtesy of Unpublished Lab experimental work, Manufacturing Research Lab, Dept. of Production Engineering, GNDEC, Ludhiana, India.)

secondary impression as discussed in the literature (Sutter *et al.,* 1993). Afterwards, this negative cast is converted into a positive CAD format through the use of scanning devices and image processing software (Romero *et al.,* 2015). After obtaining the required CAD file, it was then converted into .STL format due to its suitability in print with FDM system (Figure 14.7). Before the printing was started, the .STL format again converted into .CMB file wherein the geometrical model was split into a number of slices.

14.2.6 Printing of PD

Finally, the printing of PD and test specimens for wear analysis (please refer to Figure 14.8) were carried out on Accucraft 250i FDM system at 0° build orientation, 50% infill and ± 45° raster orientation.

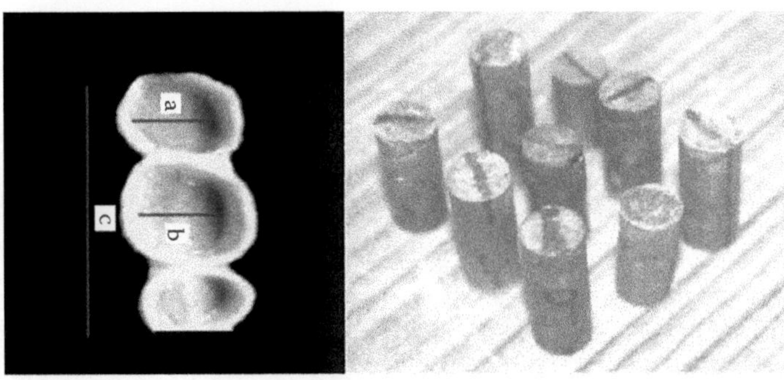

(1) (2)

FIGURE 14.8 (1) N-HAp-reinforced nylon–based PD and (2) test specimens for wear. analysis. (Courtesy of Unpublished Lab experimental work, Manufacturing Research Lab, Dept. of Production Engineering, GNDEC, Ludhiana, India.) Note: a, b, c are the measured dimensions of the PD.

14.3 RESULTS AND DISCUSSION

14.3.1 Tensile Strength

The tensile strength of the filaments developed was tested in order to check whether these filaments are able to use for prototyping or not. Since in FDM printers, the filament wire has been pulled with the help of gear arrangements during the prototyping hence poor tensile strength may result in failure of a material for prototyping. The tensile strength of the feedstock filaments (as per ASTM-638 standard) was tested on computerized UTM system. The results of the tensile test are summarized in Table 14.2. It has been found that the tensile strength of the first two filaments of Table 14.2 was higher as compared to the OEM filament. However, the tensile strength of the third filament is also sufficient to run it in FDM system. Here, it is to highlight that with an increase in the %wt. of n-HAp particles in the polymeric composite, the tensile strength reduces. This might be due to poor bonding of n-HAp with nylon matrix. In the present study, a single screw extruder was used for the fixing and extrusion of alternative FDM filament which might have affected the bonding and distribution of n-HAp particles in polymer matrix. Twin screw extruders can be used in the upcoming researches for the better compounding of composite filament ingredients. (Refer to Figure 14.9.)

TABLE 14.2
Tensile Strength of the Filaments

Material	Tensile Strength (MPa)
n-HAp(5%)+Nylon(95%)	40.26
n-HAp(10%)+Nylon(90%)	25.33
n-HAp(15%)+Nylon(85%)	16.32
ABS-P430	22

Source: Courtesy of Unpublished Lab experimental data, Manufacturing Research Lab, Dept. of Production Engineering, GNDEC, Ludhiana, India.

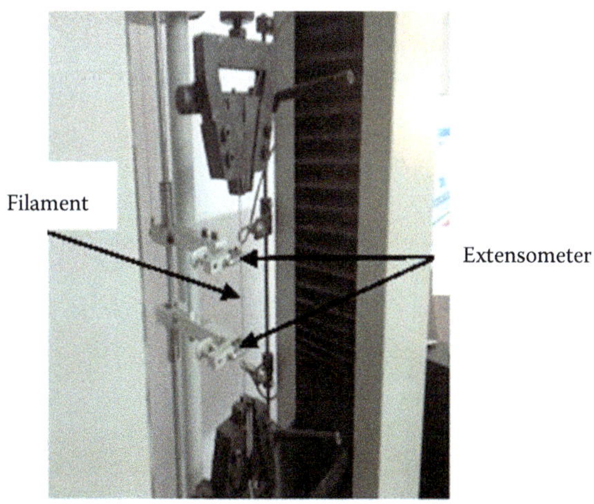

FIGURE 14.9 Tensile testing of filament. (Courtesy of Unpublished Lab experimental work, Manufacturing Research Lab, Dept. of Production Engineering, GNDEC, Ludhiana, India.) (Image courtesy: Manufacturing Research Lab, GNDEC.)

14.3.2 Wear Test

Methodologies are not established for analyzing the wear/tribological properties of composite biomaterials (Hussein *et al.*, 2015). Generally, tribo tests are used for finding out the wear properties of metallic materials and numerous methods are available (such as: block-on-disc, pin-on-disc and ball-on-disc). The wear test was conducted on pin-disk type tribo-meter setup (make: Ducom TL-20) as per ASTM-G-99 standard (refer to Figure 14.10) under dry sliding conditions and room temperature. This system is capable of calculating the amount of wear in terms of length with the help of a linear variable differential transformer, which calculates the distance following by the pin during the test. Since in original pin-on-disc machine, the test specimens, generally are cylindrical/rectangular pins, are rubbed against a hardened disc of 65 HRC. In the present case, the test material is a polymeric composite which might not wear during the test, due to its low coefficient of friction. Many researchers have highlighted that polymers and polymers composites formed a thin protective film during the test (Nagaraju *et al.*, 2011). So, in order to tackle this problem a silicon carbide emery paper (600 grit size) was pasted on the steel disc prior to the test as shown in Figure 14.10(1). The test was conducted as per Taguchi L9 orthogonal array at variable n-HAp proportion, normal load and speed of rotation. Table 14.3 shows the results of wear.

The results of Table 14.3 were used for calculation of signal/noise (S/N) ratio for the wear and for plotting the trends of input process variables for wear (in terms of S/N ratio). Figure 14.11 shows the

FIGURE 14.10 Wear testing equipment. (Courtesy of Unpublished Lab experimental work, Manufacturing Research Lab, Dept. of Production Engineering, GNDEC, Ludhiana, India.) (Image courtesy: Manufacturing Research Lab, GNDEC.)

TABLE 14.3
Results of Wear

Exp. No.	n-HAp (%wt.)	Normal Load (kg)	Speed of Rotation (rpm)	Wear (μm)	S/N Ratio (db)
1	5	1	150	159	–44.0279
2	5	1.5	200	344	–50.7312
3	5	2	250	530	–54.4855
4	10	1	200	113	–41.0616
5	10	1.5	250	136	–42.6708
6	10	2	150	167	–44.4543
7	15	1	250	112	–40.9844
8	15	1.5	150	84	–38.4856
9	15	2	200	106	–40.5061

Source: Courtesy of Unpublished Lab experimental data, Manufacturing Research Lab, Dept. of Production Engineering, GNDEC, Ludhiana, India.

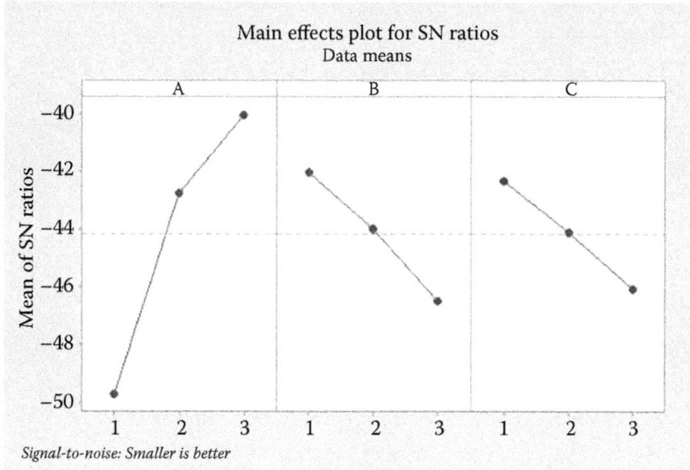

FIGURE 14.11 S/N ratio of input process parameters for wear. *Note: A, B, C are %wt. of n-HAp, Normal load and Speed of rotation respectively.* (Courtesy of Unpublished Lab experimental work, Manufacturing Research Lab, Dept. of Production Engineering, GNDEC, Ludhiana, India.)

S/N ratio plot of input process variables for wear of the composite pin. Tables 14.4 and 14.5 show the analysis of variance (ANOVA) and response table of S/N ratio for wear, respectively.

From Figure 14.11, it has been found that with an increase in the %wt. of n-HAp particles in the polymeric composite filament the wear resistance property of the same has been improved. HAp is generally categorized as a ceramic material consisted of higher wear resistance property due to which the polymeric composites with higher concentration of n-HAp particle resulted into enhanced resistance against scratch, wear and abrasion. In case of a parameter "normal load," it has been found that at higher load the composite specimens were heavily worn out. This might be due to the detachment of n-HAp particles from the polymeric material under higher values of load. Oxidation, delamination and adhesion can also be the reasons behind this trend as observed by other researchers (Xue *et al.,* 1997; Aigbodion *et al.*, 2012). In case of parameter "speed of rotation," it has been observed that at a high speed of rotation the wear of the samples was higher as compared to the test conducted at lower speeds. The main reason behind this might be the transverse force whose magnitude increased with the speed of rotation. The optimized result, as suggested by Figure 14.11,

TABLE 14.4
ANOVA for S/N Ratio of Wear

Source	Degree of Freedom	Sum of Square	Variance	Fisher's Value	Probability	Contribution (%)
%wt. of n-HAp	2	151.94	75.97	10.39	0.088	69.90
Normal load	2	29.97	14.98	2.05	0.328	13.78
Speed of rotation	2	20.82	10.41	1.42	0.412	9.57
Error	2	14.62	7.309	–	–	6.72
Total	8	217.35	–	–	–	100

Source: Courtesy of Unpublished Lab experimental data, Manufacturing Research Lab, Dept. of Production Engineering, GNDEC, Ludhiana, India.

TABLE 14.5
Response Table of S/N Ratio to Wear Data

Levels	%wt. of n-HAp	Normal Load	Speed of Rotation
1	–49.75	–42.02	–42.32
2	–42.73	–43.96	–44.10
3	–39.99	–46.48	–46.05
Delta	9.76	4.46	3.72
Rank	1	2	3

Source: Courtesy of Unpublished Lab experimental data, Manufacturing Research Lab, Dept. of Production Engineering, GNDEC, Ludhiana, India.

is n = HAp-15% (by wt.), normal load = 1 kg and speed of rotation = 150 rpm. Further, Equation 14.2 was used for the calculation of the optimized value of wear under the effect of suggested levels.

$$\eta_{opt} = m + (m_3 - m) + (m_1 - m) + (m_1 - m) \\ = -36.0173 db \quad (14.2)$$

Where, η_{opt}, m, m_3, m_1 and m_1 is the optimum value of S/N ratio obtained, overall mean of S/N data (i.e., mean of the S/N ratios for 9 experiments of Table 14.3), S/N ratio of %wt. of n-HAp particles at level 3, S/N ration of normal load at level 1 and S/N ratio of speed of rotation at level 1, respectively. Further Equation 14.3 was used for converting the η_{opt} into wear.

$$y_{opt}^2 = 1/10_{opt}^{\eta/10} \\ = 1/10^{-36.0173/10} \\ = 63.2\ \mu m \quad (14.3)$$

In order to verify the predicted value of wear and to justify the accuracy of the analysis, a confirmatory experiment has been conducted at the suggested level of input process parameters. The experimental value resulted in a wear value of 64.5 µm, which is very close to the predicted value.

14.4 CONCLUSIONS

In the present chapter, in-house n-HAp particles have been developed by using kitchen waste eggshells, and the synthesized particles characterized using SEM and XRD. The developed n-HAp particles were used as reinforcements for the development of alternative FDM feedstock filament for dentistry applications. On the basis of the tensile and wear tests, it could be concluded that the tensile strength of the filaments decreased by increasing the %wt. of n-HAp particles in the filament proportion. When the loading was up to 10%wt. of n-HAp particles, the tensile strength of the resulting filament was higher as compared to the OEM filament material. Further increase in n-HAp contents to 15%wt. resulted in reduction of tensile strength. Poor bonding between n-HAp and nylon might be the reason behind this trend, which could be tackled up to a certain extent with the use of twin screw extruder. The wear analysis indicated that no input process parameter has significantly affected the wear of the composite test specimens and ANOVA table construed that the percentage contribution of %wt. of n-HAp particles, normal load and speed of rotation was 69.9%, 13.8% and 9.6% respectively, in close agreement with predicted results. Thus, the results of Chapter 14 highlight the suitability of this alternative FDM filament for developing patient-specific

PD/CD. Nevertheless, before the actual clinical application it is mandatory to carry out *in vitro* and *in vivo* analysis for the better simulation of material behaviour. The filament developed could be used for FDM-assisted investment casting application for the manufacturing of non-removal PD/CD in future. It is believed that the presence of n-HAp particles in the PD/CD will be helpful for cell growth and biocompatibility with the passage of time. It would be a great interest to analyze the bioactive response of such dentures in the simulated body environment. After in-depth analysis, it could be finally arrived at that the filaments developed in this study are totally suitable for 4D printing applications.

ACKNOWLEDGMENTS

The authors are thankful to Department of Science and Technology, SERB, Government of India, Ref. No. SERB/F/3572/2015-2016 for providing financial support.

REFERENCES

Abraham CM. A brief historical perspective on dental implants, Their surface coatings and treatments. *Open Dent J.* 2014; 8: 50–55.

Aigbodion VS, Hassan SB, Agunsoye JO. Effect of bagasse ash reinforcement on dry sliding wear behaviour of polymer matrix composites. *Mat. and Dgn.* 2012; 33: 322–7.

Andonović V, Vrtanoski G.V, Vrtanoski G, authors. Growing rapid prototyping as a technology in dental medicine. *Mech Eng Sci J.* 2010; 29:31–39.

Araiza EMRM, Brostow W, Castano VM, Estrada JRD, Hernandez R, Rodrıguez JR. Synthesis of hydroxyapatite from eggshells. *Materials Letters.* 1999; 41:128–34.

Attaran M. The rise of 3-D printing: The advantages of additive manufacturing over traditional manufacturing. *Business Hori.* 2017; In-Press.

Azari A, Nikzad S. The evolution of rapid prototyping in dentistry: A review. *Rapid Prototyping J.* 2009; 15:216–25.

Beuer F, Schweiger J, Edelhoff D. Digital dentistry: An overview of recent developments for CAD/CAM generated restorations. *Br Dent J.* 2008; 204: 505–11.

Bibb B, Eggbeer D, Williams R. Rapid manufacture of removable partial denture frameworks. *Rapid Prototyping J.* 2006; 12: 95–9.

Campbell SD, Cooper L, Craddock H, Hyde TP, Nattress B, Pavitt SH, Seymour DW. Removable partial dentures: The clinical need for innovation. *The J of Pros Dent.* 2017; In-Press.

Chen H, Yang X, Chen L, Wang Y, Sun Y. Application of FDM three-dimensional printing technology in the digital manufacture of custom edentulous mandible trays. *Scientific Reports.* 2006; 6: 19207.

Eden SE, Kerr WJS, Brown J. A clinical trial of light cure acrylic resin for orthodontic use. *J. of Orthodontics.* 2002; 29: 51–5.

Espalin D, Arcaute K, Rodriguez D, Medina F, Posner M, Wicker R. Fused deposition modeling of patient-specific polymethylmethacrylate implants. *Rapid Prototyping J.* 2010; 16: 164–73.

Gibson I, Cheung LK, Chow SP, Cheung WL, Beh SL, Savalani M, Lee SH. The use of rapid prototyping to assist medical applications. *Rapid Prototyping J.* 2006; 12: 53–8.

Greul M, Pintat T, Greulich M. Rapid prototyping of functional metallic parts. *Comp in Ind.* 1995; 28: 23–8.

Guo N, Leu MC. Additive manufacturing: Technology, applications and research needs. *Frontiers of Mech Eng.* 2013; 8: 215–43.

Haq RHBA. Characterization and development of polycaprolactone (PCL)/montmorillonite (MMT)/hydroxapaptite (HA) nano-composites for fused deposition modelling (FDM) process. Ph.D Thesis Report, Faculty of Mechanical and Manufacturing Engineering, Universiti Tun Hussein Onn Malaysia. 2015.

Henry PJ. Future therapeutic directions for management of the edentulous predicament. *J Pros Dent.* 1998; 79: 100–6.

Hontsu S, Yoshikawa K, Kato N, Kawakami Y, Hayami T, Nishikawa H, Kusunoki M, Yamamoto K. Restoration and conservation of dental enamel using a flexible apatite sheet. *J of the Australian Ceramic Soc.* 2011; 47: 11 3.

Hussein MA, Mohammed AS, Al-Aqeeli N. Wear characteristics of metallic biomaterials: A Review. *Materials.* 2015; *8*: 2749–68.

Janus CE, Hunt RJ, Unger JW. Survey of prosthodontic service provided by general dentists in Virginia. *J Pros. Dent.* 2007; 97: 287–91.

Jendrensen MD, Allen EP, Bayne SC, Donovan TE, Hansson TL, Klooster J *et al.* Annual review of selected dental literature: Report of the Committee on Scientific Investigation of the American Academy of Restorative Dentistry. *J Pros Dent.* 1997; 78: 54–92.

Lee JH, Lee CH. A stereolithographic resin pattern for evaluating the framework, altered cast partial removable dental prosthesis impression, and maxillomandibular relationship record in a single appointment. *J of Pros. Dent.* 2015; 114: 625–6.

Li L, Pan H, Tao J, Xu X, Mao C, Gu X, Tang R. Repair of enamel by using hydroxyapatite nanoparticles as the building blocks. *J of Mat. Chem.* 2008; 18: 4079–84.

Liu Q, Leu MC, Schmitt SM. Rapid prototyping in dentistry: Technology and application. *Int J of Adv Mfg Tech.* 2006; 29: 317–35.

Nagaraju B, Ramji K, Prashad VSRK. Studies on tribology properties of ZnO filled polymer nanocomposites. *ARPN J of Eng and App Sc.* 2011; 6: 75–82.

Negrutiu M, Sinescu C, Romanu M, Pop D, Lakatos S. Thermoplastic resins for flexible framework removable partial dentures. 2005. Available at http://www.tmj.ro/pdf/2005_number_3_833235561124423.pdf.

Peres F, Martin C. Design methods applied to the selection of a rapid prototyping resource. *Proc.of 7th IEEE International Conference on Emerging Technologies and Factory Automation*, 18–21st October, Barcelona, Spain, 2002; 1, pp. 417–22.

Ring ME. *Dentistry: An Illustrated History.* 2nd edition. Abradale Press, 1985.

Rodrigues AH, Morgano SM. An expedited technique for remaking a single complete denture for an edentulous patient. *J Pros Dent.* 2007; 98: 232–4.

Romero L, Jiménez M, Espinosa MM, Domínguez M. New design for rapid prototyping of digital master casts for multiple dental implant restorations. *PLoS ONE.* 2015; 10: e0145253.

Roy DM, Linnehan SK. Hydroxyapatite formed from coral skeletal carbonate by hydrothermal exchange. *Nature.* 1974; 247: 220–2.

Singh R, Singh S, Fraternali F. Development of in-house composite wire based feed stock filaments of fused deposition modelling for wear-resistant materials and structures. *Comp Part B.* 2016; 98: 244–9.

Strub JR, Rekow ED, Witkowski S. Computer-aided design and fabrication of dental restorations: Current systems and future possibilities. *J Am Dent Assoc.* 2006; 137: 1289–96.

Sutter F, Weber H, Sorensen J, Belser U. The new restorative concept of the ITI dental implant system: Design and engineering. *Int J of Periodontics and Restorative Dent.* 1993; 13: 409–31.

Torabi K, Farjood E, Hamedani S. Rapid Prototyping technologies and their applications in prosthodontics, a Review of Literature. *J Dent (Shiraz).* 2015; 16: 1–9.

Wu J, Wang X, Zhao X, Zhang C, Gao B. A study on the fabrication method of removable partial denture framework by computeraided design and rapid prototyping. *Rapid Prototyping J.* 2012; 18: 318–23.

Xu L, Zhu Z, Chen G, Qu C. Effect of loading and sliding velocity on tribological behaviors of aramid fiber reinforced PA1010 composites. *Indl Lubr and Trib.* 2010; 62(1): 46–51.

Xue Q, Foret F, Dunayevskiy YM, Zavracky PM, McGruer NE, Karger BL. Multichannel microchip electrospray mass spectrometry. *Analytical chemistry.* 1997: 69(3):426–30.

Youssef A, Hollister SJ, Dalton PD. Additive manufacturing of polymer melts for implantable medical devices and scaffolds. *Biofabrication.* 2017; 9(1): 012002.

Zein I, Hutmacher DW, Tan KC, Teoh SH. Fused deposition modeling of novel scaffold architectures for tissue engineering applications. *Biomaterials.* 2002; 23:1169–85.

Index

Page numbers followed by f and t indicate figures and tables, respectively.

A

ABC transporter (ATP binding cassettes transporter), 70–71
Acetanilide, 204
Acetobacteracea, 152
Acetobacter xylinum, 152
Acetobacter xylinus, 152
Acinetobacter johnsonii, 125
Acrylamide, 149
Activator of transcription-3 pathways, 240
Acute myeloid leukaemia (AML), 207
Acute wounds, 221, 222f
Additive manufacturing (AM) techniques, 264
ADEPT, *see* Antibody-directed enzyme prodrug therapy (ADEPT)
Adriamycin, 88t, 90
Aggregatibacteractinomycetemcomitans biofilm, 232
Amine-functionalized NPs, 115
Amine group, 116f
Amine group-involved coupling, 115
Amine-terminated dendrimers (PAMAM-NH_2), 123
Amphiphilic gadolinium (II) chelate, 10
Amphipilic block polymer, 214f
Amphotericin B, 134
Ampicillin, 134
Anaemia of chronic disease (ACD), 241
Analyte–AuNP interactions, 2
Annexin V-Cy3 staining, 3
Antiapoptotic molecules, 240–241
Antibodies, 2f, 3, 206–207
  antigen, 102
Antibody-directed enzyme prodrug therapy (ADEPT), 211, 211f, 212
Antibody–drug conjugates (ADCs), 206–207
Anticancer bioactives, 207
Antifilarial drug, 78
Antigen antibodies, 102
Anti-HLA-DR Fab' L fragments, 89t
Anti-infective agents, 121
Anti-inflammatory drugs, 125, 157, 225, 241, 242
Antimicrobial peptide dendrimers, 124f
Antimicrobials used in OSM, 228t
Antineoplastic therapy
  approaches for targeted, 206f
  enzyme-activated prodrugs
    antibody-directed enzyme prodrug therapy, 212
    gene-directed enzyme prodrug therapy (GDEPT), 212
  linker in prodrugs, role, 205–206
  lipid-based prodrugs, 214–216
  nanostructured prodrugs for cancer therapeutics, 215t–216t
  polymeric prodrugs
    N-(2-hydroxypropyl) methacrylamide (HPMA), 210
    PEG–drug conjugates, 210
    polyglutamic acid (PGA), 210–211
  prodrugs, 205
  stimuli-sensitive prodrugs, 212–214
  targeting-ligand conjugated prodrugs
    antibody–drug conjugate, 206–207
    aptamer–drug nanoconjugates, 209
    folic acid–drug conjugates, 209–210
    peptide–drug conjugates, 207–209
Anti-TAG-72 humanized antibody (HuCC49 ΔCH2), 212
Apoptosis, 240
Aptamer–drug nanoconjugates, 209
Aqueous-free drug, 91
Arginine-glycine-aspartic acid peptide ligands, 9
Arg-8-vasopressin (VP) peptide, 187
Artificial homopolymers, 191t
Aspirin, 204
ASTM F138, 22
Atomic absorption spectroscopy (AAS), 160
Atomic layer deposition, 33
ATP-binding cassette (ABC), 69
  half transporters, 71
ATP hydrolysis, 69
Atrazine (2-chloro-4-ethylamino-6-isopropylamino-1,3,5-triazine), 159
Auger electron spectroscopy (AES), 27
Autoantibodies, 240
Avidin, 88t

B

Bacillus subtilis, 125
Bacteria, 123, 125, 128, 131, 134, 135, 148
Bacterial cellulose–functionalized MWNTs, 11t
Bacterial cellulose nanofiber, 153
Bacterial cellulose/polyaniline/SWNTs, 11t
BALY, 123, 124f
Benign tumors, 99
Benzene sulfonamide–modified dendrimers, 125
β-tricalcium phosphate (β-TCP), 156
Bioactive implant, 18
Bioactivity, 168
Biocompatibility, 167–168, 248, 249
Bioconjugated aqueous dispersible QDs, 114f
Bioconjugation, 107
  covalent coupling, 115
    amine group-involved coupling, 115
    carboxylic group involved coupling, 115
    thiol group-involved coupling, 115
  direct absorption, 107–115
Biodegradability, 169
Biodegradable polymers, 55
Biofilms
  formation, 224–225, 225f
  localized disruption of

Aggregatibacteractinomycetemcomitans biofilm, 232
antimicrobials used in OSM, 228t
approaches utilized to achieve targeted heating, 229t–230t
ciprofloxacin concentrations, 230f
cyclic di-GMP (c-di-GMP), 228
Enterococcus faecalis (E. coli) biofilms, 232
extracorporeal shock waves (ESWs), 232
fiber-based Q-switched Nd:YAG shockwave laser probes, 231–232
HIFU–induced local warming, 228
high-intensity focused ultrasound (HIFU) transducers, 228
indocyanin green–based photosensitizer (ICG–based PS), 232
low temperature–sensitive liposomes (LTSLs), 228, 231f
oral therapy, antibiotics for, 228t
photodynamic therapy (PDT), 232
photothermal therapy (PT), 232
Pseudomonas aeruginosa PA01 biofilm, 227
reactive oxygen species (ROS), 232
S. aureus survival, 231f
scanning electron microscopy (SEM), 230
superparamagnetic iron oxide nanoparticles (SPIONs), 227–228
Bioinert implant, 18
Biological sensors, 10
Biomarkers of cancer, 107
Biomaterials, 17, 139–162
Biomedical applications
carbon nanotubes (CNTs), 11t
ceramic nanoparticles (CENPs), 8t
gadolinium nanoparticles (GdNPs), 5t
graphene in, 55t
silica nanoparticles (SiNPs), 8t
of superparamagnetic iron oxide nanoparticles (SPIONs), 249f
titanium oxide nanoparticles (TiONPs), 8t
Biomimetics
mineralization, 59
network of fibers, 176
tri-component fibrous scaffold, 177
Biomimicry, 168
Biomolecular sensors, 106
Bioprecursor prodrugs, 205
Biopsy, 98, 102
Biorecognition, 10, 22
Biotin, 88t
Biotin-coated liposomes, 90
Biotolerant implant, 18
Biphasic organic–inorganic composite fibers, 176
Bipolaons, 146–147
Blood–brain barrier (BBB), 189
Blood protein testing, 100
Blood tests, cancer diagnosis, 100
Bone, 8, 11, 18
hierarchical structure nature, 18f
Bone morphogenic proteins (BMP-2), 177
Bovine serum albumin (BSA), 3, 168, 189, 192
Breast cancer, 6, 98
Breast tomosynthesis, 102
Brominated polystyrene (PBrS)/poly(ethylene oxide) (PEO), 30f
Brownian relaxation, 255–256

C

Caco-2 cell monolayers, 73
Cadmium (Cd)-based nanoparticles, 9
Calcium phosphates, 176
Calcium phosphate nanoparticles (CaPNPs), 196–197
Calcium silicate (CaO-SiO_2) coating, 31
CalichDMH, 207
Camptothecin, 209
Cancer
biomarkers, 107, 113t
breast, 6, 98
chemotherapy, 91
colon, 99
colorectal, 98
defined, 98–99
imaging techniques, 100
biopsy, 102
computerized tomography, 100–101
fluoroscopy, 101
histochemistry, 102
immunohistochemistry, 102–103
magnetic resonance imaging, 101
mammography, 102
nuclear scan (radionuclide scan), 101
positron emission tomography, 101
ultrasound imaging, 101
lab test, 100
blood tests, 100
urine tests, 100
liver, 98
lung, 98
markers, 104
nanotechnology, 103
carbon nanotubes (CNTs), 104
gold nanoparticles, 104
iron oxide nanoparticles, 103–104
nanocantilevers, 106–107
nanowires, 107
quantum dot-doped silica nanoparticles, 106
quantum dots (QDs), 104–106
utility, 108t–112t
pathophysiology, 99
pathophysiology of, 98–99
physical examination, 100
stomach, 98
therapy, 249, 257
urine tests, cancer diagnosis, 100
Candida albicans, 131
Candida glabrata, 125
Candida lipolytica, 125
Carbon-based nanomaterials, 10
Carbonic anhydrase enzymes, 125
Carbon nanofibers, 140f, 151
and Ni foam, 151–152
Carbon nanohorns (CNHs), 10
Carbon nanotubes (CNTs), 10, 51, 104, 147
biomedical application, 11t
nanofiber-supported platinum by, 147–148

Carbosilane, 122
  dendrimers, 132
Carboxylated MWNTs, 11t
Carboxylic group, 114f
  involved coupling, 115
Carboxymethyl-β-cyclodextrin decorated chitosan (CMCD-g-CS) NPs, 192
Carrier-linked prodrugs, 205
Cationic liposomes, 90
CD36, 70
CdTe QDs, 105f
Cefotaxime, 87t, 134
Cell-penetrating peptide (CPP), 192
Cell separation, 6
Cellular attachment, 59, 166, 169, 175
Cellular propagation, 167
Cellular response and nanoarchitecture, 174
  fiber alignment, 179–181
  fiber composition, 175–176
  fiber diameter, 174–175
  mechanical properties, 177–179
  organic–bioinorganic hybrid fibers, 176–177
Cellulose, 10, 149, 152
Cellulose fibers, 150, 150f
  with antibacterial Ag-loading Nano SiO_2, 149–151
Ceramic, 7
Ceramic nanoparticles (CENPs), 7
  biomedical application, 8t
Challenges for protein delivery, 186–187
Chitosan, 10, 59, 131, 156, 190–192
Chloramphenicol palmitate, 204
Chloramphenicol sodium succinate, 204
Chronic wounds, 222, 222f
  complications, 223f
  microenvironment, 222
Circulating tumor cell tests, 100
Citrullination, 240
Coaxial electrospinning, 171
Cobalt alloys, 27
Cobalt-based alloys, 20–21
Co-Cr-Mo alloys, 22
Coelectrospun, 176
Collagens, 167
Colloidal agents, 191t
Colloidal inorganic NPs, 107
Colon cancer, 99; *see also* Cancer
Colorectal cancer, 98; *see also* Cancer
Complete blood count (CBC), 100
Composite, 265, 271–274
Compritol 888 ATO, 81
Computer-aided designing (CAD), 264
Computer-aided detection, 102
Computer-aided manufacturing (CAM), 264
Computerized tomography, 100–101
Conductive polymers, 147
Copolymers, 191t
Core-shell, 249, 257
Core-shell fibers, 171–172, 172f
Corrosion, 25, 27f
Cotton fibers, 148
Coupling efflux transporters, 71–72
Covalent coupling, 115
  amine group-involved coupling, 115
  carboxylic group involved coupling, 115
  thiol group-involved coupling, 115
Cremophor, 73
Cryptococcus neoformans, 125
Curcumin, 82t
Cyclic di-GMP (c-di-GMP), 228
Cycloheximide (CHX), 84
CYP450 3A, 84
Cytochrome P450, 69
Cytotoxicity, 5, 7, 123, 167, 195, 242

D

Decomplexation, 7
Dendrimer-based nanohybrids with antimicrobial activity, 130f
Dendrimer entrapped α-tocopheryl succinate AuNPs, 110t
Dendrimer–membrane interactions, 123
Dendrimer-modified chitosan, 131
Dendrimers, 121–122
  anti-infective nanomedicines, 126t–128t
  antimicrobial activity, 122–123, 125
    nanohybrids, formation of, 129–131
    surface engineering, 125–129
  antimicrobial agents, delivery, 135
  chemical structure, 133f
  mechanisms, 122
    electrostatic interaction with membrane, 122–123
    membrane fluidity on dendrimers–membrane interactions, 123–124
    pathozymes, inhibition of, 125
    surface groups on dendrimers–membrane interactions, 123
  novel, 131
    carbosilane dendrimers, 132
    PAMAM dendrimers, 132
    peptide dendrimers, 131–132
    triazine dendrimers, 132–134
  structure of, 122f
Dental implants, 25
Dentistry, 264, 265, 274
Dentures, 264
Deposition, 265
  drug, 232
  electrochemical, 249
  silanization, 33–35
Detoxification of biological fluids, 6
Dextran NPs, 192f
Diagnosis, osteomyelitis (OSM), 225–226
Diagnostic methods, 100–103
1,4-Diazabicyclo[2.2.2]octane (DABCO), 134
Difluorodeoxyuridine (dFdU), 214
Digital mammography, 102
1,3-Dilinoyl-2-octanoyl-sn-glycerol (C18:2-C8:0-C18:2) (LML), 79
N-Dimethylformamide (DMF), 159
1,3-Dioctanoyl-2-linoleyl-sn-glycerol (C8:0-C18:2-C8:0) (MLM), 79
Dioleoyl phosphatidylethanolamine (DOPE), 188
1,4-Dioxane, 146
1,2-Dipalmitoyl-sn-glycero-3-phos-phocholine (DPPC), 123
Direct absorption, 107–115

Direct digital manufacturing, 264
Disease-modifying drugs (DMARDs), 241
Diuretics, 71
DNA-decorated SWNTS, 11t
DNA repair genes, 99
3D NF-gelatin scaffolds, 146t
'Double jeopardy' theory, 71
Doxorubicin (DOX), 195, 210, 213
3D printing, 57–58
Drug conjugate
  antibody, 206–207
  folic acid, 209–210
  PEG, 210
  peptide, 207–209
Drug delivery
  hyperthermia-mediated, 257, 257f
  lipid nanoparticles, 242
Drugs
  absorption, 68f
  carriers, polymers used for development of nanoparticles, 191t
  delivery, 154
  lymphatic transport of, 75t–77t
DTAP-PEG-mannose liposomes (DPM-LP), 90

E

EAAT1 (SLC1A3), 71
E-beam lithography, 30
Efavirenz, 82t
Eggshell, 266, 267
Elastins, 145, 167
Electrochemical anodization, 35–42, 36
Electromagnetic radiation, 227
Electrospinning, 56–57, 143–144, 166, 169–171, 170f
  core-shell fibers, 171–172
  equipment, 144f
  Janus fibers, 172–173
  nanofibers, 141–143, 141f
Electrospun fibers, 57, 156f
Electrospun nanofibers, 154
Electrospun nanopolyacrylonitrile, 155
Electrospun polymer/GO fibrous scaffolds, 57f
El-Naschie's E-infinity theory, 141
Emulsions/microemulsions/smedds/snedds, 73–80
Enhanced permeation and retention (EPR) effect, 210
Enterobacteriaceae, 224
Enterococcus faecalis (E. coli) biofilms, 232
  E. coli (ATCC 8739), 149
Enterocytes, 67
Enzyme-activated prodrugs
  antibody-directed enzyme prodrug therapy, 212
  gene-directed enzyme prodrug therapy (GDEPT), 212
Escherichia coli, 123, 125, 128, 131
Ethanol, 148
Ethoxylated castor oil (Cremophor® EL), 72
1-ethyl-3-(3-dimethylaminopropyl)carbodiimide (EDC), 115, 146
Ethylene gas, 151
Etoposide, 83t
Etoposide-loaded tripalmitin (ETP-SLN), 81
Extended spectrum beta-lactamases (ESBLs)-*E. coli*, 128
Extracellular matrix (ECM) proteins, 176
Extracorporeal shock waves (ESWs), 232
Extrusion, 267, 268, 269f, 271

F

Fabrication, 1–3
Fatigue limit, alloys, 24
FDA approved therapeutic monoclonal antibodies for cancer therapy, 207, 208t
FDM, *see* Fused deposition modelling (FDM)
Fetal bovine chondrocytes (FBC), 156
F98glioma cells, 5
Fiber
  alignment, 179–181
  composition, 175–176
  core-shell, 171–172
  cotton, 148
  diameter, 174–175
  electrospun, 57
  nanocellulose, 154
  nanostructured zinc oxide cotton, 148–149
  natural cellulose, 149
  organic–bioinorganic hybrid, 176–177
  silica aerogel, 155
Fiber-based insulation, 154
Fiber-based Q-switched Nd:YAG shockwave laser probes, 231–232
Fiber–hydrogel composite, 167
Fibronectins, 167, 177f
Fibro-porous composite nanobiomaterials, 165–167
  bioactivity, 168
  biocompatibility, 167–168
  biodegradability, 169
  biomimicry, 168
  cellular response and nanoarchitecture, 174
    fiber alignment, 179–181
    fiber composition, 175–176
    fiber diameter, 174–175
    mechanical properties, 177–179
    organic–bioinorganic hybrid fibers, 176–177
  electrospinning methods, 169–171
    core-shell fibers, 171–172
    Janus fibers, 172–173
Fibrous-porous materials, 171
Fibrous scaffolds, 175
  production, 169–171
First pass metabolism, 67–68, 79, 84
FITC–cytochrome C, 195
FLICE inhibitory protein, 240
Fluorescence quenching, 2
Fluorescent gold nanocluster (FANC), 3
  human endothelial cells and endothelial progenitor cells, 4t
Fluorescent QDs, 105f
Fluoroscopy, 101
5-fluorouracil (5-FU), 172
Focused ion beam (FIB) lithography, 30
Folic acid–drug conjugates, 209–210
Formalin-fixed paraffin sections, 102
Fourier transform infrared spectroscopy (FTIR), 148, 153
FOXO1 pathways, 244

Freund's complete adjuvant (FCA), 189
FTIR vibrational spectroscopy, 5
Fullerenes, 51; *see also* Carbon nanotubes (CNTs)
Functionalized conjugated, and ^{64}Cu-labeled SPIONS, 108t
Functionalized quantum dots probe, 112t
Fused deposition modelling (FDM), 264–265, 265f

G

Gadolinium (Gd^{3+}) ion, 3–5
Gadolinium nanoparticles (GdNPs), 3–6
  biomedical application, 5t
Gadonanotubes, 10
β-Galactosidase enzyme, 188
Gastrointestinal tract (GIT), 186
GDEPT, *see* Gene-directed enzyme prodrug therapy (GDEPT)
Gelatin, 146
Gelatin-modified PET nanofibers, 158
Geldanamycin prodrug (17-AG-C2_Gal), 212
Gelucire®, 73
Gemcitabine, 214
Gemcitabine-5′-monophospahte, 6
Gemtuzumab ozogamicin, 207
Gene-directed enzyme prodrug therapy (GDEPT), 211, 211f, 212
Gene prodrug activation therapy (GPAT), 212
Glucanacetobacter xylinum ATCC 10245, 152–154, 153
Gluconobacter xylinum, 152
Glucose-regulated protein 78 (GRP78), 209
Glucose transporter inhibitors, 71
Glutaraldehyde (GA), 34
Glycerol monostearate, 81
Glyceryl monooleate (GMO), 85
Glycosaminoglycan (GAGs), 166
Gold-coated SPIONS, 110t
Gold nanoparticles (AuNPs), 2–3, 104, 196
  bioconjugates, 3
Gold nanorods (GNR), 109t
GPAT, *see* Gene prodrug activation therapy (GPAT)
Graphene, 51, 52f
  applications, 54f
  biological effects of, 56f
  in biomedical applications, 55t
  physical properties, 53t
  toxicity of, 53
Graphene-based magnetic plasmonic nanocomposite, 111t
Graphene oxide (GO), 51, 53
Graphene oxide quantum dots (GOQD), 51
Graphene-polymer nanocomposites, 51–52, 53–54
  properties
    physical properties, 52–53
    surface chemistry, 52–53
    toxicity and antimicrobial properties, 53
  scaffolds
    biomedical applications of graphene materials, 54
    3D printing, 57–58
    electrospinning, 56–57
    hydrogel scaffolds, 58–60
    polymer/graphene scaffolds, 54–55
    state-of-the-art fabrication techniques, 55–56
Graphene quantum dots (GQD), 51
Graphene/SWNTs, 11t
G3-S16-sulphated carbosilane dendrimer, 132, 133f
Guanidinium-cholesterol, 188
Gut associated lymphoid tissue (GALT), 186

H

Halofantarine, 78, 79
N-Hap-reinforced nylon–based PD, 270f
Hepatic first pass effect, 67
HepG2 cells, 3
HIFU–induced local warming, 228
High-density lipoprotein, 71
High intensity focused ultrasound (HIFU) transducers, 228
Highly active antiretroviral therapy (HAART), 84
Hip and knee, implant for complete substitution of, 19f
Histochemistry, 102
Histopathological examination, 98
HIV-1, 132
HIV/AIDS, 67, 79, 86, 91, 92
HM-PEG-conjugated prodrug, 210
HT-29 colon cancer cell lines, 3
Human body fluids, 20
Human colon carcinomas, 9
Human epidermal growth factor receptor 2 (HER2)-positive breast cancer cell, 207
Humanized anti-CD22 mAb, 207
Human mesenchymal stem cell (hMSC), 58, 176; *see also* Mesenchymal stem cells (MSCs)
Human skin, 168
Hydrogels, 58, 59
Hydrophilicity, 52
Hydroxyapatite (HAp), 7, 10, 26f, 265
  crystals, 18
  microspheres, 59
N-(2-hydroxypropyl) methacrylamide (HPMA), 210
Hyperthermia, 6
  cellular mechanism, 254–255, 254f
  concept, 253
  future perspectives, 257
  hyperthermia-mediated drug delivery, 257, 257f
  regional, 254
  thermal mechanism, 255–256, 255f, 256f
  types, 254
Hypoglycemic effect of EE–NPs, 192

I

IgG, 88t
IgG_1, 87t
Imaging techniques, 100
  biopsy, 102
  computerized tomography, 100–101
  fluoroscopy, 101
  histochemistry, 102
  immunohistochemistry, 102–103
  magnetic resonance imaging, 101
  mammography, 102
  nuclear scan (radionuclide scan), 101
  positron emission tomography, 101
  ultrasound imaging, 101
Immunoassays, 6
Immunohistochemistry, 102–103

Implants
  for complete substitution of hip and knee, 19f
  failure, reasons, 19f
  incompatible, 18
Incompatible implant, 18
Indinavir, 89t
Indocyanin green–based photosensitizer (ICG–based PS), 232
Inductively coupled plasma atomic emission spectroscopy (ICP-AES), 160
Inductively coupled plasma mass spectrometry (ICP-MS), 160
Infectious diseases, 101, 121, 134, 135
Inflammation, 239, 240–242, 244
Inflammatory mediators, 240
In-house nano-hydroxyapatite particles
  additive fabrication, 264
  additive manufacturing (AM) techniques, 264
  experimental procedures
    alternative FDM filament, 268
    egg shell residual, 267
    Hap-reinforced nylon filament selection, 267–268, 267f
    methodology, 266, 266f
    PD printing, 270
    standard triangulation language (.STL) file, 268–270, 270f
  fused deposition modelling (FDM), 264–265, 265f
  hydroxyapatite (HAp), 265
  polycaprolactone (PCL)-filled montmorillonite (MMT), 265
  pulsed laser deposition (PLD) system, 265–266
  quality function deployment (QFD) analysis, 265
  tensile strength, 271, 271f
  wear test, 272–274, 272f, 272t, 273f
Inkjet-based system, 264
Inorganic nanocarriers, 195
  calcium phosphate nanoparticles (CaPNPs), 196–197
  gold nanoparticles, 196
  silica nanoparticles, 195–196
In situ polymerization, 32
Integrated QD, 10
Intelligent systems, 1
Interaperitonial administration, 79
Interleukin-2 (IL-2), 188
Interpolymer complex-based sensor, 6
Intestinal lymphatic system, 67
Investment casting, 275
Iodoacetamides, 115
Iron oxide/magnetic nanoparticles (IONPs), 5t, 6–7
  decorated fullerene (C_{60}), 110t
  decorated surface of fullerene, 111t
Iron oxide nanomaterials, 6
Iron oxide nanoparticles, 103–104

J

Janus fibers, 172–173
Jeffamine®-based dendrimers, 134
Jell-O®, 169

K

Klebsiella pneumoniae, 128

L

Label-free measurement technologies, 106
Lab test, cancer diagnosis, 100
  blood tests, 100
  urine tests, 100
Laminins, 167
Langmuir-Blodgett deposition, 32–35, 35f
Layer-by-layer assembly approach, 35
Layered manufacturing, 264
LED209, 125
Leishmania donovani chagasi, 125
LHRH, *see* Lutenizing hormone releasing hormone (LHRH) protein conjugates
LifeStyles® Dual Protect™ brand, 132
Ligand
  antibody–drug conjugate, 206–207
  aptamer–drug nanoconjugates, 209
  folic acid–drug conjugates, 209–210
  peptide–drug conjugates, 207–209
Linker in prodrugs, role, 205–206
Linoleic acid (LA), 85
Lipid, 67–92
Lipid-based formulations, 72–73
Lipid-based prodrugs, 214–216
Lipid digestion, 73
Lipid–drug conjugate (LDCs) NPs, 193
Lipid formulation classification system (LFCS), 73, 74t
Lipid nanoparticles, 81–85
LIPIODAL®, 78
Lipolysaccharides (LPS), 131
Lipophilic 2-chlorocabobenzoxy terminal groups, 123
Lipophilic drug delivery, 85
Liposomal formulation of adriamycin (LP-AD), 90
Liposomes, 85–92, 187–188
Lithography process, 30
Live cell imaging of GFP-expressing MSCs, 178f
Liver cancer, 98; *see also* Cancer
L-lysine-based dendrimers, 122
Long-chain triglycerides (LCT), 67, 78–80
Lopinavir, 82t
Low temperature–sensitive liposomes (LTSLs), 228, 231f
Lung cancer, 98; *see also* Cancer
Lutenizing hormone releasing hormone (LHRH) protein conjugates, 209
Lymphatic capillaries, 68
Lymphatic drug transport system, 67–69, 75t–77t
  biological transporters, 69–70
    ABC transporter (ATP binding cassettes transporter), 70–71
    coupling efflux transporters, 71–72
    SLC transporters (solute carrier transporter), 71
  fluid composition and flow, 69f
  lipid nanoparticle as medium to enhance, 82t–85t
  liposomes as nanocarrier, 87t–89t
  target drugs *via* lymphatic system, 72
    emulsions/microemulsions/smedds/snedds, 73–80
    lipid-based formulations, 72–73
    lipid nanoparticles, 81–85
    liposomes, 85–92
Lymph duct cannulation, 83t
Lymph nodes, 68, 69, 78, 79, 84–86, 90–92

Lymphoid follicles, 68
Lymphotropic carrier system, 78
Lysine-based peptide dendrimer (R 121), 124f

M

Macrophages, 84, 86, 90, 135, 187, 240
Magnetic properties, SPIONs, 252–253, 252f–253f
Magnetic resonance imaging (MRI), 3, 101
Magnetite (Fe_3O_4), 103
Magnetron sputtering deposition, 28–29
  physical surface treatments, 28
Maleimides, 115
Malignant neoplasm, 99f
Malignant tumors, 99
Mammography, 102
Mannose-conjugated and muramyl dipeptide (MDP)–conjugated dendrimers, 135
Matrix metalloproteinase 2 (MMP2), 213
MC3T3-El cells, 59
MDR-*Acinetobacter baumannii*, 128
MDR-*E. coli*, 128
MDR-*Shigella flexneri*, 128
Mechanical properties, 177–179
Medium-chain triglycerides (MCT), 78
Meltblown pitch carbon nanofibers, 155
Melt flow index (MFI), 267
Melt flow indexer (MFIer), 267, 269f
Mesenchymal stem cells (MSCs), 34, 177–179
Mesoporous silica nanoparticles (MSNs), 9, 195
Metallic alloys and applications, 26f
Metallic biomaterials, 17–20
  biocompatibility, 25
  chemical surface treatments, 32–33
    atomic layer deposition, 33
    electrochemical anodization, 35–42
    layer-by-layer assembly approach, 35
    silanization and Langmuir-Blodgett deposition, 33–35
  corrosion resistance, 24
  implants, 20–22
    cobalt-based alloys, 20–21
    design necessities, 23–25
    historical progress, 22–23
    metallic implants, 25–28
    stainless steel, 20
    titanium-based alloys, 21–22
    valve metals, 21
  mechanical efficiency, 23–24
  mechanical surface treatments, 31–32
  nanotubes, 37t–42t
  osseointegration, 25
  physical surface treatments, 28
    lithography process, 30
    magnetron sputtering deposition, 28–29
    plasma spraying, 30–31
    pulsed laser deposition (PLD), 29–30
  wear resistance, 24
Metallic nanobiomaterials, 2f
Methotrexate (MTX), 81, 82t, 214
  rheumatoid arthritis, 241–242
Methoxypoly (ethylene glycol), 86
Methylene blue (MB), 79
Micrococcus luteus, 125
Microfluidic channel, 107
Microorganisms, 121, 125, 134, 135, 157
Minimal inhibitory concentration (MIC), 131
Modelling, FDM, 264–265, 265f
Modified carbon nanofibers (mCNF), 176
Monoclonal antibodies (mAbs), 206
Monolayer graphene, 52
Monoliths, 9
MPEG-phospholipid, 86
MSC chemotaxis, 177
MTT assay, 167–168
MTX-loaded NLCs, 242–243
  characterization, 243
  effective delivery vehicle, 244
  incorporation in hydrogel for transdermal delivery, 244
  preparation, 242–243
Multidrug resistance associated proteins (MdPs), 71
Multidrug resistance protein-I (MDR1), 69
Multimodal silica nanoparticles (MSNPs), 9
Multiwall carbon nanotubes (MWCNTs), 10, 176
Murine anti- TAG-72 mAb (CC49), 212
Mylotarg, 207

N

N-acetyl gamma calicheamicin dimethylhydrazide (CalichDMH), 207
Nanobiomaterials, 2, 12
  ceramic nanoparticles (CENPs), 7
  gadolinium nanoparticles (GdNPs), 3–6
  gold nanoparticles (AuNPs), 2–3, 4t
  iron oxide/magnetic nanoparticles (IONPs), 6–7
  metallic, 2f
  organic/carbon-based nanoparticles (CNPs), 10–11
  semiconductor-based nanoparticles/quantum dots (SMNPs), 9–10
  silica nanoparticles (SiNPs), 9
  titanium oxide nanoparticles (TiONPs), 7–9
Nanobiomaterial-based therapies, 11
Nanocantilevers, 106–107
Nanocarriers, 98; *see also* Cancer; Lymphatic drug transport system
  used for protein delivery, 186f
Nanocarrier-mediated therapeutic protein delivery, 185–186
  inorganic nanocarriers, 195
    calcium phosphate nanoparticles (CaPNPs), 196–197
    gold nanoparticles, 196
    silica nanoparticles, 195–196
  limitations and challenges, 186–187
  particulate carriers, 189
    polymeric nanoparticles, 189–193
    protein/peptide-loading in SLNs, 193–195
    solid lipid nanoparticles (SLNs), 193
  vesicular nanocarriers
    liposomes, 187–188
    niosomes, 188–189
    protein and peptide delivery using niosomes, 189
Nanocellulose, 152
  fibers, 154
Nanocomposites, 1, 35

Nanofibers, 140–143, 141–143, 141f
applications
drug delivery, 154
mercury detection, 160–161
noise absorbers for aircrafts, 161
nonwoven thermal insulation, 154–155
protective clothes, 159–160
3 tissue engineering, 155–159
characterization, 140
synthesis techniques, 143
carbon nanofibers and Ni foam, 151–152
carbon nanotube/nanofiber-supported platinum by, 147–148
cellulose fibers with antibacterial Ag-loading Nano SiO_2, 149–151
electrospinning, 143–144
glucanacetobacter xylinum ATCC 10245, 152–154
nanofibrous gelatin scaffolds, 146
nanostructured zinc oxide cotton fibers, 148–149
poly(ethylene terephthalate) nanofiber, 147
self-assembling biomaterials for tissue engineering, 144–145
template synthesis of polypyrrole nanofibers, 146–147
Nanofibrous gelatin scaffolds, 146
Nanohybrid, 59, 129–130
Nanohydroxyapatite (nHA), 176
Nanomedicine, 126t–128t, 232–235
Nanoparticles, 148–151, 158
Nanophotothermolysis, 3
Nanostructured lipid carriers (NLCs), 242
Nanostructured prodrugs for cancer therapeutics, 215t–216t
Nanostructured zinc oxide cotton fibers, 148–149
Nanotechnology, 139–140
Nanotechnology, cancer diagnosis, 103
carbon nanotubes (CNTs), 104
gold nanoparticles, 104
iron oxide nanoparticles, 103–104
nanocantilevers, 106–107
nanowires, 107
quantum dot-doped silica nanoparticles, 106
quantum dots (QDs), 104–106
Nanotechnology-based prodrug conjugates
enzyme-activated prodrugs
antibody-directed enzyme prodrug therapy, 212
gene-directed enzyme prodrug therapy (GDEPT), 212
linker in prodrugs, role, 205–206
lipid-based prodrugs, 214–216
nanostructured prodrugs for cancer therapeutics, 215t–216t
polymeric prodrugs
N-(2-hydroxypropyl) methacrylamide (HPMA), 210
PEG–drug conjugates, 210
polyglutamic acid (PGA), 210–211
prodrugs, 205
stimuli-sensitive prodrugs, 212–214
targeting-ligand conjugated prodrugs
antibody–drug conjugate, 206–207
aptamer–drug nanoconjugates, 209
folic acid–drug conjugates, 209–210
peptide–drug conjugates, 207–209
Nanowires, 106, 107
Nanowoven silk fibroin nanofibers, 157
Naphthalene di-sulfonic acid, 132
Natural biopolymers, 55
Natural cellulose fibers, 149
Natural protein fibers (wool), 155
Near-infrared light region (NIR), 104
Nephrogenic system fibrosis, 5
Neural stem cells (NSCs), 175
Neurotransmitter-reuptake inhibitors, 71
NHA gradient-scaffold, 177
N-hydroxysuccinimide (NHS), 115, 146
Ni foam, 151
NIH3T3 cells, 180
Niosomes, 188–189
9-Nitro-camptothecin (9-NC), 91
Non-integrated QD, 10
Nonsteroidal antiinflammatory drugs (NSAIDs)
rheumatoid arthritis, 241
N1PC1 (Niemann Pick C1-like), 70
Nuclear factor kappa-B(NF-kB), 240, 244
Nuclear scan (radionuclide scan), 101
Nucleic acids, 2f, 3
Numeric control (NC) machining, 264
Nylon6, 147

O

O-dealkylation process, 204
Odecyltrimethylammonium bromide (DTAB), 147
Ofloxacin, 134
Oleic acid microemulsion, 79
Oncogenes, 99
Optimization, 20, 206
Oral insulin delivery, 191
Oral therapy, antibiotics for, 228t
Organic anion transporter I (OAT1), 69
Organic–bioinorganic hybrid fibers, 176–177
Organic/carbon-based nanoparticles (CNPs), 10–11
Organic collagen proteins, 18
Orthopaedic surgery, 17
Osteomyelitis (OSM)
animal models of, 232, 233t–234t
antimicrobial therapy to treat, 226–227, 227t
biofilm formation, 224–225, 225f
diabetes mellitus patients, 223
diagnosis, 225–226
local tissue invasion, 224–225
nanomedicines for treating, 232–235
pathogens, 224
periostium infection, 224f
progression of, 225, 226f
quorum sensing, 224
species in bone wounds, 224
OVA antigen, 87t
Ovalbumin (OVA), 195

P

Paclitaxel (PC), 85
1-Palmitoyl-2-oleoyl-sn-glycero-3-phosphocholine (POPC), 123
PAMAM dendrimers, 130, 132, 136
PAMAM-OH dendrimers, 123

Paraffin, 146
Particulate carriers, 189
 polymeric nanoparticles, 189–193
 protein/peptide-loading in SLNs, 193–195
 solid lipid nanoparticles (SLNs), 193
Pathophysiology of cancer, 98–99
Patient-specific dentures, 264
PCL electrospun nanofibers, 56
PCL-Gelatin nanofibers, 180
PCL-nHA-collagen, 177
Peceol®, 73
PEG–drug conjugates, 210
PEG-15-hydroxystearate, 73
PEG-oxidized multi-walled carbon nanotubes modified with ligand, 111t
PEG-ppTAT-DOX, 213
 drug delivery mechanisms of, 213f
Pegylated IL-2 (PEG-IL-2), 188
Pellicle, 152
Pendimethalin (N-(1-ethylpropyl)-3,4-dimethyl-2,6-dinitrobenzenamine), 159
Penicillin-G, 134
Peptides, 2f, 3, 122, 207, 209
Peptide dendrimers, 131–132
 BALY, 124f
 SB041, 124f
Peptide–drug conjugates, 207–209
Peptide nanofibers (PNF), 59
Peptidic-targeting molecules, 207–209
Peyer's patches, 68
P-glycoprotein (P-gp), 69, 71
 efflux transporter, 71–72
Pharmaceutical formulation approach, 72
Phenacetin, 204
Phosphate, 122
Phosphatidylinositol 3 kinase/Akt-1, 240
Photodynamic therapy (PDT), 232
Photooxidation, 8
Photosensitizer-loaded Gold Vesicles (GV-Ce6), 109t
Photothermal therapy (PT), 232
pH-sensitive drug delivery systems, 172
pH-sensitive polymeric micellar system, 214
Physical examination, cancer diagnosis, 100
Physiological environment, 25–28
Pirarubicin, 78
Plasma spraying, 30–31, 31f
Plasmodium falciparum, 125
Platelet-derived growth factor (PDGF), 177
Platinum-folic acid-CNT complex, 11t
Plausible engineered nanoparticles, 103f
PLCL-nHA-gelatin, 177
PLGA nanoparticles functionalized with aptamer associated QDs, 111t
PLGA-nHA-collagen, 177
Plurol® Oleique CC497, 72
PMSA, *see* Prostate membrane specific antigen (PMSA)
Polarons, 146–147
Polyacrylamide, 150
Poly(amidoamine) (PAMAM), 122
Polycaprolactone (PCL), 156, 168
Polycaprolactone (PCL)-filled montmorillonite (MMT), 265
Polydimethylsiloxane [PDMS], 34
Poly(ε-caprolactone) (PCL), 55, 191
 nanofibrous membranes, 174
Polyester, 155
Polyethylene glycol (PEG), 9, 141
Polyethylene glycol (PEG)-coated supramagnetic iron oxide–poly(gallol) nanoparticle complex, 6
Poly(ethylene terephthalate) (PET) nanofibers, 147
Polyglutamic acid (PGA), 210–211
Poly(glycerol sebacate) (PGS), 158
Polyglycolide, 55
Polylactide-co-glycolide (PLGA), 58
Poly-L-lysine dendrimer, 132
Polymers, 10, 18, 51, 53, 54, 125, 141, 147
 artificial, 190
 biodegradable, 55
 copolymers, 191t
 coupling, 55
 hydrophilic, 58
 synthetic, 177
 used for nanoparticles, 191t
Polymer conjugated SPIONs (HA-PEG10- SPIONs), 108t
Polymer/graphene nanocomposites, 53–54, 54f
Polymer/graphene scaffolds, 60t
Polymeric nanoparticles, 189–193, 190f
Polymeric nanoparticulate system (Pt-NP-Apt), 209
Polymeric prodrugs
 N-(2-hydroxypropyl) methacrylamide (HPMA), 210
 PEG–drug conjugates, 210
 polyglutamic acid (PGA), 210–211
Polymer nanofibers, 141
Poly(N-isopropylacrylamide) (PNIPAAm), 58
Poly(oxypropylene) triamine (Jeffamine®), 134
Poly(propylenamine) (PPI), 122
Polypyrrole/silica nanocomposite material, 147t
Polysaccharides, 55, 191t
Polysorbates, 73
Polystyrene, 141
Polyvinyl alcohol (PVA), 55, 141
Polyvinyl butyral (PVB), 141
Polyvinyl chloride (PVC), 141
Poly(vinylpyrrolidone) (PVP), 141, 173
Porous, 165–167
Porphyromonas gingivalis, 125, 240
Positron emission tomography (PET), 101
Potassium permanganate ($KMnO_4$), 149
Potent drugs, 67
PPI dendrimers conjugated MWCNTs-COOH, 129
Principal component analysis (PCA), 5
Pristine graphene, 53
Prodrugs
 antineoplastic therapy, 205
 approach, 72
 linker in, 205–206
 nanostructures possess, 216
 schematic representation of, 204f
Proinflammatory cytokines, 240–242
Prostate membrane specific antigen (PMSA), 206–207
Proteins, 2f, 3, 55, 191t
 microarrays, 30
 and peptide delivery using niosomes, 189
 peptide-loading in SLNs, 193–195
Protein tyrosine kinase 7 (PTK7), 209
Proteolytic enzymes, 186
Proto-oncogenes, 99
Prototyping, 264
Prulifloxacin, 134

Pseudomonas, 224
Pseudomonas aeruginosa, 125, 128
PA01 biofilm, 227
Pulsed laser deposition (PLD), 29–30, 265–266

Q

QD, *see* Quantum dots (QDs)
QseC, 128
Quality function deployment (QFD) analysis, 265
Quantum dots (QDs), 9–10, 104–106
doped silica nanoparticles, 106
embedded Si nanosphere, 112t
with Fe3O4-filled CNTs, 112t
surface-coated, 112t
Quorum sensing, 224

R

Radioactive iodine, 9
Radiolabelled, 83t
Radionuclide scan, 101
Radiosensitizers, 5
Raloxifene, 80, 82t
Raloxifene HCl (RHCl), 84
Reactive oxygen species (ROS), 232
Redox-responsive linkage, 214f
Reduced graphene oxide (rGO), 51, 53
Regenerative medicine, 166–168, 176, 180, 185
Rheumatoid arthritis
causes, 240
disease-modifying drugs (DMARDs), 241
drugs and vaccines, 241
inflammatory disorder, 239
lipid nanoparticles as drug delivery carriers, 242
methotrexate (MTX), 241–242
molecular, 240–241
MTX-loaded NLCs
characterization, 243
effective delivery vehicle, 244
incorporation in hydrogel for transdermal delivery, 244
preparation, 242–243
nonsteroidal antiinflammatory drugs (NSAIDs), 241
pathophysiology, 240
$(RW)_{4D}$ peptide dendrimer, 124f

S

Saccharomyces cerevisiae, 125
Salmonella paratyphi, 128
Salmonella typhimurium, 128
Saquinavir, 79
SB041, 131
Scaffolds
biomedical applications of graphene materials, 54
3D printing, 57–58
electrospinning, 56–57
hydrogel scaffolds, 58–60
nanofibrous gelatin, 146
polymer/graphene scaffolds, 54–55
state-of-the-art fabrication techniques, 55–56
Scanning electron microscopy (SEM), 153, 230
Secretable enterocyte lipoproteins (chylomicrons), 67
Selective laser sintering (SLS), 264
Self-assembling amphiphilic block copolymers, 212–213
Self-assembling biomaterials for tissue engineering, 144–145
Self-assembling peptide systems, 145f
Self-micro emulsifying drug delivery systems (SMEDDS), 78
Self-nanoemulsifying self-nano suspension loaded systems (SNESNS), 80
Semiconducting nanowires, 107
Semiconductor-based nanoparticles/quantum dots (SMNPs), 9–10
Semiconductor electron-beam lithography, 106
Sequentially-transmitted intestinal epithelial cells, 67
Short-chain triglycerides (SCT), 78
Signal transducer, 240
Silanization deposition, 32–35
and Langmuir-Blodgett deposition, 33–35
Silica aerogel fibers, 155
Silica film, 147
Silica nanoparticles (SiNPs), 9, 195–196
biomedical application, 8t
Silica nanotubes, 9
Silk fibroin nanofibers, 157
Simivastatin, 78
Single-walled carbon nanotubes (SWNTs), 10
Skin, 221SLC transporters (solute carrier transporter), 71
SLNs, *see* Solid lipid nanoparticles (SLNs)
SMCC-activated amine functionalized nanoparticle coupling, 116f
SNEDDS formulations, 80
Sodium hydroxide (NaOH), 148, 149
Sol-gel methods, 32
Solid freeform fabrication, 264
Solid lipid nanoparticles (SLNs), 81, 193, 194t, 242
Solute-carrier (SLC) super families, 69
Sonodynamic therapy, 9
Soy protein isolate (SPI), 176, 195
Specific absorption rate (SAR), 256
SPIONs, *see* Superparamagnetic iron oxide nanoparticles (SPIONs)
SRB-2 (Scavenger receptor B-1), 70
Stainless steel, 20, 25–27
Standard triangulation language (.STL) file, 264
Staphylococcus aureus (ATCC 6538), 131, 149, 224
Stearic acid, 81
Stem cell differentiation, 168, 174–178, 207
Stereolithography (SLA), 264
Stimuli-sensitive prodrugs, 212–214
Stomach cancer, 98; *see also* Cancer
Sulfo-NHS, 115
Sulfosuccinimidyl-4-(maleimidomethyl) cyclohexane-1-carboxylate (sulfo-SMCC), 115
Superparamagnetic iron oxide nanoparticles (SPIONs), 104, 227–228
biomedical applications of, 249f
materials used for coating, 250t
nanoparticles and their stages in clinical use, 248t
physicochemical properties of
magnetic properties, 252–253, 252f–253f
routes of synthesis of, 251t
shape, 251

size, 250
surface chemistry and charge, 251–252
stability, 249
synthesis of, 249–250
Surface-coated and ligand conjugated liposome containing drug and QD, 112t
Surface-coated bismuth-iron oxide nanocomposite (BIONs), 108t
Surface-coated QDs, 112t; *see also* Quantum dots (QDs)
Surface-engineered dendrimers, 129f
Surface-engineered iron oxide nanoparticles, 7
Surface-enhanced Raman spectroscopy (SER), 27
Surface functionalized (AuNP), 109t
Surface hydrophobicity, 69
Surface mechanical attrition treatment (SMAT), 32, 32f
Surface modified Fe_3O_4@Au core/shell nanostars, 110t
Surface plasmon resonance, 2
Surface properties, 196, 253
Surfactant, 73, 74t–77t
Synoviocytes, apoptosis of, 240
Synthesis
hydroxyapatite from eggshell waste, 267
of superparamagnetic iron oxide nanoparticles (SPIONs), 249–250
Synthetic macromolecule dendrimers, 135
Synthetic polymers, 55

T

Targeted drug delivery, 6, 72, 98, 195, 232, 249
Targeting-ligand conjugated prodrugs
antibody–drug conjugate, 206–207
aptamer–drug nanoconjugates, 209
folic acid–drug conjugates, 209–210
peptide–drug conjugates, 207–209
^{99}Tc-HMPAO radiolabelled compritol-SLN, 84
Teflon-coated spinneret, 173
Template synthesis of polypyrrole nanofibers, 146–147
Tensile strength, 271, 271f
Tetracycline, 134
Theranostic, 3, 6, 7, 110t, 215t
Thermally cross-linked SPION (TCL-SPION), 108t
Thermally induced phase separation (TIPS) technique, 146
Thermocouples, 155
Thin film by magnetron sputtering, 29f
Thiol group-involved coupling, 115
Three-dimensional (3-D) image, 101
Three-dimensional printing, 264
Tissue engineering, 54–60, 156f, 166–167
Tissue/organ repair, 54
Tissue repair, 6
Titanium alloys, 27
Titanium-based alloys, 21–22
Titanium dioxide, 7
Titanium oxide nanoparticles (TiONPs), 7–9
biomedical application, 8t
Tobramycin, 83t
Tobramycin loaded solid lipid nanoparticles (Tb-SLN), 85
Toxicity, 248, 249, 251
Transducer, 101
Transmission electron microscopy, 172
Trastuzumab, 207
Trastuzumab-amide-MCC-DM1, 207
Triazine dendrimers, 132–134
Tricarboxylato dendrimers, 134
Triglycerides (TG), 67, 79
Trimethylaluminum [Al $(CH_3)_3$] (TMA), 33
Tripterine, 82t
Tristearin, 81
Triton X-100, 73
Trivalent peptide dendrimer with metallocene, 124f
Trypanosoma cruzi, 125
Tumor suppressor genes, 99
Tumour-specific antigens, 206–207

U

Ultrasound imaging, 101
Uric acid transporter inhibitors, 71
Urine tests, cancer diagnosis, 100; *see also* Cancer
UV-visible (UV-Vis) spectroscopy, 148

V

Val-Cit motif, 209
Valve metals, 21
Vancomycin, 134
Vasoactive intestinal peptide (VIP), 189
VDEPT, *see* Virus-directed enzyme/prodrug therapy (VDEPT)
Vesicles, 86, 187
Vesicular nanocarriers
liposomes, 187–188
niosomes, 188–189
protein and peptide delivery using niosomes, 189
Vibrio cholerae, 125
Vinpocetine (VCP)-SMEDDS, 79
Virtual prototyping, 264
Virus-directed enzyme/prodrug therapy (VDEPT), 212
Vitamin E analogue (α-TEA), 91
Vitronectin, 177f

W

Water-soluble polyanionic carbosilane dendrimers, 132
Water-soluble tricarboxylato dendrimers, 134
Wear test, in-house nano-hydroxyapatite particles, 272–274
Whey protein isolate (WPI), 195
W/O-G emulsion, 78
Wound healing, 157
Wounds types
acute wounds, 221, 222f
chronic wounds, 222, 222f

X

X-ray diffraction, 147, 149, 153
X-ray films, 100
X-ray photoelectron spectroscopy (XPS), 27

Z

Zidovudine (ZVD), 86, 87t, 90
Zinc oxide (ZnO), 148
Zinc sulphate, 148

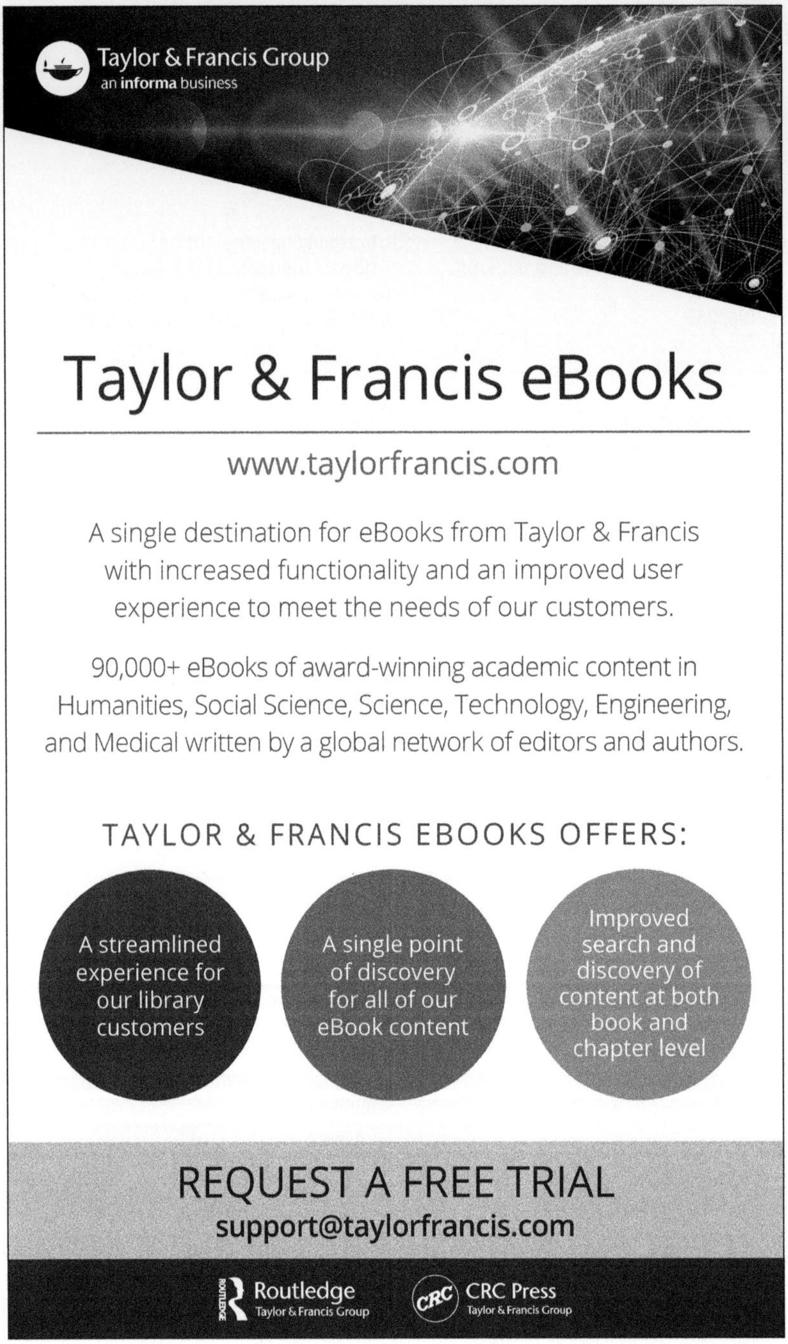
Taylor & Francis Group
an informa business
Taylor & Francis eBooks
www.taylorfrancis.com
A single destination for eBooks from Taylor & Francis with increased functionality and an improved user experience to meet the needs of our customers.
90,000+ eBooks of award-winning academic content in Humanities, Social Science, Science, Technology, Engineering, and Medical written by a global network of editors and authors.
TAYLOR & FRANCIS EBOOKS OFFERS:
A streamlined experience for our library customers
A single point of discovery for all of our eBook content
Improved search and discovery of content at both book and chapter level
REQUEST A FREE TRIAL
support@taylorfrancis.com
Routledge
Taylor & Francis Group
CRC Press
Taylor & Francis Group